Land-Rover
Parts Catalogue
Series II and IIA
Bonneted Control Models

December 1968

The Rover Company Limited
Solihull, Warwickshire, England

Part No. 605957

Service Department
Solihull, Warwickshire
Telephone: 021-743 4242
Telegrams: Rovrepair Solihull
Telex: 33-156

Parts Department
P.O. Box 79
Cardiff
Great Britain
Telephone: Cardiff 33681
Telegrams: Rovparts, Cardiff
Telex: 49-359

London Service Depot
Seagrave Road
Fulham, London SW6
Telephone:
Administration and Appointments: 01-385 1221
Reception: 01-385 7721
Parts Department: 01-385 6231
Telegrams: Rovrepair, Wesphone, London

Please note that all prices and specifications are
subject to alteration without notice

Always quote type and number of vehicle
when ordering

By Appointment to
Her Majesty
Queen Elizabeth II

Manufacturers
of Motor Cars and
Land-Rovers

By Appointment to
Her Majesty
Queen Elizabeth
the Queen Mother

Suppliers
of
Motor Cars and
Land-Rovers

The Rover Company Limited

THE "LAND-ROVER 88"

THE "LAND-ROVER 109"

Land-Rover
Parts Catalogue
Series II and IIA
Bonneted Control Models

December 1968

Glossary of American-English Automotive Part Terms listed under the following sub-headings: Body parts, brake parts, chassis parts, electrical equipment, instruments, motor and clutch parts, rear axle and transmission parts, steering parts, tools and accessories, transmission parts and tyres.

Body parts

USA	ENGLISH
Bumper guard	Overrider
Cowl	Scuttle
Dashboard	Facia panel
Door post	Door pillar
Door stop	Check strap
Door vent	Quarter light
Fender	Wing
Firewall	Bulkhead
Hood	Bonnet
License plate	Number plate
Rear seat back or backrest	Rear seat squab
Rocker panel	Valance
Skirt	Apron
Tailgate	Tailboard
Toe pan	Toe board
Top	Hood
Trunk	Boot
Windshield	Windscreen
Wheelhouse or housing	Wheel arch

Brake parts

USA	ENGLISH
Parking brake	Hand brake

Chassis parts

USA	ENGLISH
Muffler	Exhaust silencer
Side rail	Side member
Spring clamp	U-bolt

Electrical equipment

USA	ENGLISH
Back up lamp	Reverse lamp
Dimmer switch	Dip switch
Dome lamp	Roof lamp
Gas pump or fuel pump	Petrol pump
Generator	Dynamo
Ignition set	Ignition harness
Parking lamp	Side lamp
Rear lamp	Tail lamp
Spark plug	Sparking plug
Turn signal	Trafficator
Voltage regulator	Control box

Instruments

USA	ENGLISH
Tachometer	Rev-counter

Motor and clutch parts—Engine and clutch parts

USA	ENGLISH
Carburetor	Carburetter
Clutch throwout bearing	Clutch release bearing
Cylinder crankcase	Cylinder block
Hose clamp	Hose clip
Pan	Sump
Piston or wrist pin	Gudgeon pin

Rear axle and transmission parts

USA	ENGLISH
Axle shaft	Half shaft
Drive shaft	Propeller shaft
Grease fitting	Grease nipple
Ring gear and pinion	Crown wheel and pinion

Steering parts

USA	ENGLISH
Control arm	Wishbone
King pin	Swivel pin
Pitman arm	Drop arm
Steering idler	Steering relay
Steering knuckle	Stub axle
Steering post	Steering column
Tie bar or track bar	Track rod

Tools and accessories

USA	ENGLISH
Antenna	Aerial
Crank handle	Starting handle
Wheel wrench	Wheel brace
Wrench	Spanner

Transmission parts—Gearbox parts

USA	ENGLISH
Auxiliary gearbox	Transfer box
Counter shaft	Layshaft
Gear shift lever	Gear lever
Output shaft	Main shaft
Parking lock	Parking brake
Shift bar	Selector rod
Transmission casing	Gearbox housing

Tyres

USA	ENGLISH
Tire	Tyre
Tread	Track

COMMENCING VEHICLE NUMBERS

PETROL MODELS, 4 CYLINDER

	1958 88* Series II	1959 88† Series II	1960 88† Series II	1961 88† Series II	88† Series IIA†
Home, RHStg	141800001	141900001	141000001	141100001	24100001 A
Export, RHStg	142800001	142900001	142000001	142100001	24200001 A
Export, RHStg, CKD	143800001	143900001	143000001	143100001	24300001 A
Export, LHStg	144800001	144900001	144000001	144100001	24400001 A
Export, LHStg, CKD	145800001	145900001	145000001	145100001	24500001 A

	1958 109† Series II	1959 109† Series II	1960 109† Series II	1961 109† Series II	109† Series IIA
Home, RHStg	151800001	151900001	151000001	151100001	25100001 A
Export, RHStg	152800001	152900001	152000001	152100001	25200001 A
Export, RHStg, CKD	153800001	153900001	153000001	153100001	25300001 A
Export, LHStg	154800001	154900001	154000001	154100001	25400001 A
Export, LHStg, CKD	155800001	155900001	155000001	155100001	25500001 A

From March 1965 onwards, Vehicle numbers prior to this date are the same as for 88

	Series IIA†
Home, RHStg, 88 Station Wagon	31500001 B
Export, RHStg, 88 Station Wagon	31600001 B
Export, RHStg, CKD, 88 Station Wagon	31700001 B
Export, LHStg, 88 Station Wagon	31800001 B
Export, LHStg, CKD, 88 Station Wagon	31900001 B

	1959† 109† Series II	1960† 109† Series II	1961 109† Series II	109†† Series IIA†
Home, RHStg, 109 Station Wagon	161900001	161000001	161100001	26100001 A
Export, RHStg, 109 Station Wagon	162900001	162000001	162100001	26200001 A
Export, RHStg, CKD, 109 Station Wagon	163900001	163000001	163100001	26300001 A
Export, LHStg, 109 Station Wagon	164900001	164000001	164100001	26400001 A
Export, LHStg, CKD, 109 Station Wagon	165900001	165000001	165100001	26500001 A

* Fitted with 2.0 litre engine   † Fitted with 2¼ litre engine   †† Fitted with 2.6 litre engine

PETROL MODELS, 6 CYLINDER

	109†† Series IIA††
Home, RHStg	34500001 D
Export, RHStg	34600001 D
Export, RHStg, CKD	34700001 D
Export, LHStg	34800001 D
Export, LHStg, CKD	34900001 D

	109†† Series IIA††
Home, RHStg, 109 Station Wagon	35000001 D
Export, RHStg, 109 Station Wagon	35100001 D
Export, RHStg, CKD, 109 Station Wagon	35200001 D
Export, LHStg, 109 Station Wagon	35300001 D
Export, LHStg, CKD, 109 Station Wagon	35400001 D

†† Fitted with 2.6 litre engine

DIESEL MODELS

	1958 88* Series II	1959 88† Series II	1960 88† Series II	1961 88† Series II	88† Series IIA
Home, RHStg	146800001	146900001	146000001	146100001	27100001 A
Export, RHStg	147800001	147900001	147000001	147100001	27200001 A
Export, RHStg, CKD	148800001	148900001	148000001	148100001	27300001 A
Export, LHStg	149800001	149900001	149000001	149100001	27400001 A
Export, LHStg, CKD	150800001	150900001	150000001	150100001	27500001 A

	1958 109* Series II	1959 109* Series II	1960 109* Series II	1961 109* Series II	109* Series IIA
Home, RHStg	156800001	156900001	156000001	156100001	27600001 A
Export, RHStg	157800001	157900001	157000001	157100001	27700001 A
Export, RHStg, CKD	158800001	158900001	158000001	158100001	27800001 A
Export, LHStg	159800001	159900001	159000001	159100001	27900001 A
Export, LHStg, CKD	160800001	160900001	160000001	160100001	28000001 A

From March 1965 onwards, Vehicle numbers prior to this date are the same as for 88

	Series IIA†
Home, RHStg, 88 Station Wagon	32000001 B
Export, RHStg, 88 Station Wagon	32100001 B
Export, RHStg, CKD, 88 Station Wagon	32200001 B
Export, LHStg, 88 Station Wagon	32300001 B
Export, LHStg, CKD, 88 Station Wagon	32400001 B

	1959* 109* Series II	1960* 109* Series II	1961* 109* Series II	109* Series IIA†
Home, RHStg, 109 Station Wagon	166900001	166000001	166100001	28100001 A
Export, RHStg, 109 Station Wagon	167900001	167000001	167100001	28200001 A
Export, RHStg, CKD, 109 Station Wagon	168900001	168000001	168100001	28300001 A
Export, LHStg, 109 Station Wagon	169900001	169000001	169100001	28400001 A
Export, LHStg, CKD, 109 Station Wagon	170900001	170000001	170100001	28500001 A

* Fitted with 2.0 litre engine   † Fitted with 2¼ litre engine

Vehicle and chassis number
F 778

Engine number
F 779

Gearbox number
F 780

Front and rear axle number
F 781

GENERAL EXPLANATION

This Parts catalogue is to be used for all Series II and Series IIA Land-Rover vehicles, Petrol and Diesel bonneted control models, to both United Kingdom and Export specifications. However, Land-Rovers supplied to some territories are fitted with special equipment, part number information for the parts peculiar to these vehicles will be found in Supplements issued to the countries concerned.

The part number quoted is always common to all the models covered by the section in question unless otherwise stated, in which case differences are noted in the Remarks column.

As the list covers both Home and Export models, reference is made throughout the text to the 'left-hand' (LH) and 'right-hand' (RH) sides of the vehicle, rather than to 'near-side' and 'off-side'. The 'left-hand side' is that to the left hand when the vehicle is viewed from the rear; similarly, 'left-hand steering' (LHStg) models are those having the driving controls on the left-hand side, again when the vehicle is viewed from fne rear.

Assemblies are printed in capitals, and all the parts comprising that assembly are shown by indenting the alignment of the description column, the indentation reverting to the original alignment when the assembly is complete. The indentation is clearly shown by the numbers 1, 2, 3, 4 at the top of each page.

An example, taken from Page 15, is given below:

```
    1 2 3 4
    ROCKER COVER, TOP, ASSEMBLY
        Tappet clearance plate
        Drive screw fixing plate
        Joint washer for top rocker cover
```

It will be seen that the 'ROCKER COVER, TOP, ASSEMBLY' includes the 'tappet clearance plate' and 'drive screw fixing plate' (indentation 1), while the 'joint washer', by reverting to alignment with 'ROCKER', is excluded from the assembly.

In all cases where the entire assembly is required, quote only the assembly number.

Where applicable, proprietary part numbers are listed in this Parts Catalogue. The first two or three letters of these part numbers are an abbreviation of the manufacturer's name as follows:

Prefix:				
AC	— AC Delco		KLG	— KLG Ltd.
BB	— Borg and Beck		LU	— Lucas
CAV	— CAV Ltd.		SM	— Smith
GI	— Girling		SU	— SU
JA	— British Jaeger		SX	— Solex

When ordering parts, always:

(1) Quote a part or assembly number, giving the name of the part and the vehicle number for which it is required.

(2) Order from the nearest authorised Rover Distributor or Dealer.

(3) Specify how despatch is to be made.

No. de pont AV et AR
F 781

No. de boîte de vitesses
F 780

No. de moteur
F 779

No. de véhicule et de châssis
F 778

INSTRUCTIONS GENERALES

Ce Catalogue des pièces de rechange doit être utilisé pour rechercher le numéro des pièces désirées pour toutes les Land-Rover, commande normale, de la Série II et IIA munis du moteur Diesel ou à essence, spécifications 'Home' (Marché intérieur) et 'Export' (Marché extérieur). Toutefois, les véhicules pour certains pays comportent des équipements spéciaux, dont les références sont indiquées dans les additifs fournis pour ces pays.

Le numéro de la pièce est commun à tous les modèles groupés sous la section en question, à moins que la pièce ne soit de désignation spéciale. En pareil cas, les différences sont notées dans la colonne des observations.

Dans ce catalogue, nous avons recours à la spécification droite et gauche, droit et gauche étant définis par rapport à la position qu'occupe le conducteur lorsque celui-ci est assis au volant de son véhicule.

Les ensembles sont imprimés en lettres majuscules et toutes les pièces comprises dans un ensemble sont indiquées en déplaçant la désignation de ces pièces vers la droite par rapport à la désignation de l'ensemble. Lorsque cet ensemble est complet, la désignation de la pièce suivante est ramenée à l'alignement d'origine, c'est-à-dire vers la gauche. L'alignement des désignations est clairement montré par les numéros 1, 2, 3, 4, en haut de chaque page.

Un exemple pris à la Page 15 est donné ci-dessous:

```
    1 2 3 4
    ROCKER COVER, TOP, ASSEMBLY
        Tappet clearance plate
        Drive screw fixing plate
        Joint washer for top rocker cover
```

L'on remarquera que l'ensemble 'ROCKER COVER, TOP, ASSEMBLY' comprend les pièces 'Tappet clearance plate' et 'Drive screw fixing plate' (colonne 1), tandis que la pièce suivante 'Joint washer for top rocker cover' est exclue de l'ensemble, la désignation de celle-ci étant ramenée en ligne avec l'ensemble 'ROCKER COVER, TOP, ASSEMBLY'.

Lorsqu'un ensemble complet est désiré, il suffira d'indiquer le numéro de l'ensemble.

Toutes les fois qu'il a été possible, les numéros de pièces des différentes marques ont été portés dans ce catalogue. Les deux premières lettres de ces numéros représentent l'abréviation du nom du constructeur, comme suit:

Préfixe:				
AC	— AC Delco		KLG	— KLG Ltd.
BB	— Borg and Beck		LU	— Lucas
CAV	— CAV Ltd		SM	— Smith
GI	— Girling		SU	— SU
JA	— British Jaeger		SX	— Solex

Dans la passation des commandes, il est indispensable de:

(1) Donner le numéro de la pièce ou de l'ensemble, la désignation complète de chaque pièce ainsi que le numéro du véhicule auquel la pièce est destinée.

(2) Spécifier le mode d'expédition désiré.

(3) Placer la commande avec l'agent Rover le plus proche.

ANORDNUNG DER FAHRZEUG- UND GRUPPENNUMMERN

Dieser Ersatzteilkatalog ist für alle für den Inland- und Exportmarkt hergestellten Normallenker Land-Rover der Serie II und IIA mit Diesel- und Benzinmotor zu verwenden. Für gewisse Länder wird der Land-Rover mit Sonderausrüstung geliefert. Ersatzteilangaben sind in den entsprechenden Nachträgen zu finden.

Die Ersatzteilnummern sind, soweit nicht anders angegeben, stets allen, in dem betreffenden Abschnitt angeführten Modellen gemeinsam. Ausnahmen sind in der Rubrik 'Remarks' vermerkt.

Der Text dieses Katalogs, der Inland- und Exportmodelle umfasst, nimmt Bezug auf die linke (LH) und rechte (RH) Seite des Fahrzeuges. Diese Bezeichnungen beziehen sich auf die Seiten des Wagens, wenn dieser von gesteuerte und RHStg rechtsgesteuerte Modelle.
der Rückseite betrachtet wird. Dementsprechend bedeutet LHStg links-

Aus mehreren Einzelheiten bestehende, grössere Ersatzteilsätze sind in Grossbuchstaben gedruckt und alle Teile eines solchen Satzes sind im Text eingerückt angeführt. Das Ende dieses Satzes ist dadurch angegeben, dass das nächstfolgende Ersatzteil wieder auf die ursprüngliche Spaltenbreite vorgerückt ist. Die Einrückung ist durch die am Kopf jeder Seite befindlichen Ziffern 1, 2, 3, 4 deutlich angegeben.

Dies ist durch das folgende, Seite 15 entnommene Beispiel illustriert:

    1 2 3 4
ROCKER COVER, TOP, ASSEMBLY
    Tappet clearance plate
    Drive screw fixing plate
    Joint washer for top rocker cover

Das bedeutet, dass der Satz 'ROCKER COVER, TOP, ASSEMBLY' eine 'tappet clearance plate' und 'drive screw fixing plate' mit einschliesst (Rubrik 1), während der 'joint washer' wie schon sein Vorrücken auf die Höhe von 'ROCKER' anzeigt, nicht im Satze inbegriffen ist.

Beim Bestellen eines grösseren Ersatzteilsatzes gebe man nur die Nummer des kompletten Ersatzteilsatzes an.

Wo eigentumsrechtlich geschützte Ersatzteile verwendet werden, sind diese im Ersatzteilkatalog dadurch gekennzeichnet, dass die ersten zwei Buchstaben eine Abkürzung des Namens der Herstellerfirma darstellen:

Schlüsselbuchstaben: AC — AC Delco   KLG — KLG Ltd.
    BB — Borg and Beck LU — Lucas
    CAV — CAV Ltd.   SM — Smith
    GI — Girling   SU — SU
    JA — British Jaeger SX — Solex

Bei Bestellungen beachte man stets folgendes:

(1) Folgende Einzelheiten sind anzugeben: Die Nummer und Bezeichnung des Ersatzteiles oder Ersatzteilsatzes, und die Nummer des Fahrzeuges, für welches diese Teile gebraucht werden.

(2) Man bestelle vom nächsten bevollmächtigten Rover Händler oder Vertreter.

(3) Man gebe die nötigen Versandvorschriften.

Motornummer

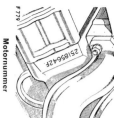
Fahrzeug- und Chassisnummer

Getriebenummer

Vorder- und Hinterachsnummer

Número do motor

Número de viatura e de chassis

Número do motor

Número do caixa velocidades

Número de eixo dianteira e trasero

(vi)

ALLGEMEINE BEMERKUNGEN

INSTRUCOES GERAIS

SITUAÇÃO DOS NÚMEROS DE UNIDADE E DE VEÍCULO

Este catálogo de Peças sobressalentes destina-se aos veículos Land-Rover modelos série II e IIA com motor Diesel o de gasolina com direção normal, tanto para os modelos do mercado Británico, como para os de exportação. Contudo, os modelos Land-Rover fornecido a alguns territórios, estão dotados com equipamento especial, cujos detalhes de Peça No. das peças especiais para estes veículos acharam-se nos Suplementos expedidos a estes territórios.

O número indicado para cada peça é sempre comum a todos os modelos abrangidos pela parte do catálogo em questão, a não ser que se especifique o contrário, em cujo caso as excepções são anotadas na coluna de 'Remarks' (Observações).

Como este catálogo se refere igualmente a modelos do Mercado Interno e de Exportação, faz-se sempre referência no texto aos lados esquerdo (LH) e direito (RH) do veículo, sendo o primeiro o que fica do lado da mão esquerda quando se olha para o veículo da parte traseira. Os veículos com volante à esquerda (LHStg) são aqueles que têm os comandos da direcção do lado esquerdo, também quando se olha para o veículo da parte de trás.

Os conjuntos são impressos em letras maiúsculas e todas as peças que fazem parte de um conjunto são indicadas escrevendo os respectivos nomes afastados um espaço para a direita do alinhamento de coluna descritiva; logo que o conjunto esteja completo, deixa de dar-se esse espaço e volta-se al alinhamento original. As mudanças de alinhamento são facilmente observadas por meio dos números 1, 2, 3, 4, escritos no topo de cada página.

Como um exemplo, tome-se a página 15, onde poderá ler-se:

    1 2 3 4
ROCKER COVER, TOP, ASSEMBLY
    Tappet clearance plate
    Drive screw fixing plate
    Joint washer for top rocker cover

Notar-se-á que o 'ROCKER COVER, TOP, ASSEMBLY' inclui o 'Tappet clearance plate' e a 'Drive screw fixing plate' (coluna 1), ao passo que a 'joint washer', já escrita por debaixo do alinhamento original da palavra 'ROCKER', não pertence a esse conjunto.

Em todos os casos em que se pretenda um conjunto completo, é suficiente indicar apenas o número do conjunto.

Onde resulta aplicável, os números de peças con nome de propriedade exclusiva são indicados neste catálogo de peças, as duas primeiras letras dos números destas peças são uma abreviação dos nomes dos fabricantes, ou seja:

Prefixo: AC — AC Delco   KLG — KLG Ltd
    BB — Borg and Beck LU — Lucas
    CAV — CAV Ltd   SM — Smith
    GI — Girling   SU — SU
    JA — British Jaeger SX — Solex

Ao fazer a encomenda de peças sobressalentes, deve sempre:

(1) Indicar-se o número da peça ou o número do conjunto, dando ao mesmo tempo o nome da peça e o número do veículo a que se destina;

(2) Dirigir-se a encomenda ao Distribuidor Rover mais próximo;

(3) Especificar-se como o despacho deve ser feito.

(vii)

UBICACION DE NUMEROS DE COCHE Y UNIDAD

F-781 — **Número del eje delantero y trasero**

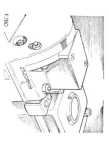

F-780 — **Número de caja velocidades**

F-779 — **Número de motor**

F-778 — **Número de coche y de chasis**

INSTRUCCIONES GENERALES

Este catálogo de Piezas de repuesto debe ser empleado para todos los vehículos Land-Rover modelos Serie II y IIA dotados del motor Diesel o de gasolina, con dirección normal, tanto en los modelos para el mercado del reino unido, como para la exportación. Sin embargo, los vehículos Land-Rover que se suministran a algunos territorios, están dotados con equipo especial, cuyos detalles de Pieza No. de las piezas especiales para estos vehículos figuran en nuestros Suplementos expedidos a dichos territorios.

El número de pieza citado es siempre común a todos los modelos a que se hace referencia en la sección, a menos que se indique de otra forma, en cuyo caso, dicha diferencia está anotada en la columna titulada 'Remarks' (Observaciones).

Como la lista se refiere a modelos Nacionales y de Exportación, la referencia en el texto se hace a vehículos con volante a la izquierda (LH) y volante a la derecha (RH). El costado izquierdo del vehículo es aquél que queda situado a la mano izquierda cuando se ve el vehículo desde su parte posterior; los vehículos con volante a la izquierda (LHStg) son aquellos que tienen los mandos de la dirección a la izquierda, también cuando el vehículo se ve desde la parte posterior.

Los conjuntos van impresos en mayúsculas y todas las piezas que comprenden dicho conjunto quedan indicadas por espaciados o desplazamientos de la alineación de la columna descriptiva; el espaciado vuelve a la alineación primitiva cuando el conjunto queda terminado. El espaciado se aprecia con toda claridad por la numeración 1, 2, 3, 4 en la parte superior de cada página.

Como ejemplo tomemos la Página 15 en la que se indica:

```
1 2 3 4
ROCKER COVER, TOP, ASSEMBLY
    Tappet clearance plate
    Drive screw fixing plate
    Joint washer for top rocker cover
```

Se podrá observar que el 'ROCKER COVER, TOP, ASSEMBLY' incluye el 'Tappet clearance plate' y el 'Drive screw fixing plate' para la cubierta delantera (espaciado 1), mientras que la 'joint washer', por volver a la alineación con la palabra ROCKER, queda excluida del conjunto.

En todos los casos donde se requiere un conjunto completo, cítese solamente el número del conjunto.

Donde resulta posible, se indican los iniciales y los números de los fabricantes ó propietarios. Las primeras dos letras de los números de estas piezas son abreviaciones del nombre del fabricante, ó sea:

Prefijo:			
AC	— AC Delco	KLG	— KLG Ltd
BB	— Borg and Beck	LU	— Lucas
CAV	— CAV Ltd	SM	— Smith
GI	— Girling	SU	— SU
JA	— British Jaeger	SX	— Solex

Al pedir piezas de repuesto, siempre:

(1) Se citará el número de la pieza o número del conjunto, dando el número de la pieza y número del vehículo para el cual se precisa.

(2) Colóquese el pedido con el Distribuidor Rover más próximo.

(3) Especifíquese como se ha de efectuar el envío.

INDEX TO SECTIONS

INDEX

CYLINDER BLOCK, PETROL ENGINE, 2 LITRE, Series II

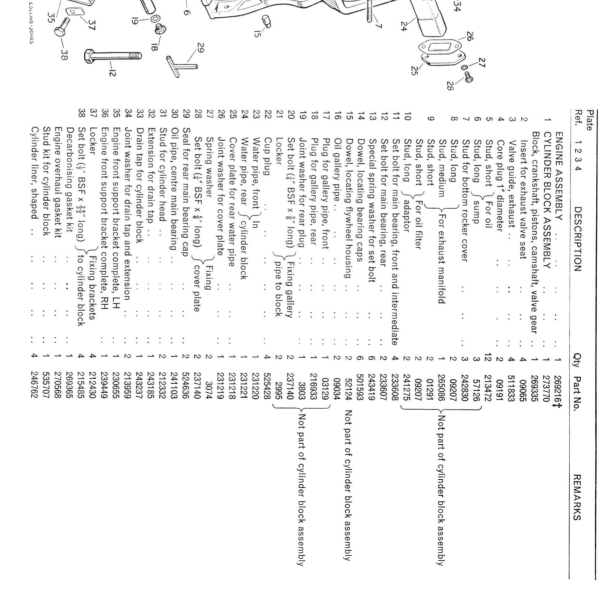

C700.

COLLINS-JONES

CYLINDER BLOCK, PETROL ENGINE, 2 LITRE, Series II

Plate Ref.	1 2 3 4	DESCRIPTION	Qty	Part No.	REMARKS
		ENGINE ASSEMBLY	1	269216†	
1		CYLINDER BLOCK ASSEMBLY	1	273770	
2		Block, crankshaft, pistons, camshaft, valve gear	1	269335	
3		Insert for exhaust valve seat	4	09065	
4		Valve guide, exhaust	4	511833	
5		Core plug 1" diameter	12	09191	
6		Stud, short } For oil	2	213472	
7		Stud, short } sump	3	57126	
8		Stud for bottom rocker cover	2	242830	
9		Stud, long }	2	09207	} For exhaust manifold Not part of cylinder block assembly
10		Stud, medium }	2	265086	
11		Stud, short }	2	01291	
12		Stud, short } For oil filter	2	09207	} Not part of cylinder block assembly
13		Stud, long } adaptor	2	241275	
14		Set bolt for main bearing, front and intermediate	6	233608	
15		Set bolt for main bearing, rear	6	243607	
16		Special spring washer for set bolt	2	243419	
17		Dowel, locating bearing caps	2	501593	Not part of cylinder block assembly
18		Dowel, locating flywheel housing	2	52124	
19		Oil gallery pipe	1	09034	
20		Plug for gallery pipe, front	1	216933	
21		Plug for gallery pipe, rear	1	03129	
22		Joint washer for rear plug	1	3803	
23		Set bolt (¼" BSF x ⅝" long) } Fixing gallery pipe to block	2	237140	
24		Locker	2	2995	
25		Cup plug	4	525428	
26		Water pipe, front } In	2	231220	
27		Water pipe, rear } cylinder block	2	231221	
28		Cover plate for rear water pipe	1	231218	
29		Joint washer for cover plate	1	231219	
30		Spring washer }	2	3074	
31		Set bolt (¼" BSF x ⅝" long) } Fixing cover plate	2	237140	
32		Seal for rear main bearing cap	1	524636	
33		Oil pipe, centre main bearing	2	241103	
34		Stud for cylinder head	1	212332	
35		Extension for drain tap	1	243185	
36		Drain tap for cylinder block	1	243237	
37		Joint washer for drain tap and extension	2	213959	
38		Engine front support bracket complete, RH	1	230655	
		Engine front support bracket complete, LH	1	239449	
		Set bolt (¼" BSF x 9/32" long) } Fixing brackets	4	212430	} to cylinder block
		Locker	4	215485	
		Engine overhaul gasket kit	1	269365	
		Decarbonising gasket kit	1	270568	
		Stud kit for cylinder block	1	535707	
		Cylinder liner, shaped	4	246762	

* Asterisk indicates a new part which has not been used on any previous Rover model

† Engine assembly does not include carburetter and fittings

C701.

CRANKSHAFT, PETROL ENGINE, 2 LITRE, Series II

Plate Ref.	1 2 3 4	DESCRIPTION	Qty	Part No.	REMARKS
1		CRANKSHAFT ASSEMBLY, STD	1	241893	
		CRANKSHAFT ASSEMBLY, .010" US	1	242255	
		CRANKSHAFT ASSEMBLY, .020" US	1	243256	
		CRANKSHAFT ASSEMBLY, .030" US	1	243257	} Home only
		CRANKSHAFT ASSEMBLY, .040" US	1	243258	
2		Dowel for flywheel	1	6395	
3		Main bearing, front and rear, Std	2	523306	
		Main bearing, front and rear, .010" US	2	523307	
		Main bearing, front and rear, .020" US	2	523308	
		Main bearing, front and rear, .030" US	2	523309	
		Main bearing, front and rear, .040" US	2	523310	
4		Main bearing, centre, Std	2	523311	
		Main bearing, centre, .010" US	2	523312	
		Main bearing, centre, .020" US	2	523313	} Pairs
		Main bearing, centre, .030" US	2	523314	
		Main bearing, centre, .040" US	2	523315	
5		Thrust washer for crankshaft, Std	1	523316	
		Thrust washer for crankshaft, .0025" OS	1	523317	
		Thrust washer for crankshaft, .005" OS	1	523318	
		Thrust washer for crankshaft, .075" OS	1	523319	
		Thrust washer for crankshaft, .010" OS	1	523320	
6		Chainwheel on crankshaft	1	59054	
7		Key locating chainwheel	1	52015	
8		Oil thrower on crankshaft	1	59053	
9		VIBRATION DAMPER ASSEMBLY	1	215935	
		DAMPER FLYWHEEL AND BUSH ASSEMBLY	1	215936	
10		Bush for flywheel	1	236289	
11		Driving flange	1	09017	
12		Rubber disc for damper	1	03017	
14		Retaining plate	2	03018	
15		Set screw ($\frac{1}{4}$" BSF × $\frac{9}{16}$" long) fixing plate to flywheel	6	03093	
16		Starting dog	1	503665	
17		Lock-washer for starting dog	1	03092	
18		Crankshaft oil retainer	1	275807	
19		Dowel, lower $\Big\}$ Fixing retainer to cylinder	2	519064	
20		Dowel, upper	2	246464	
21		Spring washer $\Big\}$ block and rear	6	3074	
22		Bolt ($\frac{1}{4}$" BSF × $\frac{7}{8}$" long) $\Big\}$ bearing cap	6	237142	

* Asterisk indicates a new part which has not been used on any previous Rover model

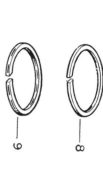
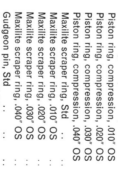

CONNECTING ROD AND PISTON, PETROL ENGINE, 2 LITRE, Series II

C723.

COLLINS-JONES

Plate Ref.	1 2 3 4	DESCRIPTION	Qty	Part No.	REMARKS
1		CONNECTING ROD ASSEMBLY	4	265737	
2		Gudgeon pin bush	4	273163	
3		Special bolt ⎱ Fixing connecting	8	07884	
4		Castle nut (⅜" BSF) ⎰ rod cap	8	2102	
5		Split pin fixing connecting rod cap ...	8	2392	
6		Connecting rod bearing halves, Std	4	523301	
		Connecting rod bearing halves, .010" US	4	523302	
		Connecting rod bearing halves, .020" US	4	523303	
		Connecting rod bearing halves, .030" US	4	523304	
		Connecting rod bearing halves, .040" US	4	523305	
7		PISTON ASSEMBLY, STD	4	278372	
		PISTON ASSEMBLY, .010" OS ...	4	278373	
		PISTON ASSEMBLY, .020" OS ...	4	278374	
		PISTON ASSEMBLY, .030" OS ...	4	278375	
		PISTON ASSEMBLY, .040" OS ...	4	278376	
8		Piston ring, compression, Std ...	8	231155	
		Piston ring, compression, .010" OS	8	236173	
		Piston ring, compression, .020" OS	8	236174	
		Piston ring, compression, .030" OS	8	236175	
		Piston ring, compression, .040" OS	8	236176	
9		Maxilite scraper ring, Std ...	4	234817	
		Maxilite scraper ring, .010" OS ...	4	236183	
		Maxilite scraper ring, .020" OS ...	4	236184	
		Maxilite scraper ring, .030" OS ...	4	236185	
		Maxilite scraper ring, .040" OS ...	4	236186	
10		Gudgeon pin, Std	4	264569	
		Gudgeon pin, .001" OS	4	267257	
		Gudgeon pin, .003" OS	4	267258	
11		Circlip for gudgeon pin	8	235100	

CYLINDER HEAD, PETROL ENGINE, 2 LITRE, Series II

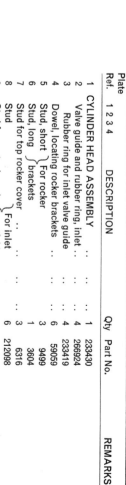

B 494

CYLINDER HEAD, PETROL ENGINE, 2 LITRE, Series II

Plate Ref.	1 2 3 4	DESCRIPTION	Qty	Part No.	REMARKS
1		CYLINDER HEAD ASSEMBLY	1	233430	
2		Valve guide and rubber ring, inlet ...	4	266924	
3		Rubber ring for inlet valve guide ...	4	233419	
4		Dowel, locating rocker brackets ...	6	59059	
5		Stud, short } For rocker brackets	3	9499	
6		Stud, long }	1	3604	
7		Stud for top rocker cover ...	3	6316	
8		Stud } For inlet	6	212098	
9		Stud for centre branch manifold	4	243531	
10		Rear end cover for cylinder head ...	1	213811	
11		Joint washer for rear end cover ...	1	213810	
12		Spring washer ...	4	3075	
13		Set bolt (⁵⁄₁₆" BSF x ⅝" long) } Fixing rear end cover	4	237161	
14		Thermostat switch for choke warning light ...	1	213574	
15		Spring washer ...	2	3073	
16		Set bolt (2 BA x 1¾" long) } Fixing switch	2	237119	
17		Cylinder head gasket ...	1	263666	
18		Spring washer ...	2	3077	
19		Special nut ...	2	212702	
20		Special set bolt, short } Fixing cylinder head to cylinder block	4	237345	
21		Special set bolt, medium	4	212334	
22		Special set bolt, long	6	212333	
23		ROCKER COVER, TOP, ASSEMBLY	1	246340	
24		Tappet clearance plate ...	1	213245	
25		Drive screw fixing plate ...	4	78001	
26		Joint washer for top rocker cover ...	1	267781	
27		Fibre washer } Fixing top	3	3055	
28		Special nut } rocker cover	3	279330	
29		Breather filter for top cover, AC 7222264 ...	1	230996	
30		Set screw fixing breather ...	1	231144	

* Asterisk indicates a new part which has not been used on any previous Rover model

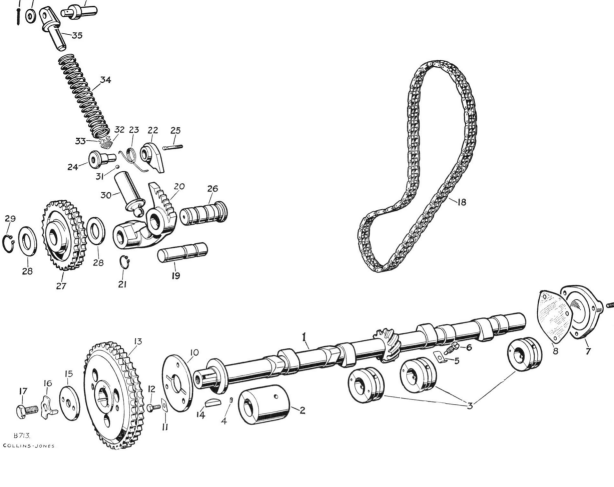

B713.
COLLINS-JONES

CAMSHAFT AND TENSIONER MECHANISM, PETROL ENGINE, 2 LITRE, Series II

Plate Ref. 1 2 3 4	DESCRIPTION		Qty	Part No.	REMARKS
1	Camshaft	... ...	1	511036	
2	Bearing complete for camshaft, front	...	1	09091	
3	Bearing complete for camshaft, centre and rear	...	3	233072	
4	Dowel for camshaft bearing	...	8	09208	
5	Locking plate ⎱ Fixing camshaft bearings		4	09205	
6	Special set screw ⎰ to cylinder block		4	213251	
7	Rear end cover for camshaft	...	1	219914	
8	Joint washer for cover	... ...	1	219913	
9	Set bolt (¼" BSF × ⅝" long) fixing cover to block	...	3	237140	
10	Thrust plate for camshaft	...	1	277063	
11	Locker ⎱ Fixing thrust plate		3	2500	
12	Set bolt (¼" BSF × ⅝" long) ⎰ to cylinder block		3	237140	
13	Chainwheel for camshaft	...	1	09023	
14	Key locating camshaft chainwheel	...	1	230313	
15	Retaining washer ⎱ Fixing		1	09093	
16	Locker ⎰ chainwheel		1	09210	
17	Set bolt (⅜" BSF × ¾" long) to camshaft	...	1	237187	
18	Camshaft chain (⅜" pitch × 78 links)	...	1	09156	
19	Pivot pin for chain adjuster arm	...	1	212365	
20	Adjuster arm for timing chain	...	1	212363	
21	Circlip fixing adjuster arm to pivot pin	...	1	09129	
22	Pawl for tensioner arm	...	1	213600	
23	Pawl spring	... ...	1	213601	
24	Pawl pin	... ...	1	213602	
25	Anchor pin for spring	...	1	214131	
26	Spindle for timing chain adjuster idler wheel	...	1	212364	
27	Idler wheel for timing chain	...	1	09075	
28	Steel washer ⎱ Fixing idler wheel		2	09126	
29	Circlip ⎰ to spindle		1	09128	
30	CYLINDER ASSEMBLY FOR TIMING CHAIN ADJUSTER	...	1	265331	
31	Steel ball (³⁄₁₆") for non-return valve	...	1	3746	
32	Spring retaining ball in cylinder	...	1	212339	
33	Retainer for steel ball	...	1	219489	
34	Spring ⎱ For timing		1	212340	
35	Piston ⎰ chain adjuster		1	219504	
36	Pivot pin for chain adjuster	...	1	212341	
37	Steel washer ⎱ Fixing chain adjuster piston		1	210679	
38	Split pin ⎰ piston to pivot pin		1	2555	

* Asterisk indicates a new part which has not been used on any previous Rover model

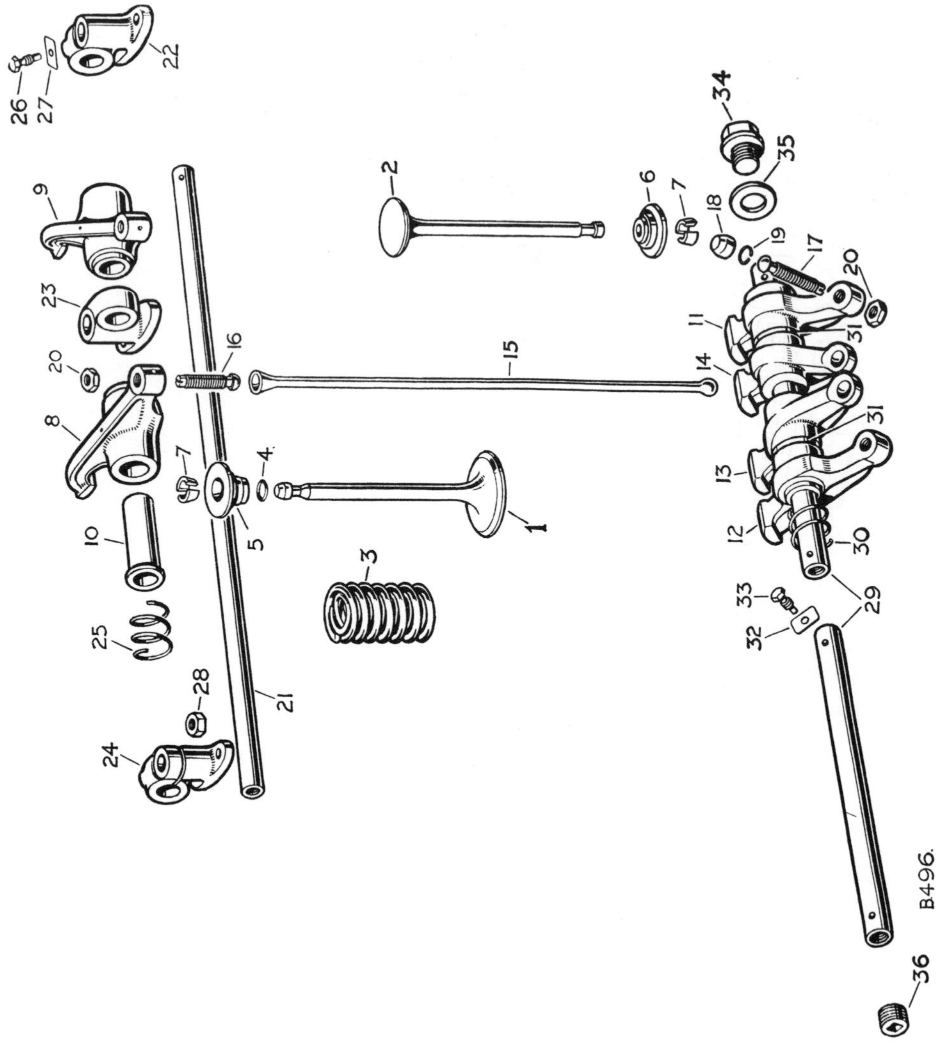

VALVE GEAR AND ROCKER SHAFTS, PETROL ENGINE, 2 LITRE, Series II

B496.

VALVE GEAR AND ROCKER SHAFTS, PETROL ENGINE, 2 LITRE, Series II

Plate Ref.	1 2 3 4	DESCRIPTION	Qty	Part No.	REMARKS
1		Inlet valve	4	233427	
2		Exhaust valve	4	264137	
3		Valve spring, inner and outer	8	273257	
4		Rubber sealing ring for valve, inlet	4	233419	
5		Valve spring cup, inlet	4	212137	
6		Valve spring cup, exhaust	4	212137	
7		Split cone for valve, pair	8	09906	
8		VALVE ROCKER ASSEMBLY, INLET, LH	2	214351	
9		VALVE ROCKER ASSEMBLY, INLET, RH	2	214350	
10		Bush for inlet valve rocker	4	214348	
11		Valve rocker, exhaust, LH	2	239547	
12		Valve rocker, exhaust, RH	2	239546	
13		Inlet cam follower, LH	2	239545	
14		Inlet cam follower, RH	2	239544	
15		Tappet push rod, inlet	4	09122	
16		Tappet adjusting screw, inlet	4	506812	
17		TAPPET ADJUSTING SCREW ASSEMBLY, EXHAUST			
18		Cap for exhaust tappet adjusting screw	4	506818	
19		Circlip fixing cap	4	212160	
20		Locknut for tappet adjusting screw	8	212161	
21		Valve rocker shaft, top	1	231057	
22		Rocker bracket, oil feed, top	1	212328	
23		Rocker bracket, intermediate, top	2	212329	
24		Rocker bracket, clamping, top	3	212330	
25		Spring for rocker shaft, top	4	502006	
26		Locating screw ⎱ For rocker shaft	1	09004	
27		Locating plate ⎰ at oil feed bracket	1	09005	
28		Nut (5⁄16" BSF) fixing rocker bracket to cylinder head	4	3490	
29		Valve rocker shaft, bottom	2	239089	
30		Spring for rocker shaft, bottom	4	500610	
31		Valve rocker shaft, bottom	4	09159	
32		Thrust washer for rocker shaft, bottom	2	09005	
33		Locating screw ⎱ Fixing bottom rocker	2	09004	
34		Locating plate ⎰ shafts to cylinder block	2	536577	
35		End plug, rear ⎱ For bottom	1	3055	
36		End plug, front ⎰ rocker shaft	1	12455	

C702.

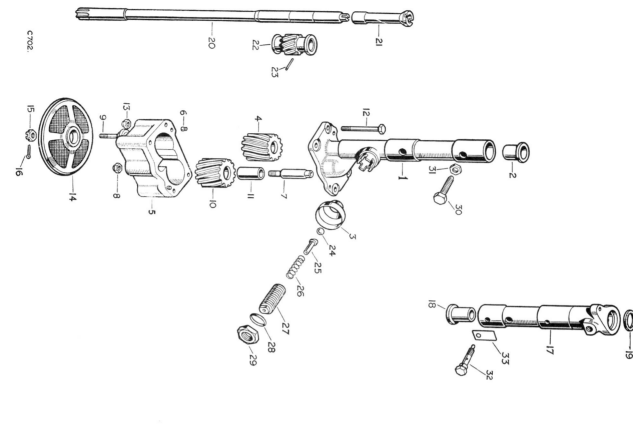

OIL PUMP, PETROL ENGINE, 2 LITRE, Series II

Plate Ref.	1 2 3 4	DESCRIPTION		Qty	Part No.	REMARKS
1		OIL PUMP ASSEMBLY	... ...	1	240651	
2		OIL PUMP BODY ASSEMBLY	...	1	501253	
3		Bush for drive shaft	...	1	212309	
4		Oil pump shield	...	1	09225	
5		Oil pump gear, driver	...	1	09027	
6		OIL PUMP COVER ASSEMBLY	...	1	234296	
7		Dowel locating body		2	52710	
8		Spindle for idler wheel		1	09050	
9		Self-locking nut ($\frac{3}{8}$") fixing spindle		1	265055	
10		Stud for oil strainer		1	234291	
11		OIL PUMP GEAR IDLER ASSEMBLY		1	212375	
12		Bush for idler gear	... ...	1	09049	
13		Bolt ($\frac{5}{16}$" BSF x 2$\frac{1}{2}$" long)	Fixing cover	4	237170	
14		Self-locking nut ($\frac{5}{16}$") to body		4	251321	
15		Oil strainer for pump		1	266900	
16		Castle nut ($\frac{1}{4}$" BSF) Fixing oil strainer		1	251600	
17		Split pin to pump		1	2422	
18		DISTRIBUTOR HOUSING ASSEMBLY		1	214047	
19		Bush for drive shaft		1	09025	
20		Cork washer for housing		1	52278	
21		Oil pump drive shaft		1	235083	
22		Drive shaft for distributor		1	267829	
23		Oil pump driving gear	...	1	212308	
24		Taper pin, fixing gear to shaft	...	1	3005	
25		Steel ball		1	01035	
26		Plunger		1	245940	
27		Spring For oil pressure		1	242648	
28		Adjusting screw release valve		1	242889	
29		Washer		1	230508	
30		Locknut		1	12028	
31		Special set screw Fixing oil pump		1	241267	
32		Locknut to cylinder block		1	3490	
33		Oil feed bolt, locating distributor housing	...	1	09052	
		Locker for bolt	... ...	1	2504	

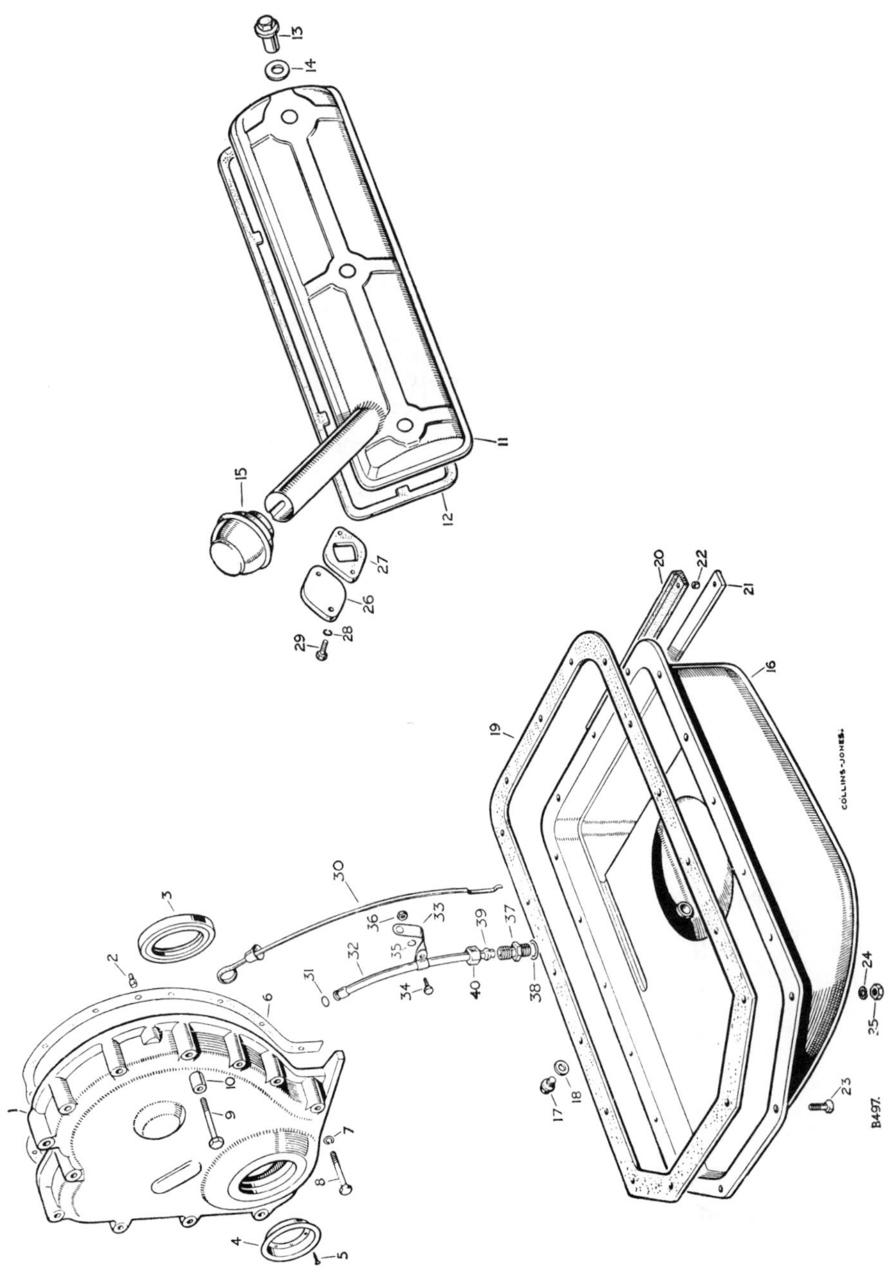
COLLINS-JONES.
B497.

FRONT COVER, SIDE COVER AND SUMP, PETROL ENGINE, 2 LITRE, Series II

Plate Ref. 1 2 3 4	DESCRIPTION	Qty	Part No.	REMARKS
1	FRONT COVER ASSEMBLY	1	236018	
2	Dowel locating front cover	2	6395	
3	Oil seal for front cover	1	213744	
4	Mud excluder	1	236011	
5	Hammer drive screw fixing excluder	8	3936	
6	Joint washer for front cover	1	09102	
7	Spring washer } Fixing front cover and water pipe to cylinder block	11	3075	
	Plain washer } Fixing front cover and water pipe to cylinder block	1	2550	
8	Set bolt ($\frac{5}{16}$" BSF x $1\frac{29}{32}$" long) } Fixing front cover and water pipe to cylinder block	10	250530	
9	Set bolt ($\frac{5}{16}$" BSF x $2\frac{1}{4}$" long) } Fixing front cover and water pipe to cylinder block	1	250531	
10	Distance piece } Fixing front cover and water pipe to cylinder block	1	59155	
11	Rocker cover, side	1	236003	
12	Joint washer for rocker cover	1	267779	
13	Special nut fixing side rocker cover	3	268543	
14	Sealing washer for nut	3	267828	
15	Oil filler cap and breather filter, AC 7964819	1	546440	
	Oil recommendation label for oil filler cap	1	272476	
16	Crankcase sump	1	246201	
17	Drain plug for crankcase sump	1	536577	
18	Washer for drain plug	1	243959	
19	Joint washer for crankcase sump	1	267780	
20	Rubber seal } For rear end of crankcase sump	1	231480	
21	Packing strip } For rear end of crankcase sump	1	231479	
22	Distance piece } For rear end of crankcase sump	3	231478	
23	Set bolt ($\frac{5}{16}$" Whit x $\frac{11}{16}$" long) } Fixing sump to cylinder block	2	215647	
24	Spring washer } Fixing sump to cylinder block	17	3075	
25	Nut ($\frac{5}{16}$" BSF) } Fixing sump to cylinder block	15	3490	
26	Cover plate for breather aperture	1	236435	
27	Joint washer } Fixing breather pipe or cover plate to cylinder block	1	214058	
28	Spring washer } Fixing breather pipe or cover plate to cylinder block	2	3075	
29	Set bolt ($\frac{5}{16}$" BSF x $\frac{3}{4}$" long) } Fixing breather pipe or cover plate to cylinder block	2	237159	
30	Oil level rod	1	237563	
31	Sealing ring for rod	1	532387	
32	Tube for oil level rod	1	236922	
33	Bracket for oil level rod tube	1	236393	
34	Bolt (2 BA x $\frac{1}{2}$" long) } Fixing bracket	1	234603	
35	Spring washer } Fixing bracket	1	3073	
36	Nut (2 BA) } Fixing bracket	1	2247	
37	Double-ended union } Fixing tube to cylinder block	1	236060	
38	Washer } Fixing tube to cylinder block	1	243958	
39	Olive } Fixing tube to cylinder block	1	236408	
40	Union nut } Fixing tube to cylinder block	1	236407	

* Asterisk indicates a new part which has not been used on any previous Rover model

C703.

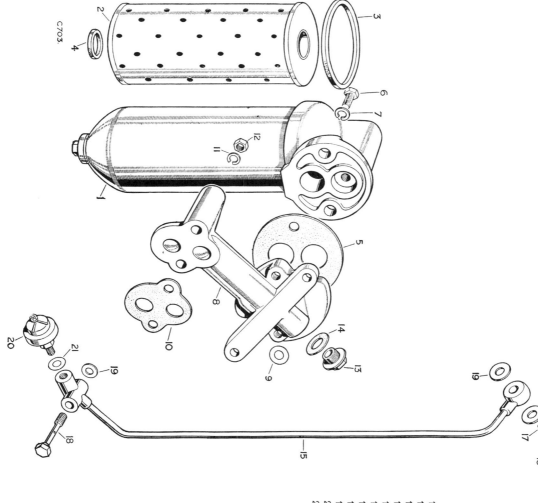

OIL FILTER AND OIL PIPES, PETROL ENGINE, 2 LITRE, Series II

Plate Ref.	1 2 3 4	DESCRIPTION	Qty	Part No.	REMARKS
1		Oil filter for engine, AC 7961133	1	513591	For filter with bottom centre bolt ⎱ Check
2		Element, AC FF 24 ...	1	246262	For filter with top centre bolt ⎰ before ordering
3		Gasket, large ...	1	246261	
4		Gasket, small, 1 9/16" diameter, AC 1530635 ...	2	277611	
5		Gasket, small, 1 5/16" diameter, AC 1531283 ⎱ For	2	516370	
		Rubber washer for centre bolt, AC 1531098 ... ⎰ oil filter	1	269889	
6		Bolt (7/16" Whit x 1 5/32" long) ⎱ Fixing filter to adaptor	1	272839	
7		Joint washer for oil filter to adaptor ... ⎰	2	240308	
		Spring washer on cylinder block	2	3077	
8		Adaptor for oil filter ...	1	241166	
9		Packing washer, top ⎱ For oil filter adaptor	2	4095	
10		Joint washer, bottom ⎰	1	268472	
11		Spring washer ⎱ Fixing adaptor to	4	3076	
12		Nut (3/8" BSF) ⎰ cylinder block	4	2827	
13		Plug ⎱ For adaptor	2	536577	
14		Joint washer ⎰ oil way	2	231577	
15		Oil pipe complete to cylinder head ...	1	267981	
16		Banjo bolt, oil pipe to cylinder head ...	1	504840	
18		Banjo bolt, oil pipe to cylinder block ...	1	233520	
19		Joint washer for banjo bolts ...	4	231575	
20		Oil pressure switch ...	1	519863	
21		Joint washer for switch ...	1	232039	

* Asterisk indicates a new part which has not been used on any previous Rover model

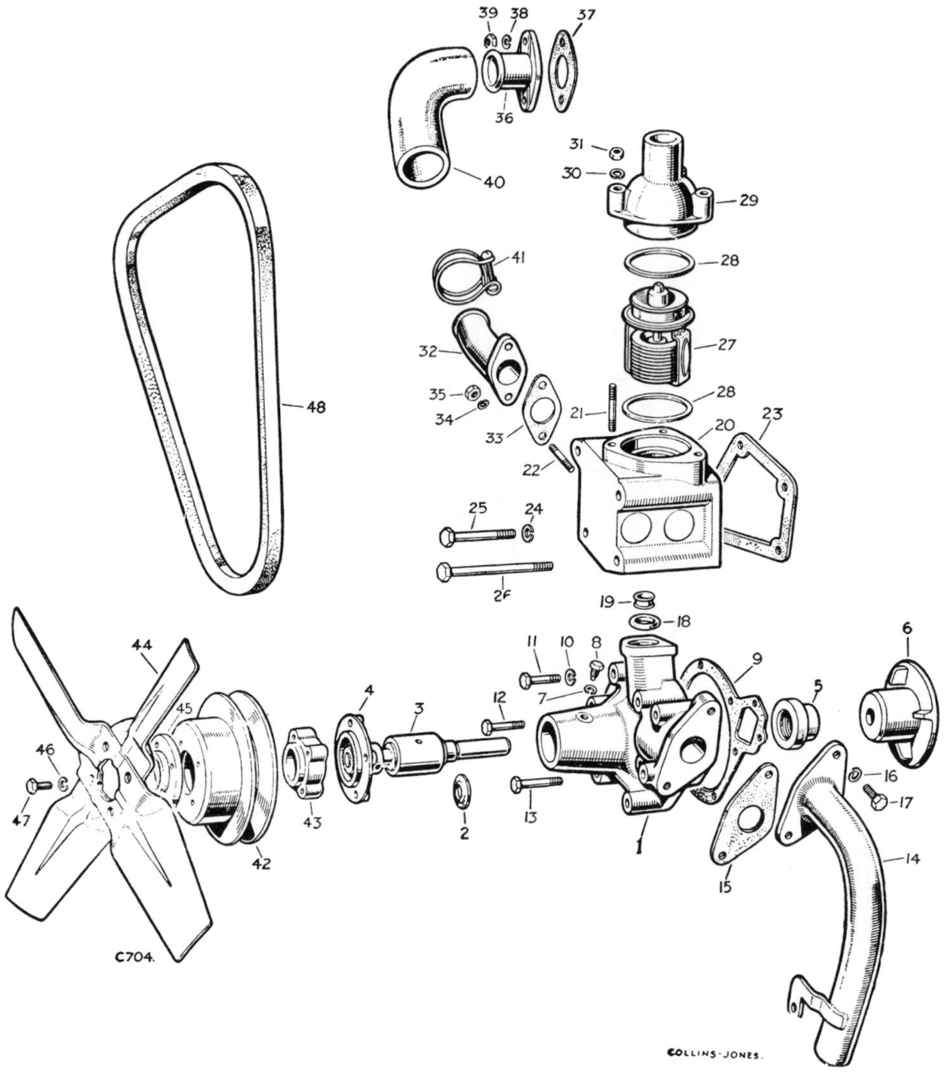
C704.
COLLINS-JONES.

WATER PUMP AND THERMOSTAT, PETROL ENGINE, 2 LITRE, Series II

Plate Ref.	1 2 3 4	DESCRIPTION	Qty	Part No.	REMARKS
		WATER PUMP ASSEMBLY	1	269974	
1		Water pump casing with washer	1	271572	
2		Water deflector washer	1	266985	
3		Pump spindle and bearing	1	213695	
4		Hub for fan	1	213698	
5		Carbon ring and seal unit	1	239855	
6		Impeller for pump	1	242329	
7		Spring washer } Locating bearing casing	1	3074	
8		Set bolt (¼" BSF x ⅞" long) } Locating bearing casing	1	211064	
9		Joint washer for water pump	1	09118	
10		Spring washer } Fixing water pump to cylinder block	8	3074	
11		Set bolt (¼" BSF x 1½" long) } Fixing water pump to cylinder block	7	250518	
13		Set bolt (¼" BSF x 2¼" long) } Fixing water pump to cylinder block	1	237150	
14		Inlet pipe for water pump	1	277508	
15		Joint washer for inlet pipe	1	213724	
16		Spring washer } Fixing pipe to water pump	2	3075	
17		Set bolt (5/16" Whit x 11/16" long) } Fixing pipe to water pump	2	215647	
18		Rubber joint ring } Connecting water pump to thermostat housing	1	09170	
19		Copper tube } Connecting water pump to thermostat housing	1	09197	
20		THERMOSTAT HOUSING ASSEMBLY	1	263670	
21		Stud for outlet pipe	3	212042	
22		Stud for inlet pipe	2	212104	
23		Joint washer for thermostat housing	1	213811	
24		Spring washer } Fixing thermostat housing to cylinder head	1	3075	
25		Set bolt (5/16" BSF x 2 19/32" long) } Fixing thermostat housing to cylinder head	1	215180	
26		Set bolt (5/16" BSF x 3 11/16" long) } Fixing thermostat housing to cylinder head	3	215191	
27		Thermostat, AC 1572249	1	515962	
28		Fibre washer for thermostat	2	01645	
29		Water outlet pipe, thermostat to radiator	1	263607	
30		Spring washer } Fixing outlet pipe to thermostat	3	3074	
31		Nut (¼" BSF) } Fixing outlet pipe to thermostat	3	2823	
32		Water inlet elbow to thermostat	1	263630	
33		Joint washer for inlet elbow	1	210419	
34		Spring washer } Fixing elbow to thermostat housing	2	3074	
35		Nut (¼" BSF) } Fixing elbow to thermostat housing	2	2823	
36		Water outlet pipe from manifold	1	242055	
37		Joint washer for outlet pipe	1	210419	
38		Spring washer } Fixing water outlet pipe to manifold	2	3074	
39		Nut (¼" BSF) } Fixing water outlet pipe to manifold	2	2823	
40		Rubber hose } Connecting water pipe to inlet elbow	1	242056	
41		Clip for hose } Connecting water pipe to inlet elbow	2	50320	
42		Fan pulley	1	230492	
43		Distance piece for fan pulley	1	501568	
44		Fan blade	1	27156R	
45		Reinforcing plate for fan blade	1	244317	
46		Spring washer } Fixing fan blade and pulley to hub	4	3074	
47		Set bolt (¼" BSF x 1" long) } Fixing fan blade and pulley to hub	4	237143	
48		Fan and dynamo belt	1	218576	

* Asterisk indicates a new part which has not been used on any previous Rover model

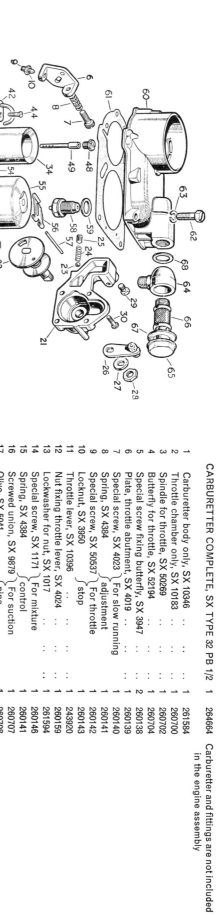

B.310.

CARBURETTER, PETROL ENGINE, 2 LITRE, Series II

CARBURETTER, PETROL ENGINE, 2 LITRE, Series II

Plate Ref.	1	2	3	4	DESCRIPTION	Qty	Part No.	REMARKS
					CARBURETTER COMPLETE, SX TYPE 32 PB 1/2	1	264664	Carburetter and fittings are not included in the engine assembly
1					Carburetter body only, SX 10346	1	261584	
2					Throttle chamber only, SX 10183	1	260700	
3					Spindle for throttle, SX 50269	1	260702	
4					Butterfly for throttle, SX 52194	1	260704	
5					Special screw fixing butterfly, SX 3947	2	260138	
6					Plate, throttle abutment, SX 4019	1	260139	
7					Special screw, SX 4023 } For slow running adjustment	1	260140	
8					Spring, SX 4384	1	260141	
9					Special screw, SX 50537 } For throttle stop	1	260142	
10					Locknut, SX 3950	1	260143	
11					Throttle lever, SX 10396	1	243920	
12					Nut fixing throttle lever, SX 4024	1	260159	
13					Lockwasher for nut, SX 1017	1	261594	
14					Special screw, SX 1171 } For mixture control	1	260146	
15					Spring, SX 4384	1	260141	
16					Screwed union, SX 9879 } For suction pipe	1	260707	
17					Olive, SX 5041	1	260708	
18					Joint washer for throttle chamber, SX 52788	1	260754	
19					Special screw and washer, SX 12493, fixing chamber to carburetter body	4	513174	
21					Starter body, SX 52778	1	261580	
22					Starter valve complete, SX 9942	1	261581	
23					Ball, SX 51728 } For starter valve	1	260155	
24					Spring, SX 51762 }	1	260156	
25					Plug retaining starter valve spring, SX 51274/1	1	270391	
26					Lever for starter, SX 9882	1	261582	
27					Special washer for lever, SX 4031/1	1	261410	
28					Nut fixing starter lever, SX 4024	1	260159	
29					Special bolt fixing starter cable, SX 12056	1	265495	
30					Special screw fixing starter body, SX 5142/12	4	262313	
31					Accelerator pump complete, SX 10146	1	261562	
					Diaphragm for pump, SX 10138	4	243394	
32					Joint washer for pump, SX 52825	1	260739	
33					Special screw fixing pump, SX 10141	4	260737	
34					Choke tube (25), SX 52846/25	1	262659	
35					Special screw fixing choke tube, SX 50362	1	260163	
36					Jet, economy (75)	1	261974	
37					Fibre washer for jet, SX 52825	1	260734	
38					Non-return valve (1.5 m/m), SX 11093/150	1	288568	
39					Fibre washer for valve, SX 52825	1	260734	
40					Jet (75) accelerator pump, SX 52200/75	1	262448	
41					Fibre washer for jet, SX 52825	1	260734	
42					Pump injector, SX 52952	1	260732	
43					Joint washer for pump, injector, SX 52735	1	261570	
44					Special screw fixing injector, SX 3947/2	1	260767	
45					Main jet (115), SX 50552/6/115	1	260173	

SEE PAGE 31 FOR ALTERNATIVE JETS

* Asterisk indicates a new part which has not been used on any previous Rover model

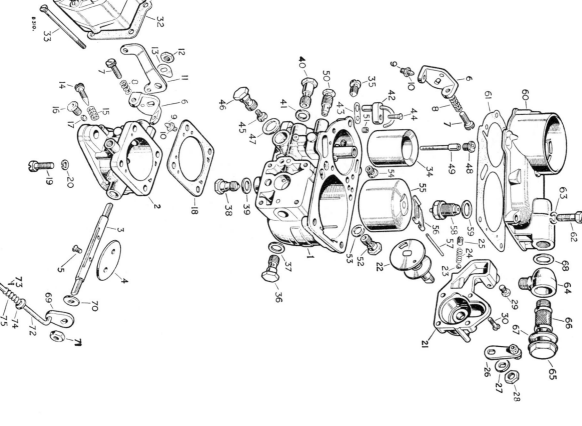

B.310.

COLLINS-JONES

CARBURETTER, PETROL ENGINE, 2 LITRE, Series II

CARBURETTER, PETROL ENGINE, 2 LITRE, Series II

Plate Ref.	1 2 3 4	DESCRIPTION	Qty	Part No.	REMARKS
46		Bolt, main jet carrier, SX 50820	1	260180	
47		Fibre washer for bolt, SX 50815	1	260181	
48		Correction jet (240), SX 51612/240	1	260266	
49		Emulsion tube, SX 52684	1	260743	
50		Pilot jet (55), SX 50797/4/55	1	260269	
51		Jet air bleed (1.5), SX 51274/15	1	260179	
52		Starter jet, petrol (135), SX 52823/135	1	260752	
53		Fibre washer for jet, SX 52825	1	260734	
54		Starter jet, air (5.5), SX 50906/55	1	260719	
55		Float, SX 51638/1	1	260745	
56		Toggle for float, SX 52180	1	260753	
57		Spindle for toggle, SX 52204	1	260747	
58		Needle valve complete, SX 51305/17	1	260164	
59		Fibre washer for valve, SX 2261	1	260165	
60		Top cover for carburetter, SX 9946	1	260721	
61		Joint washer for top cover, SX 52787	1	260749	
62		Special screw fixing top cover, SX 12869	3	512401	
63		Spring washer for screw, SX 971	3	260714	
64		Banjo union, SX 4120/5	1	260714	
65		Special bolt for union, SX 4122	1	260167	
66		Filter gauze for union, SX 4123	1	260168	
67		Fibre washer, large, SX 4124 } For	1	260169	
68		Fibre washer, small, SX 4124/1 } union	1	260170	
69		Lever for accelerator pump rod, SX 10692	1	261564	
70		Special washer for lever, SX 4031/1	1	261410	
71		Nut fixing lever to spindle, SX 4024	1	260159	
72		Control rod for accelerator pump, SX 0775	1	262312	Not part of carburetter
73		Split pin	2	3359	
74		Plain washer, SX 11051	2	260771	
75		Spring, SX 52760 } For control rod	1	240689	
		Spring, SX 2979/1	1	237774	
		Carburetter overhaul kit	1	266693	When engine governor is fitted
		Carburetter gasket kit	1	274895	

CARBURETTER SETTINGS WHICH MAY BE ADVANTAGEOUS FOR HIGH ALTITUDES OR TROPICAL CONDITIONS

Main jet (110), SX 50552/6/110		1	260274 }	3,000 to 6,000 ft
Air bleed jet (2.0), SX 51274/2.0		1	260275 }	
Main jet (107.5), SX 50552/6/107.5		1	261563 }	6,000 to 10,000 ft
Air bleed jet (2.0), SX 51274/2.0		1	260275 }	
Choke tube (26), SX 51210/26		1	262647	
Main jet (107.5), SX 50552/6/107.5		1	261563 }	10,000 to 15,000 ft
Air bleed jet (2.0), SX 51274/2.0		1	260275 }	
Correction jet (260), SX 51612/260		1	273788 }	
Main jet (110), SX 50552/6/110		1	260274	Tropical conditions

* Asterisk indicates a **new** part which has not been used on any previous Rover model

MANIFOLDS, PETROL ENGINE, 2 LITRE, Series II

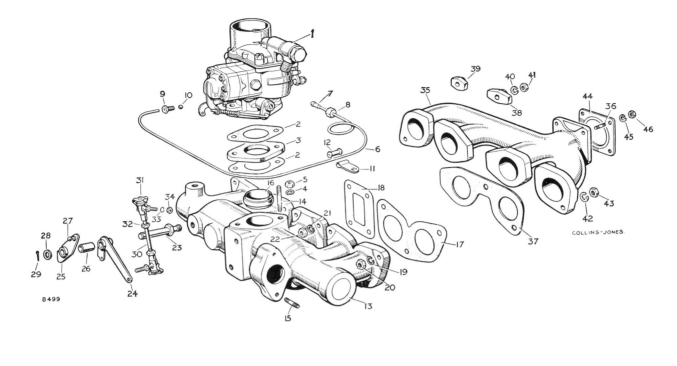

B499   COLLINS-JONES

Plate Ref.	1	2	3	4	DESCRIPTION	Qty	Part No.	REMARKS
1					CARBURETTER COMPLETE ...	1	264664	For component parts, see Pages 29 and 31
2					Joint washer for carburetter	2	212233	
3					Packing for carburetter	1	212232	
4					Spring washer } Fixing carburetter	2	3075	
5					Nut (5/16" BSF) } to manifold	2	3490	
6					Suction pipe complete, carburetter to distributor	1	214231	
7					Nipple } For suction pipe,	1	3787	
8					Nut } distributor end	1	3783	
9					Screwed union } For suction pipe,	1	260707	
10					Olive } carburetter end	1	260708	
11					Clip for suction pipe	1	214228	
12					Rubber grommet for clip	1	214229	
13					INLET MANIFOLD ASSEMBLY	1	218954	
14					Stud for carburetter	2	214054	
15					Stud for water outlet pipe	2	212104	
16					Dowel ring for carburetter	2	214019	
17					Joint washer, manifold to cylinder head	1	09179	
18					Joint washer, water branch to cylinder head	2	210447	
19					Spring washer } Fixing manifold to	6	3075	
20					Nut (5/16" BSF) } cylinder head	6	3490	
21					Spring washer } Fixing water branch	4	3074	
22					Nut (1/4" BSF) } to cylinder head	4	2823	
23					Spindle for carburetter bell crank and levers	1	240306	
24					CARBURETTER BELL CRANK ASSEMBLY	1	240181	
25					CARBURETTER RELAY LEVER ASSEMBLY	1	240180	
26					Bush for lever	2	240303	
27					Ball end for lever	1	1481	
28					Plain washer } Fixing bell crank and	1	2210	
29					Split pin } levers to spindle	1	2392	
30					ROD ASSEMBLY, BELL CRANK TO CARBURETTER	1	231267	
31					Ball joint complete	2	231261	
32					Locknut (1/4" BSF) for ball joint	2	2823	
33					Spring washer } Fixing rod to bell crank	2	3074	
34					Nut (1/4" BSF) } and carburetter	2	2823	
35					EXHAUST MANIFOLD ASSEMBLY	1	237978	
36					Stud for exhaust pipe	4	213472	
37					Joint washer, front } For exhaust	1	534100	
37					Joint washer, rear } manifold	1	534102	
38					Clamp, centre } For exhaust	2	09162	
39					Clamp, outer } manifold	2	09161	
40					Spring washer } For exhaust	3	3076	
41					Nut (3/8" BSF) } manifold clamps	3	2827	
42					Spring washer } Fixing ends of	2	3075	
43					Nut (5/16" BSF) } exhaust manifold	2	3490	
44					Joint washer for exhaust pipe	1	213358	
45					Spring washer } Fixing pipe	4	3075	
46					Nut (5/16" BSF) } to manifold	4	3771	

COLLINS-JONES.

DISTRIBUTOR AND STARTER, PETROL ENGINE, 2 LITRE, Series II

Plate Ref.	1 2 3 4	DESCRIPTION	Qty	Part No.	REMARKS
1		Distributor complete, LU 40504A	1	269240	
2		Distributor cap complete, LU 418871	1	245004	
3		Brush and spring for cap, LU 418856	1	262703	
4		Rotor arm, LU 400051	1	245005	
5		Contact breaker points, set, LU 423153	1	269987	
6		Contact breaker base plate, LU 423318	1	502282	
7		Condenser for distributor, LU 421267	1	269988	
8		Auto advance springs, set, LU 416159/S	1	245009	
9		Auto advance weight, LU 410033/S	1	262708	
10		Vacuum unit, LU 54411727	1	269989	
11		Cam for distributor, LU 496078	1	245008	
12		Shaft for distributor, LU 419689	1	245003	
13		Bush for distributor shaft, LU 419430	1	245012	
14		Driving dog for distributor, LU 420620	1	245014	
15		Clamping plate for distributor, LU 420151	1	245013	
16		Clip for cover, LU 425765	2	245015	
17		Sundry parts kit, LU 419644			
18		Cork washer for distributor housing	1	52278	
19		Special set bolt ⎫	1	215758	
		Spring washer ⎬ Fixing distributor lever	1	3074	
		Plain washer ⎭	1	3911	
20		Rubber insulator for distributor LT lead, LU 419882	1	246560	
21		Sparking plug, Lodge CLNH	4	262796	⎫ Alternatives.
		Sparking plug, Champion N8	4	512806	⎬ Check before ordering
22		Washer for sparking plug	4	40441	
23		Suppressor for sparking plug	4	240138	
24		Sparking plug cover	4	240262	
		Rubber sealing ring for cover	4	213172	
25		Cable nut, LU 410600	4	214278	
26		Washer for cable nut, LU 185015	4	214279	
		Resistor for distributor lead	1	213646	
		HT wire	As reqd	80603	
27		Starter motor complete, LU 25605	1	236287	
28		Bracket for starter, commutator end, LU 256748	1	244706	
29		Bracket, drive end, LU 256495	1	244713	
30		Armature, LU 255463	1	244715	
31		Bush, commutator end, LU 255491	1	242958	
32		Bush, pinion end, LU 270038	1	244714	
33		Pinion and sleeve, LU 255194	1	244711	
34		Spring for pinion, LU 255728	1	244712	
35		Main spring for pinion, LU 270889	1	244710	
36		Nut for pinion, LU 255851	1	244709	
37		Field coil for starter, LU 255625	1	262861	

* Asterisk indicates a new part which has not been used on any previous Rover model

COLLINS · JONES.

DISTRIBUTOR AND STARTER, PETROL ENGINE, 2 LITRE, Series II

Plate Ref.	1 2 3 4	DESCRIPTION		Qty	Part No.	REMARKS
38		Brushes for starter motor, set, LU 255659		2	260055	
39		Spring set for brushes, LU 54257221	...	1	601754	
40		Bolt for bracket, LU 255459 ...	...	2	244717	
41		Cover band, LU 255557 ...	...	1	244705	
42		Grease cap, LU 255854 ...	...	1	243095	
		Sundry parts kit, LU 256762 ...	...	1	244718	
43		Set bolt (⅜" Whit x 1" long) ...	⎫	1	215703	
		Bolt (⅜" BSF x 1½" long)	⎬ Fixing starter to	1	250543	
44		Spring washer	⎭ flywheel housing	2	3076	
45		Nut (⅜" BSF)		1	2827	

* Asterisk indicates a new part which has not been used on any previous Rover model

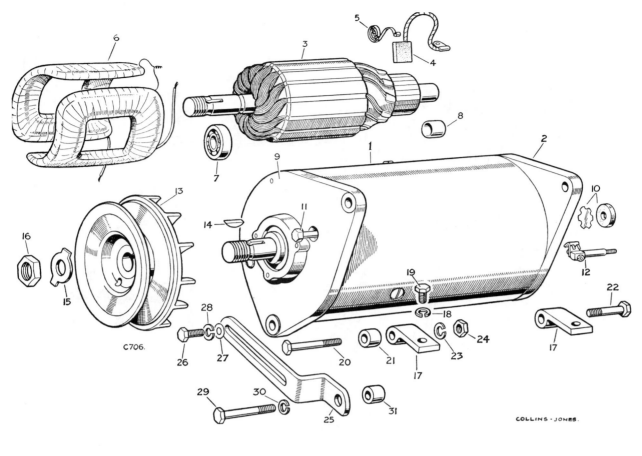

C706.

COLLINS · JONES.

DYNAMO AND FIXINGS, PETROL ENGINE, 2 LITRE, Series II

Plate Ref. 1 2 3 4	DESCRIPTION	Qty	Part No.	REMARKS
1	Dynamo complete, LU 22700, Type C40	1	510274	
2	Bracket, commutator end, LU 227702	1	242669	For dynamo type C39. Check before ordering
3	Armature for dynamo, LU 54217499	1	514194	
4	Brushes for dynamo, set, LU 227305	1	261483	
5	Spring set for brushes, LU 228159	1	261238	
6	Field coil for dynamo, LU 227291	1	262860	
7	Ball bearing for dynamo	1	529221	
8	Bush, commutator end, LU 227818	1	271614	
9	Bracket, drive end, LU 227698	1	242673	
10	Oiler for dynamo, LU 227914	1	271615	
11	Bolt for dynamo, LU 228336	2	242675	
12	Terminal for dynamo, LU 227625	1	264436	
	Sundry parts, set, LU 54211890	1	532567	
	Bracket, commutator end, LU 54211125	1	529221	For dynamo type C40. Check before ordering
	Armature for dynamo, LU 54214237	1	514195	
	Brushes for dynamo, set, LU 227541	1	514194	
	Spring set for brushes, LU 227542	1	514193	
	Field coil for dynamo, LU 227543	1	514192	
	Bolt for bracket, LU 54211097	1	514191	
	Bracket, drive end, LU 54211095	1	271614	
	Bush, commutator end, LU 227818	1	242676	
	Oiler for dynamo, LU 228336	1	514189	
	Terminal set, LU 227726	1	514190	
	Sundry parts, set, LU 54211894	2	242675	
13	Pulley for dynamo	1	236360	
14	Woodruff key (No. 5), LU 189340	1	240270	Fixing pulley to dynamo
15	Lockwasher	1	217781	
16	Special nut	1	240271	
17	Support bracket for dynamo	2	210572	
18	Spring washer	2	3075	Fixing support bracket to cylinder block
19	Set bolt (5/16" BSF x 3/4" long)	2	237325	
20	Bolt (5/16" BSF x 2 1/8" long) front	2	237333	Fixing dynamo to anchor brackets
21	Distance piece, front	1	217780	
22	Bolt (5/16" BSF x 1 1/2" long) rear	1	237329	
23	Spring washer	2	3075	
24	Nut (5/16" BSF)	2	3490	
25	Adjusting link for dynamo	1	500221	
26	Special set bolt	1	253027	Fixing adjusting link to dynamo
27	Plain washer	1	3966	
28	Spring washer	1	3075	
29	Set bolt (5/16" BSF x 2 1/4" long)	1	250531	Fixing adjusting link to cylinder block
30	Spring washer	1	3075	
31	Distance piece	1	210566	

* Asterisk indicates a new part which has not been used on any previous Rover model

FLYWHEEL AND CLUTCH, PETROL ENGINE, 2 LITRE, Series II

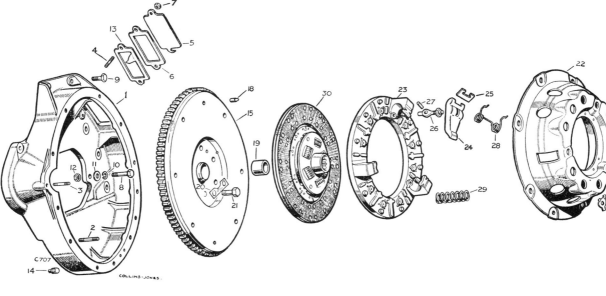

C707

COLLINS-JONES.

FLYWHEEL AND CLUTCH, PETROL ENGINE, 2 LITRE, Series II

Plate Ref.	1 2 3 4	DESCRIPTION	Qty	Part No.	REMARKS
1		FLYWHEEL HOUSING ASSEMBLY	1	246338	
2		Stud ($\frac{3}{8}$" BSF) ⎱ Fixing flywheel housing	12	3650	
3		Stud ($\frac{5}{16}$" BSF) ⎰ to bell housing	1	3200	
4		Stud ($\frac{1}{4}$" BSF) fixing inspection cover	2	3651	
5		Inspection cover plate	1	56140	
6		Joint washer for cover plate	1	50216	
7		Nut ($\frac{1}{4}$" BSF) fixing cover plate	2	2823	
8		Bolt ($\frac{3}{8}$" BSF x 1$\frac{3}{8}$" long) ⎱ Fixing	6	250542	
9		Bolt ($\frac{3}{8}$" BSF x 1$\frac{1}{8}$" long) ⎰ flywheel housing	2	237179	
10		Spring washer ⎱ to cylinder block	8	3076	
11		Plain washer	6	2219	
12		Nut ($\frac{3}{8}$" BSF)	6	2827	
13		Indicator for ignition timing	1	239820	
14		Drain plug for housing	1	3290	
15		FLYWHEEL ASSEMBLY	1	272661	
16		Ring gear for flywheel	1	506799	
17		Set bolt fixing clutch	6	237324	
18		Tab washer for set bolt	6	546197	
19		Dowel locating clutch cover plate	2	502116	
20		Bush for primary pinion	1	08566	
21		Special set bolt ⎱ Fixing flywheel	6	526161	
		Locker ⎰ to crankshaft	3	210211	
22		CLUTCH ASSEMBLY, BB 45693/41	1	236684	With black clutch springs
23		Cover plate for clutch, BB 45481	1	231888	
24		Pressure plate for clutch, BB 42652	1	231880	
25		Release lever for clutch, BB 48234	3	242993	
26		Strut for release lever, BB 42606	3	231884	
27		Eyebolt and nut for release lever, BB 48508	3	242996	
28		Pin for release lever, BB 42604	3	231885	
29		Anti-rattle spring for release lever, BB 47688	3	231883	
		Clutch spring (black), BB 40200	9	231881	For early type clutch
		Clutch spring (cream), BB 44633	9	275301	Part of 236684 Late type clutch ⎱ Check before ordering
30		Clutch plate complete, BB 47626/104	1	275811	
		Lining package for clutch plate, BB 46627	1	261921	Ferodo lining ⎱ Alterna-
		Lining package for clutch plate, BB KL75012	1	517026	Mintex lining ⎰ tives

Ferodo clutch

Not part of engine assembly

* Asterisk indicates a new part which has not been used on any previous Rover model

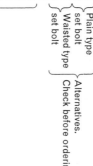

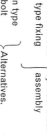

C746.

COLLINS-JONES.

CYLINDER BLOCK, PETROL ENGINE, 2¼ LITRE, Series II and IIA

Plate Ref. 1 2 3 4	DESCRIPTION	Qty	Part No.	REMARKS
	ENGINE ASSEMBLY	1	531746	Series II models
	ENGINE ASSEMBLY	1	600974	Series IIA models up to engine suffix 'J' inclusive
1	ENGINE ASSEMBLY	1	605567*	Series IIA models from engine suffix 'K' onwards
	CYLINDER BLOCK ASSEMBLY	1	276908	Series II models
	CYLINDER BLOCK ASSEMBLY	1	541891	Series IIA models
	Cylinder block, crankshaft, pistons and camshaft	1	601316	Series IIA models
	Cylinder block, crankshaft, pistons and camshaft	1	512412	
2	Core plug, 1⅛" diameter	2	09191	} Alternatives. Check before ordering
2	Core plug, 1⅛" diameter	2	250840	
3	Core plug, 1" diameter	1	524765	
4	Core plug, 1½" diameter	2	555428	
5	Cup plug, 1⅜" diameter	3	247208	Series II models } Not part of cylinder block assembly
6	Cup plug	1	247869	Series IIA models
7	Stud — For front cover and water pump	1	514527	Early type fixing
8	Stud	3	252622	
	Stud for distributor adaptor	1	252621	
	Stud for sump	1	213700	
9	Set bolt for main bearing	1	247127	Plain type set bolt } Alternatives. Check before ordering
10	Special spring washer for set bolt	6	536577	
11	Set bolt for main bearings	6	243959	Waisted type set bolt
	Special plain washer for set bolt	6	247861	
10	Dowel, locating main bearing and bearing cap	6	273166	
11	Dowel, locating flywheel housing	6	243968	
12	Dowel for timing chain adjuster	2	527269	
13	Plug for oil gallery, front	2	247965	
14	Plug for oil gallery, rear	1	537279	
15	Joint washer for rear plug	1	538608	
16	Plug for tappet feed gallery pipe, front	1		Not part of cylinder block assembly
17	Plug for tappet feed gallery pipe, rear	1		
18	Joint washer for plug	2		
19	Plug for redundant heater hole	1		
20	Plug for tappet feed hole	1		
22	Packing for rear main bearing cap	2		
23	Drain tap for cylinder block	1		
25	Engine front support bracket complete, LH	1	268635	Up to engines numbered 151023739
26	Engine front support bracket complete, RH	1	271975	
25	Engine front support bracket complete, LH	1	554437	From engines numbered 151023740 onwards
26	Engine front support bracket complete, RH	1	516133	
27	Locker	4	212430	} Fixing brackets to cylinder block
	Spring washer	4	3078	Alternative to locker
28	Set bolt (½" UNF x 1" long)	3	255085	
28	Set bolt (½" UNF x 1⅛" long)	1	255086	Alternative to locker
	Decarbonising gasket set	1	525857	
	Engine overhaul gasket set	1	525856	
	Stud kit for cylinder block	1	600245	
	Cylinder liner	4	503160	

* Asterisk indicates a new part which has not been used on any previous Rover model
† Engine assembly does not include ancillaries

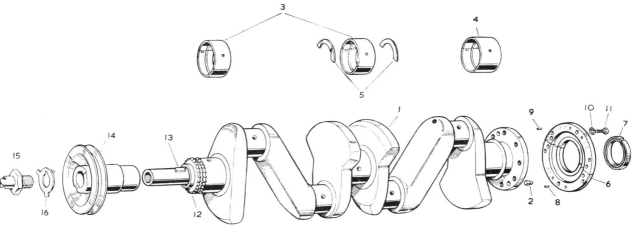

F 216

CRANKSHAFT, PETROL ENGINE, 2¼ LITRE, Series II and IIA

Plate Ref.	1 2 3 4	DESCRIPTION	Qty	Part No.	REMARKS
1		CRANKSHAFT ASSEMBLY, STD ...	1	514526	
		CRANKSHAFT ASSEMBLY, .010″ US	1	269912	
		CRANKSHAFT ASSEMBLY, .020″ US	1	269913	Series II models
		CRANKSHAFT ASSEMBLY, .030″ US	1	269914	
		CRANKSHAFT ASSEMBLY, .040″ US	1	269915	
		CRANKSHAFT ASSEMBLY, STD ...	1	527167	
		CRANKSHAFT ASSEMBLY, .010″ US	1	525852	
		CRANKSHAFT ASSEMBLY, .020″ US	1	525853	Series IIA models
		CRANKSHAFT ASSEMBLY, .030″ US	1	525854	
		CRANKSHAFT ASSEMBLY, .040″ US	1	525855	
2		Dowel for flywheel	2	265779	
3		Main bearing, front and centre, Std	2	523321	
		Main bearing, front and centre, .010″ US	2	523322	
		Main bearing, front and centre, .020″ US	2	523323	Series II models
		Main bearing, front and centre, .030″ US	2	523324	
		Main bearing, front and centre, .040″ US	2	523325	
		Main bearing, front and centre, Std	2	518748	
		Main bearing, front and centre, .010″ US	2	518749	
		Main bearing, front and centre, .020″ US	2	518750	Series IIA models
		Main bearing, front and centre, .030″ US	2	518751	
		Main bearing, front and centre, .040″ US	2	518752	
4		Main bearing, rear, Std	2	523326	
		Main bearing, rear, .010″ US	2	523327	
		Main bearing, rear, .020″ US	2	523328	Series II models
		Main bearing, rear, .030″ US	2	523329	
		Main bearing, rear, .040″ US	2	523330	
		Main bearing, rear, Std	2	518753	
		Main bearing, rear, .010″ US	2	518754	
		Main bearing, rear, .020″ US	2	518755	Series IIA models
		Main bearing, rear, .030″ US	2	518756	
		Main bearing, rear, .040″ US	2	518757	
5		Thrust washer for crankshaft, Std	1	523331	
		Thrust washer for crankshaft, .0025″ OS	1	523332	
		Thrust washer for crankshaft, .005″ OS	1	523333	Series II models
		Thrust washer for crankshaft, .0075″ OS	1	523334	
		Thrust washer for crankshaft, .010″ OS	1	523335	
		Thrust washer for crankshaft, Std	1	518758	
		Thrust washer for crankshaft, .0025″ OS	1	518759	
		Thrust washer for crankshaft, .005″ OS	1	518760	Series IIA models
		Thrust washer for crankshaft, .0075″ OS	1	518761	
		Thrust washer for crankshaft, .010″ OS	1	518762	
		Bearing set for crankshaft, Std	1	533979	
		Bearing set for crankshaft, .010″ US	1	533980	Series
		Bearing set for crankshaft, .020″ US	1	533981	IIA } Sets include main bearings,
		Bearing set for crankshaft, .030″ US	1	533982	models } thrust washers and
		Bearing set for crankshaft, .040″ US	1	533983	} connecting rod bearings

Pairs

* Asterisk indicates a new part which has not been used on any previous Rover model

CRANKSHAFT, PETROL ENGINE, 2¼ LITRE, Series II and IIA

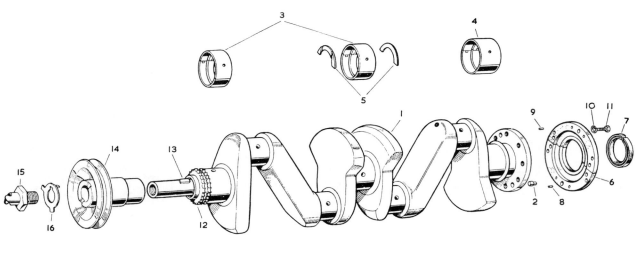

F 216

CRANKSHAFT, PETROL ENGINE, 2¼ LITRE, Series II and IIA

Plate Ref.				DESCRIPTION				Qty	Part No.	REMARKS
1	2	3	4							
				REAR BEARING OIL SEAL ASSEMBLY	..		..	1	542494	
6				Rear bearing split seal halves			..	2	523240	
7				Oil seal complete for rear bearing		..		1	542492	
				Silicone grease, MS4, 1 oz tube	..	..		1	270656	
8				Dowel, lower ⎫ Fixing retainer or				2	519064	
9				Dowel, upper ⎬ seal to cylinder				2	246464	
10				Spring washer ⎭ block and rear				10	3074	
11				Bolt (¼" UNF x ⅞" long) ⎰ bearing cap				10	255208	
12				Chainwheel on crankshaft			..	10	235770	
13				Key locating chainwheel and pulley		..		2	568333	
14				Crankshaft pulley			..	1	247676	Series II models
				Crankshaft pulley			..	1	524206	Series IIA models up to engine suffix 'J' inclusive
				Crankshaft pulley			..	1	564375 *	Series IIA models from engine suffix 'K' onwards
15				Starting dog	..	..	..	1	503665	
16				Lockwasher for starting dog	..	..		1	247771	

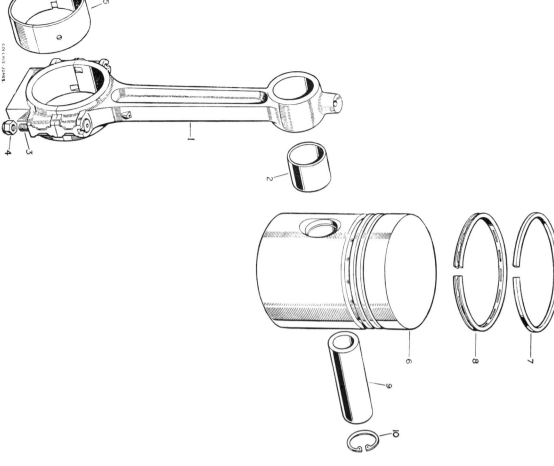

C748
COLLINS JONES.

Ref.	1 2 3 4	DESCRIPTION	Qty	Part No.	REMARKS
1		CONNECTING ROD ASSEMBLY	4	518809	Series II models
1		CONNECTING ROD ASSEMBLY	4	527164	Series IIA models
2		Gudgeon pin bush ...	4	528004	
		Complete set of bolts and nuts for connecting rod ...	1	522607	
3		Special bolt ...	8	518468	Series II models } Fixing connecting rod cap
4		Self-locking nut (⅜" UNF)	8	277390	
3		Special bolt ...	8	519440	Series IIA models } Fixing connecting rod cap
4		Self-locking nut (⅜" UNF)	8	277390	
5		Connecting rod bearing, Std	4	523336	} Series II models (Pairs)
		Connecting rod bearing, .010" US	4	523337	
		Connecting rod bearing, .020" US	4	523338	
		Connecting rod bearing, .030" US	4	523339	
		Connecting rod bearing, .040" US	4	523340	
		Connecting rod bearing, Std	4	524151	} Series IIA models (Pairs)
		Connecting rod bearing, .010" US	4	524152	
		Connecting rod bearing, .020" US	4	524153	
		Connecting rod bearing, .030" US	4	524154	
		Connecting rod bearing, .040" US	4	524155	
6		PISTON ASSEMBLY, STD, GRADE 'Z'	4	500638	
		PISTON ASSEMBLY, STD, GRADE 'A'	4	500639	
		PISTON ASSEMBLY, STD, GRADE 'B'	4	500640	
		PISTON ASSEMBLY, STD, GRADE 'C'	4	500641	
		PISTON ASSEMBLY, STD, GRADE 'D'	4	500642	
		PISTON ASSEMBLY, .010" OS	4	500140	
		PISTON ASSEMBLY, .020" OS	4	500141	
		PISTON ASSEMBLY, .030" OS	4	500142	
		PISTON ASSEMBLY, .040" OS	4	500143	
7		Piston ring, compression, Std ...	8	278286	
		Piston ring, compression, .010" OS	8	500136	
		Piston ring, compression, .020" OS	8	500137	
		Piston ring, compression, .030" OS	8	500138	
		Piston ring, compression, .040" OS	8	500139	
8		Piston ring, scraper, Std ...	4	247512	
		Piston ring, scraper, .010" OS	4	500132	
		Piston ring, scraper, .020" OS	4	500133	
		Piston ring, scraper, .030" OS	4	500134	
		Piston ring, scraper, .040" OS	4	500135	
9		Gudgeon pin, Std ...	4	265169	
10		Circlip for gudgeon pin	8	265175	

* Asterisk indicates a new part which has not been used on any previous Rover model

CYLINDER HEAD, PETROL ENGINE, 2¼ LITRE, Series II and IIA

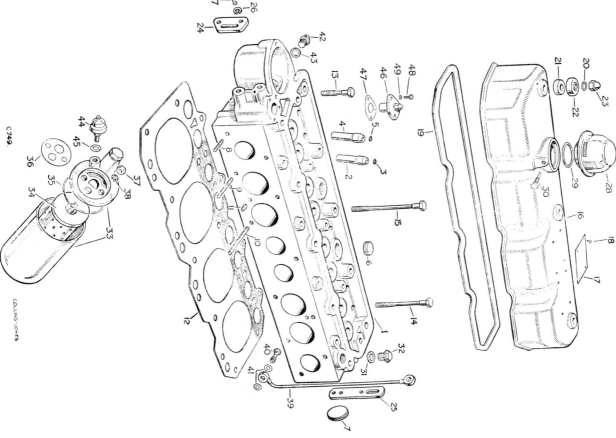

C749

COLLINS-JONES

CYLINDER HEAD, PETROL ENGINE, 2¼ LITRE, Series II and IIA

Plate Ref. 1 2 3 4	DESCRIPTION	Qty	Part No.	REMARKS
1	CYLINDER HEAD ASSEMBLY	1	279573	Series II models
1	CYLINDER HEAD ASSEMBLY	1	564167	Series IIA models up to engine suffix 'J' inclusive
	CYLINDER HEAD ASSEMBLY	1	566703*	Series IIA models from engine suffix 'K' onwards
2	Valve guide, inlet, and sealing ring	4	500144	
3	Sealing ring for inlet valve guide	4	247186†	With 'O' ring type seal.
4	Valve guide, exhaust and sealing ring	4	500145	Up to engine suffix 'H' inclusive
5	Sealing ring for exhaust valve guide	4	233419*†	
	Valve guide, inlet, and oil seal	4	605759*	
	Oil seal for inlet valve guide	4	554727*†	With lip type oil seal.
	Valve guide, exhaust, and oil seal	4	605760*†	From engine suffix 'K' onwards
	Oil seal for exhaust valve guide	4	554728*†	
6	Core plug, 7/8" diameter	1	250830	Alternatives.
7	Cup plug, 1 5/16" diameter	4	525497	Check before ordering.
8	Core plug, 1 1/8" diameter	4	09191	
	Core plug, 1" diameter	2	230250	
11	Stud, short, for manifold	2	247144	
	Stud for manifold flange, rear	5	252623	
12	Dowel for cylinder head	2	213700	
13	Cylinder head gasket	1	558160	
14	Special set bolt, short	4	279648	
15	Special set bolt, medium	9	279649	
	Special set bolt (6¼" long)	5	279650	Fixing cylinder head to cylinder block
16	Special set bolt (4⅛" long)	5	554621	
	ROCKER COVER, TOP, ASSEMBLY	1	274172	Series II models
	ROCKER COVER, TOP, ASSEMBLY	1	524846	Series IIA models
17	Tappet clearance plate	1	247634	
18	Drive screw fixing plate	4	78001	
19	Lifting bracket for engine, front	1	274173	
	Lifting bracket for engine, rear	1	525131	
20	Sealing washer	3	232038	
21	Rubber washer	3	506069	
22	Cover for rubber washer	3	247624	Fixing top rocker cover
23	Special nut	3	247121	
24	Special set bolt	3	247714	
25	Joint washer for top rocker cover	1	3075	
26	Spring washer	4	255026	Fixing brackets to cylinder head
27	Set bolt (5/16" UNF x ¾" long)	4	276501	
28	Breather filter for engine, AC 7223381	1	268887	Not part of engine assembly
29	Sealing ring for filter	1	247631	
30	Sealing ring for breather filter	1	515291	
	Sealing washer for set screw fixing breather filter	1	232037	
31	Joint washer	1	243959	For heater valve hole in cylinder head
32	Plug (⅜" BSP)	1	536577	For heater valve hole in cylinder head

* Asterisk indicates a new part which has not been used on any previous Rover model
† Not included in cylinder head assembly

CYLINDER HEAD, PETROL ENGINE, 2¼ LITRE, Series II and IIA

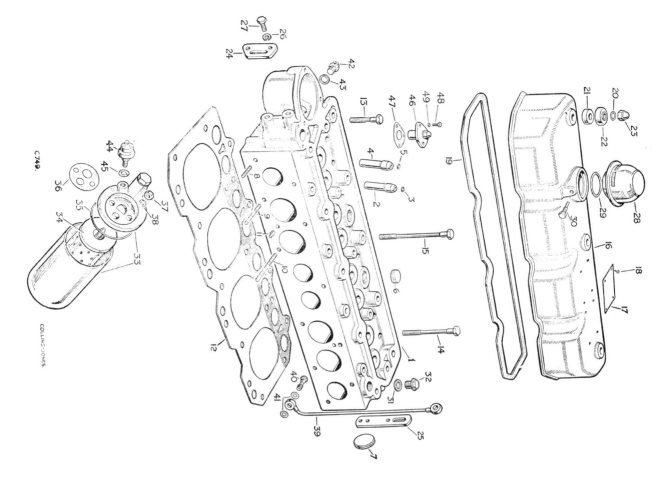

C749.

COLLINS-JONES

CYLINDER HEAD, PETROL ENGINE, 2¼ LITRE, Series II and IIA

Plate Ref.	1 2 3 4	DESCRIPTION	Qty	Part No.	REMARKS
33		Engine oil filter, AC 7965053	1	537229	Short type
34		Element for filter, AC FF 50, overall length 6¹³⁄₁₆″	1	248863	For long type filter }} Alternatives. Check before ordering
35		Element for filter, AC 72, overall length 4⅝″	1	541403	For short type filter
		Gasket for filter, AC 1530250	1	272539	
36		Rubber washer for centre bolt, AC 1531098	1	269889	
37		Set bolt (⁷⁄₁₆″ UNF x 1½″ long) }} Fixing filter to cylinder block	1	272839	
38		Spring washer	2	255068	
39		Oil pipe complete to cylinder head	2	3077	
40		Banjo bolt fixing oil pipe	1	275679	
41		Joint washer for banjo bolts	2	504840	
42		Plug for thermometer hole in cylinder head	4	232039	
		Water temperature transmitter	1	536577	Up to engine suffix 'H' inclusive
		Water temperature transmitter	1	560794 *	From engine suffix 'J' onwards
43		Adaptor for water temperature transmitter	1	568457 *	
44		Joint washer for plug or adaptor	1	231577	
		Oil pressure switch, AC 7954238	1	519863	Series II models
		Oil pressure switch, AC 7954237	1	519864	Series IIA models
45		Joint washer for switch	1	232039	
46		Shakeproof washer for switch terminal	1	70884	
47		Thermostat switch for mixture warning light	1	545010	
48		Joint washer for switch, ¹⁄₁₆″ thick	1	236022	Alternatives. Check before ordering
		Joint washer for switch, ⅛″ thick	1	535703	
		Set bolt (2 BA x ⅝″ long) }} Fixing switch to head	3	237119	Alternatives. Check before ordering
		Set bolt (2 BA x ⁷⁄₁₆″ long) }}	3	251002	
49		Spring washer	3	3073	
		Valve guide oil control kit	1	605761 *	To convert early engines with 'O' ring type valve guide seals to lip type oil seals

* Asterisk indicates a new part which has not been used on any previous Rover model

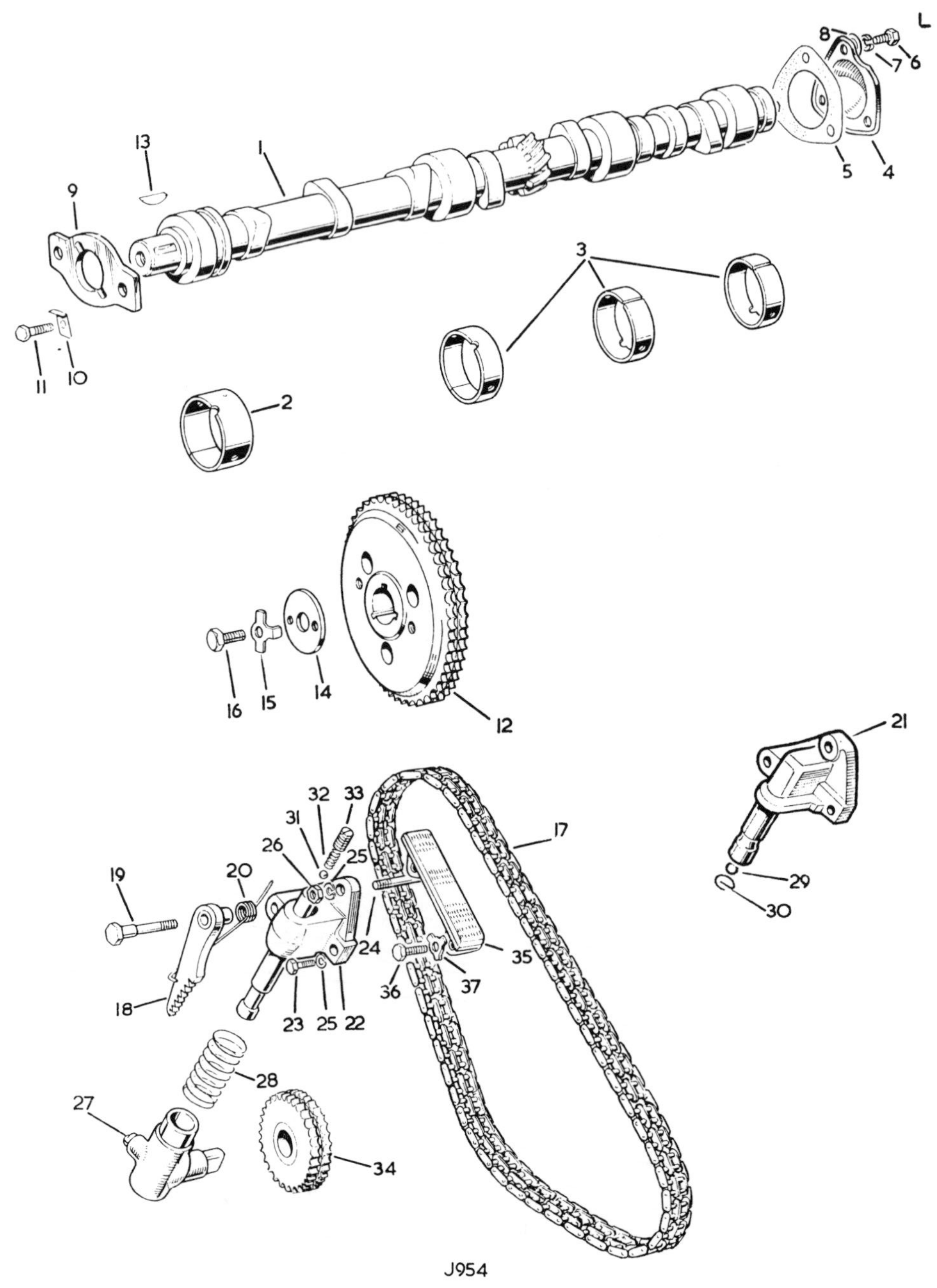

CAMSHAFT AND TENSIONER MECHANISM, PETROL ENGINE, 2¼ LITRE, Series II and IIA

J954

CAMSHAFT AND TENSIONER MECHANISM, PETROL ENGINE, 2¼ LITRE, Series II and IIA

Plate Ref.	Description	Qty	Part No.	Remarks
1	Camshaft	1	247709	
2	Bearing complete for camshaft, front	1	519054	
3	Bearing complete for camshaft, centre and rear	3	519055	
4	Rear end cover for camshaft	1	538073	
5	Joint washer for cover	1	247070	
6	Set bolt (¼" UNF x ⅞" long) } Fixing cover to block		252208	Alternative fixings. Check before ordering
6	Set bolt (¼" UNF x ¾" long)		252206	
7	Spring washer		3074	
8	Plain washer		10882	
9	Thrust plate for camshaft	1	538535	
10	Locker	2	2995	
11	Set bolt (¼" UNF x ¾" long) } Fixing thrust plate to cylinder block	2	255207	
12	Chainwheel for camshaft	1	568474	
13	Key locating camshaft chainwheel	1	230313	
14	Retaining washer } Fixing chainwheel to camshaft	1	09093	
15	Locker	1	09210	
16	Set bolt (⅜" UNF x ⅞" long)	1	255046	
17	Camshaft chain (⅜" pitch x 78 links)	1	09156	
18	Ratchet for timing chain adjuster	1	546026	
19	Special bolt fixing ratchet and piston to block	1	247199	
20	Spring for chain adjuster ratchet	1	267451	
21	Piston for timing chain adjuster	1	247912	Up to engines numbered 151017422
22	Piston for timing chain adjuster	1	515323	From engines numbered 151017423 onwards
23	Set bolt (5/16" UNF x 1¼" long) } Fixing piston to cylinder block	2	256220	1 off on late type piston
24	Stud	1	247144	Late type piston
25	Spring washer	2	3075	
26	Nut (5/16" UNF)	2	254831	Late type piston
27	Cylinder for timing chain adjuster	1	277388	
28	Spring for chain tensioner	1	233326	
29	Steel ball (5/16") for non-return valve	1	3739	Up to engines numbered 151017422
30	Retainer for steel ball	1	233328	
31	Steel ball (5/16") for non-return valve	1	3739	From engines numbered 151017423 onwards
32	Spring retainer for steel ball	1	515321	
33	Plug for spring retainer	1	515325	
34	Idler wheel for timing chain	1	286067	
35	Vibration damper for timing chain	1	275234	From engines numbered 151017423 onwards
36	Set bolt (¼" UNF x ½" long) } Fixing damper to cylinder block	2	255204	
37	Locking plate	2	557523	

* Asterisk indicates a new part which has not been used on any previous Rover model

J955

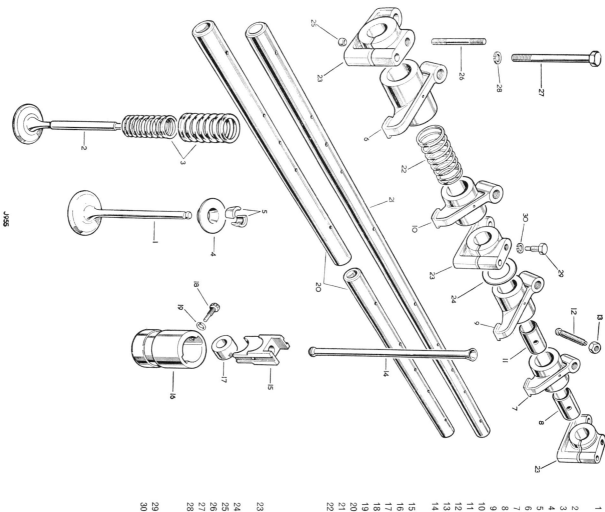

VALVE GEAR AND ROCKER SHAFTS, PETROL ENGINE, 2¼ LITRE, Series II and IIA

Plate Ref.	1 2 3 4	DESCRIPTION				Qty	Part No.	REMARKS
1		Inlet valve	…	…	…	4	277593	
2		Exhaust valve			…	4	525372	Up to engine suffix 'J' inclusive
2		Exhaust valve			…	4	557967*	From engine suffix 'K' onwards
3		Valve spring, inner and outer				8	568550	
4		Valve spring cup	…	…	…	8	268292	
5		Split cone for valve, halves				16	268293	
6		Valve rocker, inlet, LH				2	512207	
7		Valve rocker, exhaust, LH				2	512208	
8		Bush for exhaust valve rocker				4	247614	
9		Valve rocker, inlet, RH				2	512205	
10		Valve rocker, exhaust, RH				2	512206	
11		Bush for inlet valve rocker				4	247614	
12		Tappet adjusting screw				8	506814	
13		Locknut for tappet adjusting screw				8	254881	
14		Tappet push rod				8	546798	
15		Tappet, tappet guide, roller and set bolt assembly				8	507829	
15		Tappet	…	…	…	8	507026	
16		Tappet guide	…	…	…	8	502473	
17		Roller	…	…	…	8	517429	
18		Special set bolt			…	8	507025	
19		Copper washer for tappet guide set bolt				8	232038	
20		Valve rocker shaft			…	2	274125	Series II models
21		Valve rocker shaft			…	1	554070	Series IIA models
22		Spring for rocker shaft				4	247040	
23		Rocker bracket, inner				3	274068	
23		Rocker bracket, inner				1	274069	Series II models
23		Rocker bracket, rear				1	503746	Series IIA models
23		Rocker bracket, front				5	524838	'H' inclusive
23		Rocker bracket	…	…	…	5	554602	Series IIA models. Up to engine suffix
								Series IIA models. From engine suffix 'J' onwards
24		Washer for rocker bracket				6	525389	Series II models
25		Locating dowel for rocker bracket				3	277956	Series IIA models
26		Stud in rocker bracket for rocker cover				5	502656	4 off on early Series IIA models
27		Set bolt (5/16" UNF x 2¾" long) Fixing rocker bracket to cylinder head				5	525500	
28		Spring washer				5	3075	
								Fixing rocker bracket to cylinder head
28		Spring washer				5	3075	
29		Locating screw } For rocker shaft				2	247730	Series II
29		Locating screw } at oil feed bracket				2	3075	models
30		Locating screw } For rocker shaft				1	525390	Series IIA.
30		Spring washer } at bracket				1	3075	2 off on early models

* Asterisk indicates a new part which has not been used on any previous Rover model

OIL PUMP, PETROL ENGINE, 2¼ LITRE, Series II and IIA

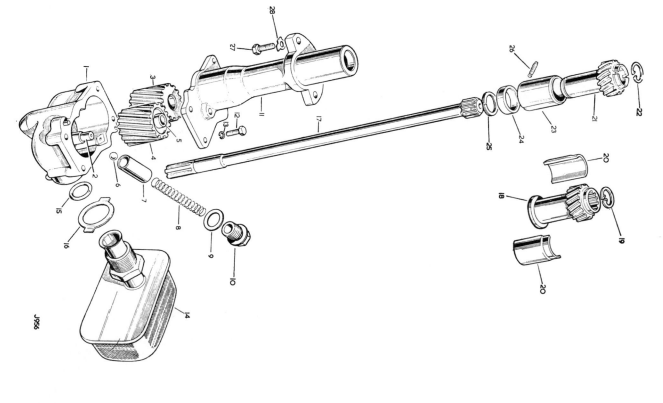

OIL PUMP, PETROL ENGINE, 2¼ LITRE, Series II and IIA

Plate Ref. 1 2 3 4	DESCRIPTION	Qty	Part No.	REMARKS
	OIL PUMP ASSEMBLY	1	247662	1958-59
	OIL PUMP ASSEMBLY	1	513640	1960 onwards
1	Oil pump body	1	247663	1958-59
2	Oil pump body	1	513641	1960 onwards
	Spindle for idler gear ...	1	502209	1960 onwards
3	Oil pump gear, driver ...	1	247659	1958-59
3	Oil pump gear, driver ...	1	240555	1960 onwards
4	Oil pump gear, idler ...	1	237884	1958-59
4	Oil pump gear, idler ...	1	278109	1960 onwards
5	Bush for idler gear	1	214995	1960 onwards
6	Steel ball	1	3748	
7	Plunger	1	273711	For oil pressure release valve
8	Spring	1	564456	
9	Washer, inside diameter 55/64"	1	243970	Alternatives. Check before ordering
9	Washer, inside diameter 49/64"	1	232044	Alternatives. Check before ordering
10	Plug	1	549909	
11	Oil pump cover	1	247658	1958-59
11	Oil pump cover	1	513639	1960 onwards
12	Set bolt (5/16" UNF x 7/8" long) } Fixing cover to body	4	255227	
13	Spring washer } Fixing cover to body	4	3075	
14	Oil filter for pump	1	247664	
15	Sealing ring } Fixing oil filter	1	244488	
16	Lockwasher } to oil pump	1	244487	
17	Drive shaft for oil pump ...	1	247739	1958-59
18	VERTICAL DRIVE SHAFT ASSEMBLY	1	511680	1960 onwards
19	Circlip for drive shaft ...	1	502266	With split bush
20	Bush for drive shaft gear ...	1	247742	With split bush
	VERTICAL DRIVE SHAFT ASSEMBLY	1	247653	Up to engine suffix 'F' inclusive
20	Bush for drive shaft gear ...	1	530175	
21	VERTICAL DRIVE SHAFT ASSEMBLY	1	541181	From engine suffix 'G' onwards With one-piece bush
22	Circlip for drive shaft ...	1	247742	
23	Bush for drive shaft gear ...	1	522745	
24	Thrust washer for drive shaft	1	530178	
25	Retaining ring for washer and bush	1	530179	
26	Locating screw for drive shaft bush	1	524769	
27	Set bolt (5/16" UNF x 1" long) } Fixing oil pump	2	255228	
28	Lockwasher } to cylinder block	2	247665	

H4O4

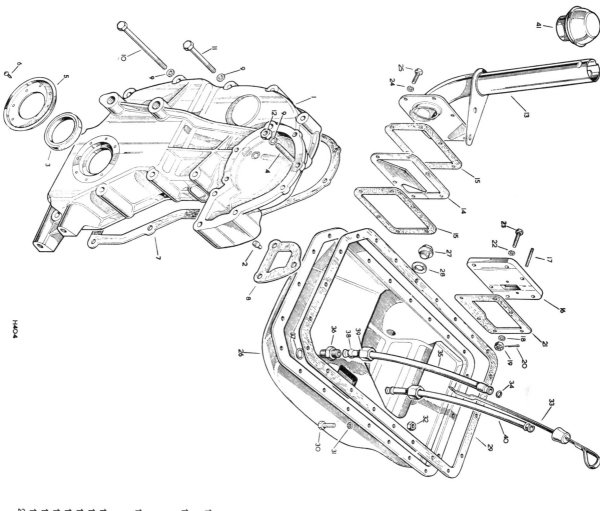

FRONT COVER, SIDE COVERS AND SUMP, PETROL ENGINE, 2¼ LITRE, Series II and IIA

Plate Ref. 1 2 3 4	Description	Qty	Part No.	Remarks
1	FRONT COVER ASSEMBLY	1	510805	Series II models
	FRONT COVER ASSEMBLY	1	514451	Series IIA models. Up to engine suffix 'G' inclusive
	FRONT COVER ASSEMBLY	1	554541	Series IIA models. From engine suffix 'H' onwards
	FRONT COVER ASSEMBLY	1	529155	Series IIA models with Lucas 2AC type 12 volt AC/DC generator. Up to engine suffix 'G' inclusive
	FRONT COVER ASSEMBLY	1	554843	Series IIA models with Lucas 2AC type 12 volt AC/DC generator. From engine suffix 'G' inclusive
2	Dowel locating front cover	2	6395	
3	Oil seal for front cover	1	213744	
	Stud } Fixing water pump	1	247595	For alloy type front cover } Series II models
	Stud } water pump	3	247159	3 off with cast iron type front cover } models
4	Stud fixing water pump	5	252500	Series IIA models. Up to engine suffix 'G' inclusive
5	Mud excluder	1	247766	
6	Hammer drive screw fixing excluder	8	78001	
7	Joint washer for front cover	1	538039	
8	Joint washer at water inlet	1	538038	
	Spring washer	15	3075	
	Set bolt (5/16" UNF x 3" long) } Fixing front cover and water pump to cylinder block	9	256030	7 off on Series IIA models
	Bolt (5/16" UNF x 3¼" long)	2	256031	Series IIA models
	Set bolt (5/16" UNF x 2" long)	3	256226	1 off on Series IIA models with Lucas 2AC type 12 volt AC/DC generator } Up to engine suffix 'G' inclusive
	Set bolt (5/16" UNF x 3½" long)	2	256232	
9	Nut (5/16" UNF)	15	254831	
10	Spring washer	2	3075	
	Set bolt (5/16" UNF x 3" long)	2	256030	
11	Set bolt (5/16" UNF x 2" long) } Fixing front cover and water pump to cylinder block	7	256226	From engine suffix 'H' onwards } Series IIA models
	Set bolt (5/16" UNF x 3½" long)	3	256031	
	Plain washer	1	2266	
	Nut (5/16" UNF)	2	254831	
12	Timing pointer at front cover	1	564362*	Series IIA models. From engine suffix 'K' onwards
13	Side cover and oil filler pipe for engine, front	1	510730	Not part of engine assembly
14	Baffle plate for side cover, front	1	529394	
15	Joint washer for side cover, front	2	247555	
16	SIDE COVER ASSEMBLY, REAR	1	542600	
17	Stud for fuel pump	2	500792	When Lucas 2AC type generator is fitted
18	Plain washer	2	2220	
19	Nut (5/16" UNF) } Fixing stud	2	254831	Nut and split pin type fixing } Alternatives. Check before ordering
20	Split pin	2	2422	
	Stud for fuel pump	2	542601	Plain type stud fixing

* Asterisk indicates a new part which has not been used on any previous Rover model

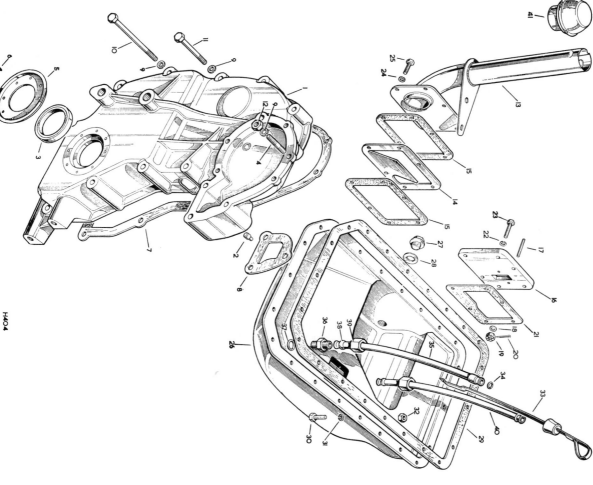

62

H4O4

FRONT COVER, SIDE COVERS AND SUMP, PETROL ENGINE, 2¼ LITRE, Series II and IIA

Plate Ref. 1 2 3 4	DESCRIPTION	Qty	Part No.	REMARKS
21	Joint washer for side cover, rear	1	247554	
22	Spring washer	6	3075	Fixing rear side
23	Set bolt (⁵⁄₁₆" UNF x 1" long)	6	255228	cover plate
24	Spring washer	7	3075	Fixing front
25	Set bolt (⁵⁄₁₆" UNF x ¾" long)	7	255226	side cover plate
26	Crankcase sump	1	529823	
27	Drain plug for crankcase sump	1	536577	
28	Washer for drain plug	1	243959	
29	Joint washer for crankcase sump	1	546841	
30	Set bolt (⁵⁄₁₆" UNF x ⅞" long)	21	255227	Fixing sump
31	Spring washer	22	3075	to
32	Nut (⁵⁄₁₆" UNF) for stud	1	254831	cylinder block
33	Oil level rod	1	274154	
34	Sealing ring for rod	1	532387	
35	Tube for oil level rod	1	504032	
36	Double-ended union	1	236060	Alternative to
37	Washer	1	243958	one-piece type
38	Olive	1	236408	oil level rod
39	Union nut	1	236407	tube
40	Tube for oil level rod	1	541860	One-piece type. Alternative to tube with separate union
41	Oil filler cap and breather filter, AC 7964819	1	546440	Not part of engine assembly
	Oil recommendation label for oil filler cap	1	272476	

* Asterisk indicates a new part which has not been used on any previous Rover model

63

WATER PUMP AND THERMOSTAT, PETROL ENGINE, 2¼ LITRE, Series II and IIA

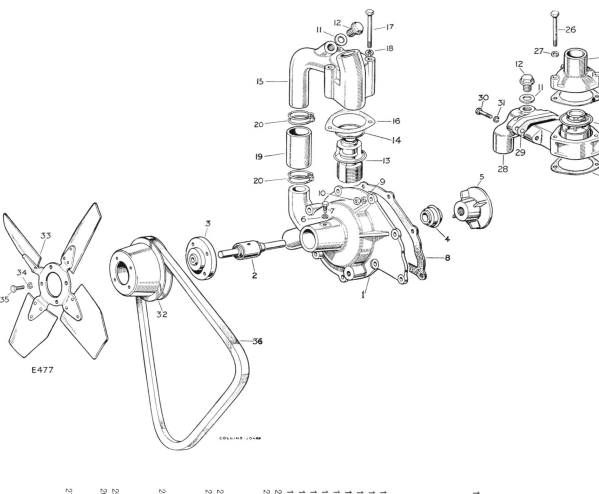

E477

COLLINS-JONES

WATER PUMP AND THERMOSTAT, PETROL ENGINE, 2¼ LITRE, Series II and IIA

Plate Ref. 1 2 3 4	Ref.	DESCRIPTION	Qty	Part No.	REMARKS
		WATER PUMP ASSEMBLY	1	501041	Series II models
		WATER PUMP ASSEMBLY	1	530477	Series IIA models
	1	Water pump casing ...	1	501039	Series II models
	1	Water pump casing ...	1	530478	Series IIA models
	2	Pump spindle and bearing ...	1	523354	
	3	Hub for fan ...	1	515729	
	4	Carbon ring and seal unit ...	1	568301	
	5	Impeller for pump ...	1	247916	
	6	Spring washer \ Locating	1	3074	
	7	Special set bolt / bearing casing	1	247078	
	8	Joint washer for water pump ...	1	247919	
	9	Spring washer \ Fixing water pump	3	3074	Series II models
	9	Spring washer /	3	254810	Series IIA
	10	Nut (¼" UNF) \ to front cover	3	3074	
	10	Nut (¼" UNF) /	3	254810	
		Bolt (⁵⁄₁₆" UNF x 3¾" long) Fixing	1	538671	
		Spring washer \ water pump	5	256233	
		Nut (¼" UNF) / to front	5	3074	
		Set bolt (¼" UNC x 1" long) cover	5	253009	From engine suffix 'H' onwards
	11	Joint washer \ For heater return	1	243959	
	12	Plug (⅜" BSP) / in water outlet pipe	1	536577	
	13	Thermostat ...	1	513465	
	14	'O' ring, thermostat to water outlet pipe	1	502832	
	15	Water outlet pipe, thermostat to radiator	1	501031	
	16	Washer for outlet pipe ...	1	247874	1958–60
	17	Set bolt (¼" UNF x 2½" long) \ Fixing outlet pipe	3	256209	
	18	Spring washer / to cylinder head	3	3074	
	19	Hose for by-pass pipe ...	1	273789	
	20	Clip for hose ...	2	603894	
	21	Thermostat, bellows type ...	1	504736	1961. Also Series IIA up to engine suffix 'C'
	22	Thermostat, wax type ...	1	532453	Series IIA. From engine suffix 'D' onwards
		'O' ring for thermostat ...	1	527235	1961
	22	Thermostat housing ...	1	516059	1961 onwards
	23	Joint washer for thermostat housing, upper	1	511957	1961. Also Series IIA. Up to engine suffix 'C'
	24	Joint washer for thermostat housing, upper	1	527110	Series IIA. From engine suffix 'D' onwards
	24	Water outlet pipe, thermostat to radiator ...	1	511956	1961. Also Series IIA up to engine suffix 'C'
	25	Joint washer for thermostat housing, lower ...	1	527109	Series IIA from engine suffix 'C'
	26	Set bolt (¼" UNF x 2½" long) \ Fixing thermostat housing and	3	256209	2 off on late models
	27	Spring washer / outlet pipe to cylinder head	1	256010	Late models
			3	3074	

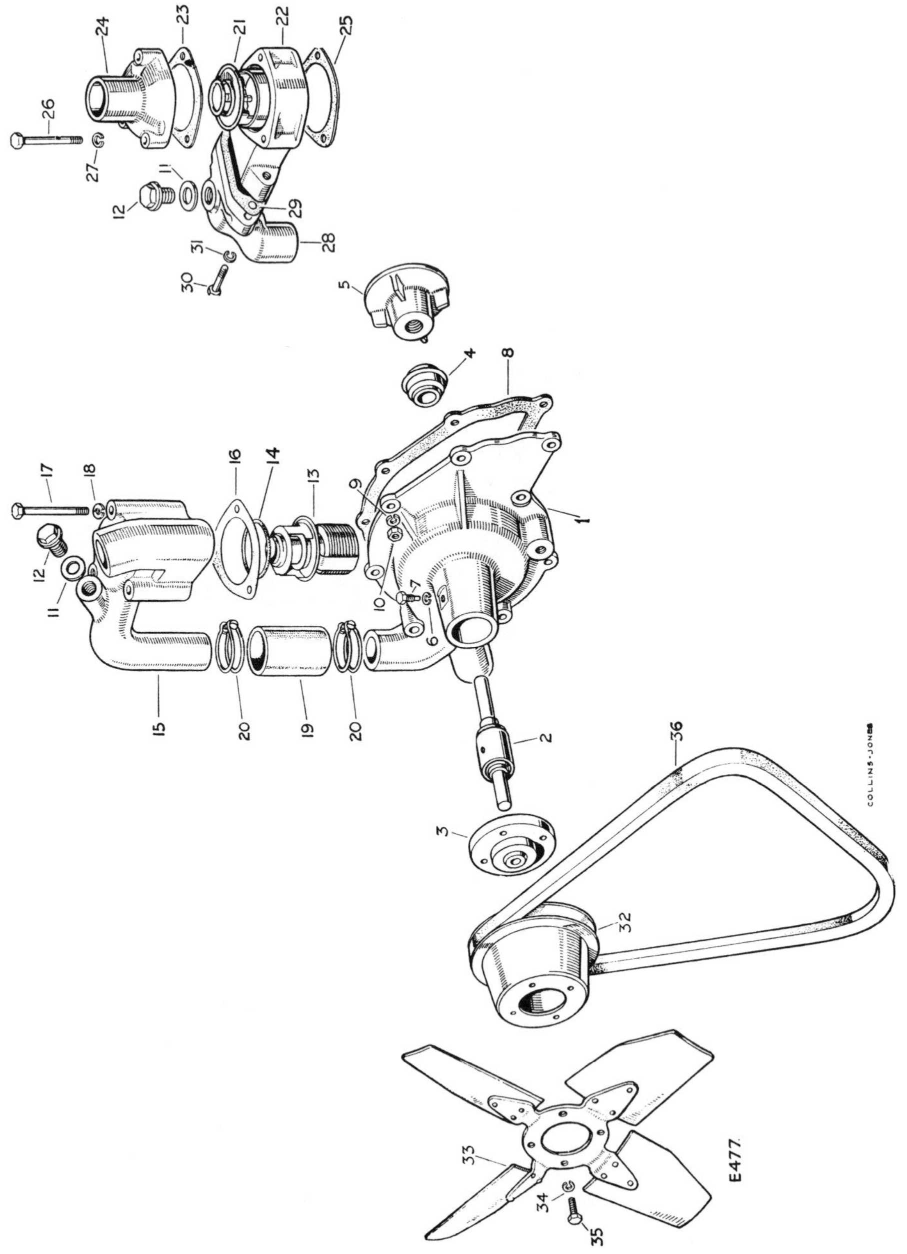

E477

COLLINS-JONES

WATER PUMP AND THERMOSTAT, PETROL ENGINE, 2¼ LITRE, Series II and IIA

Plate Ref.	1 2 3 4	DESCRIPTION	Qty	Part No.	REMARKS
28		Thermostat by-pass pipe ...	1	518818	1961 Series II models
28		Thermostat by-pass pipe ...	1	530476	Series IIA models
29		Joint washer for by-pass pipe ...	1	511958	
30		Set bolt (5/16" UNF x 1" long) ⎱ Fixing by-pass pipe to	2	255228	⎱ 1961 onwards
31		Spring washer ⎰ thermostat housing	2	3075	⎰
32		Fan pulley ...	1	247520	Series II models
33		Fan pulley ...	1	530890	Series IIA models
34		Fan blade ...	1	512018	
35		Spring washer ⎱ Fixing fan blade	4	3074	
36		Set bolt (¼" UNF x ¾" long) ⎰ and pulley to hub	4	255207	
		Fan and dynamo belt ...	1	247920	Series II models
		Fan and dynamo belt ...	1	550224	Series IIA models
		Water pump overhaul kit ...	1	530590	

Not part of engine assembly

* Asterisk indicates a new part which has not been used on any previous Rover model

E 478

COLLINS JONES

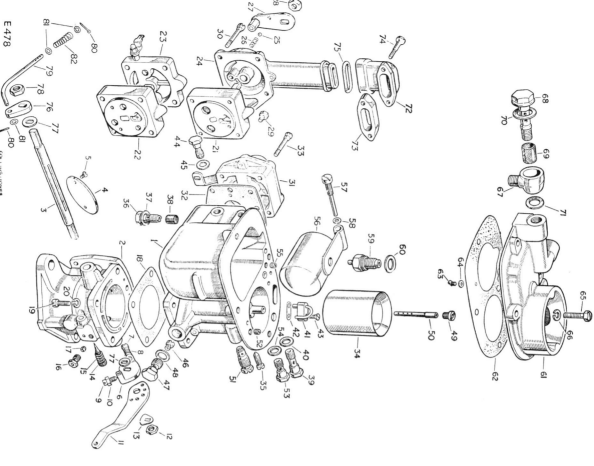

Plate Ref.	1 2 3 4	DESCRIPTION	Qty	Part No.	REMARKS
		CARBURETTER COMPLETE, SX Type 40 PA 10–5 No. 2003	1	540029†	**Without** starter heater element. Check before ordering
		CARBURETTER COMPLETE, SX Type B40 PA 10–6 No. 2004	1	540180†	**With** starter heater element. Check before ordering. *Note:* Identification, Lucar connection adjacent to starter unit
1		Carburetter body only, SX 12860	1	528037	For carburetter **with** starter element. Check before ordering } For carburetter
		Carburetter body only, SX 13077	1	513506	Alternative. Check before ordering } **without** starter element. Check before ordering
		Carburetter body only, SX 12860	1	503887	
2		Throttle chamber only, SX B 14108	1	538435	For carburetter **with** starter heater element. Check before ordering
		Throttle chamber only, SX 13326	1	524114	For carburetter **without** starter heater element. Check before ordering
		Throttle chamber only, SX B 14687	1	503891	Alternatives. Check before ordering
		Plate, throttle abutment, SX 4019	1	260139	
6		Plate, throttle abutment, SX 10191	1	260138	
5		Special screw fixing butterfly, SX 3947	2	504485	
4		Butterfly for throttle, SX 52162	1	260159	
3		Spindle for throttle, SX 50991	1	504683	
7		Special screw, SX 4023	1	260140	For slow running
8		Spring, SX 4384	1	260141	adjustment
9		Special screw, SX 50537	1	260142	For mixture
10		Locknut, SX 3950	1	260143	For throttle stop
11		Throttle lever, SX 13047	1	504683	
12		Nut fixing throttle lever, SX 4024	1	260159	
13		Lockwasher for nut, SX 12599	1	504485	
14		Special screw, SX 1171	1	260146	For mixture control
15		Spring, SX 4384	1	260141	
16		Screwed union, SX 9879	1	260707	For suction
17		Olive, SX 5041	1	260708	pipe
18		Joint washer for throttle chamber, SX 9097	1	503893	
19		Special screw and spring washer, SX 12501, fixing chamber to carburetter body	4	503895	
21		Starter complete, SX 12836	1	512402	
22		Starter complete, SX B 13942	1	527273	**Without** starter heater element. Check before ordering
		Starter valve complete, SX 12925	1	503896	For carburetter **without** starter heater element. Check before ordering
		Starter valve complete, SX B 13337	1	528039	**With** starter heater element. Check before ordering
23		Heater element for starter, SX B 13941	1	528053	For carburetter **with** starter heater element. Check before ordering
24		Cover for starter, SX 12827	1	503897	For carburetter **with** starter heater element. Check before ordering
25		Ball, SX 51728 } For	1	260155	
26		Spring, SX 51762 } starter valve	1	260156	
27		Lever for starter, SX 10508	1	503898	
28		Nut fixing starter lever, SX 4024	1	260159	

* Asterisk indicates a new part which has not been used on any previous Rover model
† Carburetter and fittings are not included in the engine assembly

COLLINS-JONES.

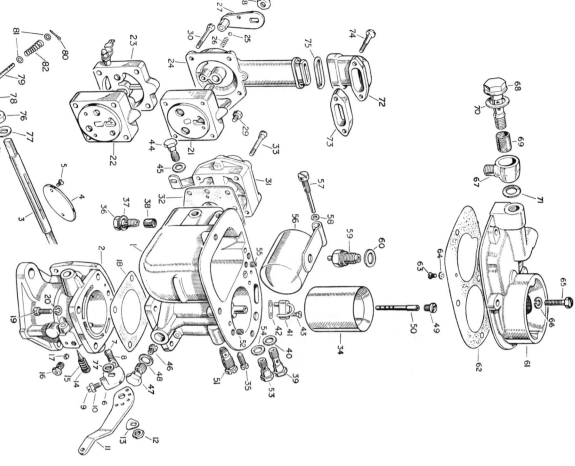

CARBURETTER, SOLEX TYPE, PETROL ENGINE, 2¼ LITRE, Series II and IIA
Up to engine suffix 'H' inclusive

Plate Ref.	1 2 3 4	DESCRIPTION	Qty	Part No.	REMARKS
29		Special bolt fixing starter cable, SX 12056	1	265495	For carburetter without starter heater element. Check before ordering
30		Special screw fixing starter body, SX 5142/12	4	262313	For carburetter without starter heater element. Check before ordering
31		Special screw fixing starter body, SX 51384/P1	4	527272	For carburetter with starter heater element. Check before ordering
		Accelerator pump complete, SX 52940/1	1	524115	For carburetter without starter heater element. Check before ordering
		Accelerator pump complete, SX B 13896	1	528041	For carburetter with starter heater element. Check before ordering
32		Diaphragm assembly for pump, SX 11314	1	503900	
33		Joint washer for pump, SX 51941/L1	1	503939	
33		Special screw fixing pump, SX 51421	4	503901	
34		Choke tube (30), SX 50459/2	4	503902	
		Choke tube (28), SX 50459/2/28	1	512105	Alternatives. Check before ordering
		Choke tube (26), SX 50459/2/26	1	542638	
35		Special screw fixing choke tube, SX 50362	1	260163	
36		Non-return valve, SX 9829	1	260765	Alternatives. Check before ordering
36		Non-return valve, SX 12505	1	513507	
37		Fibre washer for valve, SX 52825	1	260734	
38		Filter gauze for non-return valve, SX 52847	1	503903	
39		Jet (65), accelerator pump, SX 52200/65	1	503904	Alternatives. Check before ordering
		Jet (50), accelerator pump, SX 52200/50	1	260734	
40		Fibre washer for jet, SX 52825	1	503905	
41		Pump injector, SX 52950	1	261570	
42		Joint washer for pump injector, SX 52735	1	260767	
43		Special screw fixing injector, SX 3947/2	1	503906	
44		Economy jet, SX 52824/1 (blank)	1	260734	
45		Joint washer for blank jet, SX 52825/1	1	260734	
46		Main jet (135), SX 50552/6/135	1	505701	Alternatives. Check before ordering
		Main jet (125), SX 50552/6/125	1	260734	
		Main jet (120), SX 50552/6/120	1	542639	
47		Bolt, main jet carrier, SX 50820	1	260180	
48		Fibre washer for bolt, SX 50815	1	260181	
49		Correction jet (175), SX 51612/175	1	503908	Alternatives. Check before ordering
		Correction jet (185), SX 51612/2/185	1	512106	
50		Emulsion tube, SX 52043	1	503909	Alternatives. Check before ordering
		Emulsion tube, SX 52684	1	260743	
51		Pilot jet (60), SX 50797/4/60	1	260216	For carburetter without starter heater element. Check before ordering
		Pilot jet (50), SX 50797/4/50	1	260268	For carburetter with starter heater element. Check before ordering
52		Pilot jet (55), SX 50797/4/55	1	260269	For carburetter with starter heater element. Check before ordering
		Jet air bleed (1.5), SX 51274/1/150	1	260179	Alternatives. Check before ordering
		Jet air bleed (blank), SX 13334	1	542641	
		Starter jet, petrol (145), SX 52823/145	1	503910	For carburetter without starter heater element. Check before ordering
53		Starter jet, petrol (130), SX 52823/130	1	528038	For carburetter with starter heater element. Check before ordering

* Asterisk indicates a new part which has not been used on any previous Rover model

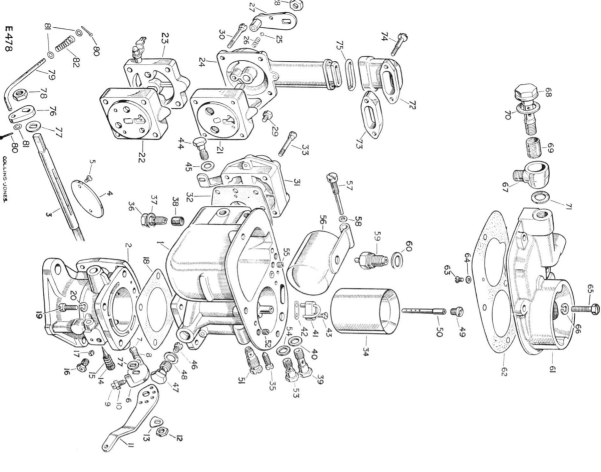

E 478

COLLINS-JONES.

CARBURETTER, SOLEX TYPE, PETROL ENGINE, 2¼ LITRE, Series II and IIA
Up to engine suffix 'H' inclusive

Plate Ref.	1 2 3 4	DESCRIPTION	Qty	Part No.	REMARKS
54		Fibre washer for jet, SX 52825	1	260734	
55		Economy jet, high speed (150), SX 51274/L1	1	503911	Alternatives.
		Economy jet, high speed 100, SX 51274/L1/100	1	512107	Check before ordering
56		Float, SX 51785	1	503912	
57		Spindle for float, SX 51786	1	503913	
58		Copper washer for spindle, SX 51287	1	503914	
59		Needle valve complete, SX 51305/7	1	260725	
60		Fibre washer for valve, SX 1019	1	601512	
61		Top cover for carburetter, SX 12861	1	503916	
62		Joint washer for top cover, SX 52732	1	503915	
63		Screw, SX 1129	1	503917	Fixing joint washer
64		Washer, SX 51149	2	503918	to top cover
65		Special screw fixing top cover, SX 9189	2	503919	
66		Spring washer for screw, SX 971	4	260714	
67		Banjo union, SX 4120/5	1	260166	
68		Special bolt for union, SX 4122	1	260167	
69		Filter gauze for union, SX 4123	1	260168	
70		Fibre washer, large, SX 4124	1	260169	For
71		Fibre washer, small, SX 4124/1	1	260170	union
72		Elbow for top cover, SX 12828	1	503920	
73		Distance piece, elbow to top cover, SX B 13934	1	527275	Not part of carburetter
74		Screw fixing elbow to top cover, SX 9698	2	503921	For carburetter with starter heater element. Check before ordering.
		Screw fixing elbow to top cover, SX 51421/3	2	260157	For carburetter without starter heater. Check before ordering
75		Rubber sealing washer, elbow to starter cover, SX 12862	1	503922	For carburetter with starter heater element. Check before ordering
76		Lever for accelerator pump rod, SX 12880	1	503923	
77		Special washer for levers, SX 4031/1	1	261410	
78		Nut fixing lever to spindle	2	260159	
79		Control rod for accelerator pump, SX 12671	1	503924	
80		Split pin	4	3359	
81		Plain washer, SX 52760	3	260771	For control rod
82		Spring, SX 53010	1	503925	
		Carburetter overhaul kit	1	507687	
		Carburetter gasket kit	1	507693	
		CARBURETTER CONVERSION KIT	1	542637	To convert early type PA 10-5A and PA 10-6 carburetters to late types with revised jet sizes

CARBURETTER SETTINGS WHICH MAY BE ADVANTAGEOUS FOR HIGH ALTITUDES

Main jet (120), SX 50552/6/120				260494	5,000 to 7,000 ft
Main jet (117.5), SX 50552/6/117.5			1	515314	7,000 to 9,000 ft
Main jet (115), SX 50552/6/115			1	260173	9,000 to 12,000 ft
Pilot jet (45), SX 50797/4/45			1	260178	12,000 ft
Main jet (112.5), SX 50552/6/112.5			1	515313	12,000 to 14,000 ft
Pilot jet (45), SX 50797/4/45			1	260178	14,000 ft
Altitude corrector, SX B 13759/130			1	531984	For all altitudes. Alternative to jets listed above

* Asterisk indicates a new part which has not been used on any previous Rover model

CARBURETTER, ZENITH TYPE, PETROL ENGINE, 2¼ LITRE, Series IIA
From engine suffix 'J' onwards

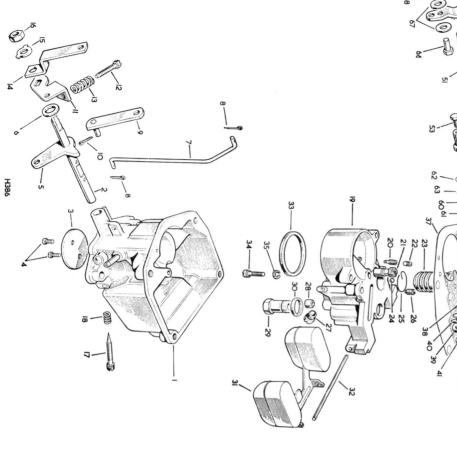

H386

CARBURETTER, ZENITH TYPE, PETROL ENGINE, 2¼ LITRE, Series IIA
From engine suffix 'J' onwards

Plate Ref.	1 2 3 4	DESCRIPTION	Qty	Part No.	REMARKS
		CARBURETTER COMPLETE, ZENITH 36 IV			
1		Carburetter main body, Zenith B17120Z	1	554149	
2		Throttle spindle, Zenith 020150	1	601834	
3		Butterfly for throttle, Zenith 018227	1	601835	
4		Special screw fixing butterfly, Zenith 010350	2	601844	
5		Floating lever on throttle spindle, Zenith 020739	1	601845	
6		Plain washer on spindle for floating lever, Zenith 08545	1	601842	
7		Interconnecting link, throttle to choke, Zenith 06893	1	601843	
8		Split pin fixing link to levers, Zenith 05889	2	601841	
9		Relay lever, throttle to accelerator pump, Zenith 020476	1	601893	
10		Split pin fixing relay lever to floating lever, Zenith 020740	1	601894	
11		Throttle stop and fast idle lever, Zenith 020738	1	601840	
12		Special screw, Zenith 016543 ⎰ For throttle	1	601837	
13		Spring, Zenith 04611 ⎱ stop	1	601846	
14		Throttle lever, Zenith B17115Z	1	601847	
15		Lockwasher, Zenith 019020 ⎰ Fixing throttle	1	601836	
16		Special nut, Zenith 016411 ⎱ levers	1	601839	
17		Volume control screw, Zenith 020458	1	601838	
18		Spring for control screw, Zenith L1177	1	601848	
19		Emulsion block, Zenith B17117Z	1	601849	
20		Pump jet (65), Zenith 020371	1	601851	
21		Plug for pump jet, Zenith 05972	1	601899	
22		Pump discharge valve, Zenith B17110Z	1	601855	
23		Piston for accelerator pump, Zenith 020164	1	601854	
24		Ball for piston, Zenith 09061	1	601856	
25		Circlip for piston, Zenith 020157	1	601852	
26		Slow-running jet (60), Zenith 017563	1	601853	
27		Main jet (125), Zenith 012571	1	601897	
28		Enrichment jet (195), Zenith 012572	1	601895	
29		Needle valve (1.75), Zenith 020173	1	601896	
30		Special washer (2 mm) for needle valve, Zenith 016622	1	601900	
31		Float, Zenith B17001Z	1	601857	
32		Spindle for float, Zenith 020224	1	601858	
33		'O' ring, emulsion block to body, Zenith 020646	1	601859	
34		Special screw, emulsion block to body, Zenith 020196 ⎰ Fixing emulsion	1	601850	
35		Spring washer, Zenith L1474 ⎱ block to body	1	601886	
36		Top cover for carburetter, Zenith B17113Z	1	601887	
37		Gasket for top cover, Zenith 17313Z	1	601860	
38		Ventilation screw (3.0) for choke, Zenith 016353	1	601885	
			1	601898	

* Asterisk indicates a new part which has not been used on any previous Rover model

CARBURETTER, ZENITH TYPE, PETROL ENGINE, 2¼ LITRE, Series IIA
From engine suffix 'J' onwards

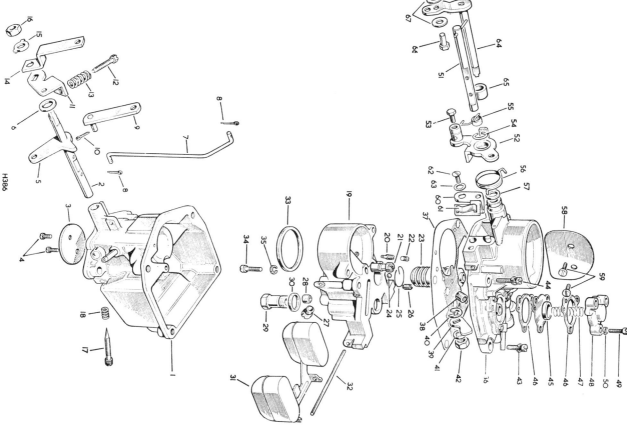

H386

CARBURETTER, ZENITH TYPE, PETROL ENGINE, 2¼ LITRE, Series IIA
From engine suffix 'J' onwards

Plate Ref.	1 2 3 4	DESCRIPTION	Qty	Part No.	REMARKS
39		Pump lever, internal, Zenith 020194	1	601876	
40		Retaining ring for pump lever, Zenith 020594	1	601878	
41		Shakeproof washer, Zenith 016412 ⎱ Fixing	1	601877	
42		Special nut, Zenith 016411 ⎰ pump lever	1	601838	
43		Screw and spring washer, short, Zenith 01810B ⎱ Fixing	1		
44		Screw and spring washer, long, Zenith 018252 ⎰ top cover to main body	2	601889	
45		Diaphragm for carburetter, Zenith 016127	1	601888	
46		Gasket for diaphragm, Zenith 010270	1	601879	
47		Spring for diaphragm, Zenith 016818	2	601880	
48		Cover for diaphragm, Zenith 016358	1	601881	
49		Screw, Zenith 07087 ⎱ Fixing diaphragm	1	601882	
50		Spring washer, Zenith 04692 ⎰ cover	3	601883	
51		Spindle and pin for choke lever, Zenith 020743	3	601884	
52		Lever and swivel for choke, Zenith 020737	1	601885	
53		Screw for choke lever swivel, Zenith 012754	1	601861	
54		Circlip fixing choke lever to top cover, Zenith 015127	1	601884	
55		Spring, small, Zenith B17122Z ⎱ For choke	1	601863	
56		Spring, large, Zenith 018462 ⎰ lever	1	601866	
57		Plain washer for choke spindle, Zenith 018697	1	601882	
58		Butterfly for choke, Zenith 020377	1	601869	
59		Special screw fixing butterfly, Zenith L1942	1	601867	
60		Bracket and clip for choke cable, Zenith 020742	2	601868	
61		Clip for choke bracket, Zenith B17626	1	601870	
62		Special screw, Zenith 09200 ⎱ Fixing choke	1	601871	Part of 601870
63		Shakeproof washer, Zenith 07734 ⎰ bracket to top cover	1	601872	
64		Spindle and lever for accelerator pump, Zenith 020486	1	601873	
65		Spacing washer for pump spindle, Zenith 020491	1	601874	
66		Pin, Zenith 016260 ⎱ Fixing relay	1	601875	
67		Plain washer, Zenith 016707 ⎰ lever to pump lever	2	601890	
68		Split pin, Zenith 03408	1	601892	
		Carburetter overhaul kit, Zenith MRK 3082	1	601891	
		Carburetter gasket kit, Zenith 237Z	1	605093	

CARBURETTER SETTINGS WHICH MAY BE ADVANTAGEOUS FOR HIGH ALTITUDES

Main jet (120)	...	...	1	605683*	5,000 to 7,000 ft
Main jet (117.5)	...	...	1	605684*	7,000 to 9,000 ft
Main jet (115)	...	...	1	605685*	9,000 ft
Slow running jet (55)	...	...	1	605687*	12,000 ft
Main jet (112.5)	...	...	1	605686*	12,000 to 14,000 ft
Slow running jet (55)	...	...	1	605687*	14,000 ft

* Asterisk indicates a new part which has not been used on any previous Rover model

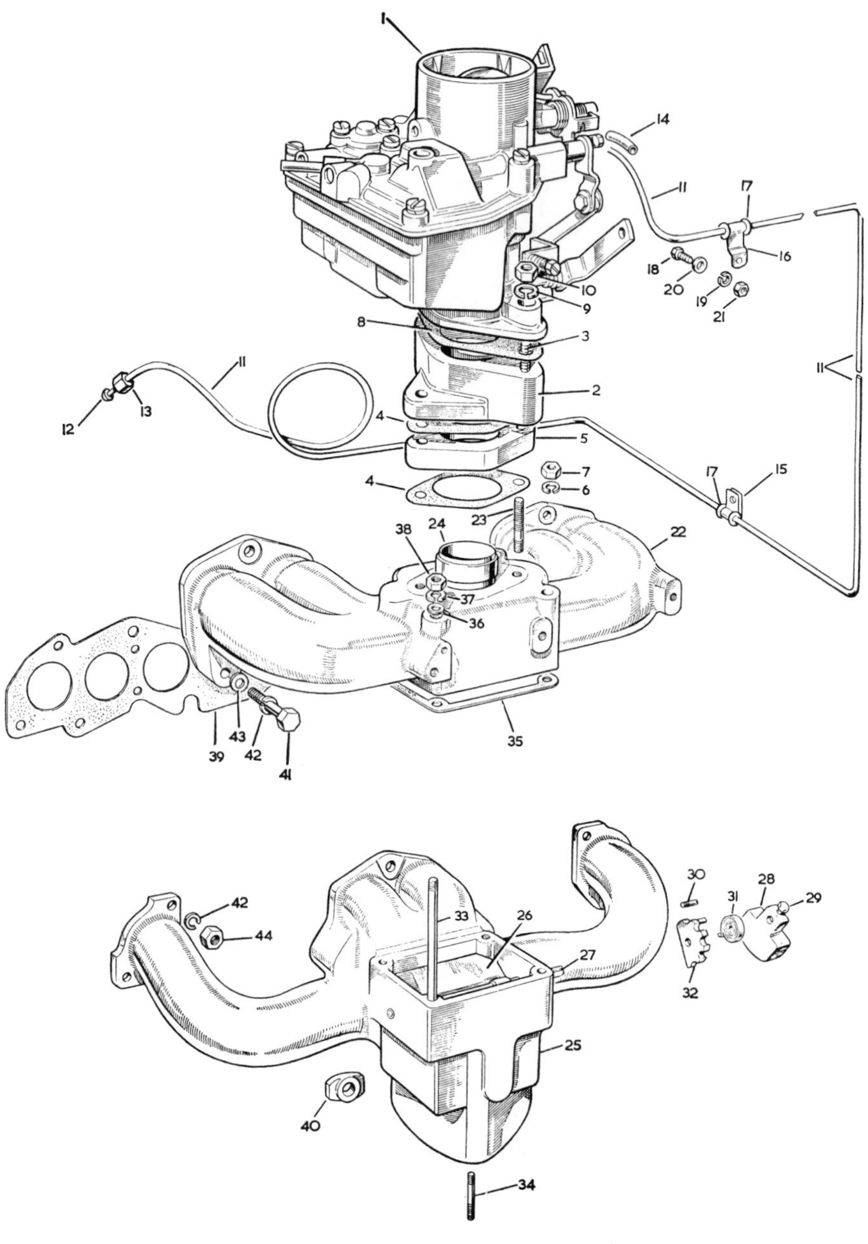

MANIFOLDS, PETROL ENGINE, 2¼ LITRE, Series II and IIA

MANIFOLDS, PETROL ENGINE, 2¼ LITRE, Series II and IIA

Plate Ref. 1 2 3 4	DESCRIPTION	Qty	Part No.	REMARKS
1	Carburetter complete, Solex type	2	278163	See page 69 } Up to engine suffix 'H' inclusive
2	Joint washer for carburetter	1	278162	
3	Packing for carburetter	2	3076	
	Spring washer } Fixing carburetter	2	254812	
	Nut (⅜" UNF) }	2	254812	
1	Carburetter complete, Zenith 36 IV type	1	554175	See page 69 } From engine suffix 'J' onwards
1	Carburetter complete, Zenith 36 IV type	1	252514	See page 75
2	Adaptor for carburetter	1	254812	
3	Stud for carburetter	2	278163	
4	Joint washer for adaptor and packing piece	1	278162	
5	Packing for adaptor	1	3076	
6	Spring washer } Fixing adaptor and packing	2	254812	
7	Nut (⅜" UNF) }	2	554163	
8	Joint washer for carburetter } to inlet manifold	2	3075	
9	Spring washer } Fixing carburetter	2	254831	
10	Nut (5/16" UNF) } to adaptor	2	527349	
	Suction pipe, complete, carburetter to distributor	1	554145	Up to engine suffix 'H' inclusive
11	Suction pipe complete, carburetter to distributor	1	514580	From engine suffix 'J' onwards
14	Rubber tube, suction pipe to carburetter	1	514228	
	Olive } Suction pipe, carburetter end	1	566865	
	Screwed union }	1	260708	Up to engine suffix 'H' inclusive
13	Nut } carburetter end	1	260707	
12	Nipple } Suction pipe, carburetter end	1	3783	From engine suffix 'J' onwards
13	Nut }	1	3787	
16	Clip, suction pipe to manifold stud	1	255206	
15	Clip for suction pipe	1	214229	
16	Clip, suction pipe to cylinder head stud	2	512646	
17	Clip, suction pipe to cylinder head stud	1	278161	
	Rubber grommet for clip	3	252529	
18	Bolt (¼" UNF x ⅝" long) } Fixing clip	1	278598	
19	Spring washer }	1	254810	
20	Plain washer }	1	3840	
21	Nut (¼" UNF) }	1	3074	
22	INLET MANIFOLD ASSEMBLY	1	501422	
23	Stud for carburetter or adaptor	2	501175	
24	Liner for inlet manifold	1	247811	
25	Exhaust manifold	1	247819	1958-60
26	Exhaust manifold	1	279168	1961 onwards
27	Spindle for butterfly valve	1	279169	
28	Butterfly valve	1	237119	
29	Counterbalance weight	1	247818	
30	Set screw (2 BA x ⅜" long) fixing weight to spindle	1	536646	
31	Stop pin for adjusting plate	1	574048	
32	Bi-metal spring for butterfly	1	530135	
33	Adjusting plate and pin for spring	3	256426	
	Stud, 3¼" long } For inlet manifold	1	256429	
	Stud, 3⅞" long }	1		
	Special set bolt } Fixing inlet to exhaust manifold	2	256426	
	Special set bolt }	2	256429	LHStg

Not part of engine assembly

* Asterisk indicates a **new** part which has not been used on any previous Rover model

MANIFOLDS, PETROL ENGINE, 2¼ LITRE, Series II and IIA

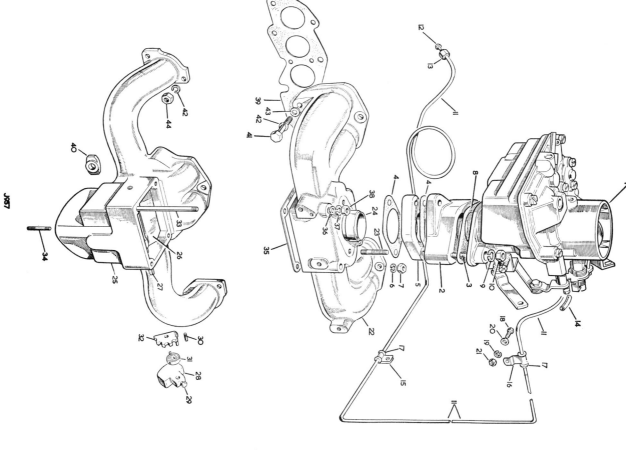

J87

MANIFOLDS, PETROL ENGINE, 2¼ LITRE, Series II and IIA

Plate Ref. 1 2 3 4	DESCRIPTION	Qty	Part No.	REMARKS
34	Stud for exhaust pipe	3	568664	
35	Joint washer ⎱	1	247824	
36	Plain washer ⎰ Fixing inlet to	4	2550	
37	Spring washer ⎱ exhaust manifold	4	3075	
38	Nut (5⁄16″ UNF) ⎰	4	254821	
39	Joint washer, inlet and exhaust manifold to cylinder head	4	274171	Metal and fibre type. Up to engines numbered 25283642J except Switzerland 25282861J Switzerland
	Joint washer, inlet manifold to cylinder head	2	564307†	Corrugated tin plate type. From engines numbered 25283643J onwards except Switzerland 25282862J onwards Switzerland
40	Clamp for manifold	2	247836	Not part of engine assembly
41	Set bolt (5⁄16″ UNF x 2¼″ long) ⎱	2	256029	
42	Set bolt (5⁄16″ UNF x 2″ long) ⎰ Fixing	2	256226	
42	Spring washer ⎱ manifold to	11	3075	
43	Plain washer ⎰ cylinder head	4	2550	
44	Nut (5⁄16″ UNF)	7	254831	

* Asterisk indicates a new part which has not been used on any previous Rover model
† With corrugated tin plate inlet manifold joint washer, the exhaust manifold and cylinder head are in metal-to-metal contact

DISTRIBUTOR AND STARTER MOTOR, PETROL ENGINE, 2¼ LITRE, Series II and IIA

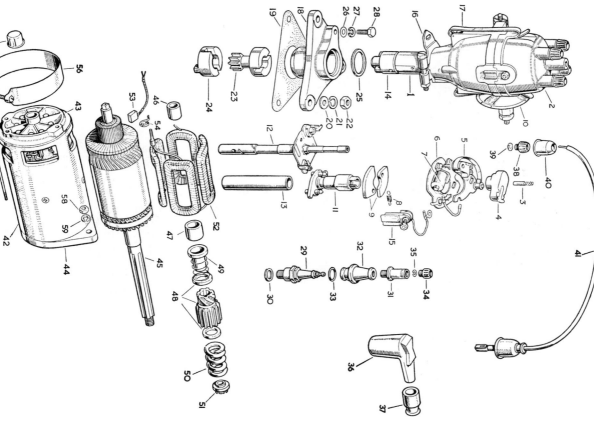

J968

COLLINS-JONES.

DISTRIBUTOR AND STARTER MOTOR, PETROL ENGINE, 2¼ LITRE, Series II and IIA

Plate Ref.	1 2 3 4	DESCRIPTION	Qty	Part No.	REMARKS
1		DISTRIBUTOR COMPLETE, LU 40944	1	546695	
2		Distributor cap complete, LU 5441451 3	1	600328	
3		Brush and springs for cap, LU 418856	1	262703	
4		Rotor arm, LU 400051	1	245005	
5		Contact breaker points set, LU 423153	1 set	269987	
6		Contact breaker base plate, LU 422318	1	502282	
7		Condenser for distributor, LU 423871	1	269988	
8		Auto advance springs set, LU 54410187	1 set	502285	
9		Auto advance springs set, LU 54411631	1 set	600476	
10		Auto advance weight, LU 410033/S	1	262708	
11		Auto advance weight, LU 54413073	1	539572	
		Vacuum unit, LU 54411598	1	502286	
11		Cam for distributor, LU 496078	1	245008	
12		Cam for distributor, LU 491072	1	600475	
		Shaft for distributor, LU 419689	1	269990	
		Shaft and action plate for distributor, LU 54410290	1	539575	
13		Bush for distributor shaft, LU 419430	1	245012	
14		Bush for distributor shaft, LU 421998	1	600334	
		Driving dog for distributor, LU 410601	1	245014	
15		Terminal bush and lead, LU 422238	1	502283	
		Terminal bush and lead, LU 54413549	1	600329	
16		Clamping plate for distributor, LU 420151	1	245003	
17		Clip for cover, LU 421824	2	502287	
		Sundry parts kit, LU 419644		245015	
18		Adaptor for distributor	1	247672	
19		Adaptor for distributor	1	549610	
20		Joint washer, adaptor to cylinder block	1	247212	
21		Plain washer	3	3966	
22		Spring washer	3	3075	
		Nut (5/16" UNF)	3	254831	
23		Set bolt (5/16" UNF x 7/8" long) Fixing adaptor to cylinder block	3	255227	
24		Driving shaft for distributor, top	1	247806	
		Distributor drive coupling	1	549611	
25		Cork washer for distributor housing	1	52278	
26		Plain washer Fixing distributor to adaptor	1	3911	
27		Spring washer	1	3074	
28		Set bolt (¼" UNC x 9/16" long)	1	253205	

Remarks column (grouped):
- For distributor LU 40944
- For distributor LU 40609
- For distributor LU 40944
- For distributor LU 40609
- For distributor LU 40944
- For distributor LU 40609
- For distributor LU 40944
- For distributor LU 40944
- Up to engine suffix 'F'
- From engine suffix 'G' onwards
- Alternative to set bolt type fixing
- Alternative to stud and nut type fixing
- Spline type shaft. Up to engine suffix 'F'
- Blade type coupling. From engine suffix 'G' onwards

Not part of engine assembly

DISTRIBUTOR AND STARTER MOTOR, PETROL ENGINE, 2¼ LITRE, Series II and IIA

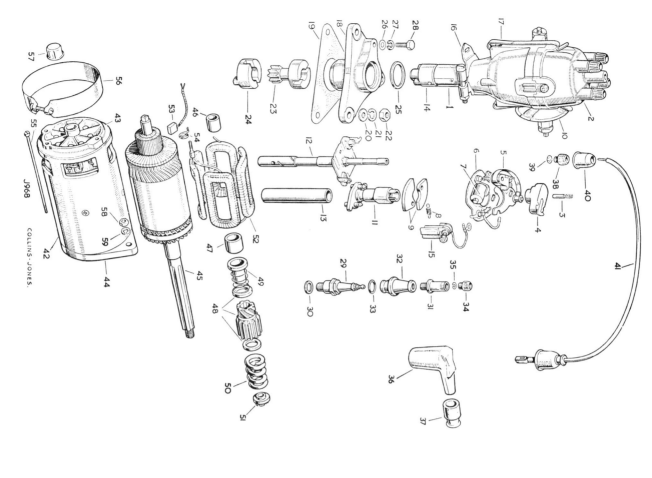

J968

COLLINS-JONES.

DISTRIBUTOR AND STARTER MOTOR, PETROL ENGINE, 2¼ LITRE, Series II and IIA

Plate Ref. 1 2 3 4	DESCRIPTION	Qty	Part No.	REMARKS
29	Sparking plug, Lodge CLNH	4	262796	Alternatives. Check before ordering
29	Sparking plug, Champion N8	4	512806	
30	Joint washer for sparking plug	4	40441	
31	Suppressor for sparking plug	4	240138	
32	Sparking plug cover	4	214262	Alternative to 6 lines below. Check before ordering
33	Sealing ring for plug cover	4	213172	
34	Cable nut, LU 410600	9	214278	
35	Washer for cable nut, LU 185015	9	214279	
35	Suppressor and cover for sparking plug	4	513694	Lodge type — Alternative to 5 lines above. Check before ordering
36	Suppressor and cover for sparking plug	4	558045	Champion type
37	Rubber boot for suppressor	4	240408	
38	Cable nut for distributor, LU 410600	5	214278	
39	Washer for cable nut, LU 185015	5	214279	
40	Rubber boot for distributor cable nut	5	507001	
41	Lead, HT, distributor to coil	1	80603	
	HT Wire	As reqd	568766 *	For ignition coil with 'push-in' type connection Check before ordering
	STARTER MOTOR COMPLETE, LU 25605	1	236287	
42	Bracket, commutator end, LU 256748	1	244706	
43	Bracket, drive end, LU 256495	1	244713	
44	Armature, LU 255463	1	244715	
45	Bush, commutator end, LU 255491	1	242958	
46	Bush, pinion end, LU 270038	1	244714	
47	Pinion and sleeve, LU 255194	1	244711	
48	Spring for pinion, LU 255728	1	244712	
49	Main spring for pinion, LU 270889	1	244710	
50	Nut for pinion, LU 255851	1	244709	
51	Field coil for starter, LU 256578	1	262861	
52	Brushes for starter motor, set, LU 256659	2	260055	
53	Spring set for brushes, LU 54257221	2	601754	
54	Bolt for bracket, LU 255459	2	244717	
55	Cover band, LU 255557	1	244705	
56	Grease cap, LU 255854	1	243095	
57	Sundry parts kit, LU 256762	1	244718	
58	Nut (⅜" UNF)	2	254812	Fixing starter motor to flywheel housing — Not part of engine assembly
59	Spring washer	2	3076	
	Fan disc washer	1	512305	

* Asterisk indicates a new part which has not been used on any previous Rover model

DYNAMO AND FIXINGS, PETROL ENGINE, 2¼ LITRE, Series II and IIA

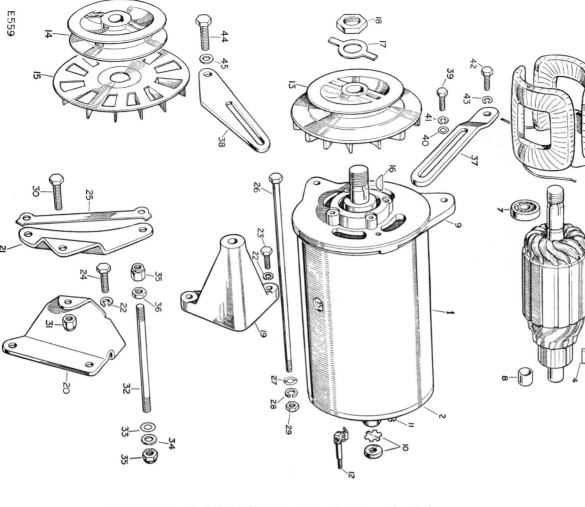

DYNAMO AND FIXINGS, PETROL ENGINE, 2¼ LITRE, Series II and IIA

Plate Ref.	1 2 3 4	DESCRIPTION	Qty	Part No.	REMARKS
1		Dynamo complete, LU 22700, type C40 ...	1	551569	
2		Bracket, commutator end, LU 227702 ...	1	242669	For dynamo type C39
2		Bracket, commutator end, LU 54211125 ...	1	514195	For dynamo type C40
3		Armature for dynamo, LU 5421 1499 ...	1	514194	For dynamo type C40
4		Brushes for dynamo, set, LU 227305 ...	1	261483	For dynamo type C39
4		Brushes for dynamo, set, LU 227541 ...	1	514193	For dynamo type C40
5		Spring set for brushes, LU 228159 ...	1	261238	For dynamo type C39
5		Spring set for brushes, LU 227542 ...	1	514192	For dynamo type C40
6		Field coil for dynamo, LU 227291 ...	1	262860	For dynamo type C39
6		Field coil for dynamo, LU 227543 ...	1	514191	For dynamo type C40
7		Ball bearing, front ...	1	529221	
8		Bush, commutator end, LU 227818 ...	1	271614	For dynamo type C39
9		Bracket, drive end, LU 227698 ...	1	242673	For dynamo type C39
9		Bracket, drive end, LU 54211095 ...	1	514190	For dynamo type C40
10		Oiler for dynamo, LU 227914 ...	1	271615	For dynamo type C39
10		Oiler for dynamo, LU 54211097 ...	2	514189	For dynamo type C40
11		Bolt for bracket, LU 228336 ...	2	242675	
12		Terminal, set, LU 227625 ...	1	264436	For dynamo type C39
12		Terminal, set, LU 54216878 ...	1	532566	For dynamo type C40
13		Sundry parts, set, LU 227726 ...	1	242676	For dynamo type C39
13		Sundry parts, set, LU 54211894 ...	1	532567	For dynamo type C40
14		Pulley for dynamo ...	1	247209	Series IIA models
15		Fan for dynamo ...	1	522913	Series II models
16		Distance washer for fan ...	2	554055	Series IIA models
17		Woodruff key ...	1	533860	
18		Lockwasher ⎫ Fixing pulley	1	1664	
19		Spring washer ⎭ to dynamo	1	217781	Alternatives
			1	547604	
20		Special nut ...	1	3466	
21		Anchor bracket for dynamo ...	1	247085	Series II models
22		Support bracket for dynamo ...	1	277999	Series IIA models
23		Steady bracket for dynamo ...	1	278000	
24		Spring washer ...	2	3075	
25		Set bolt (⁵⁄₁₆" UNF x 1⅛" long) ⎫ Fixing anchor or	2	255029	Series II models
26		Set bolt (⁵⁄₁₆" UNF x ⅝" long) ⎭ support bracket	2	255225	Series IIA models
27		Locking plate for dynamo bolts ...	1	501198	Not required when front cover timing pointer is fitted
28		Bolt (⁵⁄₁₆" UNF x 7" long)	1	273271	⎫ Series IIA models
29		Shim washer	3	4148	⎭
30		Spring washer	1	3075	Series II models
31		Nut (⁵⁄₁₆" UNF)	1	254831	Not part of engine assembly
		Bolt (⁵⁄₁₆" UNF x 1⅜" long) ⎫ Fixing dynamo	1	256421	Series IIA models
		Self-locking nut (⁵⁄₁₆" UNF) ⎭ to front support	1	252211	

E559.

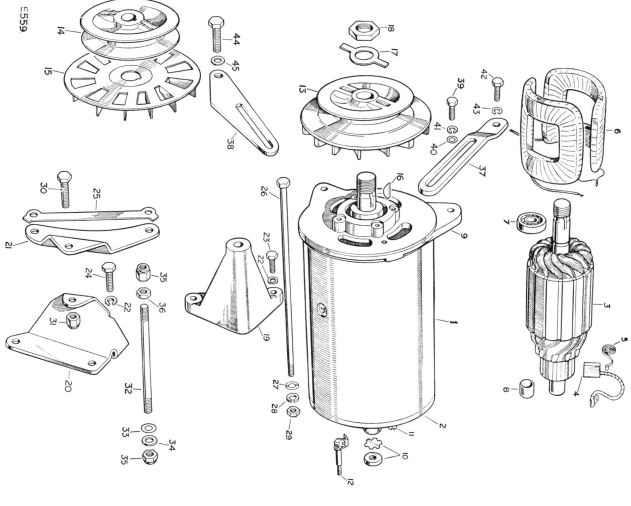

DYNAMO AND FIXINGS, PETROL ENGINE, 2¼ LITRE, Series II and IIA

Plate Ref.	1 2 3 4	DESCRIPTION	Qty	Part No.	REMARKS
32		Special stud	1	526225	
33		Shim washer	As reqd	4148	
34		Plain washer	2	2550	Series IIA models
35		Self-locking nut (5/16″ UNF)	2	252161	
36		Locknut (5/16″ UNF)	1	254821	
37		Adjusting link for dynamo	1	247523	Series II models
38		Adjusting link for dynamo	1	514472	Series IIA models
39		Special bolt	1	253027	
40		Plain washer } Fixing	1	2266	
41		Spring washer } adjusting link	1	3075	
42		Plain washer } Fixing	1	253025	Series II models
		} link to			
43		Spring washer } adjusting	1	3075	} Series IIA models
44		Set bolt (5/16″ UNC x 5/8″ long) } link to	1	256031	} models
45		Bolt (5/16″ UNF x 3¼″ long) } front cover	1	2266	
		Plain washer	1	2266	Not part of engine assembly
		Dynamo complete, LU 22764, type C40-T	1	551360	Early type
		Dynamo complete, LU 22756, type C40-T	1	605968*	Late type
		Bracket, commutator end, LU 54216363	1	605055	
		Cover for shaft, commutator end, LU 54216152	1	605056	For dynamo LU 22764
		Ball bearing, sealed, commutator end, LU 54160219	1	605057	
		Armature for dynamo, LU 54216228	1	605058	For dynamo LU 22764
		Armature for dynamo, LU 54218417	1	605969*	For dynamo LU 22756
		Brushes for dynamo, set, LU 54216361	1	605060	
		Retaining clip for brush, LU 54215103	2	605061	
		Spring set for brushes, LU 54216362	2	605062	
		Field coil for dynamo, LU 227543	1	605063	
		Ball bearing, sealed, drive end, LU 54160139	1	242675	
		Bracket, drive end, LU 54216225	1	532567	
		Bolt for bracket, LU 228336	2	550336	For use with sealed } Optional
		Sundry parts, set, LU 54211894	1		bearing type dynamo } equipment
		Fan for dynamo, LU 54214561	1		} Optional
		Distance washer for fan	1	550281	} equipment

* Asterisk indicates a new part which has not been used on any previous Rover model

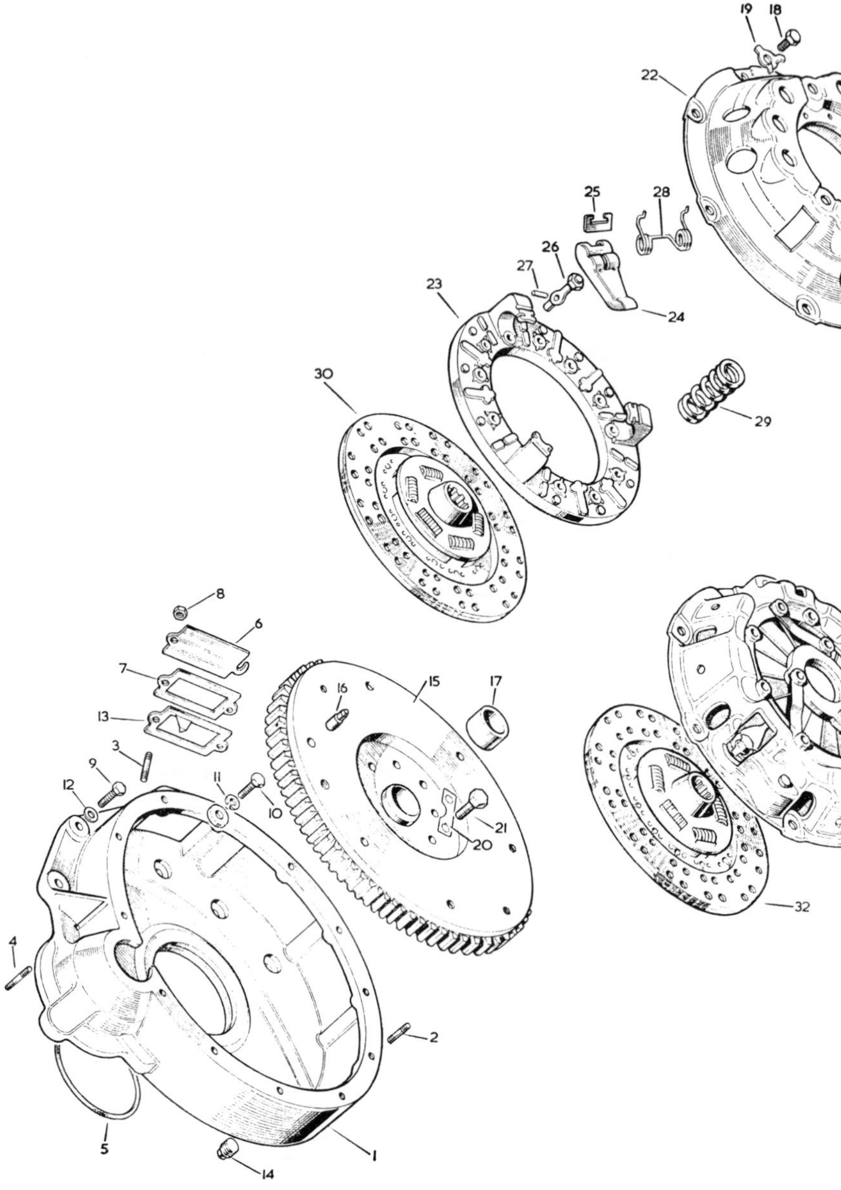

H407

Plate Ref.	1 2 3 4 DESCRIPTION	Qty	Part No.	REMARKS
1	FLYWHEEL HOUSING ASSEMBLY	1	247526	Up to engine suffix 'J' inclusive
	FLYWHEEL HOUSING ASSEMBLY	1	564394*	From engine suffix 'K' onwards
2	Stud ($\frac{3}{8}''$ UNF) fixing flywheel housing to bell housing	12	247145	
3	Stud ($\frac{1}{4}''$ UNF) fixing inspection cover	2	247146	Up to engine suffix 'J' inclusive
4	Stud ($\frac{3}{8}''$ UNF) for starter motor	2	247145	
5	Sealing ring for flywheel housing	1	246169	
6	Inspection cover plate	1	56140	Up to engine suffix 'J' inclusive (6–8)
7	Joint washer for cover plate	2	50216	
8	Nut ($\frac{1}{4}''$ UNF) fixing cover plate	2	254810	
9	Bolt ($\frac{3}{8}''$ UNF x $1\frac{3}{4}''$ long) Fixing flywheel housing to cylinder block (9–12)	2	256043	Up to engine suffix 'J' inclusive
	Bolt ($\frac{3}{8}''$ UNF x $1\frac{3}{8}''$ long)	2	256040	From engine suffix 'K' onwards
10	Bolt ($\frac{3}{8}''$ UNF x $1\frac{1}{8}''$ long)	6	255248	
11	Spring washer	8	3076	
12	Plain washer	6	2219	
13	Indicator for engine timing	1	272784	Up to engine suffix 'J' inclusive
14	Drain plug for housing	1	3290	
	Stowage bracket for drain plug	1	276511	
15	FLYWHEEL ASSEMBLY	1	600243	
	Ring gear for flywheel	1	506799	
16	Dowel locating clutch cover plate	2	502116	3 off on vehicles with $9\frac{1}{2}''$ clutch
17	Bush for primary pinion	1	08566	
18	Set bolt fixing clutch	6	255427	
19	Tab washer for set bolt	6	546197	Alternatives
	Spring washer for set bolt	6	3075	
20	Locker } Fixing flywheel to crankshaft	4	526161	
21	Special set bolt	8	247135	
	CLUTCH ASSEMBLY, BB 45693/41	1	236684	9″ type clutch with black clutch springs
22	Cover plate for clutch, BB 45481	1	231888	
23	Pressure plate for clutch, BB 42652	1	231880	
24	Release lever for clutch, BB 51022	3	534358	
25	Strut for release lever, BB 42606	3	231884	
26	Eyebolt and nut for release lever, BB 48508 ..	3	242996	
27	Pin for release lever, BB 42604	3	231885	
28	Anti-rattle spring for release lever, BB 47688 ..	3	231883	
29	Clutch spring (yellow and light green), BB 44780	9	243593	For early type clutch } Check before ordering
	Clutch spring (black), BB 44633	9	275301	Part of 236684 For late type clutch
30	Clutch plate complete, BB 47626/161	1	504519	For 9″ type clutch
	Lining package for clutch plate, BB 46627 ..	1	261921	Ferodo lining } Alternatives
	Lining package for clutch plate, BB KL 75012 ..	1	517026	Mintex lining
31	CLUTCH ASSEMBLY, BB 75302/11	1	540700	For $9\frac{1}{2}''$ type clutch. America Dollar Area. From engines numbered 25163284 onwards. Optional equipment for all other Areas
32	Clutch plate complete, BB 51226/12	1	539827	
	Lining package for clutch plate, BBSSC 75073 ..	1	600388	

Not part of engine assembly (CLUTCH ASSEMBLY 236684 to 600388)

* Asterisk indicates a new part which has not been used on any previous Rover model

CYLINDER BLOCK, 2.6 LITRE 6-CYLINDER PETROL MODELS

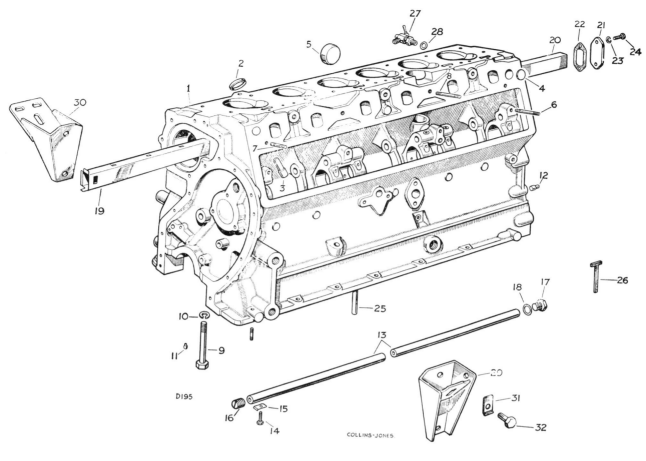

D195

COLLINS-JONES

CYLINDER BLOCK, 2.6 LITRE 6-CYLINDER PETROL MODELS

Plate Ref.	1 2 3 4	DESCRIPTION	Qty	Part No.	REMARKS
		ENGINE ASSEMBLY, 7.8:1 compression ratio	1	605104	
		ENGINE ASSEMBLY, 7:1 compression ratio	1	605103	Optional
1		CYLINDER BLOCK ASSEMBLY, less studs	1	605157	
2		Insert for exhaust valve seat	6	09065	
3		Valve guide, exhaust	6	511833	
4		Cup plug	2	525497	
5		Cup plug	5	525428	
6		Plug for redundant heater tapping	1	527269	
7		Stud for side rocker cover	4	252627	
8		Stud, short } For exhaust manifold	5	252622	
8		Stud, long } manifold	5	252638	
9		Special set bolt for main bearings	14	272749	
11		Dowel locating bearing caps	14	530354	
12		Dowel locating flywheel housing	2	52124	
13		Oil gallery pipe	2	272451	
14		Set bolt ($\frac{1}{4}$" UNF x $\frac{5}{8}$" long) } Fixing gallery pipes	2	255206	
15		Locker }	2	2995	
16		Plug for gallery pipe, front	1	279413	
17		Plug for gallery pipe, rear	1	279415	
18		Joint washer for rear plug	1	243970	
19		Water pipe in block, front	1	563047	
20		Water pipe in block, rear	1	272452	
21		Cover plate for rear water pipe	1	231218	
22		Joint washer for cover plate	1	231219	
23		Spring washer	2	3074	
24		Bolt ($\frac{1}{4}$" UNF x $\frac{5}{8}$" long) } Fixing cover plate	2	255206	
25		Stud for sump	1	252621	
26		Oil return pipe	1	275836	
27		Seal for rear bearing cap	2	524636	
28		Drain tap for cylinder block	1	538608	Taper thread type
		Joint washer for drain tap	1	213959	For early type tap with straight thread
30		Support bracket for engine, LH	1	562654	
29		Support bracket for engine, RH	1	239449	
31		Lockwasher } Fixing support brackets	4	212430	
32		Set bolt ($\frac{1}{2}$" UNF x $\frac{7}{8}$" long) } to cylinder block	4	255084	
		Stud and dowel kit for cylinder block	1	535708	
		Engine overhaul gasket kit for cylinder block	1	605106	

* Asterisk indicates a new part which has not been used on any previous Rover model

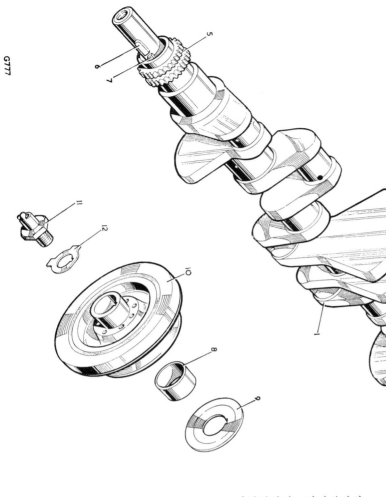

CRANKSHAFT, 2.6 LITRE 6-CYLINDER PETROL MODELS

Plate Ref.	1 2 3 4	DESCRIPTION	Qty	Part No.	REMARKS
1		CRANKSHAFT ASSEMBLY, STD	1	541910	
2		Dowel for flywheel	1	6395	
3		Main bearing, Std	7	600161	
		Main bearing, .010" US	7	600162	
		Main bearing, .020" US	7	600163	Pairs
		Main bearing, .030" US	7	600164	
		Main bearing, .040" US	7	600165	
4		Thrust washer for crankshaft, Std	1	600177	
		Thrust washer for crankshaft, .0025" OS	1	600174	
		Thrust washer for crankshaft, .005" OS	1	600175	
		Thrust washer for crankshaft, .0075" OS	1	600176	Pairs
		Thrust washer for crankshaft, .010" OS	1	600178	
		Thrust washer for crankshaft, .0125" OS	1	600179	
		BEARING SET FOR CRANKSHAFT, STD ...	1	600181	
		BEARING SET FOR CRANKSHAFT, .010" US	1	600182	Sets include
		BEARING SET FOR CRANKSHAFT, .020" US	1	600183	main bearings,
		BEARING SET FOR CRANKSHAFT, .030" US	1	600184	thrust washers
		BEARING SET FOR CRANKSHAFT, .040" US	1	600185	and connecting rod bearings
5		Chainwheel on crankshaft	1	564463	
6		Key, front, locating vibration damper	1	541921	
7		Key, rear, locating chainwheel	1	542623	
8		Oil seal ring for crankshaft	1	542622	
9		Oil thrower on crankshaft	1	546324	
10		VIBRATION DAMPER AND PULLEY ASSEMBLY ...	1	564155	Not part of engine assembly
11		Starting dog	1	530343	
12		Lockwasher for starting dog	1	542494	
13		Crankshaft oil retainer and seal assembly ...	1	542492	
14		Oil seal assembly	1	546333	
15		Silicone grease	1	270656	
15		Rear bearing split seal halves	2	523240	
16		Dowel, lower	2	519064	Fixing retainer to cylinder block
17		Dowel, upper	2	246464	
18		Spring washer	10	3074	and rear bearing cap
19		Bolt (¼" UNF x ⅞" long)	10	255208	

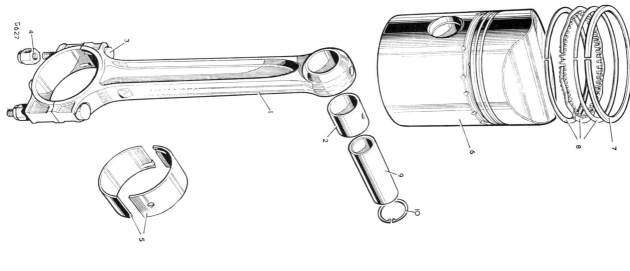

CONNECTING ROD AND PISTON, 2.6 LITRE 6-CYLINDER PETROL MODELS

Plate Ref. 1 2 3 4	DESCRIPTION	Qty	Part No.	REMARKS
1	CONNECTING ROD ASSEMBLY	6	524492	
2	Gudgeon pin bush	12	273163	
3	Special bolt } Fixing connecting	12	518100	
4	Special locking nut } rod cap	12	272317	
5	Connecting rod bearing, Std ...	6	523341	
	Connecting rod bearing, .010" US	6	523342	
	Connecting rod bearing, .020" US	6	523343	
	Connecting rod bearing, .030" US	6	523344	
	Connecting rod bearing, .040" US	6	523345	Check before ordering
6	PISTON ASSEMBLY, STD, GRADE 'Z'	6	537263	Engine with
	PISTON ASSEMBLY, STD, GRADE 'A'	6	537264	7.8:1 compression ratio.
	PISTON ASSEMBLY, STD, GRADE 'B'	6	537265	Check before ordering
	PISTON ASSEMBLY, STD, GRADE 'C'	6	537266	
	PISTON ASSEMBLY, STD, GRADE 'D'	6	537267	
	PISTON ASSEMBLY, .010" OS	6	537268	
	PISTON ASSEMBLY, .020" OS	6	537269	
	PISTON ASSEMBLY, .030" OS	6	537270	
	PISTON ASSEMBLY, .040" OS	6	537271	
	PISTON ASSEMBLY, STD, GRADE 'Z'	6	536267	Engine with
	PISTON ASSEMBLY, STD, GRADE 'A'	6	536268	7:1 compression ratio
	PISTON ASSEMBLY, STD, GRADE 'B'	6	536269	Check before ordering
	PISTON ASSEMBLY, STD, GRADE 'C'	6	536270	
	PISTON ASSEMBLY, STD, GRADE 'D'	6	536271	
	PISTON ASSEMBLY, .010" OS	6	536272	
	PISTON ASSEMBLY, .020" OS	6	536273	
	PISTON ASSEMBLY, .030" OS	6	536274	
	PISTON ASSEMBLY, .040" OS	6	536275	
7	Piston ring, compression, Std	12	231155	
	Piston ring, compression, .010" OS	12	236173	
	Piston ring, compression, .020" OS	12	236174	
	Piston ring, compression, .030" OS	12	236175	
	Piston ring, compression, .040" OS	12	236176	
8	Duaflex scraper ring, Std ...	6	554620	
	Duaflex scraper ring, .010" OS	6	554789	
	Duaflex scraper ring, .020" OS	6	554790	
	Duaflex scraper ring, .030" OS	6	554791	
	Duaflex scraper ring, .040" OS	6	554792	
9	Gudgeon pin, Std	6	264569	
	Gudgeon pin, .001" OS ...	6	267257	
	Gudgeon pin, .003" OS ...	6	267258	
10	Circlip for gudgeon pin ...	12	235100	

* Asterisk indicates a new part which has not been used on any previous Rover model

OIL PUMP AND OIL PIPES, 2.6 LITRE 6-CYLINDER PETROL MODELS

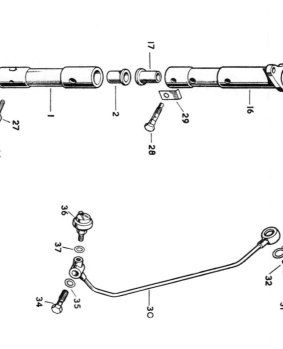

OIL PUMP AND OIL PIPES, 2.6 LITRE 6-CYLINDER PETROL MODELS

Plate Ref. 1 2 3 4	DESCRIPTION	Qty	Part No.	REMARKS
1	OIL PUMP ASSEMBLY	1	564334	
1	OIL PUMP BODY ASSEMBLY	1	542396	
2	Bush for drive shaft	1	212209	
3	Oil pump gear, driver	1	240555	
4	OIL PUMP COVER ASSEMBLY	1	564335	
5	Dowel locating body	2	52710	
6	Spindle for idler wheel	1	502209	
7	Stud for oil strainer	1	564217	
8	OIL PUMP IDLER GEAR ASSEMBLY	1	278109	
9	Bush for idler gear	1	214995	
10	Oil pump shield	1	09225	
11	Set bolt ($\frac{5}{16}$" UNF x $\frac{7}{8}$" long) } Fixing cover	4	255227	
	Spring washer } to body	4	3075	
13	Oil strainer	1	266900	
	Extension piece, strainer to oil pump	1	564216	
14	Castle nut } Fixing strainer	1	254950	
15	Split pin } to pump cover	1	2556	
16	DISTRIBUTOR HOUSING ASSEMBLY	1	274084	
17	Bush for drive shaft	1	521583	
18	Cork washer for distributor housing	1	52278	
19	Drive shaft for distributor	1	267829	
20	Oil pump and distributor gear and shaft assembly	1	515969	
	(phosphor-bronze gear)			
21	Steel ball	1	01035	
22	Plunger	1	245940	
23	Spring	1	504997	
24	Retaining cap } For oil pressure release valve	1	504995	
25	Joint washer	1	243971	
26	Special set screw } Fixing oil pump	1	274086	
27	Locknut ($\frac{5}{16}$" UNF) } to cylinder block	1	254861	
28	Oil feed bolt locating distributor housing	1	274928	
29	Locker for bolt	1	2504	
30	Oil pipe to cylinder head	1	558302	
31	Banjo bolt for oil pipe	1	557782	
32	Banjo bolt fixing oil pipe to cylinder head	1	265038	
33	Meter plug for banjo bolt	1	231421	
34	Joint washer for banjo bolt	2	231577	
35	Banjo bolt fixing oil pipe to cylinder block	1	233520	
36	Banjo bolt fixing oil pipe to cylinder block	1	232039	
36	Oil pressure switch complete	1	519664	
37	Joint washer for switch	1	232039	

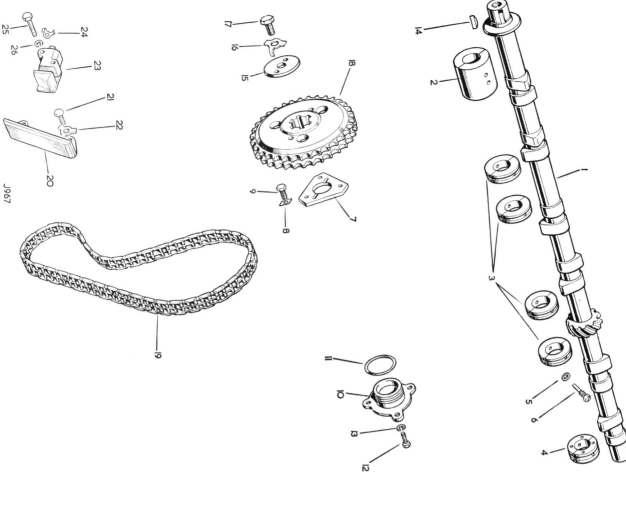

J967

Plate Ref.	1 2 3 4	DESCRIPTION		Qty	Part No.	REMARKS
1		Camshaft	...	1	523139	
2		Bearing, front		1	274116	
3		Bearing, intermediate	For camshaft	4	274115	
4		Bearing, rear		1	274117	
5		Spring washer	Fixing camshaft	6	3075	
6		Special set screw	bearings to block	6	274118	
7		Thrust plate for camshaft	...	1	502266	
8		Locker	Fixing thrust	1	2500	
9		Set bolt (¼" UNF x ⅝" long)	plate to block	3	255206	
10		Sealing plate for camshaft	...	1	530481	
11		Joint washer for sealing plate	...	1	276541	
12		Set bolt (¼" UNF x ⅝" long)	Fixing sealing plate	3	255206	
13		Spring washer	to cylinder block	3	3074	
14		Key locating chainwheel	...	1	230313	
15		Retaining washer	Fixing	1	09093	
16		Locker	chainwheel	1	09210	
17		Set bolt (⅜" UNF x ⅞" long)	to camshaft	1	255046	
18		Chainwheel for camshaft	...	1	563145 *†	
19		Camshaft chain	...	1	266662	
20		Vibration damper for timing chain	...	1	275234	
21		Locker	Fixing damper to	2	255204	
22		Set bolt (¼" UNF x ½" long)	cylinder block	2	557523	
23		Tensioner for timing chain	...	1	266661	
24		Tab washer for tensioner	...	1	504443	
25		Set bolt (¼" UNF x 1⅜" long)	Fixing tensioner	2	256202	
26		Spring washer	to cylinder block	2	3074	

* Asterisk indicates a new part which has not been used on any previous Rover model
† Also supplied as replacement for early type chainwheel with separate hub and fixings

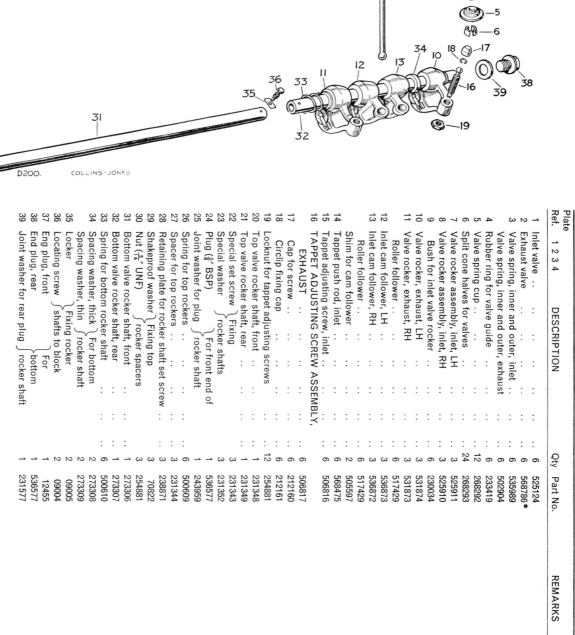

D200.   COLLINS·JONES

VALVE GEAR AND ROCKER SHAFTS, 2.6 LITRE 6-CYLINDER PETROL MODELS

Plate Ref. 1 2 3 4	DESCRIPTION	Qty	Part No.	REMARKS
1	Inlet valve	6	525124	
2	Exhaust valve	6	568786	*
3	Valve spring, inner and outer, inlet ...	6	535989	
4	Rubber ring for valve guide ...	6	502904	
5	Valve spring cup	12	233419	
6	Split cone halves for valves ...	24	268292	
7	Valve rocker assembly, inlet, LH ...	3	268293	
8	Valve rocker assembly, inlet, RH ...	3	525911	
9	Bush for inlet valve rocker ...	6	525910	
10	Valve rocker, exhaust, LH ...	3	230034	
11	Valve rocker, exhaust, RH ...	3	531874	
	Roller follower	3	531873	
12	Inlet cam follower, LH ...	6	536872	
13	Inlet cam follower, RH ...	6	536873	
14	Shim for cam follower ...	2	505597	
	Tappet push rod, inlet ...	6	568475	
15	Tappet adjusting screw, inlet ...	6	508816	
16	TAPPET ADJUSTING SCREW ASSEMBLY, EXHAUST	6	506817	
17	Cap for screw	6	212160	
18	Circlip fixing cap ...	6	212161	
19	Locknut for tappet adjusting screws	12	254881	
20	Top valve rocker shaft, front ...	1	231348	
21	Top valve rocker shaft, rear ...	1	231349	
22	Special set screw } Fixing	3	231351	
23	Special washer } rocker shafts	3	231352	
24	Plug (⅜" BSP) } For front end of	1	536577	
25	Joint washer for plug } rocker shaft	1	243959	
26	Spring for top rockers ...	6	500609	
27	Spacer for top rockers ...	3	231344	
28	Retaining plate for rocker shaft set screw	3	238871	
29	Shakeproof washer } Fixing top	3	70822	
30	Nut (5/16" UNF) } rocker spacers	3	254881	
31	Bottom valve rocker shaft, front ...	1	273306	
32	Bottom valve rocker shaft, rear ...	1	273307	
33	Spring for bottom rocker shaft	6	500610	
34	Spacing washer, thick } For bottom	2	273309	
	Spacing washer, thin } rocker shaft	2	273308	
35	Locker } Fixing rocker	2	09005	
36	Locating screw } shafts to block	2	09004	
37	Eng plug, front } For	1	12455	
38	End plug, rear } bottom	1	536577	
39	Joint washer for rear plug } rocker shaft	1	231577	

FRONT COVER, SIDE COVER AND SUMP, 2.6 LITRE 6-CYLINDER PETROL MODELS

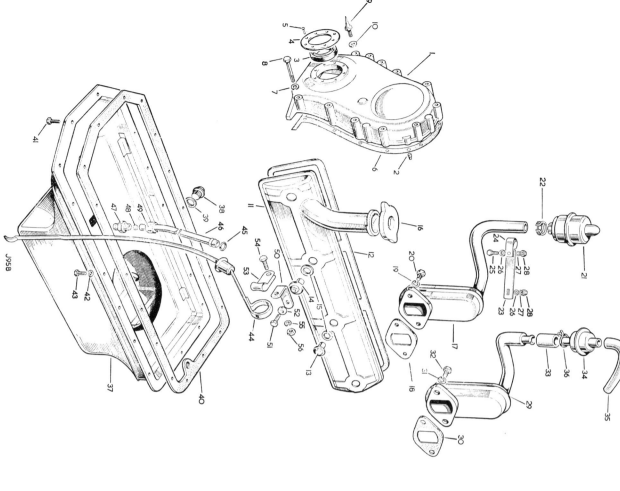

J95B

FRONT COVER, SIDE COVER AND SUMP, 2.6 LITRE 6-CYLINDER PETROL MODELS

Plate Ref.	1 2 3 4	DESCRIPTION		Qty	Part No.	REMARKS
1		FRONT COVER ASSEMBLY	...	1	554924	
2		Dowel locating front cover		2	6395	
3		Oil seal for front cover		1	516028	
4		Mud excluder		1	542073	
5		Drive screw fixing excluder		8	78001	
6		Joint washer for front cover		1	272835	
7		Spring washer for front cover		10	3075	
8		Set bolt ($\frac{5}{16}$" UNF x 1$\frac{7}{8}$" long) } Fixing front cover to cylinder block		10	256025	
9		Timing pointer at front cover	...	1	564163	
10		Spring washer for timing pointer		1	3074	
11		Rocker cover, side, and oil filler		1	542425	
12		Joint washer for rocker cover	...	1	274817	
13		Special nut		3	274091	
14		Special union nut		1	554831	
15		Rubber sealing washer	} Fixing side cover	4	267828	
16		Oil filler cap, AC 2571943		1	546254	
17		Breather pipe for engine crankcase		1	541291	
18		Joint washer for breather pipe		1	214058	
19		Spring washer		2	3075	
20		Set bolt ($\frac{5}{16}$" UNF x $\frac{3}{4}$" long) } Fixing breather pipe to cylinder block		2	255226	
21		Filter for breather pipe, AC 7222485		1	518145	Up to engines numbered 34500469B
22		Clip fixing filter to breather pipe		1	50309	
23		Steady bracket for breather pipe		1	554917	
24		Clip for breather pipe		1	518146	34600024A
25		Bolt ($\frac{1}{4}$" UNF x $\frac{5}{8}$" long) } Fixing breather pipe bracket and clip		1	255206	
26		Plain washer		2	3840	
27		Spring washer		2	3074	
28		Nut ($\frac{1}{4}$" UNF)		2	254810	
29		Breather pipe for engine crankcase		1	566957 *	
30		Joint washer for breather pipe		1	214058	
31		Spring washer		2	3075	
32		Set bolt ($\frac{5}{16}$" UNF x $\frac{3}{4}$" long) } Fixing breather pipe to cylinder block		2	255226	From engines numbered 34500470B
33		Hose, breather pipe to flame trap		1	569958 *	34600025A onwards
34		Flame trap	...	1	603330	
35		Hose, flame trap to carburetter		1	569959 *	
36		Clip for hoses		3	554260 *	
37		Crankcase sump	...	1	564215	
38		Drain plug for sump	...	1	536577	
39		Joint washer for plug	...	1	243959	
40		Joint washer for sump	...	2	546457	
41		Set bolt ($\frac{5}{16}$" UNC x $\frac{7}{8}$" long) } Fixing sump to cylinder block		2	253027	
42		Spring washer		19	3075	
43		Set bolt ($\frac{5}{16}$" UNF x $\frac{7}{8}$" long) } Fixing sump to cylinder block		16	255227	
		Nut ($\frac{5}{16}$" UNF)		1	254831	

* Asterisk indicates a new part which has not been used on any previous Rover model

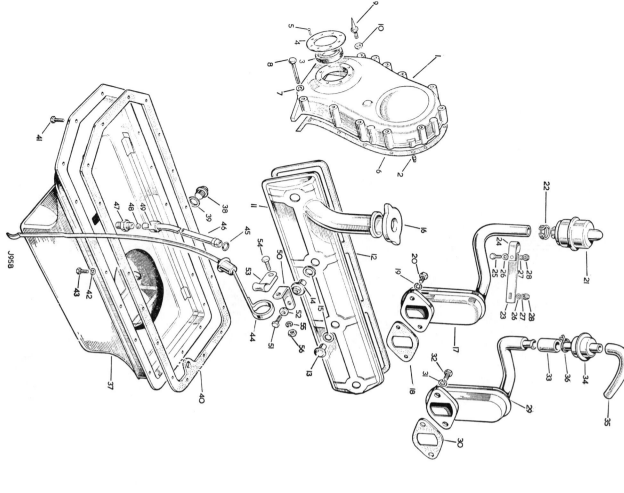

J95B

FRONT COVER, SIDE COVER AND SUMP, 2.6 LITRE 6-CYLINDER PETROL MODELS

Plate Ref. 1 2 3 4	DESCRIPTION		Qty	Part No.	REMARKS
44	Oil level rod	... ...	1	554834	
45	Sealing ring for oil level rod	...	1	532387	
46	Tube for oil level rod	... ...	1	554832	
47	Adaptor for tube	... ...	1	541266	
48	Olive	... ...	1	236408	Fixing tube
49	Union nut ($\frac{3}{8}$" BSP)		1	236407	to adaptor
50	Bracket, oil level tube to side cover	...	1	554833	
51	Set bolt ($\frac{1}{4}$" UNF x $\frac{9}{16}$" long)		1	255005	Fixing bracket to side cover
52	Sealing washer	... ...	2	232037	special union nut on side cover
53	Clip for oil level tube	...	1	554852	
54	Bolt ($\frac{1}{4}$" UNF x $\frac{5}{8}$" long)		1	255206	Fixing clip and tube to bracket
55	Spring washer	... ...	1	3074	
56	Nut ($\frac{1}{4}$" UNF)	... ...	1	254810	

* Asterisk indicates a new part which has not been used on any previous Rover model

J959

CYLINDER HEAD, 2.6 LITRE 6-CYLINDER PETROL MODELS

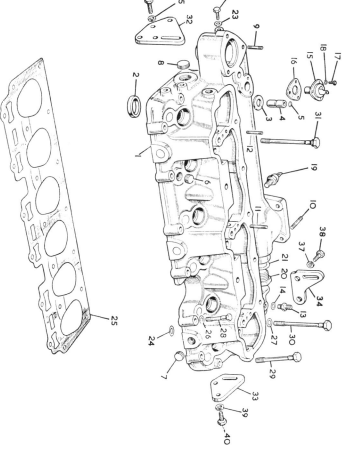

CYLINDER HEAD, 2.6 LITRE 6-CYLINDER PETROL MODELS

Plate Ref.	1 2 3 4	DESCRIPTION	Qty	Part No.	REMARKS
1		CYLINDER HEAD ASSEMBLY	1	564202	
2		Insert for inlet valve seat	6	266321	
3		Washer for valve spring	6	230062	
4		Valve guide, inlet	6	504169	
5		Stud for rocker cover, long	6	233419	
6		Stud for rocker cover, short	6	230251	
7		Double-ended union for servo pipe	4	210492	
8		Joint washer for union	1	09191	
9		Rubber ring for valve guide, inlet	2	230250	
10		Core plug, $\frac{11}{16}$" diameter	4	252501	
11		Core plug, $\frac{7}{8}$" diameter	4	252515	
12		Core plug, 1" diameter	3	252497	
13		Core plug, 1$\frac{3}{8}$" diameter	2	506047	
14		Stud for water outlet pipe	1	506046	
15		Stud for carburetter	1	513171	
16		Stud for carburetter relay bracket	1	243958	
17		Thermostat switch for choke warning light	1	545010	
18		Joint washer for thermostat switch	1	236022	
19		Set bolt (2 BA x $\frac{7}{16}$" long)	1	251002	Fixing switch
20		Water temperature transmitter	1	3073	
21		Plug for heater connection hole	1	560794	
22		Joint washer for plug	1	278164	For hole in thermostat housing
23		Set bolt ($\frac{1}{4}$" UNC x $\frac{7}{16}$" long)	1	243972	
24		'O' ring for oil return hole in cylinder head	1	253003	
25		Cylinder head gasket	1	547820	
26		Plain washer, small	6	2210	
27		Plain washer, large	11	3843	
28		Special set bolt ($\frac{3}{8}$" UNF x 1$\frac{13}{32}$" long)	6	274093	Fixing cylinder head to block
29		Special set bolt ($\frac{7}{16}$" UNF x 2$\frac{1}{2}$" long)	3	274094	
30		Special set bolt ($\frac{7}{16}$" UNF x 4$\frac{3}{32}$" long)	8	274095	
31		Special set bolt ($\frac{7}{16}$" UNF x 4$\frac{31}{32}$" long)	3	276400	
32		Front lifting bracket	1	568363	
33		Rear lifting bracket at rear face	1	500628	
34		Rear lifting bracket at rear side face, RH	1	564206	
35		Spring washer	2	253026	
36		Set bolt ($\frac{5}{16}$" UNC x $\frac{3}{4}$" long)	2	3075	Fixing front bracket
37		Spring washer	2	253026	
38		Set bolt ($\frac{5}{16}$" UNC x $\frac{3}{4}$" long)	2	3075	Fixing rear bracket
39		Spring washer	2	3076	
40		Set bolt ($\frac{3}{8}$" UNC x $\frac{3}{4}$" long)	2	253045	Fixing bracket at side face / at rear face

* Asterisk indicates a new part which has not been used on any previous Rover model

J959

CYLINDER HEAD, 2.6 LITRE 6-CYLINDER PETROL MODELS

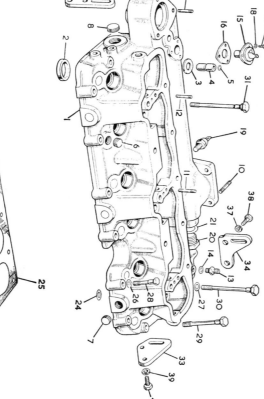
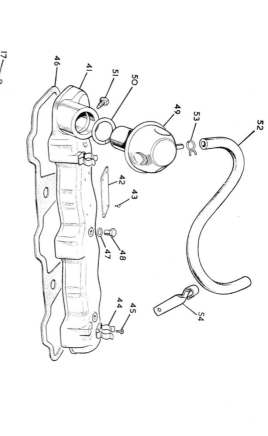

CYLINDER HEAD, 2.6 LITRE 6-CYLINDER PETROL MODELS

Plate Ref.	1 2 3 4	DESCRIPTION	Qty	Part No.	REMARKS
41		ROCKER COVER ASSEMBLY, TOP	1	563002	
42		Tappet clearance plate	1	274100	
43		Hammer drive screw for plate	4	3767	
44		Clip for ignition wire carrier	2	231569	Early models with ignition wire carrier
45		Drive screw fixing clip	2	77919	
46		Joint washer for rocker cover	1	267668	
47		Sealing washer	3	231576	
48		Special nut } Fixing top rocker cover	3	274089	
49		Engine breather filter, top, AC 7971724	1	547605 ††	
50		Engine breather filter, top, AC 7971799	1	546203 ††	
51		Sealing ring for breather filter	1	268887	
52		Special set screw fixing filter	1	515291	
53		Sealing washer for set screw	1	232037	
52		Hose, top breather filter to carburetter elbow	1	568391 *††	Not part of engine assembly
53		Clip, fixing hose to breather filter and elbow	2	546210 ††	
54		Clip, securing hose to carburetter body	1	568392 *††	
54		Gasket set, decarbonising	1	605105	

* Asterisk indicates a new part which has not been used on any previous Rover model
† Up to engines numbered 345004698, 346000024A
†† From engines numbered 345004708, 346000025A onwards

G780

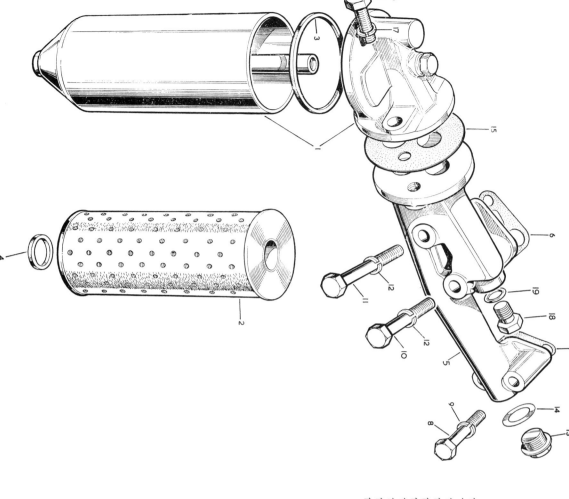

ENGINE OIL FILTER AND ADAPTOR, 2.6 LITRE 6-CYLINDER PETROL MODELS

Plate Ref.	1 2 3 4	DESCRIPTION		Qty	Part No.	REMARKS
1		Oil filter for engine, AC 7961133	... ...	1	513591	
2		Element, AC FF24	...	1	246262	Not part of engine assembly
3		Gasket, large, AC 1530764	...	1	246261	
4		Gasket, small, lower, AC 7961054	...	1	605169	
5		Adaptor for oil filter	... ...	1	554925	
6		Joint washer, front ⎱ For oil filter		1	274609	
7		Joint washer, rear ⎰ adaptor		1	274104	
8		Set bolt (5/16″ UNF x 1 7/8″ long)	⎱ For oil filter	2	256025	
9		Spring washer		2	3075	
10		Set bolt (7/16″ UNF x 3″ long) ⎱ Fixing	adaptor to	1	256068	
11		Set bolt (7/16″ UNF x 3½″ long) ⎰ cylinder block		1	256070	
12		Spring washer		2	3077	
13		Plug (3/8″ BSP) ⎱ For adaptor		1	563121	Part of
14		Joint washer ⎰ oil way		1	231577	adaptor
15		Joint washer		1	272839	
16		Set bolt (7/16″ UNC x 1¼″ long) ⎱ Fixing oil filter	to adaptor	2	253068	
17		Spring washer		2	3077	
18		Plug (3/8″ BSF) ⎱ Blanking oil pressure		1	557498	Part of
19		Joint washer ⎰ transmitter hole		1	232039	adaptor

* Asterisk indicates a new part which has not been used on any previous Rover model

WATER PUMP AND THERMOSTAT, 2.6 LITRE 6-CYLINDER PETROL MODELS

H907

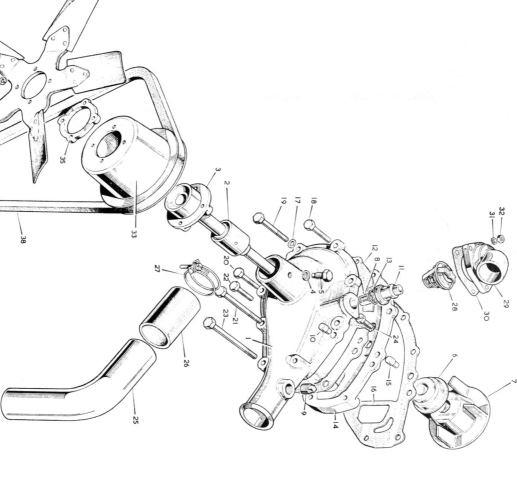

WATER PUMP AND THERMOSTAT, 2.6 LITRE 6-CYLINDER PETROL MODELS

Plate Ref. 1 2 3 4	DESCRIPTION	Qty	Part No.	REMARKS
1	WATER PUMP ASSEMBLY ...	1	564197	
2	Spindle and bearing complete ...	1	523354	
3	Hub for fan blade ...	1	564270	
4	Spring washer } Locating bearing	1	3074	
5	Special set bolt } in casing	1	247078	
6	Carbon ring and seal ...	1	568301	} Not part of engine assembly
7	Impeller for pump ...	1	563037	
8	Tube for thermostat by-pass ...	1	564164	
9	Plug (3/8" BSP) for heater adaptor hole in casing	1	3291	
10	Dowel for water pump casing ...	2	6395	
11	Connector, by-pass to water pump ...	1	564165	
12	Joint washer for water pump ...	1	564157	
13	Rubber seal, pump to cylinder head ...	2	09170	
14	Adaptor for water pump ...	1	563154	
15	Dowel for adaptor ...	2	6395	
16	Joint washer for adaptor ...	2	563038	
17	Spring washer ...	10	3074	
18	Set bolt (1/4" UNF x 1" long) ...	3	255009	
19	Set bolt (1/4" UNF x 1 3/8" long) ...	3	255004	
20	Set bolt (1/4" UNF x 1 1/8" long) } Fixing water	2	256004	
21	Set bolt (1/4" UNF x 2 1/4" long) } pump and	1	256200	
22	Set bolt (1/4" UNF x 1 7/8" long) } adaptor to block	1	256208	
23	Set bolt (1/4" UNF x 2 3/4" long) ...	2	255208	
24	Special 'Wedgelok' screw ...	3	256410	
25	Inlet pipe for water pump ...	3	563050	
26	Hose, water inlet pipe to pump ...	1	552517	
27	Clip, hose to inlet pipe ...	1	562915	} Not part of engine assembly
	Clip, hose to pump ...	2	50324	
	...		50319	
28	Thermostat. wax type ...	1	563122	
29	Outlet pipe to radiator ...	1	558195	
30	Joint washer for water outlet pipe ...	1	276510	
31	Spring washer } Fixing pipe to	3	3074	
32	Nut (1/4" UNF) } cylinder head	3	254810	
33	Pulley for fan ...	1	564267	
34	Fan blade ...	1	564247	
35	Distance piece for fan blade ...	1	274737	
36	Spring washer } Fixing pulley	4	3075	
37	Set bolt (1/4" UNF x 1 1/4" long) } and blade to hub	4	256001	
38	Fan belt ...	1	564268	
	Water pump overhaul kit ...	1	605716 *	

* Asterisk indicates a new part which has not been used on any previous Rover model

CARBURETTER, SU TYPE, 2.6 LITRE 6-CYLINDER PETROL MODELS

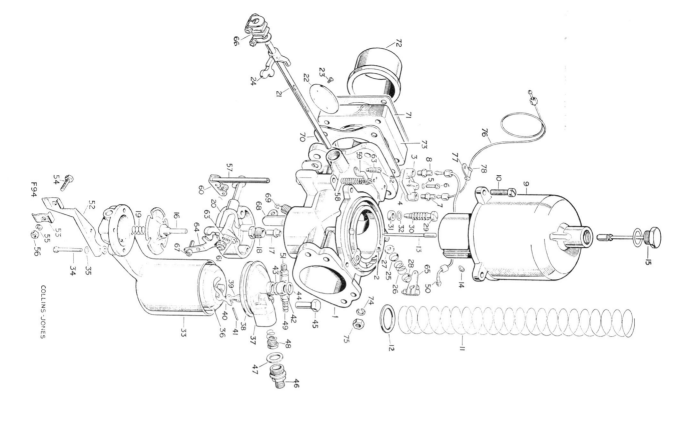

F94

COLLINS-JONES

CARBURETTER, SU TYPE, 2.6 LITRE 6-CYLINDER PETROL MODELS

Plate Ref.	1 2 3 4	DESCRIPTION	Qty	Part No.	REMARKS
		CARBURETTER, SU AUD 247			
1		Carburetter body, SU AUC 8262 ...	1	564209	Carburetter and fittings are not included in engine assembly. †See footnote for applicability
3		Adaptor, ignition and weakening device, SU AUC 2043 ...	1	512458	
4		Gasket for adaptor, SU AUC 2014 ...	1	504082	
5		Shakeproof washer, SU LWN 403 } Fixing adaptor	2	274950	
6		Screw, SU AUC 2175 ...	2	601594	
7		Union for ignition pipe, SU AUC 4490 ...	1	262493	
8		Union for economiser pipe, SU AUC 4831 ...	1	274963	
9		Suction chamber and piston complete, SU AUC 8073 ...	1	245295	
10		Special screw fixing suction chamber, SU AUC 2175 ...	1	274954	
11		Spring for piston (yellow), SU AUC 1167 ...	3	262493	
12		Thrust washer for suction chamber, SU AUC 3071 ...	3	262443	
13		Needle, SS, SU AUD 1334 ...	1	262492	
14		Special screw fixing needle, SU AUC 2057 ...	1	245292	
15		Oil cap complete, SU AUC 8102 ...	1	278897	
16		Jet complete, SU AUC 8155 ...	1	245261	
17		Jet bearing, SU AUC 2001 ...	1	274956	
18		Jet screw, SU AUC 2002 ...	1	274957	
19		Jet spring, SU AUC 2006 ...	1	274958	
20		Jet housing complete, SU AUD 9708 ...	1	274959	
21		Throttle spindle, SU AUC 5137 ...	1	605162	
22		Throttle butterfly, SU AUC 1295 ...	1	512459	
23		Screw for throttle butterfly, SU AUC 1358 ...	2	600351	
24		Throttle stop, SU AUC 2023 ...	1	262481	
25		Taper pin for throttle stop lever, SU AUC 2106 ...	1	274963	
		Gland washer for throttle spindle, brass, SU AUC 2096 ...	2	601592	
26		Spring for throttle spindle gland, SU AUC 2097 ...	2	274964	
27		Gland washer for throttle spindle, langite, SU AUC 2098 ...	2	274965	
28		Retainer cap for gland washer, SU AUC 2010 ...	2	274966	
29		Slow-running adjusting valve, SU AUC 2028 ...	1	274967	
30		Gland spring for slow running, SU AUC 2027 ...	1	274968	
31		Gland washer for slow running, rubber, SU AUC 2029 ...	1	274969	
32		Brass washer for slow running, SU AUC 2030 ...	1	274970	
33		Float chamber, SU AUC 2009 ...	1	274971	
34		Bolt, SU AUC 2110 } Fixing float chamber	4	274972	
35		Shakeproof washer, SU LWN 403	4	274973	
36		Float, SU AUC 1123 ...	1	601594	
37		Lid for float chamber, SU AUC 2283 ...	1	41710	
38		Joint washer for float chamber lid, SU AUC 1147 ...	1	601524	
39		Needle valve and seat, SU AUD 9096 ...	1	261980	
40		Lever for float, SU AUD 2285 ...	1	262827	
41		Pin for lever, SU AUC 1152 ...	1	601522	

* Asterisk indicates a new part which has not been used on any previous Rover model
† All items on this page up to engines numbered 345004698, 34600024A

CARBURETTER, SU TYPE, 2.6 LITRE 6-CYLINDER PETROL MODELS

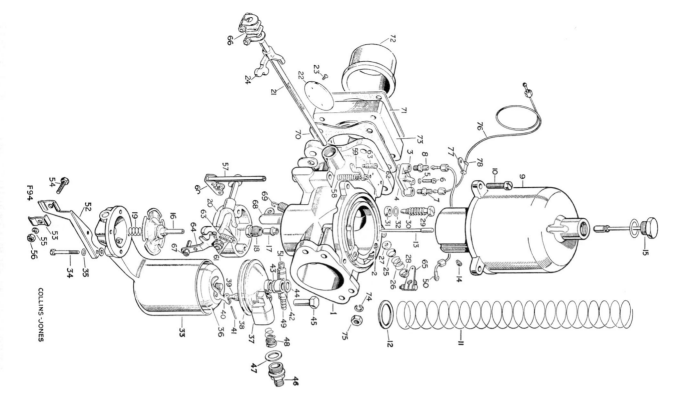

COLLINS-JONES

F94

CARBURETTER, SU TYPE, 2.6 LITRE 6-CYLINDER PETROL MODELS

Plate Ref. 1 2 3 4	DESCRIPTION	Qty	Part No.	REMARKS
42	Banjo, SU AUC 4700 ...	1	245279	† See footnote for applicability
43	Fibre washer for banjo, SU AUC 1928	1	261981	
44	Aluminium washer for banjo, SU AUC 1557	1	262427	On float chamber
45	Cap nut fixing banjo, SU AUC 1867	1	245277	
46	Double-ended union for carburetter, SU AUC 1290	1	245278	
47	Washer for union, SU AUC 2141 ...	1	232006	
48	Filter and spring for carburetter body, SU AUC 2139	1	262446	
49	Economiser union for rubber tube, SU AUC 4788	1	245258	
50	Pipe for economiser, SU AUC 8237	1	512460	
51	Union for economiser pipe, SU AUC 8166	1	245266	
57	Sliding rod, roller and cam shoe, SU AUC 8964	1	536796	
58	Spring for sliding rod, SU AUC 2020	1	274979	
59	Top plate, SU AUC 2019	1	274980	
60	Stop screw, bottom, SU AUC 4790	1	601521	
61	Stop screw, top, SU AUC 4790 ...	1	601521	
62	Spring for stop screw, SU AUC 2451	1	262489	
63	Cold-start lever, SU AUC 4996	2	262489	
64	Lever for throttle return spring, SU AUC 8209	1	513991	
65	Bolt (2 BA x ¾" long) Fixing lever	1	501374	
	Spring washer } to throttle	1	250961	
66	Nut (2 BA) spindle	1	3073	
	Throttle lever for carburetter, SU AUC 8420	1	505796	
67	Bolt (2 BA x ¾" long) Fixing lever	1	2247	
	Spring washer } to carburetter	1	250961	
68	Nut (2 BA) spindle	1	3073	
	Ball end for carburetter lever ...	1	2247	
69	Spring washer } Fixing	1	273964	
70	Nut (2 BA) ball end	1	3073	
71	Bracket for throttle return spring	1	2247	
68	Throttle return spring ...	1	505787	
69	Joint washer for carburetter ...	1	505786	
70	Joint washer for carburetter ...	1	511690	
71	Joint washer for distance piece	1	511652	

CARBURETTER, SU TYPE, 2.6 LITRE 6-CYLINDER PETROL MODELS

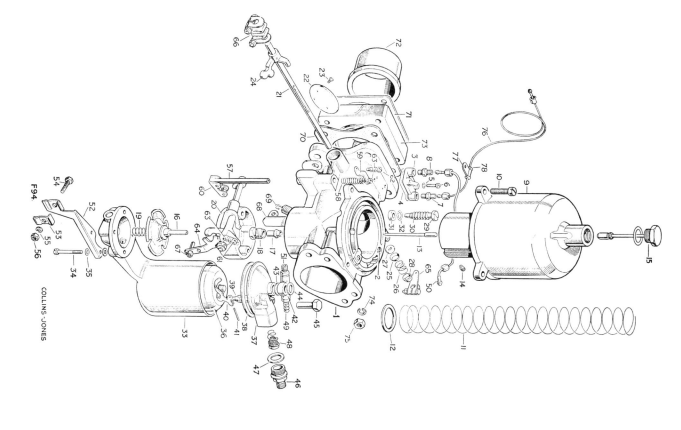

F94.

COLLINS -JONES

CARBURETTER, SU TYPE, 2.6 LITRE 6-CYLINDER PETROL MODELS

Plate Ref. 1 2 3 4	DESCRIPTION	Qty	Part No.	REMARKS
				†See footnote for applicability
72	Liner for manifold	1	505612	
73	Distance piece for carburetter	1	505613	
74	Spring washer ⎱ Fixing carburetter and distance	4	3075	
75	Nut (5/16" UNF) ⎰ piece to cylinder head	4	254831	
76	Suction pipe complete	1	501769	
77	Clip for suction pipe	1	275037	
78	Rubber grommet for clip ⎱ On ⎰ engine	1	214229	
	AIR INLET ADAPTOR ASSEMBLY ...	1	557693	
	Stud ⎱ For ⎰ adaptor	1	541494	
	Stud ⎰	1	252728	
	Joint washer for adaptor	1	242375	
	Locking pin for cold start cable ...	1	560103	
	Support for cold start cable	1	563181	
	Plain washer ⎱ Fixing adaptor	1	2920	
	Spring washer and support ⎰	2	3075	
	Nut (5/16" UNF) ⎰ to carburetter	2	254831	
	Clip ⎱ Fixing cold start	1	3075	
	Bolt (10 UNF x 1/2" long) ⎱ cable to	1	41379	
	Nut (10 UNF) ⎰ support	1	257017	
	Economiser hose for carburetter ...	1	257023	
	Adaptor for economiser hose... ...	1	557694	
	Joint washer for adaptor	1	242319	
	Clip fixing hose to adaptor	1	232039	
	Clip fixing hose to adaptor and carburetter	2	546017	
			50301	

* Asterisk indicates a new part which has not been used on any previous Rover model
† All items on this page up to engines numbered 34500469B, 34600024A

J960

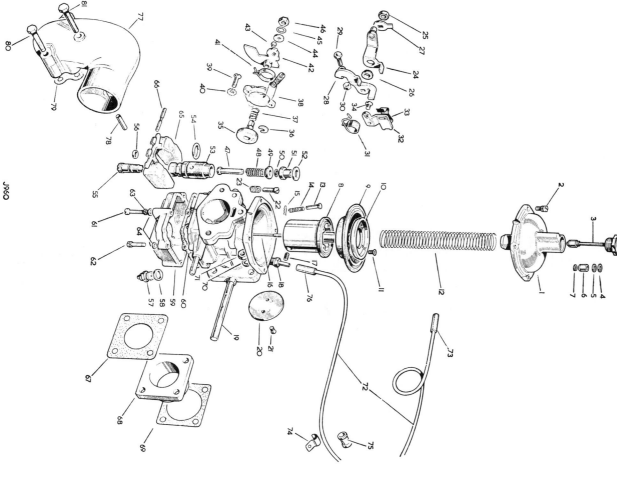

CARBURETTER, ZENITH TYPE, 2.6 LITRE 6-CYLINDER PETROL MODELS

Plate Ref.	1 2 3 4	DESCRIPTION	Qty	Part No.	REMARKS
		CARBURETTER COMPLETE, ZENITH 175 CD 2S	1	569956*	Carburetter and fittings are not
1		Top cover for carburetter, Zenith 020296..	1	605846*	included in engine assembly. †See
2		Special screw and washer fixing top cover, Zenith 019654	4	605847*	footnote for applicability
3		Damper and oil cap assembly, Zenith B17873Z	1	605848*	
4		Special washer, upper, Zenith B17632Z ..	1	605849*	
5		Special washer, lower, Zenith 017109 ..	1	605850*	
6		Bush for damper, Zenith 019640 ..	1	605850*	
7		Retaining ring, for damper, Zenith 019641	1	605851*	
8		Air valve, shaft and diaphragm assembly, Zenith 020303	1	605844*	
9		Diaphragm, Zenith B17421Z ..	1	605841*	
10		Retaining ring for diaphragm, Zenith 019870 ..	1	605842*	
11		Special screw fixing retaining ring, Zenith B18008Z	1	605843*	
12		Return spring for air valve, Zenith 020685	4	605855*	
13		Lifting pin for air valve, Zenith B16826Z	1	605820*	
14		Spring for lifting pin, Zenith B16831Z ..	1	605821*	
15		Spring clip for lifting pin, Zenith 019535	1	605822*	
16		Metering needle, Zenith B17667Z ..	1	605853*	
17		Locking screw for metering needle, Zenith 019673	1	605845*	
18		Ignition adaptor, Zenith B17086Z ..	1	605793*	
19		Throttle spindle, Zenith B17458Z ..	1	605794*	
20		Butterfly for throttle, Zenith 019608	1	605800*	
21		Special screw fixing butterfly, Zenith 010351	2	605801*	
22		Throttle stop screw, Zenith 019532 ..	1	605816*	
23		Spring for stop screw, Zenith 019508	1	605817*	
24		Throttle lever, Zenith B18168Z ..	1	605795*	
25		Special nut, Zenith L1475 ..	1	605799*	Fixing throttle
26		Special washer, Zenith B17858Z ..	1	605797*	levers
27		Tab washer, Zenith 019067 ..	1	605798*	
28		Throttle stop and fast-idle lever, Zenith B17085Z	1	605796*	
29		Special screw, Zenith B17562Z ..	1	605818*	For throttle stop
30		Locknut, Zenith 019480 ..	1	605819*	
31		Throttle return spring, Zenith B17552Z ..	1	605802*	
32		Bracket and clip for choke cable, Zenith B17088Z	1	605813*	
33		Clip for choke bracket, Zenith B17626Z ..	1	605814*	

* Asterisk indicates a new part which has not been used on any previous Rover model
† All items on this page from engines numbered 34500470B, 34600025A onwards

J96O

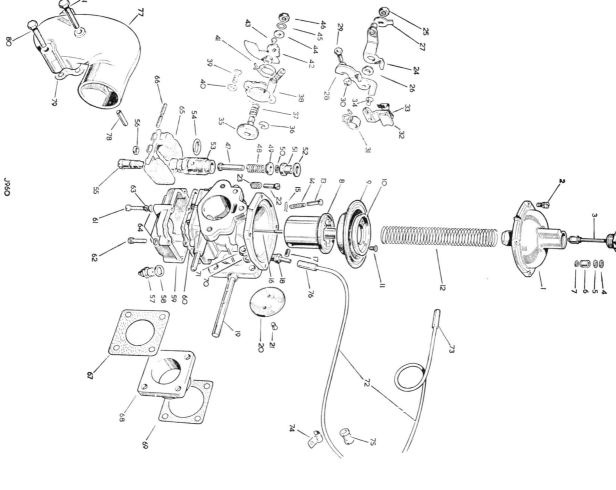

CARBURETTER, ZENITH TYPE, 2.6 LITRE 6-CYLINDER PETROL MODELS

Plate Ref.	1 2 3 4	DESCRIPTION	Qty	Part No.	REMARKS
34		Special screw fixing choke bracket, Zenith B15346	1	605815*	†See footnote for applicability
35		Cold start spindle, Zenith B18215Z	1	605804*	
36		Special washer for starter spindle, Zenith B16556Z	1	605806*	
37		Cold start spring, Zenith B16533Z	1	605805*	
38		Cover for cold start, Zenith B18228Z	1	605803*	
39		Special screw, Zenith B17269Z ⎫ Fixing	2	605811*	
40		Shakeproof washer, Zenith L1402 ⎬ cover	2	605812*	
41		Return spring for cam lever, Zenith B16934Z	1	605809*	
42		Cam lever for cold start, Zenith B17797Z	1	605810*	
43		Clamping screw for cam lever swivel, Zenith P12867	1	605808*	
44		Spacing washer, Zenith B16670Z ⎫ Fixing cam	1	605807*	
45		Shakeproof washer, Zenith 016412 ⎬ lever to cold	1	601877*	
46		Special nut, Zenith 016411 ⎭ start spindle	1	601838*	
47		Jet orifice, Zenith 019522	1	605827*	
48		Spring, Zenith B17894Z	1	605828*	
49		Guide bush, Zenith B17895Z ⎫ For jet orifice	1	605826*	
50		'O' ring, Zenith 019657 ⎭	1	605825*	
51		Bush, Zenith 019663	1	605823*	
52		Special washer, Zenith 019664	1	605824*	
53		Carrier for jet orifice, Zenith B17406Z	1	605829*	
54		'O' ring for carrier, Zenith 019658	1	605830*	
55		Adjusting screw for jet orifice, Zenith 019926	1	605832*	
56		'O' ring for adjusting screw, Zenith 016268	1	605831*	
57		Needle valve, Zenith B17855Z	1	605854*	
58		Washer for needle valve, Zenith 09619	1	605852*	
59		Float chamber, Zenith B16506Z	1	605836*	
60		Gasket for float chamber, Zenith B16518Z	1	605835*	
61		Special screw, long, Zenith B16266 ⎫ Fixing	2	605837*	
62		Special screw, short, Zenith B16265 ⎬ float	4	605838*	
63		Spring washer, Zenith 021091 ⎭ chamber	6	605839*	
64		Plain washer, Zenith 020102	6	605840*	
65		Float and arm, Zenith B16722Z	1	605833*	
66		Spindle for float, Zenith B16511Z	1	605834*	
		Gasket kit for carburetter, Zenith 235Z	1	605857*	
		Overhaul kit for carburetter, Zenith MRK 3136	1	606098*	

CARBURETTER, ZENITH TYPE, 2.6 LITRE 6-CYLINDER PETROL MODELS

J960

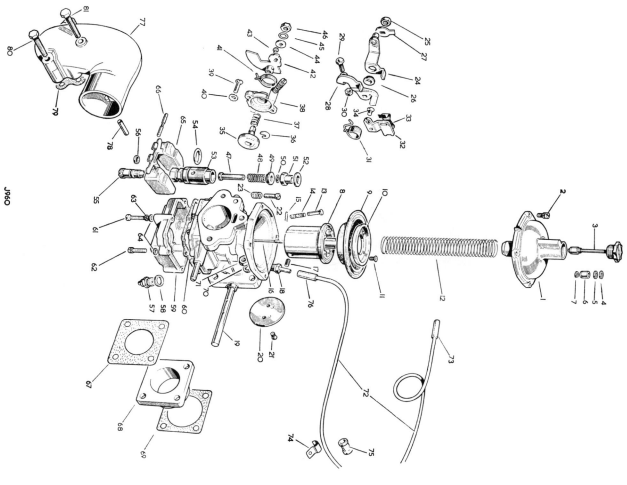

CARBURETTER, ZENITH TYPE, 2.6 LITRE 6-CYLINDER PETROL MODELS

Plate Ref.	1 2 3 4	DESCRIPTION	Qty	Part No.	REMARKS
67		Joint washer for carburetter	1	568365*	
68		Adaptor for carburetter	1	564446*	
69		Joint washer for adaptor	1	563364*	
70		Spring washer ⎫ Fixing carburetter and	4	3075	
71		Nut ($\frac{5}{16}$" UNF) ⎬ adaptor to cylinder head	4	254831	
72		Suction pipe, carburetter to distributor ...	1	564602*	For distributor with 'screw-on' connection ⎫ Alternatives. Check before ordering
73		Suction pipe, carburetter to distributor ...	1	564603*	
74		Rubber sleeve, suction pipe to distributor ...	1	566865	For distributor with 'push-on' connection ⎭
75		Clip for suction pipe	1	275037	
76		Rubber grommet for clip ⎫ On	1	214229	
77		Rubber sleeve, suction pipe to carburetter ...	1	566865	
		AIR INLET ELBOW ASSEMBLY	1	568378*	
78		Adaptor for top breather hose	1	563380*	
79		Joint washer for inlet elbow	1	564447*	
80		Set bolt ($\frac{5}{16}$" UNC x 2$\frac{1}{4}$" long) ⎫ Fixing elbow	2	254028	
81		Set bolt ($\frac{5}{16}$" UNC x 3" long) ⎭ to carburetter	1	254030*	

* Asterisk indicates a new part which has not been used on any previous Rover model
† All items on this page from engines numbered 34500470B, 34600025A onwards

EXHAUST MANIFOLD, 2.6 LITRE 6-CYLINDER PETROL MODELS

G 782

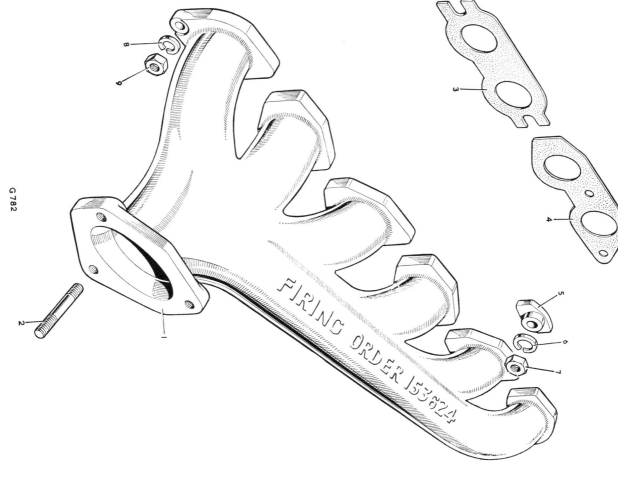

EXHAUST MANIFOLD, 2.6 LITRE 6-CYLINDER PETROL MODELS

Plate Ref.	1 2 3 4	DESCRIPTION				Qty	Part No.	REMARKS
1		EXHAUST MANIFOLD ASSEMBLY			...	1	557763	
2		Stud for exhaust pipe	...	...	...	3	252623	
3		Joint washer, front	...	...	...	1	534100	
4		Joint washer, centre	...	...	...	1	534101	
5		Joint washer, rear	...	...	...	1	534102	
5		Clamp for manifold	...	...	...	5	09161	
6		Spring washer ⎫ Fixing			...	5	3076	
7		Nut (⅜" UNF) ⎭ clamps			...	5	254812	
8		Spring washer ⎫ Fixing ends				2	3075	
9		Nut (5/16" UNF) ⎭ of manifold				2	254831	

* Asterisk indicates a new part which has not been used on any previous Rover model

DISTRIBUTOR AND STARTER, 2.6 LITRE 6-CYLINDER PETROL MODELS

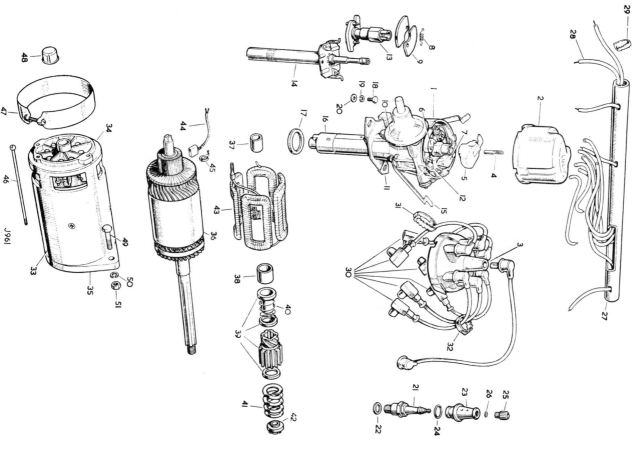

DISTRIBUTOR AND STARTER, 2.6 LITRE 6-CYLINDER PETROL MODELS

Plate Ref. 1 2 3 4	DESCRIPTION	Qty	Part No.	REMARKS
1	DISTRIBUTOR COMPLETE, LU 40923A	1	541492	Early type with side cable fixing } Alternatives. Check before ordering
			568729*	Late type with top cable fixing
	DISTRIBUTOR COMPLETE, LU 41227	1	568729*	For use in conjunction with early type distributor. Check before ordering
2	Distributor cap, LU 54412474	1	539570	Early type with side cable fixing } Alternatives. Check before ordering
			605544†*	Late type with top cable fixing
3	Distributor cap	1		Check before ordering
4	Brush and spring for cap, LU 418856	6	262703	
5	Rotor arm, LU 418726	1	262704	
6	Contact points, LU 423153	1	269987	
7	Condenser, LU 423871	1	269988	
8	Auto advance spring, set, LU 54415427	1	539579	
9	Auto advance weight, LU 54413073	1	539572	
10	Vacuum unit, LU 54413882	1	539573	
11	Clamping plate, LU 421191	1	248857	
12	Base plate for contact breaker, LU 422318	1	502282	
13	Cam, LU 54414859	1	539578	
14	Shaft and action plate, LU 54410290	1	539575	
15	Clip for cover, LU 54414490	1	539576	
16	Driving dog, LU 420620	1	539577	
	Sealing ring, LU 188639	1	539574	
17	Sundry parts kit, LU 419644	1	245014	
18	Cork washer for distributor housing	1	245015	
	Set bolt (¼" UNF × 9/16" long) } Fixing distributor to cylinder block	1	52278	
19	Spring washer	1	253205	
20	Plain washer	1	3074	
21	Sparking plug, Champion N5	6	3911	
22	Washer for plug	6	512445	
23	Cover for sparking plug	6	40441	
24	Rubber sealing ring for plug cover	6	214262	
25	Cable nut, LU 410600	6	213172	
26	Rubber boot for cable nut, LU 414965	6	214278	
27	Washer for cable nut, LU 185015	6	506679	
28	Ignition wire carrier	6	214279	
29	Sparking plug lead set	1	231234	} Not part of engine assembly
	Cable cleat securing No. 1 plug lead and coil leads	1	558321	
	Cable cleat securing Nos. 1, 2 and 3 plug leads to coil lead	1	240429	

* Asterisk indicates a new part which has not been used on any previous Rover model
† To enable latest type spark plug leads with integral covers to be fitted to vehicles with early type distributor and terminal nut type ignition coil, Part Numbers 605544, 574022, 240429, 240428 and 240428 are required

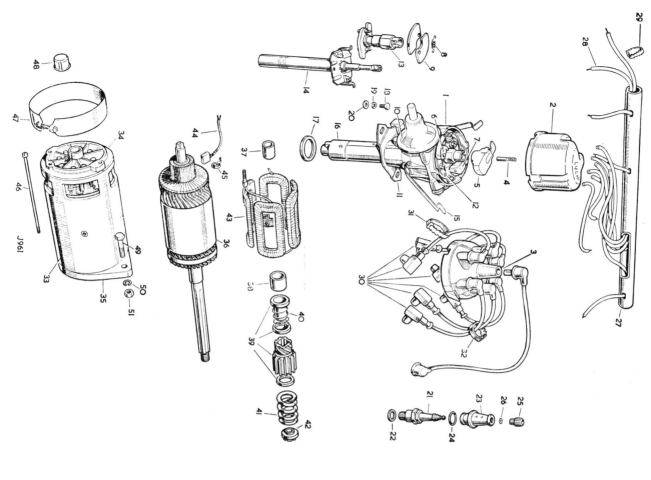

DISTRIBUTOR AND STARTER, 2.6 LITRE 6-CYLINDER PETROL MODELS

DISTRIBUTOR AND STARTER, 2.6 LITRE 6-CYLINDER PETROL MODELS

Plate Ref. 1 2 3 4	DESCRIPTION	Qty	Part No.	REMARKS
30	Spark plug lead set with integral spark plug covers	1	574076*	For use in conjunction with late type distributor with top cable fixing and coil with 'push-in' connection
31	Cable cleat ⎱ Securing plug and	4	240429	
32	Cable cleat ⎰ coil leads	1	240428	
	Spark plug lead conversion set with integral spark plug covers	1	574022†*	Alternatives. Check before ordering.
	Cable cleat ⎱ Securing plug and	4	240429†	For use in conjunction with late type distributor with top cable fixing and coil with terminal nut connection
	Cable cleat ⎰ coil leads	1	240428†	
				Not part of engine assembly
33	STARTER MOTOR COMPLETE, LU M418G 25605	1	236287	
34	Bracket for starter, commutator end, LU 256748	1	244706	
35	Bracket, drive end, LU 256495	1	244713	
36	Armature, LU 255463	1	244715	
37	Bush, commutator end, LU 255491	1	242958	
38	Bush, pinion end, LU 270038	1	244714	
39	Pinion and sleeve, LU 255194	1	244711	
40	Spring for pinion, LU 255728	1	244712	
41	Main spring for pinion, LU 270889	1	244710	
42	Nut for pinion, LU 255851	1	244709	Not part of engine assembly
43	Field coil for starter, LU 256578	1	262861	
44	Brushes for starter motor, set, LU 255659	2	260055	
45	Spring set for brushes, LU 54257221	2	601754	
46	Bolt for bracket, LU 255459	2	244717	
47	Cover band, LU 255557	1	244705	
48	Grease cap, LU 255854	1	243095	
	Sundry parts kit, LU 256762	1	244718	
49	Set bolt (³⁄₈″ Whit x 1″ long) ⎱ Fixing starter motor	1	215703	
50	Spring washer ⎱ to flywheel housing	2	250543	
51	Nut (³⁄₈″ BSF) ⎰	1	3076	
	Bolt (³⁄₈″ BSF x 1¼″ long)	2	2827	

* Asterisk indicates a new part which has not been used on any previous Rover model

† To enable latest type spark plug leads with integral covers to be fitted to vehicles with early type distributor and terminal nut type ignition coil, Part Numbers 605544, 574022, 240429 and 240428 are required

DYNAMO AND FIXINGS, 2.6 LITRE 6-CYLINDER PETROL MODELS

H9O6

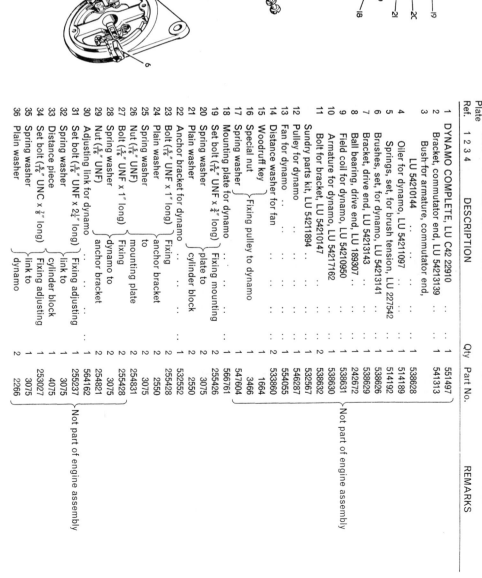

DYNAMO AND FIXINGS, 2.6 LITRE 6-CYLINDER PETROL MODELS

Plate Ref. 1 2 3 4	DESCRIPTION	Qty	Part No.	REMARKS
1	DYNAMO COMPLETE, LU C42 22910	1	551497	
2	Bracket, commutator end, LU 54213139	1	541313	
3	Bush for armature, commutator end, LU 54210144	1	538628	
4	Oiler for dynamo	1	514189	
5	Springs, set, for brush tension, LU 227542	1	514192	
6	Brushes, set, for dynamo, LU 54213141	1	538626	
7	Bracket, drive end, LU 54213143	1	538629	
8	Ball bearing, drive end, LU 189307	1	242672	
9	Field coil for dynamo, LU 54210950	1	538631	
10	Armature for dynamo, LU 54217162	1	538630	
11	Bolt for bracket, LU 54210147	2	538632	
	Sundry parts kit, LU 54211894	1	532567	
12	Pulley for dynamo	1	546287	
13	Fan for dynamo	1	554055	
14	Distance washer for fan	2	533860	
15	Woodruff key	1	1664	
16	Special nut } Fixing pulley to dynamo	1	3466	
17	Spring washer }	1	547604	
18	Mounting plate for dynamo	1	566761	
19	Set bolt ($\frac{5}{16}''$ UNF x $\frac{3}{4}''$ long) } Fixing mounting plate to cylinder block	2	255426	
20	Spring washer }	2	3075	
21	Plain washer }	2	2550	
22	Anchor bracket for dynamo	1	532552	
23	Bolt ($\frac{5}{16}''$ UNF x 1″ long) } Fixing anchor bracket to mounting plate	2	255428	
24	Plain washer }	2	2550	
25	Spring washer }	2	3075	
26	Nut ($\frac{5}{16}''$ UNF) }	2	254831	Not part of engine assembly
27	Bolt ($\frac{5}{16}''$ UNF x 1″ long) } Fixing dynamo to anchor bracket	2	255428	
28	Spring washer }	2	3075	
29	Nut ($\frac{5}{16}''$ UNF) }	2	254821	
30	Adjusting link for dynamo	1	564162	
31	Set bolt ($\frac{5}{16}''$ UNF x $2\frac{1}{4}''$ long) } Fixing adjusting link to cylinder block	1	255237	Not part of engine assembly
32	Spring washer }	1	3075	
33	Distance piece }	1	4075	
34	Set bolt ($\frac{5}{16}''$ UNC x $\frac{7}{8}''$ long) } Fixing adjusting link to dynamo	1	253027	
35	Spring washer }	1	3075	
36	Plain washer }	2	2266	

* Asterisk indicates a new part which has not been used on any previous Rover model

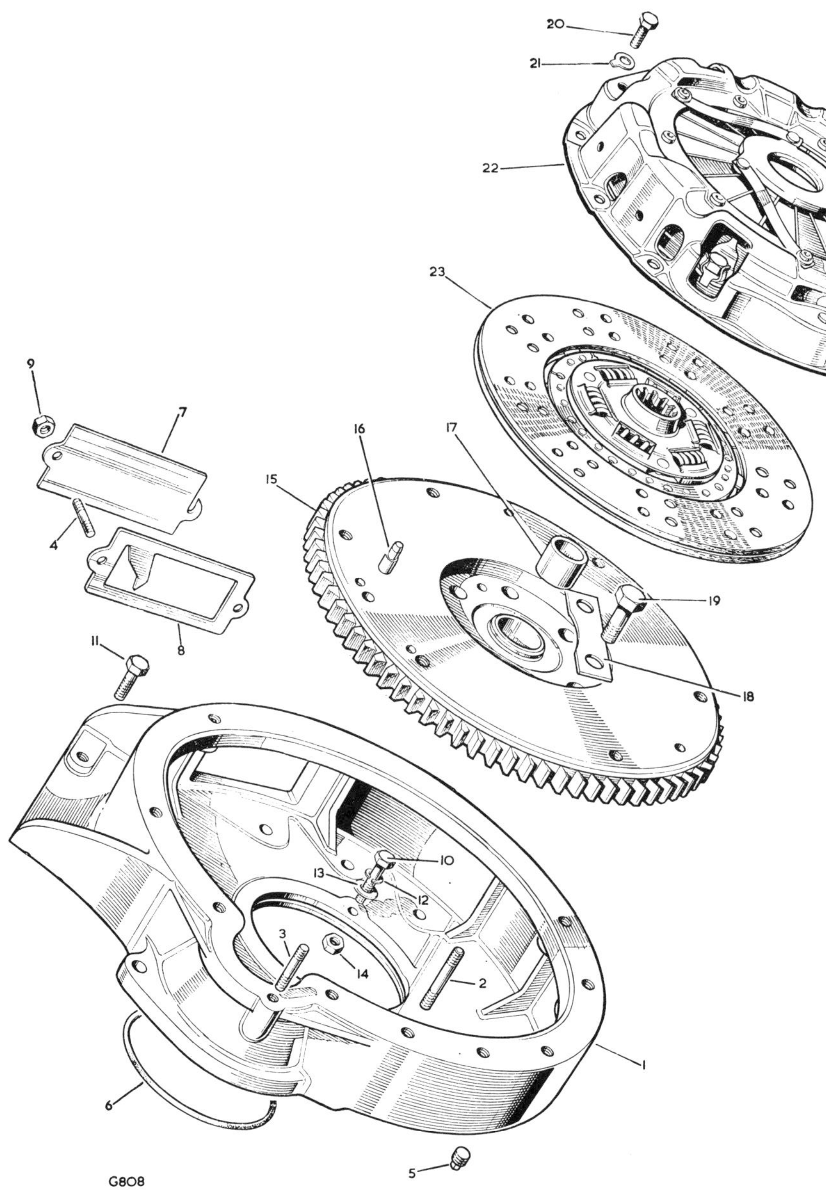

G808

Plate Ref.	1 2 3 4	DESCRIPTION	Qty	Part No.	REMARKS
1		FLYWHEEL HOUSING ASSEMBLY	1	541195	
2		Stud, short ($\frac{3}{8}$") } Fixing flywheel housing	12	3650	
3		Stud ($\frac{5}{16}$") } to bell housing	1	3200	
4		Stud ($\frac{1}{4}$") for inspection cover	2	3651	
5		Drain plug for housing	1	3290	
6		Sealing ring for flywheel housing	1	246169	
7		Inspection cover plate	1	56140	
9		Nut ($\frac{1}{4}$" BSF) fixing cover plate	2	2823	
10		Bolt ($\frac{3}{8}$" UNF x $1\frac{1}{2}$" long) } Fixing flywheel housing to cylinder block	3	256041	
11		Bolt ($\frac{3}{8}$" UNF x $1\frac{3}{8}$" long)	2	256040	
		Bolt ($\frac{3}{8}$" UNF x $1\frac{1}{8}$" long)	3	255248	
12		Spring washer	8	3076	
13		Plain washer	6	2219	
		Plain washer	2	4085	
14		Nut ($\frac{3}{8}$" UNF)	3	254812	
15		FLYWHEEL ASSEMBLY	1	541760	
		Ring gear for flywheel	1	506799	
16		Dowel for clutch cover plate	3	502116	
17		Bush for primary pinion	1	08566	
18		Locker } Fixing flywheel	3	534098	
19		Special set bolt } to crankshaft	6	247135	
20		Set bolt ($\frac{5}{16}$" BSF x $\frac{7}{8}$" long) } Fixing clutch	6	237324	Not part of engine assembly
21		Locker } to flywheel	6	546197	Not part of engine assembly
22		CLUTCH ASSEMBLY, BB 75698/28	1	571228	Not part of engine assembly
23		Clutch plate complete, BB 51226/26	1	561536	Not part of engine assembly
		Lining package for clutch plate, BB SSC 75073	1	600388	Not part of engine assembly

* Asterisk indicates a new part which has not been used on any previous Rover model

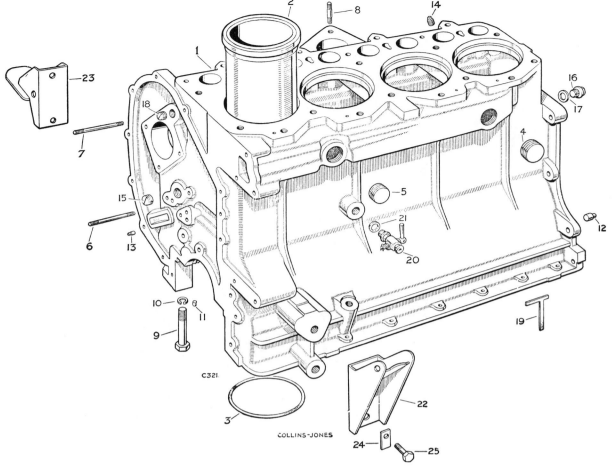

C321.

COLLINS-JONES

CYLINDER BLOCK, DIESEL ENGINE, 2 LITRE, Series II

Plate Ref.	1 2 3 4	DESCRIPTION	Qty	Part No.	REMARKS
		ENGINE ASSEMBLY	1	247796	Does not include liners and pistons
1		CYLINDER BLOCK ASSEMBLY	1	276909	Does not include liners and pistons
2		Cylinder block, crankshaft, pistons and camshaft	1	271502	
		LINER FOR CYLINDER WITH PISTONS	4	269911	For piston components see page 143
3		Sealing ring for liner	8	247746	
		Silicone grease		270656	
4		Shim for liner .002" thick	4	503747	Required for first
4		Shim for liner .004" thick	4	503748	4000 engines
5		Core plug, 1⅛" diameter	1	512412	
5		Core plug, 1" diameter	1	09191	
6		Plug for immersion heater boss	1	527269	
7		Stud For front cover	1	247208	
7		Stud and water pump	1	247869	Not part of cylinder block assembly
8		Stud for distributor pump	3	252624	
		Stud for sump	1	252621	
9		Set bolt for main bearings	6	504007	
10		Special plain washer for set bolt	6	504006	
11		Dowel, locating main bearing and bearing cap	6	501593	
12		Dowel, locating flywheel housing	2	52124	
13		Dowel for timing chain adjuster	1	213700	
15		Plug for oil gallery, front	2	247127	
16		Plug for oil gallery, rear	1	536577	
17		Joint washer for rear plug	1	243959	
18		Joint washer for plug	1	247861	Not part of cylinder block assembly
		Plug for tappet feed gallery pipe, front	1	273166	
		Plug for tappet feed gallery pipe, rear	1	243968	
		Joint washer for plug	1	247965	
19		Plug for tappet feed hole	1	537279	
		Packing for rear main bearing cap	2	243237	
20		Drain tap for cylinder block	1	213959	
21		Joint washer for drain tap	1	543823	
22		Engine front support bracket complete, LH	1	271975	
23		Engine front support bracket complete, RH	1	212430	
24		Locker	4	255085	
25		Set bolt (¼" UNF x 1" long) Fixing	3	255086	
		Set bolt (¼" UNF x 1⅛" long) brackets to cylinder block	6	272563	
		Decarbonising gasket kit	1	272564	
		Engine overhaul gasket kit	1	600245	
		Stud kit for cylinder block			
		'Hylomar' sealing compound for cylinder liner, 4 oz tube	As reqd	534244	

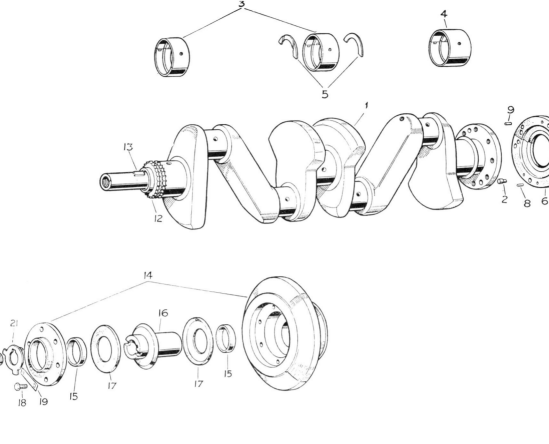

Plate Ref.	1 2 3 4	DESCRIPTION	Qty	Part No.	REMARKS
1		CRANKSHAFT ASSEMBLY, STD	1	514426	
2		Dowel for flywheel	1	265779	
3		Main bearing, front and centre, Std ...	2	523321	
4		Main bearing, rear, Std	1	523326	
5		Thrust washer for crankshaft, Std ...	1	523331	
		Thrust washer for crankshaft, .0025" OS	1	523332	
		Thrust washer for crankshaft, .005" OS	1	523333	
		Thrust washer for crankshaft, .0075" OS	1	523334	}Pairs
		Thrust washer for crankshaft, .010" OS	1	523335	
		REAR BEARING OIL SEAL ASSEMBLY ...	1	542494 †	
6		Rear bearing split seal halves ...	2	523240	
7		Oil seal complete for rear bearing ...	1	542492	
		Silicone grease, MS4, 1 oz tube ...	1	270656	
8		Dowel, lower ⎫ Fixing retainer or	2	519064	
9		Dowel, upper ⎬ seal to cylinder	2	246464	
10		Spring washer ⎪ block and rear	10	3074	
11		Bolt (¼" UNF x ⅞" long) ⎭ bearing cap	10	255208	
12		Chainwheel on crankshaft ...	1	235726	
13		Key locating chainwheel and vibration damper	2	235770	
		VIBRATION DAMPER ASSEMBLY ...	1	247154	
14		DAMPER FLYWHEEL AND BUSH ASSEMBLY	1	247094	
15		Bush for flywheel	1	236289	
16		Driving flange	1	247674	
17		Rubber disc for damper ...	2	03017	
18		Set bolt (¼" UNF x ⅝" long) ⎫ Fixing	6	255206	
19		Double tab washer ⎬ plate	3	247647	
20		Starting dog	1	503665	
21		Lockwasher for starting dog ...	1	247771	

* Asterisk indicates a new part which has not been used on any previous Rover model
† Supply complete assembly for engines numbered up to 146102060 88 and 156102153 109

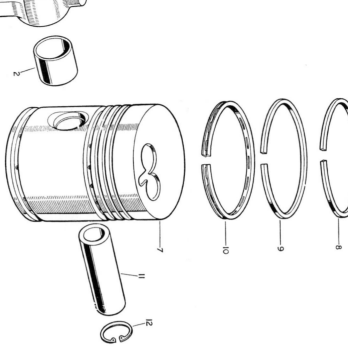

CONNECTING ROD AND PISTON, DIESEL ENGINE, 2 LITRE, Series II

Plate Ref.	1 2 3 4	DESCRIPTION	Qty	Part No.	REMARKS
1		CONNECTING ROD ASSEMBLY	4	518808	
2		Gudgeon pin bush	4	247583	
		Complete set of bolts and nuts for connecting rod	1	522607	
3		Special bolt ⎫ Fixing connecting rod cap	8	518468	⎱ Alternative fixings.
		Self-locking nut (3/8" UNF) ⎭ rod cap	8	277390	⎰ Check before ordering
4		Special bolt ⎫ Fixing connecting	8	273677	
5		Castle nut (3/8" UNF) ⎬ rod cap	8	247115	
		Split pin fixing connecting rod cap	8	2392	
6		Connecting rod bearing pairs, Std ...	4	523336	
7		PISTON ASSEMBLY AND LINER ...	4	269911	NOTE—Oversize pistons are not supplied
8		Piston ring, compression, Std, chromed ...	4	247170	
9		Piston ring, compression, Std, tapered ...	8	247171	
10		Scraper ring, Std	4	247172	
11		Gudgeon pin, Std	4	248930	
12		Circlip for gudgeon pin	8	266945	

* Asterisk indicates a new part which has not been used on any previous Rover model

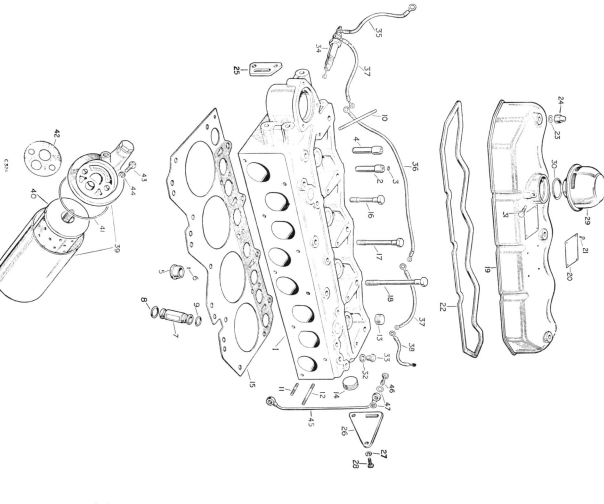

C.526

COLLINS JONES

CYLINDER HEAD, DIESEL ENGINE, 2 LITRE, Series II

Plate Ref.	1 2 3 4	DESCRIPTION	Qty	Part No.	REMARKS
1		CYLINDER HEAD ASSEMBLY	1	504588	
2		Valve guide, inlet, and sealing ring	4	272206	
3		Sealing ring for inlet valve guide	4	247186	Not included in cylinder head assembly
4		Valve guide, exhaust and sealing ring	4	275552	Not included in cylinder head assembly
		Sealing ring for exhaust valve guide	4	233419	Not included in cylinder head assembly
5		Packing washer for valve guides	20	230062	
		Hot plug in cylinder head	4	502615	
6		Peg for hot plug	4	271881	
7		Push rod tube	8	514224	
8		'O' ring for push rod tube, large	8	247877	
9		'O' ring for push rod tube, small	8	265019	
10		Stud for injector, thread length $\frac{7}{8}$"	6	247883	5 off on 1961 models } Not included in cylinder head assembly
		Stud for injector, thread length 1 $\frac{7}{8}$"	2	277996	1958-60
		Stud for injector, thread length 2 $\frac{1}{2}$"	3	521600	1961 models
11		Stud, short } For manifold	4	247144	
12		Stud, long }	2	247143	
		Cup plug	2	250830	Alternative to core plug } Check before ordering
		Core plug, $\frac{7}{8}$" diameter	2	525497	Alternative to $\frac{7}{8}$" diameter core plug
13		Core plug, $\frac{3}{4}$" diameter	3	512413	
14		Core plug, 1" diameter	7	09191	
		Core plug, 1 $\frac{3}{8}$" diameter	1	230250	
15		Shroud for injector bore	4	275143	
		Cylinder head gasket	1	535649	
16		Special set bolt, short	2	247683	
17		Special set bolt, medium	9	279649	
18		Special set bolt, long	5	247723	Fixing cylinder head to cylinder block
		Stud ($\frac{1}{2}$" UNF)	5	518466	
		Nut ($\frac{1}{2}$" UNF)	2	254824	
19		ROCKER COVER, TOP, ASSEMBLY	1	247558	
20		Tappet clearance plate	1	247634	
21		Drive screw fixing plate	4	78001	
22		Joint washer for top rocker cover	1	247606	
23		Sealing washer } Fixing top	3	232038	
24		Special nut } rocker cover	3	247121	
25		Lifting bracket for engine, front	1	247714	
26		Lifting bracket for engine, rear	1	247715	
27		Spring washers	4	3075	
28		Set bolt ($\frac{5}{16}$" UNF x $\frac{3}{4}$" long) } Fixing brackets to cylinder head	4	255226	
29		Breather filter for engine, AC 7223381	1	247631	
30		Sealing ring for filter	1	268887	
31		Special set screw fixing breather filter	1	515291	
32		Sealing washer for set screw	1	232037	
		Joint washer } For heater valve hole	1	243959	
33		Plug ($\frac{3}{8}$" BSF) } in cylinder head	1	536577	

* Asterisk indicates a new part which has not been used on any previous Rover model

CYLINDER HEAD, DIESEL ENGINE, 2 LITRE, Series II

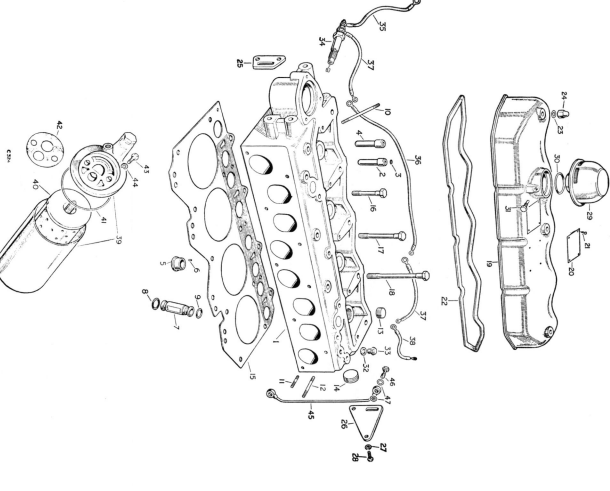

COLLINS JONES
C 324

CYLINDER HEAD, DIESEL ENGINE, 2 LITRE, Series II

Plate Ref. 1 2 3 4	DESCRIPTION	Qty	Part No.	REMARKS
34	Heater plug	4	510078	
35	Heater plug lead, No. 1 to earth	1	247953	
36	Heater plug lead, No. 2 to No. 3	1	247952	
37	Heater plug lead, No. 1 to 2 and 3 to 4 ...	2	247951	
38	Heater plug lead, No. 4 to resistor ...	1	247951	
	Shakeproof washer, large ⎱ Fixing leads to	4	77626	
	Shakeproof washer, small ⎰ heater plug	4	71082	
	Bolt (⅜" UNF × ⅝" ⎱ Fixing heater plug	4	253905	
	long) ⎰ to cylinder head			
39	Fan disc washer ⎱ earth lead to	1	537229	
	Engine oil filter, AC 7965063 ⎰ cylinder head	1	512305	
40	Element for filter, AC FF 50, overall length 6¹³⁄₁₆"	1	248863	For long type filter ⎱ Alternatives.
	Element for filter, AC 72, overall length 4⅝"	1	541403	For short type filter ⎰ Check before ordering
41	Gasket for filter, AC 1530250	1	272539	Short type
	Rubber washer for centre bolt, AC 1531098	1	269889	
42	Joint washer for filter	1	272839	
43	Set bolt (⅛" UNF × 1¼" long) ⎱ Fixing filter to	2	255068	
44	Spring washer ⎰ cylinder block	2	3077	
45	Oil pipe complete to cylinder head, length 10"	1	274500	
46	Banjo bolt fixing oil pipe	2	50840	
	Joint washer for banjo bolts	4	232039	
47	Plug for thermometer hole in cylinder head	1	278164	
	Joint washer for plug	1	243972	
	Oil pressure switch, AC 7954238	1	519863	
	Joint washer	1	232039	

* Asterisk indicates a new part which has not been used on any previous Rover model

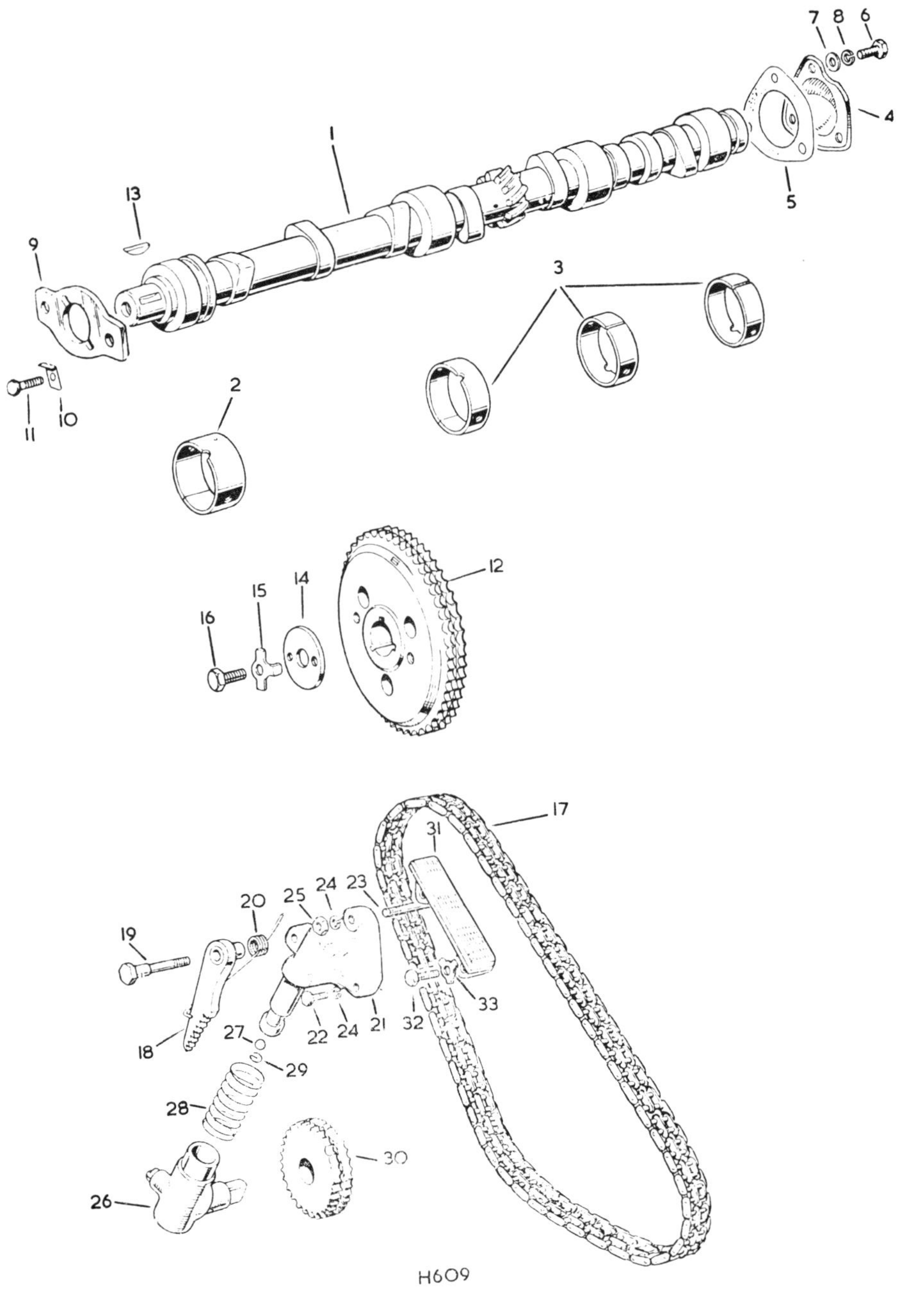

CAMSHAFT AND TENSIONER MECHANISM, DIESEL ENGINE, 2 LITRE, Series II

Plate Ref.	1 2 3 4	DESCRIPTION	Qty	Part No.	REMARKS
1		Camshaft	1	274711	
2		Bearing complete for camshaft, front	1	519054	
3		Bearing complete for camshaft, centre and rear	3	519055	
4		Rear end cover for camshaft	1	538073	
5		Joint washer for cover	1	247070	
		Set bolt ($\frac{1}{4}$" UNF x $\frac{7}{8}$" long) } Fixing cover to block	3	255208	} Alternative fixings. Check before ordering
6		Set bolt ($\frac{1}{4}$" UNF x $\frac{5}{8}$" long) } Fixing cover to block	3	255206	} Alternative fixings. Check before ordering
7		Plain washer } Fixing cover to block	4	10882	} Alternative fixings. Check before ordering
8		Spring washer } Fixing cover to block	3	3074	} Alternative fixings. Check before ordering
9		Thrust plate for camshaft	1	538535	
10		Locker } Fixing thrust plate to cylinder block	2	2995	} 4 off on early type thrust plate
11		Set bolt ($\frac{1}{4}$" UNF x $\frac{3}{4}$" long) } Fixing thrust plate to cylinder block	2	255207	} 4 off on early type thrust plate
12		Chainwheel for camshaft	1	276133	
13		Key locating camshaft chainwheel	1	230313	
14		Retaining washer } Fixing chainwheel to camshaft	1	09093	
15		Locker } Fixing chainwheel to camshaft	1	09210	
16		Set bolt ($\frac{3}{8}$" UNF x $\frac{7}{8}$" long) } Fixing chainwheel to camshaft	1	255246	
17		Camshaft chain ($\frac{3}{8}$" pitch x 78 links)	1	09156	
18		Ratchet for timing chain adjuster	1	546026	
19		Special bolt fixing ratchet and piston to block	1	247199	
20		Spring for chain adjuster ratchet	1	267451	
21		Piston for timing chain adjuster	1	247912	
22		Set bolt ($\frac{5}{16}$" UNF x $1\frac{1}{4}$" long) } Fixing piston to cylinder block	2	256220	1 off on late type fixing
23		Stud ($\frac{5}{16}$" UNF) } Fixing piston to cylinder block	1	247144	Late type fixing. Check before ordering
24		Spring washer } Fixing piston to cylinder block	2	3075	
25		Nut ($\frac{5}{16}$" UNF) } Fixing piston to cylinder block	1	254831	Late type fixing. Check before ordering
26		Cylinder for timing chain adjuster	1	277388	
27		Steel ball ($\frac{7}{32}$") for non-return valve	1	3739	
28		Spring for chain tensioner	1	233326	
29		Retainer for steel ball	1	233328	
30		Idler wheel for timing chain	1	236067	
31		Vibration damper for timing chain	1	275234	
32		Set bolt ($\frac{1}{4}$" UNF x $\frac{1}{2}$" long) } Fixing damper to cylinder block	2	255204	
33		Locking plate } Fixing damper to cylinder block	2	278769	

* Asterisk indicates a new part which has not been used on any previous Rover model

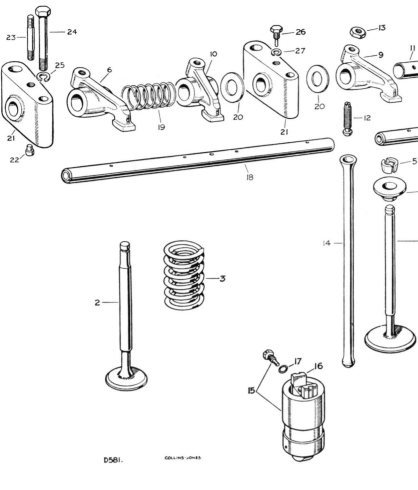

D581. COLLINS-JONES

VALVE GEAR AND ROCKER SHAFTS, DIESEL ENGINE, 2 LITRE, Series II

Plate Ref. 1 2 3 4	DESCRIPTION	Qty	Part No.	REMARKS
1	Inlet valve	4	527240	
2	Exhaust valve	4	525975	
3	Valve spring, inner and outer	8	276609	
4	Valve spring cup	8	268292	
5	Split cone for valve, halves	16	268293	
6	Valve rocker, exhaust, LH	2	274772	
7	Valve rocker, exhaust, RH	2	274773	
8	Bush for exhaust valve rocker	4	247738	
9	Valve rocker, inlet, LH	2	274774	
10	Valve rocker, inlet, RH	2	274775	
11	Bush for inlet valve rocker	4	247737	
12	Tappet adjusting screw	8	506814	
13	Locknut for tappet adjusting screw	8	254861	
14	Tappet push rod	8	546799	
15	Tappet, tappet guide, roller and set bolt assembly	8	507829	
16	Tappet	8	507026	
	Tappet guide	8	500473	
	Roller	8	517429	
17	Special set bolt	8	507025	
	Copper washer for tappet guide set bolt	8	232038	
18	Valve rocker shaft	2	274683	
19	Spring for rocker shaft	4	247040	
20	Washer for rocker shaft	8	247153	
21	Rocker bracket	5	274645	4 off on 1961 models
	Rocker bracket, centre	1	524806	1961 models
22	Locating dowel for rocker bracket	3	52710	
23	Stud for rocker bracket	3	247607	
24	Set bolt (7/16" UNF x 2" long) } Fixing rocker bracket to cylinder head	5	256226	
25	Spring washer	5	3075	
26	Locating screw } For rocker shaft at oil feed bracket	2	247730	
27	Spring washer	2	3075	

* Asterisk indicates a new part which has not been used on any previous Rover model

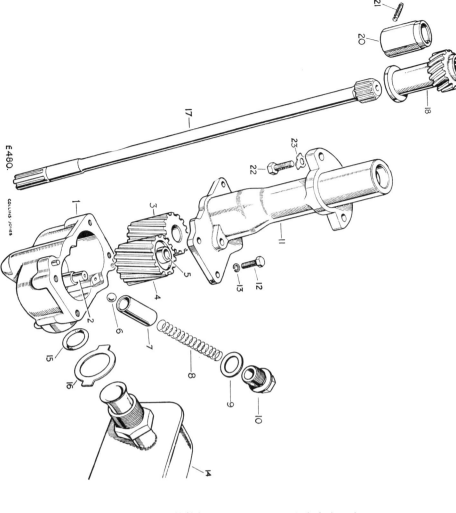

E480.

COLLINS JONES

OIL PUMP, DIESEL ENGINE, 2 LITRE, Series II

Plate Ref.	1 2 3 4	DESCRIPTION		Qty	Part No.	REMARKS
		OIL PUMP ASSEMBLY		1	247662	1958-59
		OIL PUMP ASSEMBLY		1	513640	1960-61
1		Oil pump body ...		1	247663	1958-59
2		Oil pump body ...		1	513641	1960-61
1		Spindle for idler gear		1	502209	1958-59
2		Oil pump gear, driver ...		1	247659	1960-61
3		Oil pump gear, driver ...		1	240055	1958-59
3		Oil pump gear, idler ...		1	237884	1960-61
4		Oil pump gear, idler ...		1	278109	1958-59
5		Bush for idler gear ...		1	214995	1960-61
6		Steel ball		1	3748	
7		Plunger		1	273711	
8		Spring	For oil pressure	1	564456	
9		Washer, inside diameter $\frac{55}{64}''$	release valve	1	243970	} Alternatives.
9		Washer, inside diameter $\frac{49}{64}''$		1	232044	} Check before ordering
10		Plug		1	549909	
11		Oil pump cover		1	247658	1958-59
11		Oil pump cover		1	513639	1960-61
12		Set bolt ($\frac{5}{16}''$ UNF x $\frac{7}{8}''$ long)	} Fixing cover	4	255227	
13		Spring washer	to body	4	3075	
14		Oil filter for pump		1	247664	
15		Sealing ring	} Fixing oil filter	1	244488	
16		Lockwasher	to oil pump	1	244487	
17		Drive shaft for oil pump		1	247739	1958-59
18		Drive shaft for oil pump		1	511680	1960-61
		VERTICAL DRIVE SHAFT ASSEMBLY		1	503266	
19		Circlip for drive shaft		1	247742	
20		Bush for drive shaft		1	247653	
21		Locating screw for drive shaft bush		1	524769	
22		Set bolt ($\frac{5}{16}''$ UNF x 1" long)	} Fixing oil pump	2	255228	
23		Lockwasher	to cylinder block	2	247665	

FRONT COVER, SIDE COVERS AND SUMP, DIESEL ENGINE, 2 LITRE, Series II

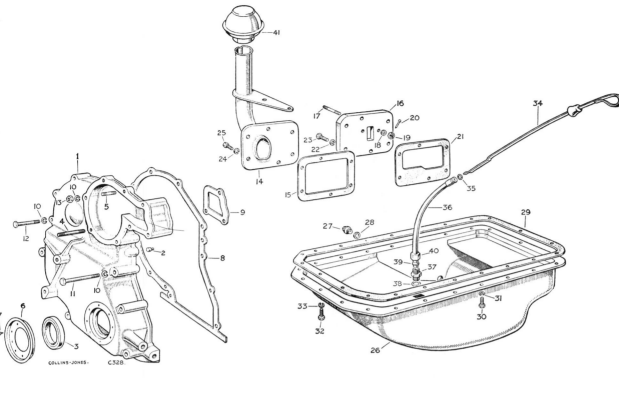

COLLINS-JONES · C328.

FRONT COVER, SIDE COVERS AND SUMP, DIESEL ENGINE, 2 LITRE, Series II

Plate Ref.	1 2 3 4	DESCRIPTION	Qty	Part No.	REMARKS
1		FRONT COVER ASSEMBLY	1	510805	
2		Dowel locating front cover ...	2	6395	
3		Oil seal for front cover ...	1	213744	
4		Stud } Fixing	1	247595	For alloy type front cover
5		Stud } water pump	1	247159	3 off with cast iron type front cover
6		Mud excluder	2	247766	
7		Drive screw fixing excluder ...	1	247766	
8		Joint washer for front cover ...	8	78001	
9		Joint washer at water inlet ...	1	538039	
10		Spring washer	1	538038	
		Spring washer	12	3075	
11		Set bolt ($\frac{5}{16}$" UNF x 3" long) } Fixing	9	256030	
12		Set bolt ($\frac{5}{16}$" UNF x 2" long) } front cover	3	256226	
		Set bolt ($\frac{5}{16}$" UNF x $2\frac{1}{4}$" long)	1	256027	
13		Distance piece	1	2920	
		Plain washer	1	3830	
		Locking plate	1	501198	
		Nut ($\frac{5}{16}$" UNF) ...	2	254831	
14		Side cover and oil filler pipe for engine, front	1	510730	
15		Joint washer for side front cover ...	5	247555	
16		SIDE COVER ASSEMBLY, REAR ...	1	542600	
17		Stud for fuel pump	2	500792	
18		Plain washer	2	2220	
19		Nut ($\frac{5}{16}$" UNF) } Fixing stud	2	254911	Nut and split pin type fixing } Alternatives.
20		Split pin	2	2422	Check before ordering
		Stud for fuel pump	2	542601	Plain type stud fixing
21		Joint washer for side cover, rear ...	1	247554	
22		Spring washer ... } Fixing	6	3075	
23		Set bolt ($\frac{5}{16}$" UNF x 1$\frac{1}{8}$" long) } rear side	5	255029	
		Set bolt ($\frac{5}{16}$" UNF x 1$\frac{1}{2}$" long) } cover plate	1	256220	
24		Spring washer ... } Fixing front	7	3075	
25		Set bolt ($\frac{5}{16}$" UNF x $\frac{3}{4}$" long) } side cover plate	7	255226	
26		Crankcase sump	1	528823	
27		Drain plug for crankcase sump ...	1	536677	
28		Washer for drain plug ...	1	243959	
29		Joint washer for crankcase sump ...	1	546841	
30		Set bolt ($\frac{5}{16}$" UNF x $\frac{7}{8}$" long) } Fixing sump to	19	255227	22 off with cast iron type front cover
31		Spring washer ... } cylinder block	19	3075	
32		Set bolt ($\frac{5}{16}$" UNC x $\frac{3}{4}$" long) } Fixing sump	3	253026	For use with alloy type front cover
33		Spring washer ... } to front cover	3	3075	
34		Oil level rod	1	274154	
35		Sealing ring for rod ...	1	532387	
36		Tube for oil level rod ...	1	504032	
37		Double-ended union } Fixing tube to cylinder	1	236060	
38		Washer } block	1	243958	
39		Olive	1	236408	
40		Union nut	1	236407	
41		Oil filler cap and breather filter, A.C 7964819	1	546440	
		Oil recommendation label for oil filler cap	1	272476	

* Asterisk indicates a new part which has not been used on any previous Rover model

FUEL INJECTION SYSTEM, DIESEL ENGINE, 2 LITRE, Series II, 1958–60

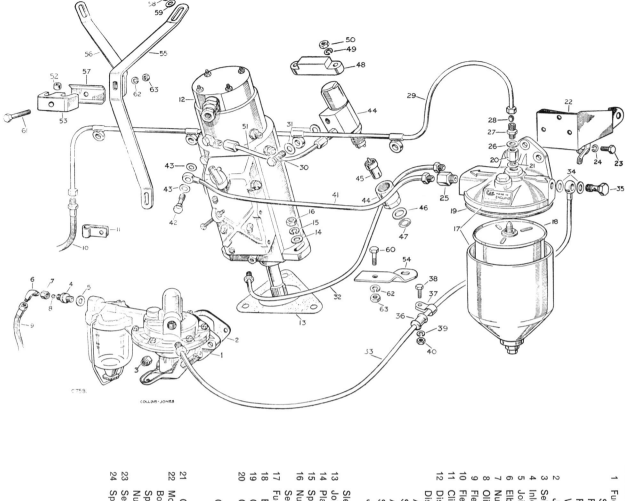

COLLINS-JONES

C759

Plate Ref.	1 2 3 4	DESCRIPTION	Qty	Part No.	REMARKS
1		Fuel pump, mechanical, AC 7950445	1	550324	With red diaphragm
		Sediment bowl, AC 1523620	1	236891	
		Retainer for bowl, AC 5592050	1	268797	
		Filter for bowl, AC 854009	1	506796	For early type pump
		Washer for bowl, AC 5593321	1	241225	
2		Joint washer, fuel pump to cylinder block	1	275565	
3		Self-locking nut fixing fuel pump	2	252161	
4		Inlet union for fuel pump	1	247929	
5		Joint washer for inlet union	1	243967	
6		Elbow for inlet union	1	50499	
7		Nut } Fixing elbow to	1	270115	
8		Olive } fuel pump	1	270105	
9		Flexible fuel feed pipe complete, leak-off to tank	1	276382	
10		Flexible fuel feed pipe complete to pump	1	276267	
11		Clip fixing flexible fuel leak-off pipe	1	276407	
12		Distributor pump, CAV DPA 3240099	1	554525	DPA 3240099 } Use with distributor
		Distributor pump, CAV DPA 3240095	1	509009	Use with distributor pump DPA 3240095
		Distributor pump, CAV DPA 3240099	1	277619	Alternatives. Check CAV number before ordering
		Joint washer for injection pipe, distributor pump	6	513617	
		Stop lever } For	1	277553	
		Accelerator control lever } distributor pump	1	276627	
		Stop lever } For	1	276407	
		Accelerator control lever } distributor pump	1	276627	
13		Sleeve for control lever stop screw, CAV 7139/184	1	247802	
14		Joint washer for distributor pump	1	546282	
		Plain washer }	3	247212	
15		Spring washer } Fixing distributor pump	3	2920	
16		Nut (5/16" UNF) } to cylinder block	3	3075	
		Sealing plugs for injection system	1 set	279702	For sealing unions in distributor pump
17		Fuel filter	1	515437	
18		Element for fuel filter	1	271479	
19		Gasket, large, for fuel filter	1	278282	
20		Check valve for filter, overall length 1¼"	1	279810	Alternative to 528108. Check before ordering
		Check valve and filter bolt, overall length 2¾" CAV 7111/269	1	528108	Alternative to 279810. Check before ordering
21		Copper washer for check valve	1	517976	
22		Mounting bracket for fuel filter	1	278517	Not part of engine assembly
		Bolt (3/8" UNF x 1⅛" long) }	3	255248	
		Spring washer } Fixing fuel filter	3	3076	
		Nut (3/8" UNF) } to bracket	3	254812	
23		Set bolt (3/8" UNF x ¾" long) } Fixing bracket to	3	255245	
24		Spring washer } cylinder head	3	3076	

FUEL INJECTION SYSTEM, DIESEL ENGINE, 2 LITRE, Series II, 1958-60

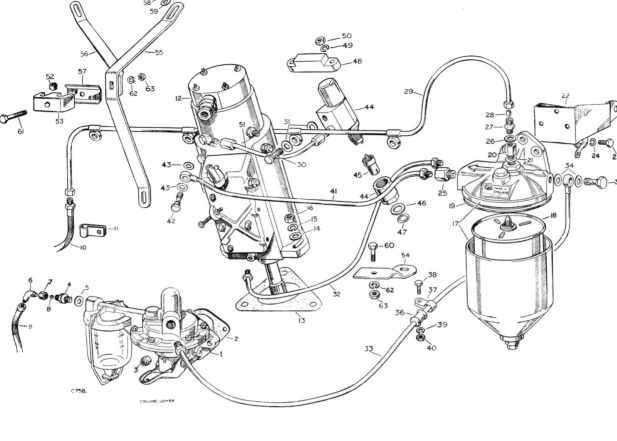

C758.

COLLINS·JONES

FUEL INJECTION SYSTEM, DIESEL ENGINE, 2 LITRE, Series II, 1958-60

Plate Ref.	1 2 3 4	DESCRIPTION	Qty	Part No.	REMARKS
25		Adaptor in filter for distributor pump and drain pipe	2	247778	
26		Joint washer } In filter for	1	243957	} Not part of engine assembly
27		Union } leak-off pipe	1	502776	
28		Restrictor for filter union	1	272396	} Not part of engine assembly
29		Leak-off pipe complete, injector to filter	1	274847	
30		Banjo bolt } Fixing leak-off pipe	4	273521	
31		Washer } to injector	8	273069	
32		Drain pipe complete, distributor pump to filter	1	273070	
33		Pipe complete, mechanical pump to filter	1	277774	
34		Joint washer } Fixing pipe	2	247808	
35		Banjo bolt } to filter	1	247774	
36		Grommet for fuel filter pipe	1	272512	
37		Clip for grommet	1	232425	} Not part of engine assembly
38		Bolt ($\frac{1}{4}$" UNF x $\frac{1}{2}$" long) } Fixing grommet	1	255206	
39		Spring washer } and clip to oil	1	3074	
40		Nut ($\frac{1}{4}$" UNF) } filler pipe	1	254810	
41		Pipe complete, filter to distributor pump	1	277556	
42		Banjo bolt } Fixing pipe to	1	275266	
43		Washer } distributor pump	2	275265	
44		Injector complete	4	273452	
45		Nozzle for injector	4		
46		Joint washer for injector, copper, CAV 5539/423	4	247726	
47		Joint washer for injector, steel	4	247179	
48		Clamping strip for injector	4	247474	
49		Spring washer } Fixing injector	8	247213	
50		Nut ($\frac{5}{16}$" UNF) } to cylinder head	8	3075	
51		Injector pipe to No. 1 cylinder	1	277766	
		Injector pipe to No. 2 cylinder	1	277767	
		Injector pipe to No. 3 cylinder	1	277768	
		Injector pipe to No. 4 cylinder	1	277769	
52		Damper for injector pipe	4	277838	
53		Shroud for damper	1	277839	
54		Bracket for damper shroud	1	277816	
55		Support strap for shroud	1	278002	
56		Steady strap for shroud	1	278021	
57		Back plate for shroud	1	277815	
58		Locknut ($\frac{5}{16}$" UNF)	4	254861	
59		Plain washer	4	2220	
60		Bolt, front ($\frac{5}{16}$" UNF x $1\frac{1}{4}$" long) } Fixing shrouds,	2	256222	
61		Bolt, rear ($\frac{5}{16}$" UNF x $1\frac{1}{2}$" long) } dampers and	1	256220	
62		Spring washer } brackets	2	3075	
63		Nut ($\frac{5}{16}$" UNF)	2	254831	

* Asterisk indicates a **new** part which has not been used on any previous Rover model

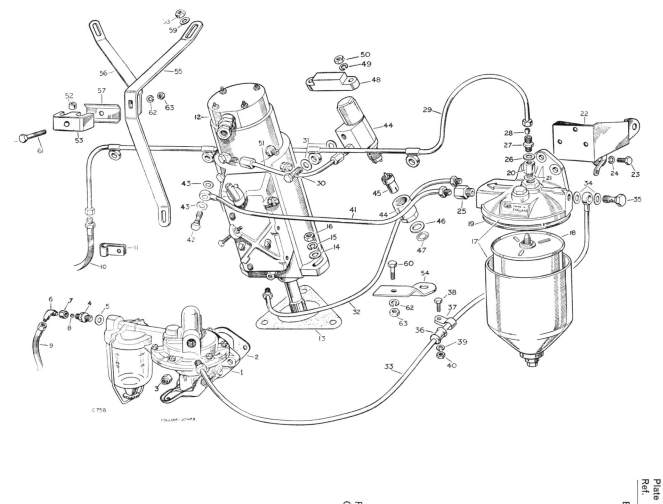

C 758

COLLINS·JONES

FUEL INJECTION SYSTEM, DIESEL ENGINE, 2 LITRE, Series II, 1958-60

Plate Ref.	1 2 3 4	DESCRIPTION	Qty	Part No.	REMARKS
		EXTRA FUEL FILTER COMPLETE ASSEMBLY ON DASH	1	502583	Standard on Export models. Optional equipment on Home models
		Rivnut for dash	2	501224	
		Fuel filter on dash	1	515437	
		Adaptor for fuel filter outlet	1	247778	
		Plug for fuel filter	1	272224	
		Air bleed plug for filter	1	509513	
		Joint washer for plug	1	517976	
		Bolt ($\frac{5}{16}$" UNF x 1$\frac{1}{8}$" long) Fixing filter to dash	3	255029	
		Spring washer	3	3075	
		Plain washer	4	2249	
		Inlet pipe complete, flexible pipe to filter	1	501251	
		Joint washer } Fixing inlet pipe	2	247808	
		Banjo bolt } to filter	1	247774	
		Flexible pipe to inlet pipe	1	276267	
		Elbow for flexible pipe	1	268660	
		Outlet pipe complete, filter to flexible pipe	1	501252	
		Clip for inlet and outlet pipes	3	50183	
		Drive screw fixing outlet pipe clip to dash	1	78006	
		Bolt (2 BA x $\frac{3}{8}$" long) } Fixing inlet pipe clips	2	237119	
		Spring washer } to dash	2	3073	
		Nut (2 BA)	2	2247	
		Repair kit for mechanical fuel pump, AC 7950706	1	600904	
		Overhaul kit for mechanical fuel pump, AC 7950707	1	600905	

* Asterisk indicates a new part which has not been used on any previous Rover model

FUEL INJECTION SYSTEM, DIESEL ENGINE, 2 LITRE, Series II, 1961

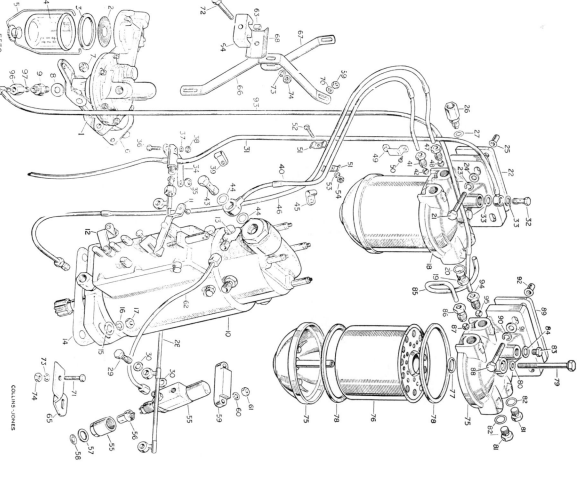

E55Q

COLLINS-JONES

Plate Ref.	1 2 3 4	DESCRIPTION	Qty	Part No.	REMARKS
1		Fuel pump, mechanical, AC 7950445	1	550324	With red diaphragm
2		Filter for bowl, AC 854009	1	506796	For early type pump
3		Washer for bowl, AC 5593321	1	241225	
4		Sediment bowl, AC 1523620	1	236891	
5		Retainer for bowl, AC 5592050	1	268797	
6		Joint washer, fuel pump to cylinder block	2	275565	
7		Self-locking nut fixing fuel pump ...	2	252161	
8		Union } inlet and outlet } For mechanical pump	2	243967	
9		Washer }	2	247929	
10		Distributor pump, CAV DPA 3240099	1	513617	For distributor
11		Accelerator control lever } pump	1	509009	
12		Stop lever }	1	554525	
13		Joint washer for injection pipe, distributor pump	1		
		end ...			Part of engine assembly
14		Sleeve for control lever stop screw, CAV 7139/184	6	247802	
15		Joint washer for distributor pump	1	546282	
		Plain washer } Fixing distributor	1	247212	
16		Spring washer } pump to	3	2920	
17		Nut ($\frac{5}{16}$" UNF) } cylinder block	3	3075	
		Washer for centre bolt, CAV 5936/188A ...	3	254831	
		Sealing plugs for injection system	1 set	279702	For sealing unions in distributor pump
18		Fuel filter, CAV V 7005/2173	1	511682	
		Element for fuel filter, CAV 7111/296 ...	1	517711	
19		Plug for filter, CAV Z 7111/312	1	517707	Not required when additional fuel filter is fitted
20		Joint washer for plug, CAV Y 5936/58T	1	517706	
21		Seal for element, small, CAV 5855/30G	1	522937	
22		Seal for element, large, CAV 5339/256B	2	522938	
23		Special centre bolt for filter, CAV 7111/291	1	522939	
24		Washer for centre bolt, CAV 5936/188A	8	522940	
25		Rivnut	2	509909	
26		Non-return valve for filter	1	505805	
27		Joint washer for non-return valve	1	517976	
28		Leak-off pipe complete	4	521584	Part of engine assembly
29		Banjo bolt } Fixing leak-off	2	254861	
30		Washer } to injector	8	273069	Part of engine assembly
31		Fuel pipe, spill return to tank ...	1	273521	
32		Banjo bolt } Fixing spill return	2	548262	
33		Joint washer } pipe to filter	2	517706	
34		Bracket for leak-off pipe	1	250961	Part of engine assembly
35		Locknut ($\frac{5}{16}$" UNF) fixing bracket to injector stud	2	254861	
36		Bolt (2 BA x $\frac{3}{4}$" long) } Fixing spill	1	558190	
		Plain washer } return pipe	1	3073	
37		Spring washer } to bracket on	1	2247	
38		Nut (2 BA) } injector stud	1	525168	
39		Clip for spill return pipe	1	552439	
		Fuel pipe, mechanical pump to filter ...	1	517531*	
		Olive for fuel pipe, pump end	1	517690	
		Nut } Fixing pipe to	1	530966	
		Olive } filter	1	270115	Early type fixing
		Olive } Fixing pipe to	1	270105	
		Nut } mechanical pump	1		Not required when additional fuel filter is fitted
		Olive }			

* Asterisk indicates a new part which has not been used on any previous Rover model

E550

COLLINS-JONES

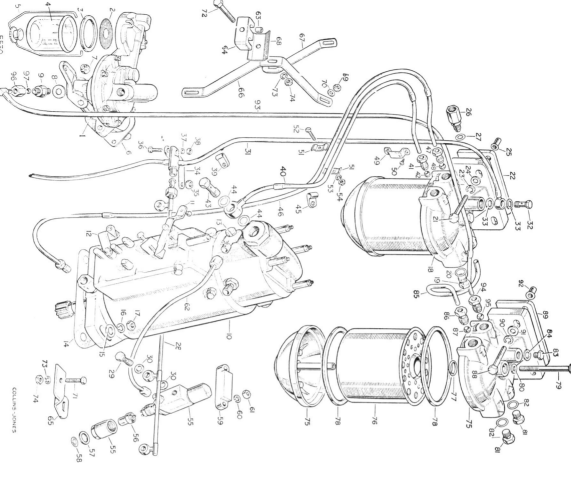

FUEL INJECTION SYSTEM, DIESEL ENGINE, 2 LITRE, Series II, 1961

Plate Ref.	1 2 3 4	DESCRIPTION	Qty	Part No.	REMARKS
40		Fuel pipe, filter to distributor pump ...	1	517685	
41		Nut ⎱ Fixing pipe	1	517690	
42		Olive ⎰ to filter	1	530966	
43		Banjo bolt ⎱ Fixing pipe to	1	275266	
44		Joint washer ⎰ distributor pump	2	275265	
45		Clip, fixing pipe to distributor pump	1	8885	
46		Fuel pipe, distributor pump return to filter ...	1	517686	
47		Nut ⎱ Fixing pipe	1	517690	
48		Olive ⎰ to filter	1	530966	
49		Clip, fixing fuel pipe to dash ...	1	517684	
50		Drive screw, fixing clip	2	78006	
51		Double pipe clip (fixing distributor pump feed and return pipes together)	2	243395	
52		Bolt (2 BA x ⅜" long) ⎱ Fixing pipes	1	234603	
53		Spring washer ⎰ and	1	3073	
54		Nut (2 BA) ⎰ clips	1	2247	
55		Injector complete, CAV 5345902	4	273452	
56		Nozzle for injector	4	247726	
57		Joint washer for injector, copper, CAV 5339/423	4	247179	
58		Joint washer for injector, steel	4	272474	
59		Clamping strip for injector	4	247213	
60		Spring washer ⎱ Fixing injector	8	3075	
61		Nut (¹⁄₁₆" UNF) ⎰ to cylinder head	8	254831	
62		Injector pipe to No. 1 cylinder ...	1	277766	
		Injector pipe to No. 3 cylinder ...	1	277768	
		Injector pipe to No. 4 cylinder ...	1	277769	
		Injector pipe to No. 2 cylinder ...	1	277767	
63		Damper for injector pipe	4	277838	
64		Shroud for damper	2	277839	
65		Bracket for damper shroud	2	277816	
66		Support strap for shroud	1	278002	
67		Steady strap for shroud	1	278021	
68		Backplate for shroud	1	277815	
69		Locknut (¹⁄₁₆" UNF) ⎱ Fixing straps to	4	254861	
70		Plain washer ⎰ injector studs	4	2220	
71		Bolt, front (¹⁄₁₆" UNF x 1¼" long) ⎱ Fixing shrouds and	1	256020	
72		Bolt, rear (¹⁄₁₆" UNF x 1½" long) ⎰ dampers to backplate, strap and support bracket	1	256222	
73		Spring washer	2	3075	
74		Nut (¹⁄₁₆" UNF)	2	254831	

Part of engine assembly

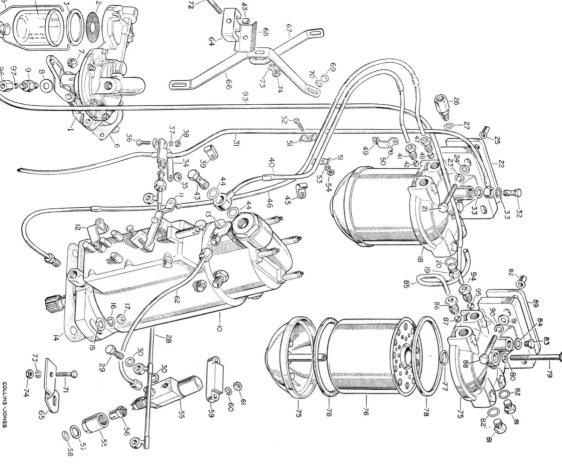

E55Q

COLLINS-JONES

Plate Ref. 1 2 3 4	DESCRIPTION	Qty	Part No.	REMARKS
	ADDITIONAL FUEL FILTER, COMPLETE ASSEMBLY	1	522756	Standard on Export models. Optional equipment on Home models
75	Fuel filter, CAV type FS	1	517682	
76	Element for fuel filter, CAV 7111/296 ..	1	517711	
77	Seal for element, small, CAV 5855/30G ..	2	522937	
78	Seal for element, large, CAV 7111/298	1	522938	
79	Special centre bolt for filter, CAV 7111/291 ..	1	522939	
80	Washer for centre bolt, CAV 5936/188A ..	2	522940	
81	Plug for fuel filter	2	517689	
82	Joint washer for plug	1	517706	
83	Plug for fuel filter, top, leak-off, CAV Z 7111/311	1	517855	
84	Joint washer for leak-off plug	1	517976	
85	Transfer pipe, extra filter to basic filter ..	2	544391	
86	Nut } Fixing pipe	2	517690	
87	Olive } to filter	2	530966	} For early type separate fixings
88	Bolt (5/16" UNF x 1 1/4" long)	2	256220	
89	Distance plate	2	544389	
90	Spring washer } Fixing filter	2	501224	
91	Plain washer } to dash	2	3075	
92	Rivnut (5/16" UNF)	1	3830	
93	Fuel pipe, mechanical pump to twin filters ..	1	552438	
94	Olive for fuel pipe, pump end	1	557531 *	} For early type pipe with separate fixings
95	Nut } Fixing pipe	1	517690	
96	Olive } to filter	1	530966	} Early type fixing
97	Nut } Fixing pipe to	1	270105	
	Olive } mechanical pump	1	270115	
	Repair kit for mechanical fuel pump, AC 7950706 ...	1	600904	
	Overhaul kit for mechanical fuel pump, AC 7950707	1	600905	

* Asterisk indicates a new part which has not been used on any previous Rover model

WATER PUMP AND THERMOSTAT, DIESEL ENGINE, 2 LITRE, Series II

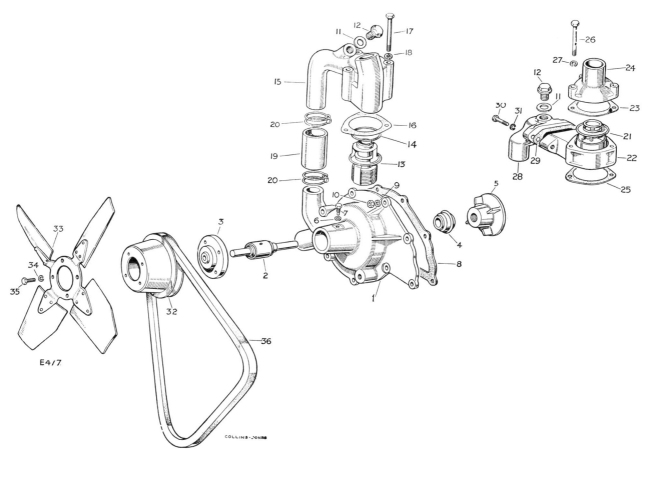

E4/7

COLLINS-JONES

WATER PUMP AND THERMOSTAT, DIESEL ENGINE, 2 LITRE, Series II

Plate Ref.	1 2 3 4	DESCRIPTION	Qty	Part No.	REMARKS
		WATER PUMP ASSEMBLY	1	501041	
1		Water pump casing	1	501039	
2		Pump spindle and bearing	1	523354	
3		Hub for fan	1	515729	
4		Carbon ring and seal unit	1	568301	
5		Impeller for pump	1	247916	
6		Spring washer	1	3074	
7		Special set bolt } Locating bearing casing	1	247078	
8		Joint washer for water pump	1	247919	
9		Spring washer	3	3074	
10		Nut (¼" UNF) } Fixing water pump	3	254810	
11		Joint washer } to front cover	3	243959	
12		Plug (⅜" BSP) in water outlet pipe } For heater return	1	536577	
13		Thermostat, AC 1572270	1	513465	
14		'O' ring, thermostat to water outlet pipe	1	502832	
15		Water outlet pipe, thermostat to outlet pipe	1	501031	1958-60
16		Washer for outlet pipe	1	247874	
17		Spring washer } Fixing outlet pipe	3	256209	
18		Set bolt (¼" UNF x 2½" long) } to cylinder head	3	273789	
19		Hose for by-pass pipe	1	50318	
20		Hose clip for by-pass pipe	2	504736	
21		Thermostat	1	516059	
22		Thermostat housing	1	511957	
23		Joint washer for thermostat housing, upper	1	511956	
24		Water outlet pipe, thermostat to radiator	1	247874	
25		Joint washer for thermostat housing, lower	1	256029	
26		Set bolt (¼" UNF x 2½" long) } Fixing thermostat housing	3	3074	1961
27		Spring washer } and outlet pipe to cylinder head	3	247520	
28		Thermostat by-pass pipe	1	518818	
29		Joint washer for by-pass pipe	1	511958	
30		Set bolt (7/16" UNF x 1" long) } Fixing by-pass pipe to thermostat housing	2	3075	
31		Spring washer	2	255228	
32		Fan pulley	1	247520	
33		Fan blade	1	271568	
34		Spring washer	4	3074	
35		Set bolt (¼" UNF x ¾" long) } Fixing fan blade and pulley to hub	4	255207	
36		Fan and dynamo belt	1	511963	
		Water pump overhaul kit	1	530590	

COLLINS-JONES

C331.

MANIFOLDS, DIESEL ENGINE, 2 LITRE, Series II

Plate Ref.	1 2 3 4	DESCRIPTION			Qty	Part No.	REMARKS
1		Inlet manifold	...	...	1	272376	
2		EXHAUST MANIFOLD ASSEMBLY	...	...	1	247651	
3		Stud for exhaust pipe	...	...	4	252621	
4		Joint washer for inlet and exhaust manifold	...	...	1	247101	
5		Clamp for manifold	...	...	4	247836	
6		Plain washer ⎱ Fixing inlet	...	...	5	2550	
7		Spring washer ⎰ and exhaust	...		9	3075	
8		Nut ($\frac{5}{16}$" UNF) ⎰ manifold			9	254831	

* Asterisk indicates a new part which has not been used on any previous Rover model

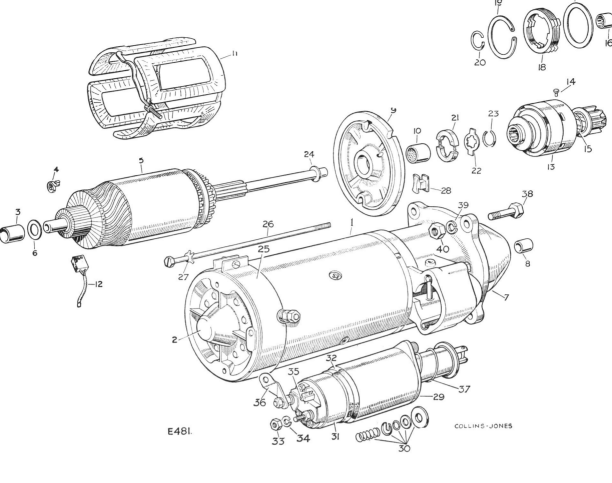

E481.   COLLINS-JONES

STARTER MOTOR AND SOLENOID, DIESEL ENGINE, 2 LITRE, Series II

Plate Ref.	1 2 3 4	DESCRIPTION	Qty	Part No.	REMARKS
1		STARTER MOTOR COMPLETE, LU M45G26189D	1	529971	
2		Bracket for starter, commutator end, LU 54250082	1	532568	
3		Bush, commutator end, LU 255491	1	242958	
4		Spring set for brushes, LU 270004	1	261239	For starter motor, LU 26147A
4		Spring set for brushes, LU 54251042	1	532569	For starter motor, LU 26189D
5		Armature, LU 272334	1	279006	For starter motor, LU 26147A
5		Armature, LU 54251813	1	532570	For starter motor, LU 26189D
6		Thrust washer, commutator end, LU 152380	1	279007	
7		Bracket for starter, drive end, LU 272745	1	279008	
8		Bush for bracket, LU 271364	1	279009	For starter motor, LU 26147A
9		Bracket for brake, LU 272700	1	270243	
10		Bush for brake bracket, LU 271733	1	270245	
		Bracket for starter, drive end, LU 54252236	1	532571	
		Bush for bracket, LU 54252800	1	532572	For starter motor, LU 26189D
		Bracket for brake, LU 54252494	1	532573	
		Bush for brake bracket, LU 54252532	1	532574	
		Pivot pin for starter motor, LU 272548	1	601689	
11		Field coil for starter, LU 272394	1	279010	
12		Brushes for starter motor, set, LU 271735	1	270225	
13		Drive assembly for starter motor, set, LU 292124	1	279001	
		Retaining ring complete for pinion, LU 295298	1	601760	
14		Rivet for pinion retaining ring, LU 291887	1	526258	
15		Return spring for starter pinion, LU 292071	1	512889	For starter motor, LU 26147A
15		Return spring for starter pinion, LU 54252496	1	532575	For starter motor, LU 26189D
16		Bush, pinion end, LU 292053	2	279012	
17		Clutch plates, set, LU 291540	1	270238	
		Clutch adjusting shim (0.006"), LU 291374	As reqd	270239	
		Clutch adjusting shim (0.005"), LU 291378	As reqd	270240	
		Clutch adjusting shim (0.004"), LU 291379	As reqd	270241	
18		Circlip retaining clutch assembly, LU 291560	1	291560	
19		Lock ring retaining clutch plates, LU 291862	1	270235	
20		Brake shoe complete with springs, LU 272701	1	279013	
21		Driving washer for brake shoe, LU 272340	1	279014	
22		Lock ring retaining brake, LU 272337	1	279015	
23		Lock washer, drive end, LU 134950	1	279016	
24		Thrust washer, drive end, LU 272396	1	279017	
25		Cover band for starter, LU 272660	1	279018	
26		Bolt for starter motor, LU 272660	2	279019	For early type bracket
27		Lockwasher for bolt, LU 188382	2	279020	For early type bracket
		Bolt for starter motor, LU 54250580	2	532568	For late type bracket
		Lockwasher for bolt, LU 272773	2	532577	Check before ordering
28		Rubber grommet in drive end and bracket, LU 272552	1	279021	
29		Solenoid for starter motor, LU 76458A	1	279022	For starter motor, LU 26147A
29		Solenoid for starter motor, LU 76490	1	532578	For starter motor, LU 26189D
30		Contact plate complete for starter solenoid, LU 54330168	1	519004	

* Asterisk indicates a new part which has not been used on any previous Rover model

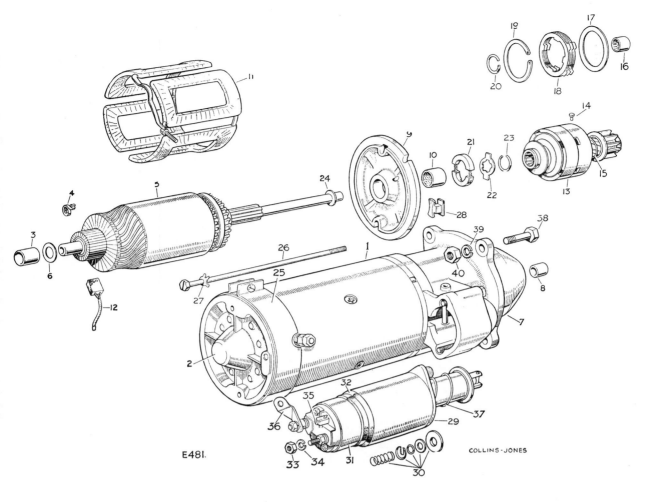

E481.

COLLINS-JONES

Plate Ref.	1 2 3 4	DESCRIPTION	Qty	Part No.	REMARKS
31		Base complete for starter solenoid, LU 54330169	1	519005	For starter motor, LU 26147A
31		Base complete for starter solenoid, LU 54640665	1	532579	For starter motor, LU 26189D
32		Gasket for starter solenoid, base, LU 765842	1	519006	
33		Terminal nut for starter solenoid, LU 170570	4	519007	For starter motor, LU 26147A
33		Terminal nut for starter solenoid, LU 156456	4	532580	For starter motor, LU 26189D
34		Terminal washer for starter solenoid, LU 185062	4	519008	
35		Terminal screw for starter solenoid, LU 121610	1	519009	
36		Terminal connector for starter motor, LU 272555	1	519010	
37		Plunger spring for starter solenoid, LU 765866	1	519011	
		Sundry parts kit, LU 271731	1	270251	
38		Bolt ($\frac{5}{16}$" UNC x 1$\frac{1}{4}$" long) ⎞ Fixing	1	253068	
39		Spring washer ⎟ starter	3	3077	
40		Nut ($\frac{5}{16}$" UNF) ⎠ motor	2	254823	

* Asterisk indicates a new part which has not been used on any previous Rover model

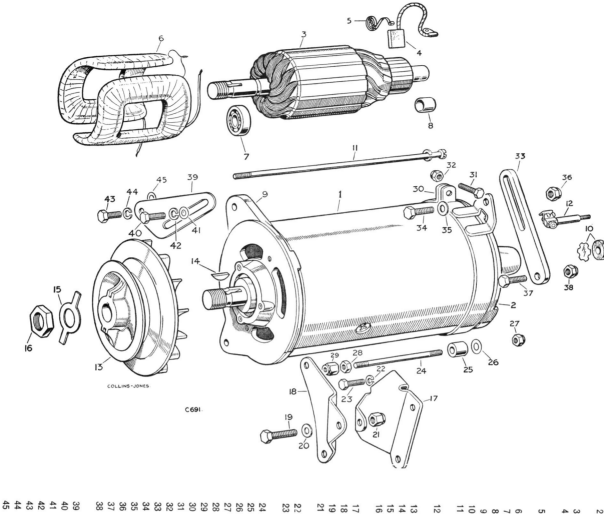

COLLINS-JONES

C 691.

Plate Ref. 1 2 3 4	DESCRIPTION	Qty	Part No.	REMARKS
1	DYNAMO COMPLETE, C 45 PV6, LU 22489	1	512249	For dynamo type PV5
2	Bracket, commutator end, LU 239266	1	263394	For dynamo type PV6
2	Bracket, commutator end, LU 54210057	1	512798	For dynamo type PV6
3	Armature for dynamo, LU 239428	1	512800	For dynamo type PV6
4	Brushes for dynamo, set, LU 238061	1	261256	For dynamo type PV6
4	Brushes for dynamo, set, LU 54212267	1	512797	For dynamo type PV6
5	Spring set for brushes, LU 238062	1	244708	For dynamo type PV5
5	Spring set for brushes, LU 54210091	1	512799	For dynamo type PV6
6	Field coil for dynamo, LU 238820	1	264435	For dynamo type PV5
7	Ball bearing, front	1	260026	
8	Bush, commutator end, LU 239263	1	263395	
9	Bracket, drive end, LU 239012	1	264431	
10	Oiler for dynamo, LU 239368	1	264433	
11	Bolt for bracket, LU 238821	2	264436	
11	Bolt for bracket, LU 272660	2	264438	
12	Terminal, LU 227625	1	272704	
13	Sundry parts, set, LU 239024	1	279019	
13	Pulley for dynamo	1	264437	Fixing pulley to dynamo
14	Woodruff key	1	278000	
15	Lockwasher	1	256421	
16	Special nut	1	252161	
17	Support bracket for dynamo	1	3075	
18	Steady bracket for dynamo support bracket	1	1664	Fixing dynamo support bracket to support and steady brackets
19	Bolt (5/16" UNF x 1 3/8" long)	1	03748	Fixing dynamo to support bracket
21	Self-locking nut (5/16" UNF)	2	3466	Fixing support bracket and steady brackets
22	Spring washer	2	4148	
23	Set bolt (5/16" UNF x 5/8" long)	2	252161	Fixing support bracket to cylinder block
24	Special stud fixing dynamo	1	272303	
25	Distance piece	1	278001	Fixing dynamo to special stud at front
26	Shim washer	As reqd	255225	
27	Self-locking nut (5/16" UNF)	1	255227	
28	Nut (5/16" UNF)	1	277544	Fixing dynamo at rear to special stud
29	Self-locking nut (5/16" UNF)	1	256221	
30	Strap for dynamo, rear	1	252211	
31	Bolt (5/16" UNF x 1 3/8" long)	1	255227	Fixing strap to dynamo
32	Self-locking nut (5/16" UNF)	1	2266	
33	Adjusting link for dynamo, rear	1	255227	Fixing strap
34	Bolt (5/16" UNF x 7/8" long)	1	277543	link to dynamo
35	Plain washer	1	252211	Fixing adjusting link to dynamo
36	Self-locking nut (5/16" UNF x 7/8" long)	1	2266	
37	Bolt (5/16" UNF x 7/8" long)	1	256221	Fixing adjusting strap
38	Self-locking nut (5/16" UNF)	1	252211	
39	Adjusting link for dynamo, front	1	252211	Fixing adjusting link to engine mounting foot
40	Special set bolt	1	274862	
41	Plain washer	1	233568	Fixing front adjusting link
42	Spring washer	1	2550	
43	Set bolt (5/16" UNC x 3/4" long)	1	3075	Fixing adjusting link to front cover
44	Spring washer	1	3075	
45	Distance washer	1	253026	

* Asterisk indicates a new part which has not been used on any previous Rover model

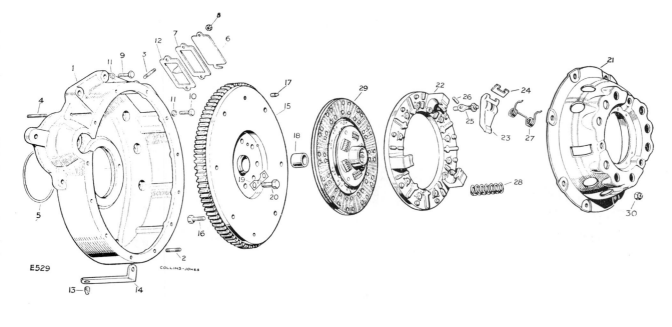

E529

COLLINS-JONES

Plate Ref.	1 2 3 4	DESCRIPTION	Qty	Part No.	REMARKS
1		FLYWHEEL HOUSING ASSEMBLY	1	277531	
2		Stud (⅜" UNF) fixing flywheel housing to bell housing	12	247145	
3		Stud (¼" UNF) fixing inspection cover	12	247146	
4		Stud (7/16" UNF) for starter motor	2	277532	
5		Sealing ring for flywheel housing	2	246169	
6		Inspection cover plate	1	56140	
7		Joint washer for cover plate	1	50216	
8		Nut (¼" UNF) fixing cover plate	2	254810	
9		Bolt (⅜" UNF x 1¾" long) } Fixing flywheel	2	256043	
10		Bolt (⅜" UNF x 1⅛" long) } housing to cylinder block	6	255248	
11		Spring washer	8	3076	
12		Indicator for engine timing	1	500597	
13		Drain plug for housing	1	3290	
14		Stowage bracket for drain plug	1	276511	
15		FLYWHEEL ASSEMBLY	1	247167	
		Ring gear for flywheel	1	510489	
16		Special fitting bolt fixing clutch cover plate	6	247166	
17		Dowel locating clutch cover plate	2	502116	
18		Bush for primary pinion	1	08566	
19		Locker } Fixing flywheel	4	526161	
20		Special set bolt } to crankshaft	8	247135	
		CLUTCH ASSEMBLY, BB 45693/41		236684	With black clutch springs
21		Cover plate for clutch, BB 45481	1	275301	For early type clutch } Check before ordering
					For late type clutch. Part of assembly 236684
22		Pressure plate for clutch, BB 42652	1	231888	
23		Release lever for clutch, BB 51022	3	231880	
24		Strut for release lever, BB 42606	3	534358	
25		Eyebolt and nut for release lever, BB 48508	3	231884	
26		Pin for release lever, BB 42604	3	242996	
27		Anti-rattle spring for release lever, BB 47688	3	231885	
28		Clutch spring (yellow and light green), BB 44780	9	243593	
28		Clutch spring (black), BB 44633	9	231883	
29		Clutch plate complete, BB 47626/104	1	275811	Ferodo lining } Alternatives.
			1	261921	Mintex lining } Check before ordering
		Lining package for clutch plate, BB 46627	1	517026	Ferodo lining }
		Lining package for clutch plate, BB KL75012	1		Mintex lining }
30		Self-locking nut (5/16" UNF) fixing cover plate	6	252211	

Not part of engine assembly

* Asterisk indicates a new part which has not been used on any previous Rover model

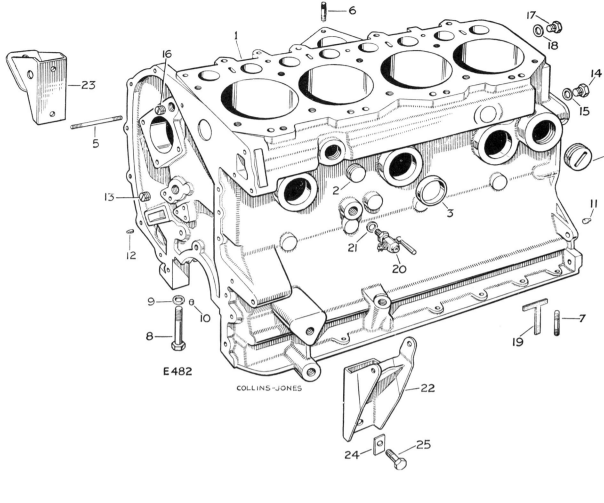

E 482

COLLINS-JONES

CYLINDER BLOCK, DIESEL ENGINE, 2¼ LITRE, Series IIA

Plate Ref. 1 2 3 4	DESCRIPTION	Qty	Part No.	REMARKS
	ENGINE ASSEMBLY	1	556831	Up to engine suffix 'J' inclusive ⎫ Models with Lucas 11 AC type 12 volt AC/DC Generator.
	ENGINE ASSEMBLY	1	605654*	From engine suffix 'K' onwards ⎭
	ENGINE ASSEMBLY	1	534085	Models with Lucas 2 AC type 12 volt AC/DC Generator. Up to engine suffix 'J' inclusive
	ENGINE ASSEMBLY	1	606082*	From engine suffix 'K' onwards
	ENGINE ASSEMBLY	1	565036	Up to engine suffix 'J' inclusive ⎫ Models with Prestolite 12 volt AC/DC generator
	ENGINE ASSEMBLY	1	605727*	From engine suffix 'K' onwards ⎭
1	CYLINDER BLOCK ASSEMBLY	1	541891	
	Cylinder block, crankshaft, pistons and camshaft			
2	Core plug, 1" diameter	2	528114	
	Core plug, 1½" diameter	2	09191	
3	Cup plug, 1⅛" diameter	2	250840	Alternatives.
4	Cup plug	3	524765	Check before ordering
	Plug for immersion heater boss	1	525428	
5	Stud for front cover and adjusting link	1	527269	
6	Stud for distributor pump	3	514527	Not part of cylinder block assembly
7	Stud for sump	1	252624	
8	Set bolt for main bearings	6	252621	
9	Special plain washer for set bolt	6	504007	
10	Dowel, locating main bearing and bearing cap	3	504006	
11	Dowel, locating flywheel housing	2	501593	
12	Dowel for timing chain adjuster	1	52124	
13	Plug for oil gallery, front	1	213700	
14	Plug for oil gallery, rear	1	247127	
15	Joint washer for rear plug	1	247861	
16	Plug for tappet feed gallery pipe, front	1	243959	
17	Plug for tappet feed gallery pipe, rear	1	536677	
18	Joint washer for plug	2	273166	Not part of cylinder block assembly
	Plug for tappet feed hole	1	243968	
19	Packing for rear main bearing cap	1	247965	
20	Drain tap for cylinder block	1	537279	
22	Engine front support bracket complete, LH	1	538608	
23	Engine front support bracket complete, RH	1	543823	
24	Locker	4	271975	
25	Set bolt (⅜" UNF x 1⅜" long)	3	212430	Fixing brackets ⎫ to cylinder block
	Set bolt (⅜" UNF x 1" long)	3	3078	Alternatives
	Spring washer	1	255085	
	Decarbonising gasket kit	1	255086	
	Engine overhaul gasket kit	1	525521	
	Stud kit for cylinder block	1	525520	
	Cylinder liner	1	600245	⎫ Alternatives
		4	503160	⎭

* Asterisk indicates a new part which has not been used on any previous Rover model

182

H6O7

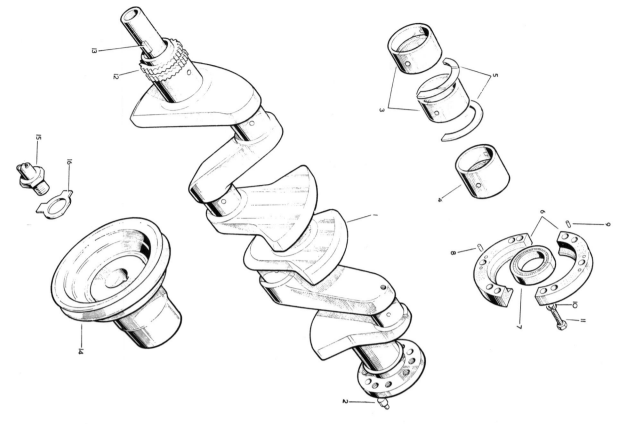

CRANKSHAFT, DIESEL ENGINE, 2¼ LITRE, Series IIA

Plate Ref. 1 2 3 4	DESCRIPTION	Qty	Part No.	REMARKS
1	CRANKSHAFT ASSEMBLY, STD ...	1	527167	
2	Dowel for flywheel	1	265779	
3	Main bearing, front and centre, Std	2	518748	
	Main bearing, front and centre, .010" US	2	518749	
	Main bearing, front and centre, .020" US	2	518750	
	Main bearing, front and centre, .030" US	2	518751	
	Main bearing, front and centre, .040" US	2	518752	
4	Main bearing, rear, Std	1	518753	
	Main bearing, rear, .010" US	1	518754	
	Main bearing, rear, .020" US	1	518755	
	Main bearing, rear, .030" US	1	518756	
	Main bearing, rear, .040" US	1	518757	
5	Thrust washer for crankshaft, Std	1	518758	
	Thrust washer for crankshaft, .0025" OS	1	518759	Pairs
	Thrust washer for crankshaft, .005" OS	1	518760	
	Thrust washer for crankshaft, .0075" OS	1	518761	
	Thrust washer for crankshaft, .010" OS	1	518762	
	Bearing set for crankshaft, Std	1	533979	
	Bearing set for crankshaft, .010" US	1	533980	Sets include main bearings, thrust washers and connecting rod bearings
	Bearing set for crankshaft, .020" US	1	533981	
	Bearing set for crankshaft, .030" US	1	533982	
	Bearing set for crankshaft, .040" US	1	533983	
	REAR BEARING OIL SEAL ASSEMBLY			
6	Rear bearing split seal halves	2	542494	
7	Oil seal complete for rear bearing ...	1	523240	
	Silicone grease, MS4, 1 oz tube	1	542492	
8	Dowel, lower — to cylinder block	2	270556	
9	Dowel, upper — and rear bearing cap	2	519064	
10	Spring washer	10	246464	
11	Bolt (¼" UNF x ⅞" long)	10	3074	
12	Chainwheel on crankshaft	1	255208	
13	Key locating chainwheel and pulley	2	235726	
14	Crankshaft pulley ...	1	235570	
15	Starting dog ...	1	524206	
16	Lockwasher for starting dog	1	503665	

184

COLLINS-JONES
KII5

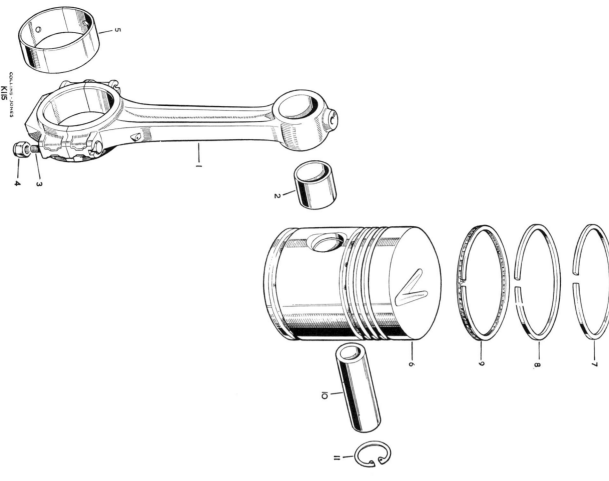

Plate Ref.	1 2 3 4	DESCRIPTION		Qty	Part No.	REMARKS
1		CONNECTING ROD ASSEMBLY	... ...	4	527169	
2		Gudgeon pin bush	... ...	4	247583	
3		Special bolt ⎱ Fixing connecting		8	519440	
4		Self-locking nut ⎰ rod cap		8	277390	
5		Connecting rod bearing, pairs, Std	...	4	524151	
		Connecting rod bearing, pairs, .010" US		4	524152	
		Connecting rod bearing, pairs, .020" US		4	524153	
		Connecting rod bearing, pairs, .030" US		4	524154	
		Connecting rod bearing, pairs, .040" US		4	524155	
6		Piston assembly, Std, Low Diameter		4	564228	
		Piston assembly, Std, High Diameter		4	564229	
		Piston assembly, .010" OS	...	4	564230	
		Piston assembly, .020" OS	...	4	564231	
		Piston assembly, .030" OS	...	4	564232	
		Piston assembly, .040" OS	...	4	564233	
7		Piston ring, compression, Std, chromed		4	568588	
		Piston ring, compression, .010" OS ⎱		4	605539	
		Piston ring, compression, .020" OS ⎱ chromed		4	605540	
		Piston ring, compression, .030" OS ⎰		4	605541	
		Piston ring, compression, .040" OS ⎰		4	605542	
8		Piston ring, compression, Std, tapered	...	8	520978	
		Piston ring, compression, .010" OS ⎱		8	520940	
		Piston ring, compression, .020" OS ⎱ tapered		8	520941	
		Piston ring, compression, .030" OS ⎰		8	520942	
		Piston ring, compression, .040" OS ⎰		8	520943	
9		Oil control ring, Std	... ...	4	564226	
		Oil control ring, .010" OS	...	4	605047	
		Oil control ring, .020" OS	...	4	605048	
		Oil control ring, .030" OS	...	4	605049	
		Oil control ring, .040" OS	...	4	605050	
10		Gudgeon pin	... ...	4	502029	
11		Circlip for gudgeon pin	...	8	266945	

CYLINDER HEAD, DIESEL ENGINE, 2¼ LITRE, Series IIA

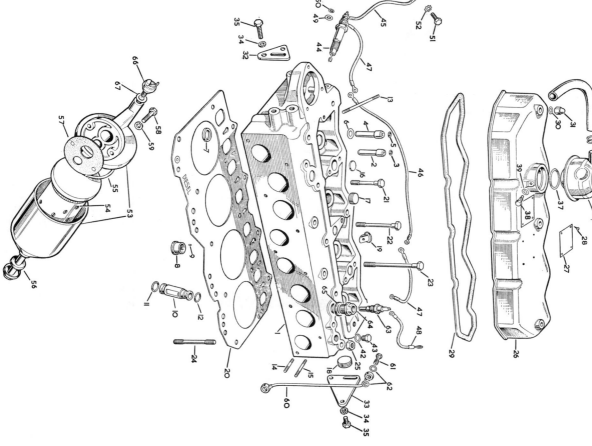

J962

CYLINDER HEAD, DIESEL ENGINE, 2¼ LITRE, Series IIA

Plate Ref. 1 2 3 4	DESCRIPTION	Qty	Part No.	REMARKS
1	CYLINDER HEAD ASSEMBLY	1	601469	Up to engine suffix 'J' inclusive
1	CYLINDER HEAD ASSEMBLY	1	605480*	From engine suffix 'K' onwards
2	Valve guide, inlet and exhaust	4	272206	
3	Sealing ring for inlet valve guide	4	247186†	
4	Valve guide, inlet and sealing ring	4	275552	
5	Valve guide, exhaust and sealing ring	4	233419†	
6	Sealing ring for exhaust valve guide	4	230062	
	Packing washer for valve guides	20	512828	
7	Insert for exhaust valve seat	4	558168	
	Oil seal for exhaust valve guide	4	605482*	With 'O' ring type oil seal. Up to engine suffix 'J' inclusive
	Valve guide, exhaust, and oil seal	4	554727*†	Up to engine suffix 'J' inclusive
	Oil seal for inlet valve guide	4	605483*	With lip type oil seal. From engine suffix 'K' onwards
	Valve guide, inlet, and oil seal	4	554728*†	From engine suffix 'K' onwards
8	Hot plug in cylinder head	4	515572	
9	Peg for hot plug	4	515573	
10	Push rod tube	8	514224	
11	'O' ring for push rod tube, large	8	271881	
12	'O' ring for push rod tube, small	8	558168	
13	Stud for injector, thread length 1 3/32"	8	531896	For plain type clamping strip for injector. Check before ordering. Up to engine suffix 'J' inclusive
	Stud for injector, thread length 2 3/32"		531895	For flanged type injector. From engine suffix 'K' onwards
	Stud for injector, thread length 1 7/8"		252624	For stepped type clamping strip for injector. Check before ordering
	Stud for injector, thread length 2½"		521600	before ordering
	Stud for injector, thread length 1 3/4"		247883	
14	Stud, short } For manifold	5	247144	
15	Stud, long }	3	247143	
16	Core plug, 7/8" diameter	2	250830	
	Cup plug	2	525497	Alternative to 7/8" diameter core plug
	Core plug, 3/4" diameter		09191	Alternative to cup plug. Check before ordering
17	Core plug, 1" diameter	1	230250	
18	Core plug, 1 3/8" diameter	1	232038	
19	Shroud for injector bore	4	512413	
20	Cylinder head gasket	1	558132	
21	Special set bolt, short	2	247683	
22	Special set bolt, medium } Fixing cylinder head	5	279649	
23	Special set bolt, long } to cylinder block	9	247723	
24	Stud (½" UNF)	2	518466	
25	Nut (½" UNF)	2	254824	
26	ROCKER COVER, TOP, ASSEMBLY	1	518719	
27	Tappet clearance plate	1	247634	
28	Drive screw fixing plate	1	78001	
29	Joint washer for top rocker cover	1	247606	
30	Sealing washer } Fixing top rocker cover	3	232038	
31	Special nut }	3	247121	
32	Special nut	3	525131	
33	Lifting bracket for engine, front	1	247715	
34	Lifting bracket for engine, rear	1	525131	
35	Spring washers } Fixing brackets	4	3075	
35	Set bolt (5/16" UNF × 3/4" long) } to cylinder head	4	255226	

* Asterisk indicates a new part which has not been used on any previous Rover model
† Not included in cylinder head assembly

CYLINDER HEAD, DIESEL ENGINE, 2¼ LITRE, Series IIA

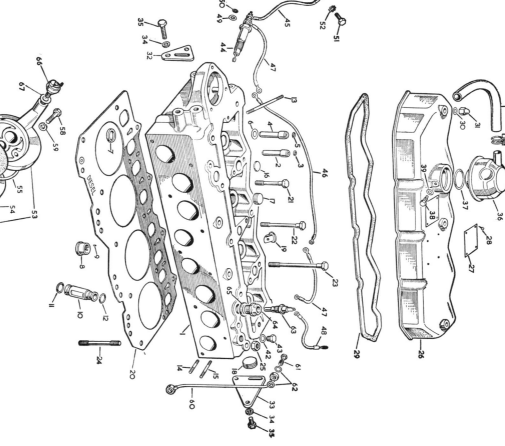

J962

CYLINDER HEAD, DIESEL ENGINE, 2¼ LITRE, Series IIA

Plate Ref. 1 2 3 4	DESCRIPTION	Qty	Part No.	REMARKS
36	Breather filter for engine, AC 7972354	1	563180	Up to engine suffix 'J' inclusive
36	Breather filter for engine, AC 7971799/1	1	574658*	From engine suffix 'K' onwards } Not part of engine assembly
37	Sealing ring for filter	1	268887	
38	Special set screw fixing breather filter	1	515291	
39	Sealing washer for set screw	1	232037	From engine suffix 'K' onwards
40	Hose, breather filter to inlet manifold	1	574655*	
41	Clip fixing hose	2	554260	
42	Clip for hose at manifold clamp	1	243959	
43	Joint washer } For heater valve hole	1	536577	
44	Plug (⅛" BSF) } in cylinder head	1	510078	
45	Heater plug	4	247953	
46	Heater plug lead, No. 1 to earth	1	247952	
47	Heater plug lead, No. 2 to No. 3	1	247951	
47	Heater plug lead, No. 1 to 2 and 3 to 4	2	529972	
48	Heater plug lead, No. 4 to resistor	1	541523	For Lucar blade type resistor } Alternatives. Check before ordering
48	Heater plug lead, No. 4, to resistor	1	541403	For terminal type resistor
49	Shakeproof washer, large } Fixing leads to	4	77626	
50	Shakeproof washer, small } heater plug	4	71082	
51	Bolt (⅜" UNF x ⅝" long) } Fixing heater plug	1	253905	
52	Fan disc washer } lead to cylinder head	1	512305	
53	Engine oil filter, AC 7965053	1	537229	Short type
54	Element for filter, AC FF 50, overall length 6¼⅛"	1	248863	For long type filter } Alternatives. Check before
54	Element for filter, AC 72, overall length 4⅝"	1	541403	For short type filter ordering
55	Gasket for filter, AC 1530250	1	272539	
56	Rubber washer for centre bolt, AC 1531098	1	269889	
57	Joint washer for filter	1	272839	
58	Set bolt (⁷⁄₁₆" UNF x 1¼" long) } Fixing filter to	2	255068	
59	Spring washer } cylinder block	2	3077	
60	Oil pipe complete to cylinder head, length 10½"	1	275679	
61	Banjo bolt fixing oil pipe	2	504840	
62	Joint washer for banjo bolts	4	232039	
63	Plug for thermometer hole in cylinder head	1	278164	Up to engine suffix 'G' inclusive
64	Water temperature transmitter	1	560794	From engine suffix 'H' onwards
65	Adaptor for water temperature transmitter	1	546177	
66	Joint washer for plug or adaptor	1	243972	
67	Oil pressure switch, AC 7954237	1	519864	
	Joint washer	1	232039	
	Valve guide oil control kit	1	605484*	To convert early engines with 'O' ring type valve guide seals to lip type oil seals

* Asterisk indicates a new part which has not been used on any previous Rover model

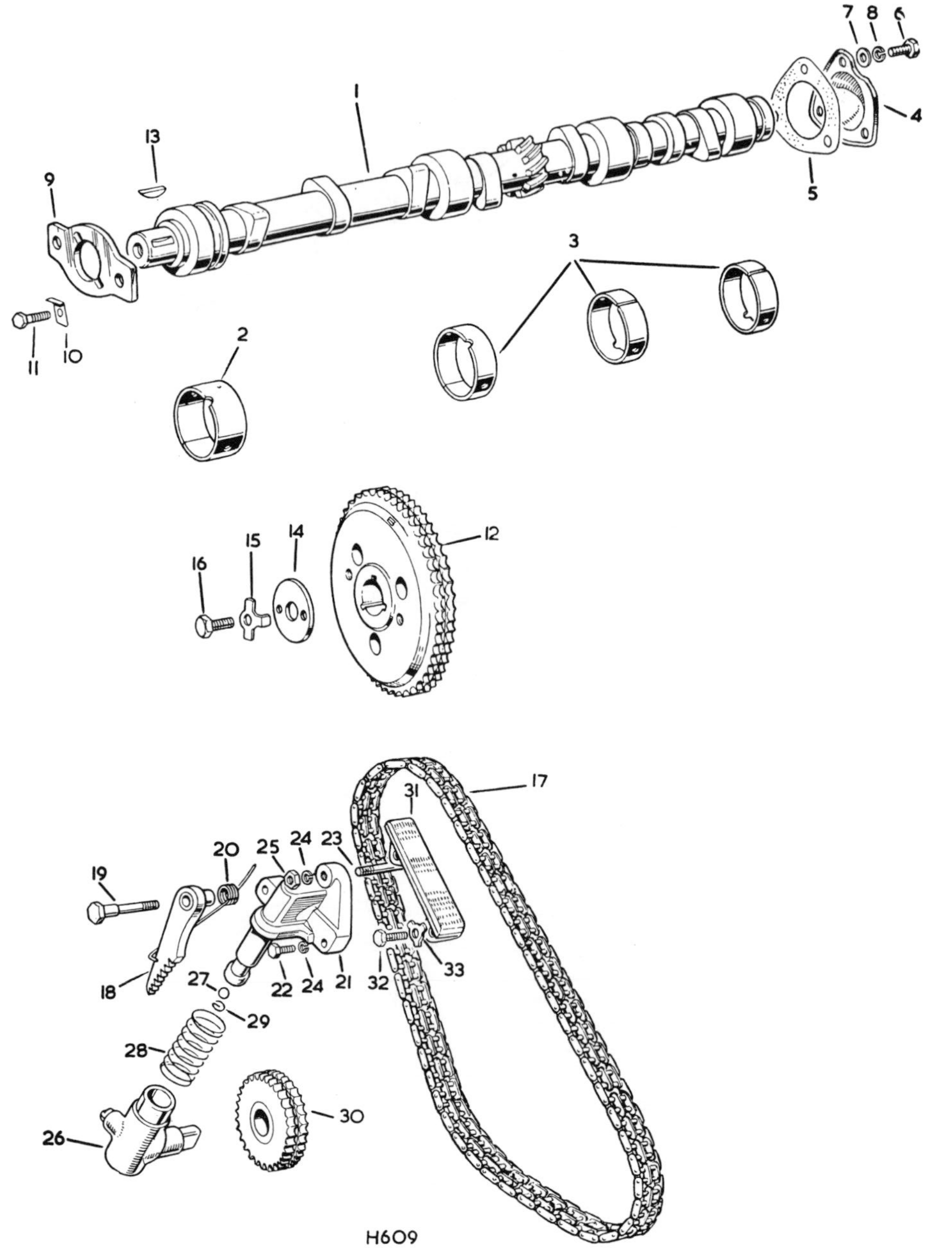

H609

CAMSHAFT AND TENSIONER MECHANISM, DIESEL ENGINE, 2¼ LITRE, Series IIA

Plate Ref.	1 2 3 4	DESCRIPTION	Qty	Part No.	REMARKS
1		Camshaft	1	274711	
2		Bearing complete for camshaft, front	1	519054	
3		Bearing complete for camshaft, centre and rear	3	519055	
4		Rear end cover for camshaft	1	538073	
5		Joint washer for cover	1	247070	
		Set bolt (¼″ UNF x ⅞″ long) } Fixing cover to block	3	255208	} Alternative fixings. Check before ordering
6		Set bolt (¼″ UNF x ⅝″ long) } Fixing cover to block	3	255206	} Alternative fixings. Check before ordering
		Set bolt (¼″ UNF x ¾″ long) } Fixing cover to block	1	255207	} Alternative fixings. Check before ordering
7		Plain washer } Fixing cover to block	4	10882	} Alternative fixings. Check before ordering
8		Spring washer } Fixing cover to block	3	3074	} Alternative fixings. Check before ordering
9		Thrust plate for camshaft	1	538535	
10		Locker } Fixing thrust plate to cylinder block	2	2995	} 4 off on early type thrust plate
11		Set bolt (¼″ UNF x ¾″ long) } Fixing thrust plate to cylinder block	2	255207	} 4 off on early type thrust plate
12		Chainwheel for camshaft	1	276133	
13		Key locating camshaft chainwheel	1	230313	
14		Retaining washer } Fixing chainwheel to camshaft	1	09093	
15		Locker } Fixing chainwheel to camshaft	1	09210	
16		Set bolt (⅜″ UNF x ⅞″ long) } Fixing chainwheel to camshaft	1	255246	
17		Camshaft chain (⅜″ pitch x 78 links)	1	09156	
18		Ratchet for timing chain adjuster	1	546026	
19		Special bolt fixing ratchet and piston to block	1	247199	
20		Spring for chain adjuster ratchet	1	267451	
21		Piston for timing chain adjuster	1	247912	
22		Set bolt ($\frac{5}{16}$″ UNF x 1¼″ long) } Fixing piston to cylinder block	1	256220	
23		Stud ($\frac{5}{16}$″ UNF) } Fixing piston to cylinder block	1	247144	
24		Spring washer } Fixing piston to cylinder block	2	3075	
25		Nut ($\frac{5}{16}$″ UNF) } Fixing piston to cylinder block	1	254831	
26		Cylinder for timing chain adjuster	1	277388	
27		Steel ball ($\frac{7}{32}$″) for non-return valve	1	3739	
28		Spring for chain tensioner	1	233326	
29		Retainer for steel ball	1	233328	
30		Idler wheel for timing chain	1	236067	
31		Vibration damper for timing chain	1	275234	
32		Set bolt (¼″ UNF x ½″ long) } Fixing damper to cylinder block	2	255204	
33		Locking plate } Fixing damper to cylinder block	2	557523	

* Asterisk indicates a new part which has not been used on any previous Rover model

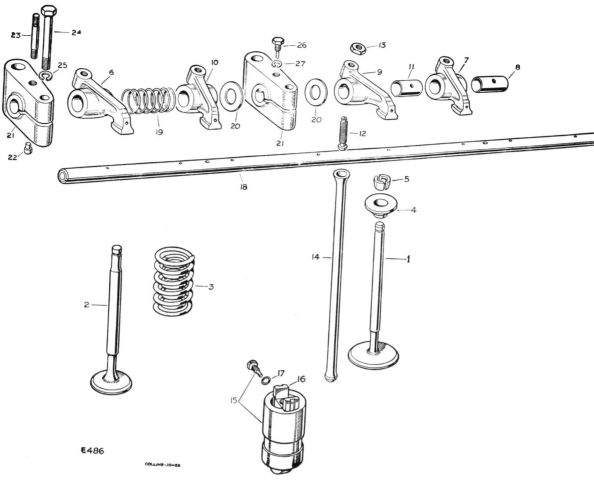

E486

COLLINS-JONES

VALVE GEAR AND ROCKER SHAFTS, DIESEL ENGINE, 2¼ LITRE, Series IIA

Plate Ref. 1 2 3 4	DESCRIPTION	Qty	Part No.	REMARKS
1	Inlet valve	4	527240	
2	Exhaust valve	4	550198	
3	Valve spring, inner and outer	8	568550	
4	Valve spring cup	8	268292	
5	Split cone for valve, halves	16	268293	
6	Valve rocker, exhaust, LH ...	2	274772	
7	Valve rocker, exhaust, RH ...	2	274773	
8	Bush for exhaust valve rocker	4	247738	
9	Valve rocker, inlet, LH ...	2	274774	
10	Valve rocker, inlet, RH ...	2	274775	
11	Bush for inlet valve rocker	4	247737	
12	Tappet adjusting screw ...	8	506814	
13	Locknut for tappet adjusting screw	8	254881	
14	Tappet push rod	8	546799	
15	Tappet, tappet guide, roller and set bolt assembly	8	507829	
16	Tappet	8	507026	
	Tappet guide	8	502473	
	Roller	8	517429	
17	Special set bolt	8	507025	
	Copper washer for tappet guide set bolt	8	232038	
18	Valve rocker shaft	1	554073	
19	Spring for rocker shaft ...	4	247040	
20	Washer for rocker shaft ...	8	247153	
21	Rocker bracket	5	523181	
22	Locating dowel for rocker bracket	3	277956	4 off on early models
23	Stud for rocker bracket ...	3	247607	
24	Set bolt (5⁄16" UNF x 2" long) Fixing rocker bracket to cylinder head	5	256226	
25	Spring washer	5	3075	
26	Locating screw For rocker shaft at bracket	1	525390	2 off on early models
27	Spring washer	1	3075	

OIL PUMP, DIESEL ENGINE, 2¼ LITRE, Series IIA

E480.

COLLINS JONES

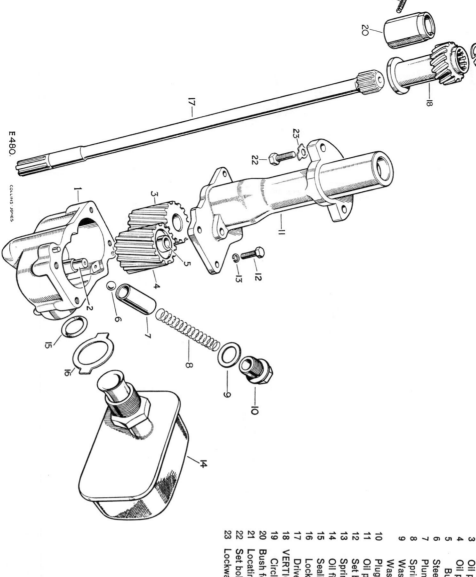

OIL PUMP, DIESEL ENGINE, 2¼ LITRE, Series IIA

Plate Ref. 1 2 3 4	DESCRIPTION		Qty	Part No.	REMARKS
	OIL PUMP ASSEMBLY				
1	Oil pump body		1	513640	
2	Spindle for idler gear		1	513641	
3	Oil pump gear, driver		1	502209	
4	Oil pump gear, idler		1	240555	
5	Bush for idler gear		1	278109	
6	Steel ball		1	214995	
7	Plunger		1	3748	
8	Spring		1	273711	
9	Washer, inside diameter $\frac{5}{16}$"	For oil pressure release valve	1	564456	
	Washer, inside diameter $\frac{43}{64}$"		1	243970	} Alternatives. Check before ordering
10	Plug		1	232044	
11	Oil pump cover		1	549909	
12	Set bolt ($\frac{5}{16}$" UNF x $\frac{7}{8}$" long)	} Fixing cover to body	4	513639	
13	Spring washer		4	255227	
14	Oil filter for pump		1	3075	
15	Sealing ring	} Fixing oil filter to oil pump	1	247664	
16	Lockwasher		1	244488	
17	Drive shaft for oil pump		1	244487	
	VERTICAL DRIVE SHAFT ASSEMBLY				
18	Drive shaft for oil pump		1	511680	
19	Circlip for drive shaft		1	503266	
20	Bush for drive shaft gear		1	247742	
21	Locating screw for drive shaft bush		1	247653	
			1	524769	
22	Set bolt ($\frac{5}{16}$" UNF x 1" long)	} Fixing oil pump to cylinder block	2	255228	
23	Lockwasher		2	247665	

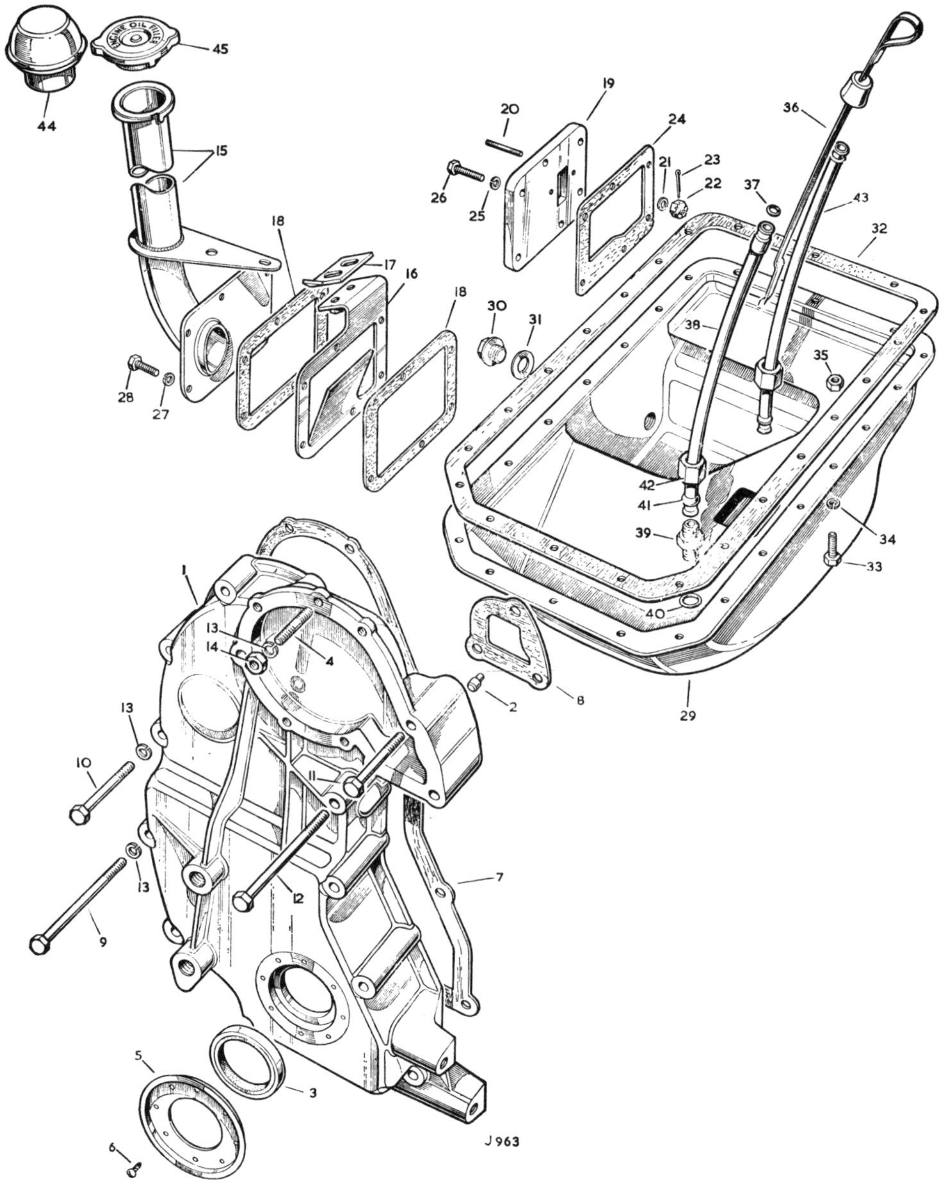

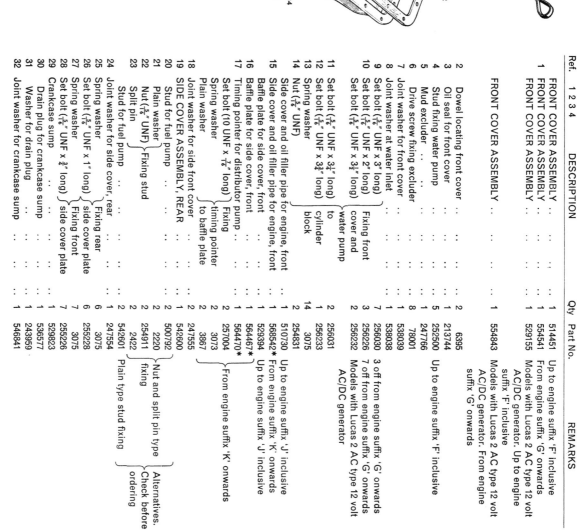

Plate Ref.	1 2 3 4	DESCRIPTION	Qty	Part No.	REMARKS
		FRONT COVER ASSEMBLY	1	514451	Up to engine suffix 'F' inclusive
1		FRONT COVER ASSEMBLY	1	554541	From engine suffix 'G' onwards
		FRONT COVER ASSEMBLY	1	529155	Models with Lucas 2 AC type 12 volt AC/DC generator. Up to engine suffix 'F' inclusive
		FRONT COVER ASSEMBLY	1	554843	Models with Lucas 2 AC type 12 volt AC/DC generator. From engine suffix 'G' onwards
2		Dowel locating front cover	2	6395	
3		Oil seal for front cover	1	213744	
4		Stud fixing water pump	5	252500	Up to engine suffix 'F' inclusive
5		Mud excluder	1	247766	
6		Drive screw fixing excluder	8	78001	
7		Joint washer for front cover	1	538039	
8		Joint washer at water inlet	1	538038	
9		Set bolt (5/16" UNF x 3" long)	7	256030	3 off from engine suffix 'G' onwards
10		Set bolt (5/16" UNF x 2" long) Fixing front	3	256226	7 off from engine suffix 'G' onwards
		Set bolt (5/16" UNF x 3½" long) cover and	2	256232	Models with Lucas 2 AC type 12 volt AC/DC generator
		water pump			
11		Set bolt (5/16" UNF x 3¼" long) to	2	256031	
12		Set bolt (5/16" UNF x 3¾" long) cylinder	1	256233	
13		Spring washer block	14	3075	
14		Nut (5/16" UNF)	2	254831	
		Side cover and oil filler pipe for engine, front	1	510730	Up to engine suffix 'J' inclusive
15		Side cover and oil filler pipe for engine, front	1	568542*	From engine suffix 'K' onwards
		Baffle plate for side cover, front	1	529394	Up to engine suffix 'J' inclusive
16		Baffle plate for side cover, front	1	564467*	
17		Timing pointer for distributor pump	1	564470*	
		Set bolt (10 UNF x 7/16" long) Fixing	2	257004	From engine suffix 'K' onwards
		Spring washer timing pointer	2	3073	
		Plain washer to baffle plate	2	3867	
18		Joint washer for side front cover	2	247555	
19		SIDE COVER ASSEMBLY, REAR	1	542600	
20		Stud for fuel pump	2	500792	
21		Plain washer	2	2220	Nut and split pin type fixing — Alternatives. Check before ordering
22		Nut (5/16" UNF) Fixing stud	2	254911	
23		Split pin	2	2422	
		Stud for fuel pump	2	542601	Plain type stud fixing
24		Joint washer for side cover, rear	1	247554	
25		Spring washer Fixing rear	6	3075	
26		Set bolt (5/16" UNF x 1" long) side cover plate	6	255228	
27		Spring washer Fixing front	7	3075	
28		Set bolt (5/16" UNF x ¾" long) side cover plate	7	255226	
29		Crankcase sump	1	529823	
30		Drain plug for crankcase sump	1	536577	
31		Washer for drain plug	1	243959	
32		Joint washer for crankcase sump	1	546841	

* Asterisk indicates a new part which has not been used on any previous Rover model

198

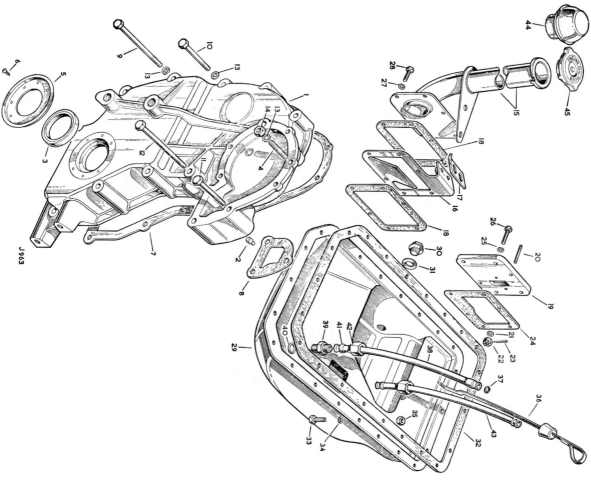

J963

Plate Ref.	1 2 3 4	DESCRIPTION		Qty	Part No.	REMARKS
33		Set bolt (⁵⁄₁₆" UNF x ⁷⁄₈" long)	Fixing sump to cylinder block	21	255227	
34		Spring washer		22	3075	
35		Nut (⁵⁄₁₆" UNF)		2	254831	
36		Oil level rod		1	274154	
37		Sealing ring for rod		1	532387	
38		Tube for oil level rod		1	504032	
39		Double-ended union		1	236060	
40		Washer		1	243958	Alternative to one-piece type oil level rod tube
41		Olive	Fixing tube to cylinder block	1	236408	
42		Union nut		1	236407	
43		Tube for oil level rod		1	541860	One-piece type. Alternative to tube with separate union. Check before ordering
44		Oil filler cap and breather filter, AC 7964819		1	546440	Up to engine suffix 'J' inclusive.
		Oil recommendation label for oil filler cap		1	272476	Not part of engine assembly
45		Oil filler cap		1	568543*	From engine suffix 'K' onwards

* Asterisk indicates a new part which has not been used on any previous Rover model

199

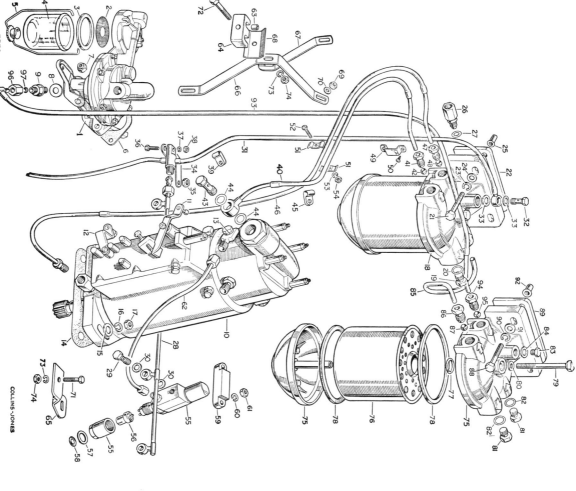

E53C

COLLINS-JONES

Plate Ref.	1 2 3 4	DESCRIPTION	Qty	Part No.	REMARKS
1		Fuel pump, mechanical, AC 7950445	1	550324	With red diaphragm
2		Filter for bowl, AC 854009	1	506796	For early type pump
3		Washer for bowl, AC 5593321	1	241225	
4		Sediment bowl, AC 1523620	1	236691	
5		Retainer for bowl, AC 5592050	1	268797	
6		Joint washer, fuel pump to cylinder block	1	275565	
7		Self-locking nut fixing fuel pump	2	252211	
8		Joint washer } For mechanical pump	2	243967	Early type
			2	247929	} fixing
9		Union } inlet and outlet	1	554527	
10		Distributor pump, CAV DPA 3248090	1	509009	
11		Accelerator control lever	1	554525	
12		Stop lever	1	219582	
13		Joint washer for injection pipe, distributor pump			
		Swivel clamp for stop lever } For	1		
		Sealing plugs for injection system } distributor	1 set	279702	For sealing unions in distributor pump
14		Sleeve for control lever stop screw, CAV 7139/184	6	247802	
15		Joint washer for distributor pump, CAV Z1/339/963	1	546282	
16		Plain washer } Fixing distributor	3	247212	
17		Spring washer } pump to	3	2920	
		Nut (5/16" UNF) } cylinder block	3	3075	
18		Fuel filter, CAV V 7005/2173	1	254831	
		Element for fuel filter, CAV 7111/296	1	517682	
19		Seal for element, small, CAV 5855/30G	1	517711	
		Seal for element, large, CAV 5339/256B	2	522937	
		Special centre bolt for filter, CAV 7111/291	1	522938	
		Washer for centre bolt, CAV 5936/188A	1	522939	
20		Plug for filter, CAV Z 7111/312	1	522940	
21		Joint washer for plug, CAV Y 5936/58T	1	517689	
22		Distance plate	1	517706	
23		Spring washer	2	256220	
24		Plain washer	2	544389	
25		Rivnut (5/16" UNF)	2	3075	
26		Non-return valve for filter	2	2249	
27		Bolt (5/16" UNF x 1¼" long)	2	501224	Not required when additional
		Joint washer for non-return valve	1	517707	fuel filter is fitted
28		Leak-off pipe complete	1	517706	
29		Banjo bolt } Fixing leak-off pipe	4	548262	
30		Washer } to injector	8	273521	} Part of engine assembly
31		Fuel pipe, spill return to injector	1	273069	
32		Banjo bolt } Fixing spill return	2	509909	
33		Joint washer } pipe to filter	4	505805	
34		Bracket for leak-off pipe	1	517976	
35		Locknut (7/16" UNF) fixing bracket to injector stud	2	521584	} Part of engine assembly
36		Bolt (2 BA x ¾" long) } Fixing spill	1	254861	
37		Plain washer } return pipe	1	250961	
		Spring washer } to bracket on	1	558190*	
38		Nut (2 BA) } injector stud	1	3073	
			1	2247	

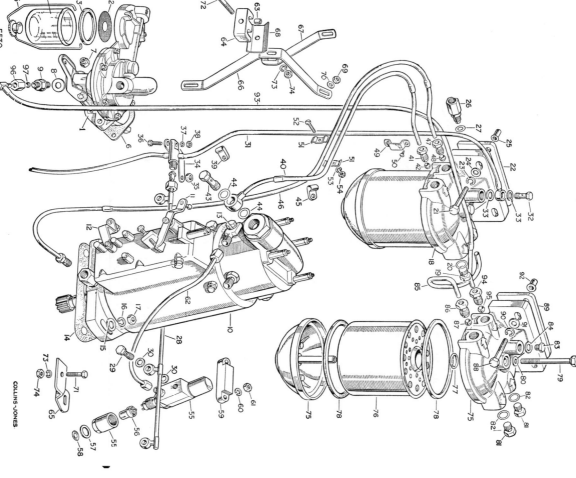

E53Q

COLLINS-JONES

Plate Ref. 1 2 3 4	DESCRIPTION	Qty	Part No.	REMARKS
39	Clip for spill return pipe	1	525168	
	Plain washer for spill pipe clip	2	10882	
	Fuel pipe, mechanical pump to filter	1	552439	
40	Olive for fuel pipe, pump end	1	557531	
	Nut } to filter	1	517690	
	Olive }	1	530966	
	Nut } Fixing pipe to mechanical pump	1	270105	Early type fixing
	Olive }	1	270115	Early type fixing
	Fuel pipe, filter to distributor pump	1	517685	Not required when additional fuel filter is fitted
41	Nut } Fixing pipe to filter	1	517690	
	Olive }	1	530966	
42	Nut } Fixing pipe to filter	1	517690	
	Olive }	1	530966	
	Banjo bolt	1	275266	
43	Clip, fixing fuel pipe to distributor pump	1	8885	
44	Joint washer distributor pump	2	275265	
45	Clip, fixing pipe to distributor pump	1	530966	
46	Fuel pipe, distributor pump return to filter	1	517686	
47	Nut } Fixing pipe to filter	1	517690	
	Olive }	1	530966	
48	Nut } Fixing pipe to filter	1	517684	
	Olive }	1	78006	
49	Clip, fixing fuel pipe to dash	2	243395	
50	Drive screw, fixing clip	2	234603	
51	Double pipe clip (fixing distributor pump feed and return pipes together)	2	3073	
52	Bolt (2 BA x ½" long) } Fixing pipes and clips	1	2247	
53	Spring washer }	1		
54	Nut (2 BA) }	1		
55	Injector complete, CAV 5345501	4	515552	
56	Nozzle for injector	4	247726	
57	Joint washer for injector	4	247179	
58	Joint washer for injector, copper	4	272474	
59	Clamping strip for injector	4	247213	Stepped type, alternative to plain type. Check before ordering
	Clamping strip for injector	4	531897	Plain type, alternative to stepped type. Check before ordering
60	Spring washer } Fixing injector	8	3075	
61	Nut (5/16" UNF) } to cylinder head	8	254831	
	Injector pipe to No. 1 cylinder	1	513926	
	Injector pipe to No. 2 cylinder	1	513928	
	Injector pipe to No. 3 cylinder	1	513929	
	Injector pipe to No. 4 cylinder	1	513927	
62	Damper for injector pipe	4	277838	
63	Damper for injector pipe	4	277839	Damper and shroud type fixing. Check before ordering
64	Shroud for damper	2	541229	Clamp and shroud type fixing. Check before ordering. Alternatives.
	Clamping plate for injector pipe grommet	4	272512	Clamp and grommet type fixing
	Grommet for injector pipe	4	277816	
65	Bracket for damper shroud	1	278002	Up to engines numbered 27105814
66	Support strap for shroud	1	278021	
67	Steady strap for shroud	1	278021	Not required with clamp and grommet type fixing
68	Backplate for shroud	2	277815	Part of engine assembly

* Asterisk indicates a new part which has not been used on any previous Rover model

E550

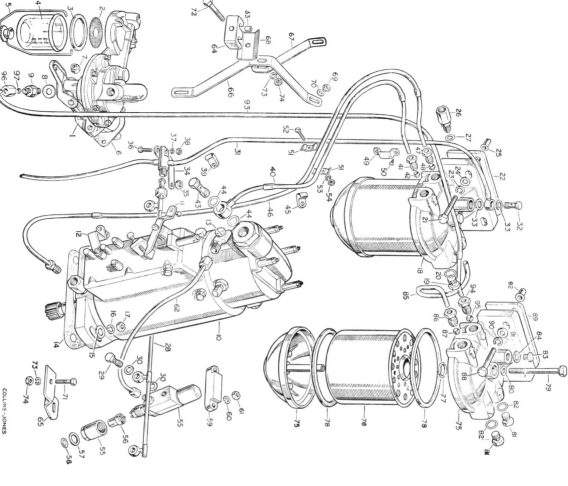

COLLINS-JONES

FUEL INJECTION SYSTEM, DIESEL ENGINE, 2¼ LITRE, Series IIA
Up to engine suffix 'J' inclusive

FUEL INJECTION SYSTEM, DIESEL ENGINE, 2¼ LITRE, Series IIA
Up to engine suffix 'J' inclusive

Plate Ref.	1 2 3 4	DESCRIPTION	Qty	Part No.	REMARKS
69		Locknut (5/16" UNF) ⎫ Fixing straps to	4	254861	
70		Plain washer ⎬ injector studs	4	2220	
71		Bolt, front (5/16" UNF x 1¼" long)	1	256220	Up to engines numbered 27105814
72		Bolt, rear (5/16" UNF x 1½" long)	1	256220	Up to engines numbered 27105814
73		Spring washer ⎫ Fixing shrouds and dampers to backplate, strap and support bracket	2	3075	
74		Nut (5/16" UNF) ⎭	2	254831	
75		Spring washer ⎫ Fixing shrouds and dampers to backplate	2	3075	
76		Nut (5/16" UNF) ⎭	2	254831	
77		Bolt (1/4" UNF x 1¼" long) ⎫ Fixing clamping plates and	2	255005	From engines numbered 27105815 onwards
78		Bolt (1/4" UNF x 7/16" long) ⎭	2	3074	
79		Spring washer ⎫ grommets	2	254810	
80		Nut (1/4" UNF)	2	254810	
81		Bolt (1/4" UNF x ½" long) ⎫ Fixing bracket for damper shroud	2	255204	Check before ordering
82		Plain washer	1	254810	Up to engines numbered 27105814
83		Spring washer ⎫ to oil filler pipe	1	3946	
84		Nut (1/4" UNF) ⎭	1	3074	
85		ADDITIONAL FUEL FILTER, COMPLETE ASSEMBLY	1	254810	
86		Fuel filter, CAV type FS	1	522756	
87		Element for fuel filter, CAV 7111/296	2	517682	
88		Seal for element, small, CAV 5855/30G	2	517711	
89		Seal for element, large, CAV 5539/256B	2	522937	
90		Special centre bolt for filter, CAV 7111/291	1	522938	
91		Washer for centre bolt, CAV 5936/188A	1	522939	
92		Plug for fuel filter	2	522940	
93		Joint washer for plug	2	517689	
94		Plug for fuel filter, top, leak-off, CAV Z7111/311	1	517706	
95		Joint washer for leak-off plug	1	517855	
96		Transfer pipe, extra filter to basic filter	1	517976	
97		Nut ⎫ Fixing pipe	2	544391	
		Olive ⎭ to filter	2	544391	
73		Bolt (5/16" UNF x 1¼" long) ⎫ Fixing	2	517690	Early type models
74		Nut ⎭ filter	2	530966	
		Distance plate	1	256220	
		Spring washer	2	544389	
		Plain washer	2	3075	
		Rivnut (5/16" UNF)	2	3830	
		Fuel pipe, mechanical pump to twin filters	1	501224	Standard on Export models. Optional equipment on Home models
		Olive for fuel pipe, pump end ⎫	2	552458	
		Nut ⎬ Fixing pipe	2	557531	
		Olive ⎭ to filter	2	517690	
		Nut ⎫ Fixing pipe to dash	1	530966	
		Olive ⎭	1	270015	Early type fixing
		Nut ⎫ Fixing pipe to	1	270105	Early type fixing
		Olive ⎭ mechanical pump	1	600904	
		Repair kit for mechanical fuel pump, AC 7950706	1	600905	
		Overhaul kit for mechanical fuel pump, AC 7950707	1		

Part of engine assembly

Alternative fixings.

* Asterisk indicates a new part which has not been used on any previous Rover model

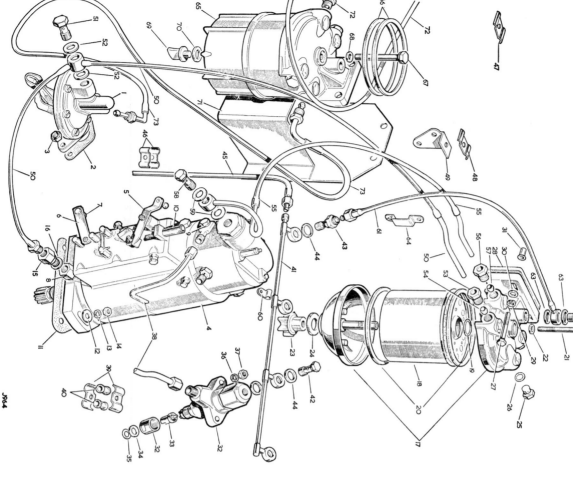

J964

FUEL INJECTION SYSTEM, DIESEL ENGINE, 2¼ LITRE, Series IIA
From engine suffix 'K' onwards

Plate Ref. 1 2 3 4	DESCRIPTION	Qty	Part No.	REMARKS
1	Fuel pump, mechanical, AC 7971192 ...	1	563146	
2	Joint washer, fuel pump to cylinder block ...	1	275565	
3	Self-locking nut, fixing fuel pump ...	2	252211	
4	Distributor pump, CAV DPA 3248760 ...	1	564495*	
5	Accelerator control lever ...	1	509009	
6	Stop lever ...	1	563525*	
7	Swivel clamp for stop lever ...	1	565526*	
8	Union, fuel pipe connection ...	1	566617*	
9	Joint washer for injection pipe, distributor pump	3	254831	
10	Sleeve for control lever stop screw, CAV 7139/184	1 set	279702	For sealing unions in distributor pump
11	Joint washer for distributor pump, CAV Z1/339/963	6	247802	
12	Plain washer ...	1	546282	} For distributor pump
13	Spring washer ...	3	247212	
14	Nut (5/16" UNF) ...	3	2920	
end		3	3075	
15	Non-return valve for distributor pump	1	564909*	
16	Joint washer for non-return valve	1	517706	
17	Fuel filter, CAV	1	563190	
18	Element for fuel filter, CAV 7111/296 ...	1	517711	
19	Seal for element, small, CAV 5855/30G ...	1	522937	} Fixing filter
20	Seal for element, large, CAV 5339/256B ...	2	522938	} to dash
21	Special centre bolt for filter, CAV 7111/291 ...	1	522939	
22	Washer for centre bolt, CAV 5936/188A ...	1	522940	
23	Nylon drain plug for filter, CAV ...	1	605012	
24	Rubber seal for drain plug, CAV ...	1	605013	
25	Plug for filter, CAV Z 7111/312 ...	2	517689	
26	Joint washer for plug, CAV Y 5936/58T ...	2	517706	
27	Bolt (5/16" UNF x 1 1/8" long) ...	2	255029	
28	Distance plate ...	1	544389	
29	Spring washer ...	2	3075	
30	Plain washer ...	2	3830	
31	Rivnut (7/16" UNF) ...	2	501224	
32	Injector complete, CAV D 5385001 ...	4	564332*	
33	Nozzle for injector ...	4	247726	
34	Joint washer for injector, copper ...	4	247179	
35	Joint washer for injector, steel ...	4	272474	
36	Spring washer) Fixing injectors to	8	3075	
37	Nut (5/16" UNF) } cylinder head studs	8	254831	
38	Injector pipe to No. 1 cylinder ...	1	563165*	
	Injector pipe to No. 3 cylinder ...	1	563167*	Part of
	Injector pipe to No. 4 cylinder ...	1	563168*	engine
	Injector pipe to No. 2 cylinder ...	1	563166*	assembly
39	Clamping plate for injector pipe grommet ...	4	541229	
40	Grommet for injector pipe) Fixing clamping	4	272512	
	Spring washer } plates and	2	3074	
	Nut (¼" UNF)) grommets	2	254810	

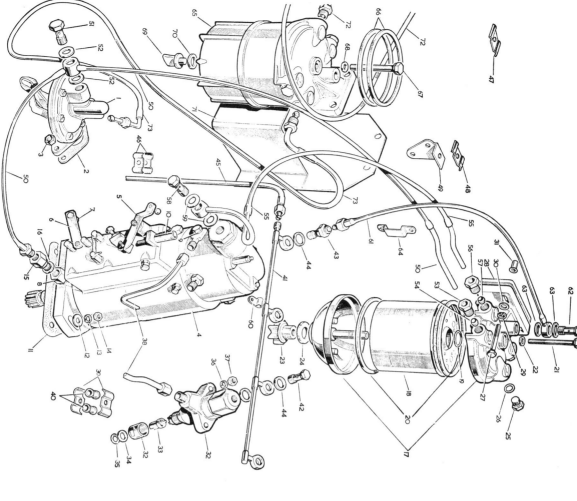

J964

FUEL INJECTION SYSTEM, DIESEL ENGINE, 2¼ LITRE, Series IIA
From engine suffix 'K' onwards

Plate Ref.	1 2 3 4	DESCRIPTION		Qty	Part No.	REMARKS
41		Spill rail pipe complete		1	564386*	
42		Banjo bolt for No. 1, 2 and 3 injectors	⎫ Fixing spill	3	273521	
43		Banjo bolt for No. 4 injector	⎬ rail pipe to	1	563195*	⎫ Part of engine assembly
44		Joint washer for banjo bolt	⎭ injectors	8	273069	⎭
45		Fuel pipe, spill return to tank		1	564907*	88 and 109
		Fuel pipe, spill return to tank		1	564962*	109 Station Wagon
		Fuel pipe, tank to mechanical pump ...		1	564898	88 and 103
		Fuel pipe, tank to mechanical pump ...		1	552436	109 Station Wagon
46		Double clip		2	243395	⎫ Except when
		Double clip		1	234603	88 and 109
		Double clip		1	250969	109 Station Wagon ⎭ sediment or is fitted
		Bolt (2 BA x ½" long)	⎫ Clamping	1	3073	
		Bolt (2 BA x 2" long)	⎬ feed and return	1	2247	
		Spring washer	⎬ pipes together			
47		Nut (2 BA)	⎭			
		Double clip		1	509412	
		Double clip	⎫ Fixing feed and return pipes	1	72626	88 and
		Drive screw	⎬ to chassis sidemember	1	509415	109
		Double clip	⎭	3	72626	Station Wagon
		Drive screw		3	509415	Station Wagon
48		Double clip for feed and return pipes	...	3	509412	
49		Bracket for clip		1	270297	
		Bolt (2 BA x ½" long)	⎫ Fixing clip and	3	234603	
		Plain washer	⎬ bracket to	3	3902	
		Spring washer	⎬ chassis crossmember	3	3073	
		Nut (2 BA)	⎭	3	2247	
		Double clip	⎫ Fixing pipes to	1	509412	
		Bracket for clip	⎬ check strap mounting	1	552118	
		Bolt (2 BA x ⅝" long)	⎬ in chassis	1	250960	109 Station Wagon
		Spring washer		1	3073	
		Nut (2 BA)		1	2247	
50		Fuel pipe, mechanical pump and distributor pump to filter ...		1	564899*	
51		Banjo bolt ⎫ Fixing fuel pipe		1	564889*	
52		Joint washer ⎬ to mechanical pump		2	231576	
53		Nut ⎫ Fixing pipe		1	517690	
54		Olive ⎬ to filter		1	530566	
55		Fuel pipe, filter to distributor pump		1	564902*	
56		Nut ⎫ Fixing pipe		1	517690	
57		Olive ⎬ to filter		1	530966	
58		Banjo bolt ⎫ Fixing pipe to		1	275266	
59		Joint washer ⎬ distributor pump		2	275265	
60		Clip ⎫ Fixing fuel		1	8885	
		Spring washer ⎬ pipe to		1	3073	
		Nut (10 UNF) ⎭ distributor pump		1	257023	

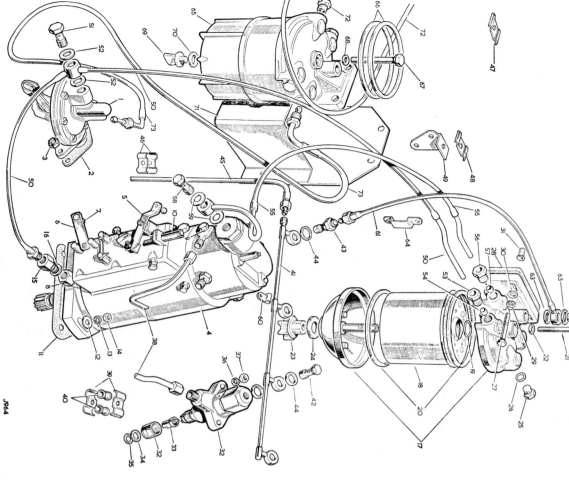

J964

FUEL INJECTION SYSTEM, DIESEL ENGINE, 2¼ LITRE, Series IIA
From engine suffix 'K' onwards

Plate Ref. 1 2 3 4	DESCRIPTION	Qty	Part No.	REMARKS
61	Fuel pipe filter to spill rail at No. 4 injector	1	564905*	
62	Banjo bolt } Fixing fuel pipe	1	505805	
63	Joint washer } to filter	2	517976	
64	Double clip } Fixing fuel pipes	1	517684	
	Drive screw } to bulkhead	2	78006	
	SEDIMENTOR COMPLETE ASSEMBLY	1	605560*	88 and 109
	SEDIMENTOR COMPLETE ASSEMBLY	1	605561*	109 Station Wagon
65	Sedimentor	1	562748	
66	Seal for sedimentor	2	522938	
67	Special centre bolt for sedimentor ...	1	522939	
68	Washer for centre bolt	1	522940	
69	Drain plug for sedimentor	1	605010	
70	Rubber seal for drain plug	1	605011	
71	Mounting bracket for sedimentor	1	569585*	
72	Bolt (⅜" UNF x 1" long) } Fixing sedimentor	2	255247	
	Plain washer } to	2	3833	
	Spring washer } mounting bracket	2	3076	
	Nut (⅜" UNF) }	2	254812	
73	Screw (10 UNF x ¾" long) } Fixing mounting	4	78384	
	Plain washer } bracket to chassis	4	3816	
	Spring washer }	4	3073	
	Rivnut (10 UNF)	4	532848	
72	Plain pipe, tank to sedimentor	1	564936*	88 and 109
	Fuel pipe, tank to sedimentor	1	564963*	109 Station Wagon
73	Fuel pipe, sedimentor to mechanical pump ...	1	564937*	
	Double clip } Fixing fuel pipe to	1	509412	109 Station
	Drive screw } chassis sidemember	1	72626	Wagon
	Repair kit for mechanical fuel pump, AC 7950706 ...	1	600904	
	Overhaul kit for mechanical fuel pump, AC 7950707	1	600905	

Standard on
Export models.
Optional
equipment on
Home models

* Asterisk indicates a new part which has not been used on any previous Rover model

FUEL INJECTION SYSTEM, DIESEL ENGINE, 2¼ LITRE, Series IIA
From engine suffix 'K' onwards

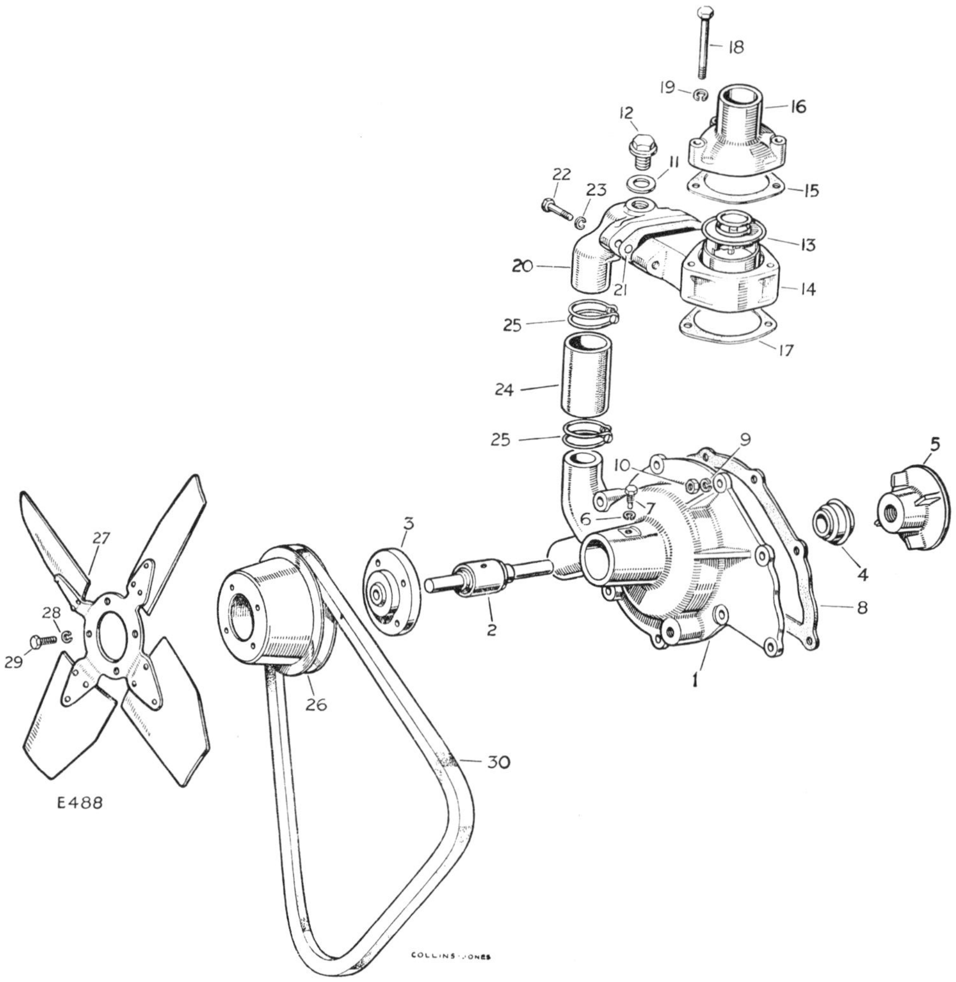

WATER PUMP AND THERMOSTAT, DIESEL ENGINE, 2¼ LITRE, Series IIA

Plate Ref. 1 2 3 4	DESCRIPTION	Qty	Part No.	REMARKS
	WATER PUMP ASSEMBLY	1	530477	
1	Water pump casing	1	530478	
2	Pump spindle and bearing	1	523354	
3	Hub for fan	1	566688	
4	Carbon ring and seal unit	1	568301	
5	Impeller for pump	1	247916	
6	Spring washer } Locating bearing casing	1	3074	
7	Special set bolt } Locating bearing casing	1	247078	
8	Joint washer for water pump	1	538671	
9	Spring washer } Fixing water pump to cylinder block	5	3074	
10	Nut (¼" UNF) } Fixing water pump to cylinder block	5	254810	Up to engine suffix 'F' inclusive
	Set bolt (¼" UNC x 1" long) } Fixing water pump to cylinder block	5	253009	From engine suffix 'G' onwards
11	Joint washer } For heater return in thermostat by-pass pipe	1	243959	
12	Plug (⅜" BSP) } For heater return in thermostat by-pass pipe	1	536577	
13	Thermostat, bellows type	1	504736	Up to engine suffix 'B'
	Thermostat, wax type	1	532453	From engine suffix 'C' onwards
	'O' ring for thermostat	1	527235	From engine suffix 'C' onwards
14	Thermostat housing	1	516059	
15	Joint washer for thermostat housing, upper	1	511957	Up to engine suffix 'B'
	Joint washer for thermostat housing, upper	1	527110	From engine suffix 'C' onwards
16	Water outlet pipe, thermostat to radiator	1	511956	Up to engine suffix 'B'
	Water outlet pipe, thermostat to radiator	1	527109	From engine suffix 'C' onwards
17	Joint washer for thermostat housing, lower	1	247874	
18	Set bolt, ¼" UNF x 2½" long } Fixing thermostat housing and outlet pipe to cylinder head	3	256209	
19	Spring washer } Fixing thermostat housing and outlet pipe to cylinder head	3	3074	
20	Thermostat by-pass pipe	1	530476	
21	Joint washer for by-pass pipe	1	511958	
22	Set bolt (5/16" UNF x 1" long) } Fixing by-pass pipe to thermostat housing	2	255228	
23	Spring washer } Fixing by-pass pipe to thermostat housing	2	3075	
24	Hose for by-pass pipe	1	273789	
25	Hose clip for by-pass pipe	2	603894	
26	Fan pulley	1	530890	
27	Fan blade	1	515090	
28	Spring washer } Fixing fan blade and pulley to hub	4	3074	
29	Set bolt (¼" UNF x ¾" long) } Fixing fan blade and pulley to hub	4	255207	
30	Fan and dynamo belt	1	550224	
	Water pump overhaul kit	1	530590	

* Asterisk indicates a new part which has not been used on any previous Rover model

COLLINS-JONES

L65

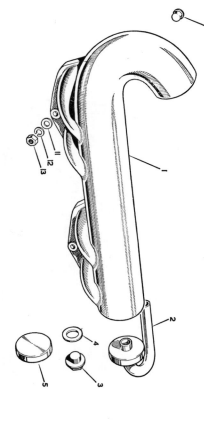

MANIFOLDS, DIESEL ENGINE, 2¼ LITRE, Series IIA

Plate Ref. 1 2 3 4	DESCRIPTION	Qty	Part No.	REMARKS
1	Inlet manifold	1	550263	Up to engine suffix 'J' inclusive
	Inlet manifold	1	574661 ✱	From engine suffix 'K' onwards
2	Connecting tube and plug, hose to manifold	1	574656 ✱	
3	Plug for inlet manifold	1	10713	Hexagon head, brass type plug
4	Joint washer for plug	1	243960	
5	Plug for inlet manifold, ⅞" outside diameter	1	524765	Cup type, Early models.
6	Plug for inlet manifold, 1¼" outside diameter	1	525428	steel plug Alternatives.
7	Plug for redundant hole at manifold elbow	1	574664 ✱	Check before ordering
	EXHAUST MANIFOLD ASSEMBLY	1	247651	Up to engine suffix 'C'
	EXHAUST MANIFOLD ASSEMBLY	1	536514	From engine suffix 'D' onwards
8	Stud for exhaust pipe	4	252621	
9	Joint washer for inlet and exhaust manifold	1	247101	
10	Clamp for manifold	5	247836	Fixing inlet
11	Plain washer	4	2550	and exhaust manifold
12	Spring washer	9	3075	
13	Nut (⁵⁄₁₆" UNF)	9	254831	

✱ Asterisk indicates a new part which has not been used on any previous Rover model

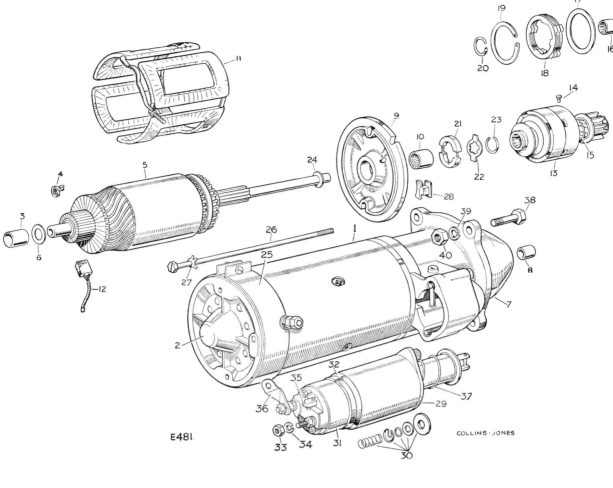

E481.   COLLINS-JONES

STARTER MOTOR AND SOLENOID, DIESEL ENGINE, 2¼ LITRE, Series IIA
Up to engine suffix 'J' inclusive

Plate Ref.	1 2 3 4	DESCRIPTION	Qty	Part No.	REMARKS
1		STARTER MOTOR COMPLETE, LU 26189D	1	529971	
2		Bracket, commutator end, LU 54250082	1	532568	
3		Bush, commutator end, LU 255491	1	242958	
4		Spring set for brushes, LU 54259524	1	532569	
5		Armature, LU 54251818	1	532570	
6		Thrust washer, commutator end, LU 152380	1	279007	
7		Bracket for starter, drive end, LU 54252532	1	532571	
8		Bush for bracket, LU 54252236	1	532572	
9		Field coil for starter, set, LU 272394	1	601689	
10		Bush for brake bracket, LU 54252494	1	532574	
11		Bracket for starter motor, LU 54252800	1	532573	
12		Bush for starter, LU 272548	1	279010	
13		Pivot pin for starter motor, LU 272394	1	270225	
14		Drive assembly for starter motor, LU 292124	1	279011	
15		Return spring for starter pinion, LU 271735	1	601760	
16		Bush, pinion end, LU 292053	1	526258	
17		Retaining ring complete for pinion, LU 295298	1	532575	
		Rivet for pinion retaining ring LU 291887	2	270238	
18		Clutch plates, set, LU 291540	1	270012	
19		Circlip retaining clutch assembly, LU 291560	1	270235	
20		Lock ring retaining clutch plates, LU 291862	1	270239	
21		Brake shoe complete with springs, LU 291374 As reqd	1	279013	
22		Clutch adjusting shim (0.006"), LU 291378 As reqd		279014	
23		Clutch adjusting shim (0.005"), LU 272701		279015	
		Clutch adjusting shim (0.004"), LU 291379 As reqd		270241	
		Driving washer for brake shoe, LU 272340	1	279016	
24		Lock ring retaining brake, LU 272337	1	279018	
25		Cover band for starter, LU 272396	1	605199	
26		Bolt for starter motor, LU 54257814	2	532577	
27		Lockwasher for bolt, LU 272273	2	279021	
28		Rubber grommet in drive end and bracket, LU 272552	1	532578	
29		Solenoid for starter motor, LU 76490	1	532580	
30		Contact plate for solenoid, LU 54330168	1	519004	
31		Base for solenoid, LU 54642436	1	532579	
32		Gasket for starter solenoid base, LU 765842	1	519006	
33		Terminal nut for starter solenoid, LU 156456	4	519008	
34		Terminal washer for starter solenoid LU 185062	4	519009	
35		Terminal screw for solenoid, LU 121610	1	519010	
36		Terminal connector for starter motor, LU 272555	1	519011	
37		Plunger spring for solenoid, LU 765866	1	270251	
		Sundry parts kit, LU 271731	1	253068	
38		Bolt (7/16" UNC x 1¼" long)	1	256064	
39		Spring washer	3	3077	
40		Nut (7/16" UNF)	2	254823	
		STARTER MOTOR COMPLETE, LU 26259	1	560870*	Waterproof type.
		Set bolt (7/16" UNF x 2" long)		560874*	Optional equipment.
		Terminal eyelet for starter cable		560875*	Up to engine suffix 'J' inclusive.
		Insulating boot for starter terminal, LU 54931746	1	575082*	Fitted as standard from engine suffix 'K' onwards.
		Clip for starter cable			See page 219 for details of components

* Asterisk indicates a new part which has not been used on any previous Rover model

STARTER MOTOR AND SOLENOID, DIESEL ENGINE, 2¼ LITRE, Series IIA
From engine suffix 'K' onwards

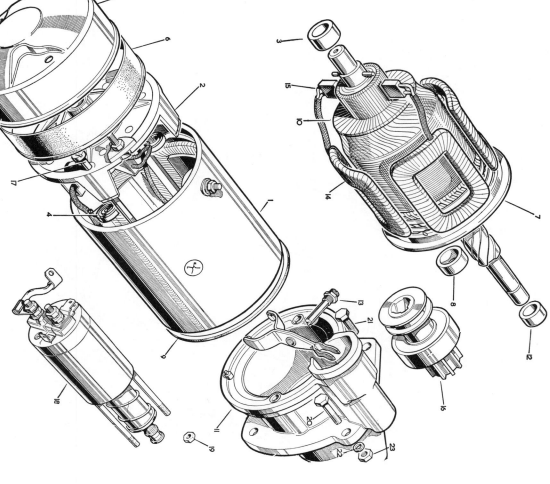

STARTER MOTOR AND SOLENOID, DIESEL ENGINE, 2¼ LITRE, Series IIA
From engine suffix 'K' onwards

Plate Ref.	1 2 3 4	DESCRIPTION	Qty	Part No.	REMARKS
1		STARTER MOTOR COMPLETE, LU 26259	1	560870*	
2		Bracket, commutator end, LU 54258084	1	605613*	
3		Bush, commutator end, LU 54256132	1	605614*	
4		Spring set for brushes, LU 54259524	1	532569	
5		Cover for starter, commutator end, LU 54259704	1	605612*	
6		Sealing ring for cover, commutator end, LU 54240484	1	605701*	
7		Intermediate bracket, LU 54258129	1	605619*	
8		Bush for bracket, LU 54256151	1	605620*	
9		Sealing ring for intermediate bracket, LU 54241202	1	605702*	
10		Armature, LU 54255952	1	605621*	
11		Bracket for starter, drive end, LU 54242247 ..	1	605617*	
12		Bush for bracket, LU 54255966	1	605618*	
13		Pivot pin for starter motor, LU 54257740 ..	1	605615*	
14		Field coil for starter, LU 272394	1	279010	
15		Brushes for starter motor, set, LU 271735 ..	1	270225	
16		Drive (roller clutch) for starter, LU 54259688	1	605616*	
17		Bolt for starter motor, LU 272660	2	279019	
18		Solenoid for starter motor, LU 76800	1	605703*	
		Spindle and contact plate for starter solenoid, LU 54641902	1	605653	
19		Base complete for starter solenoid, LU 54642190	1	605704*	
		Special nut for starter solenoid, LU 156396 ..	2	605705*	
20		Sundry parts kit, LU 54240082	1	605622*	
21		Bolt (⁷⁄₁₆" UNC x 1¼" long)	1	253068	
22		Set bolt (⁷⁄₁₆" UNF x 2" long) Fixing	1	256064	
23		Spring washer starter motor	3	3077	
		Nut (⁷⁄₁₆" UNF)	2	254823	

* Asterisk indicates a new part which has not been used on any previous Rover model

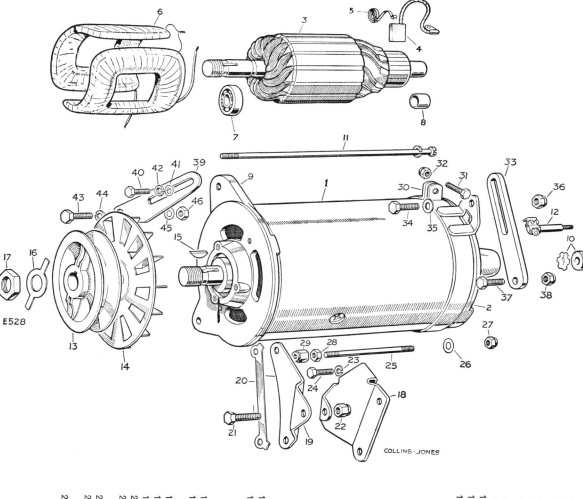

E528

COLLINS · JONES

DYNAMO AND FIXINGS, DIESEL ENGINE, 2¼ LITRE, Series IIA

Plate Ref.	1 2 3 4	DESCRIPTION	Qty	Part No.	REMARKS
1		DYNAMO COMPLETE, C40/1, LU 22700	1	551569	
2		Bracket, commutator end, LU 5421125	1	514195	
3		Armature for dynamo, LU 5421499	1	514194	
4		Brushes for dynamo, set, LU 227541	1	514193	
5		Spring set for brushes, LU 227542	1	514192	
6		Field coil for dynamo, LU 227543	1	514191	
7		Ball bearing, front	1	529221	
8		Bush, commutator end, LU 227818	1	271614	
9		Bracket, drive end, LU 5421095	1	514190	
10		Oiler for dynamo, LU 5421097	1	514189	
11		Bolt for bracket, LU 228336	2	242675	
12		Terminal, LU 5421687B	1	532566	
		Sundry parts, set, LU 5421894	1	532567	
		DYNAMO COMPLETE, C40/T, LU 22764	1	551360	Early type
		Dynamo complete, LU 22756, type C40-T	1	605963 *	Late type
		Bracket, commutator end, LU 5421363	1	605055	For dynamo LU 22764
		Cover for shaft, commutator end, LU 54216152	1	605056	For dynamo LU 22764
		Ball bearing, sealed, commutator end, LU 54160219	1	605057	
		Armature for dynamo, LU 5421628	1	605058	For dynamo LU 22764
		Armature for dynamo, LU 5421417	1	605969 *	For dynamo LU 22756
		Brushes for dynamo, set, LU 5421361	1	605059	
		Retaining clip for brush, LU 54215103	2	605060	
		Spring set for brushes, LU 54216362	1	605061	
		Field coil for dynamo, LU 227543	1	514191	
		Ball bearing, sealed, drive end, LU 54160139	1	605062	
		Bracket, drive end, LU 54216225	1	605063	
		Bolt for bracket, LU 228336	2	242675	
		Sundry parts, set, LU 5421894	1	532567	
13		Pulley for dynamo	1	522913	
14		Fan for dynamo	1	554055	
		Fan for dynamo	1	533860	For use with sealed bearing type dynamo
		Distance washer for fan	2	550336	
		Distance washer for fan	1	550281	bearing type dynamo
15		Woodruff key	1	1664	
16		Lockwasher ⎫ Fixing pulley	1	217781	
17		Special nut ⎭	1	547604	
18		Spring washer ⎫ Fixing dynamo	1	3466	
19		Support bracket for dynamo to dynamo	1	277999	
20		Steady bracket for dynamo support bracket	1	278000	
21		Locking plate for steady bracket	1	501198	Optional equipment
22		Bolt (5/16" UNF x 3/8" long) ⎫ Fixing dynamo	1	256421	
23		Self-locking nut (5/16" UNF) ⎫ to support and	1	252211	Optional equipment
24		Spring washer ⎬ steady brackets	2	3075	
		⎫ Fixing support bracket to			
		Set bolt (5/16" UNF x 5/8" long) ⎬ cylinder block	2	255225	

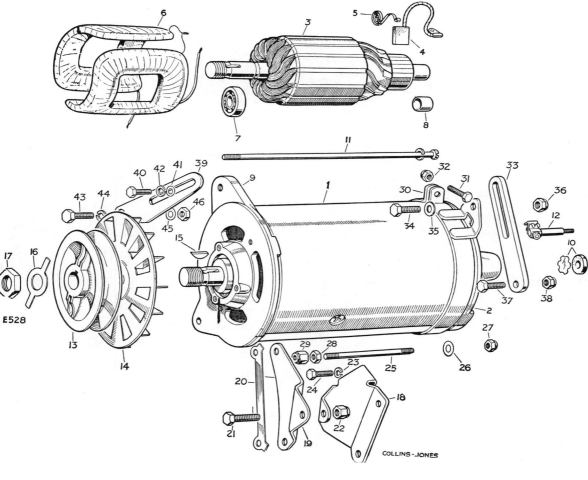

COLLINS-JONES

E528

DYNAMO AND FIXINGS, DIESEL ENGINE, 2¼ LITRE, Series IIA

Plate Ref.	1 2 3 4	DESCRIPTION	Qty	Part No.	REMARKS
25		Special stud fixing dynamo	1	526225	
26		Shim washer	As reqd	4148	
27		Self-locking nut (5/16" UNF) } Fixing dynamo	1	252211	
28		Nut (5/16" UNF) } to special stud at rear	1	254821	
29		Self-locking nut (5/16" UNF) Fixing dynamo to special stud at front	1	252211	
30		Strap for dynamo, rear, diameter 3 1/16"	1	536679	Alternatives. Check before ordering
31		Bolt (5/16" UNF x 1⅜" long) } Fixing strap	1	256221	
32		Self-locking nut (5/16" UNF) } to dynamo	1	252211	
33		Adjusting link for dynamo, rear	1	524210	
34		Bolt (5/16" UNF x ⅞" long) } Fixing adjusting	1	255027	
35		Plain washer } link to dynamo	1	2266	
36		Self-locking nut (5/16" UNF) }	1	252211	
37		Bolt (5/16" UNF x ⅞" long) Fixing adjusting link to engine	1	255027	
38		Self-locking nut (5/16" UNF) mounting foot	1	252211	
39		Adjusting link for dynamo, front	1	514472	
40		Special set bolt } Fixing front	1	253027	
41		Plain washer } adjusting link	1	2550	
42		Spring washer } to dynamo	1	3075	
43		Set bolt (5/16" UNC x ¾" long) } Fixing adjusting	1	253026	
44		Spring washer } link to front cover	1	3075	
45		Distance washer }	1	2266	
46		Nut (5/16" UNF)	1	254831	

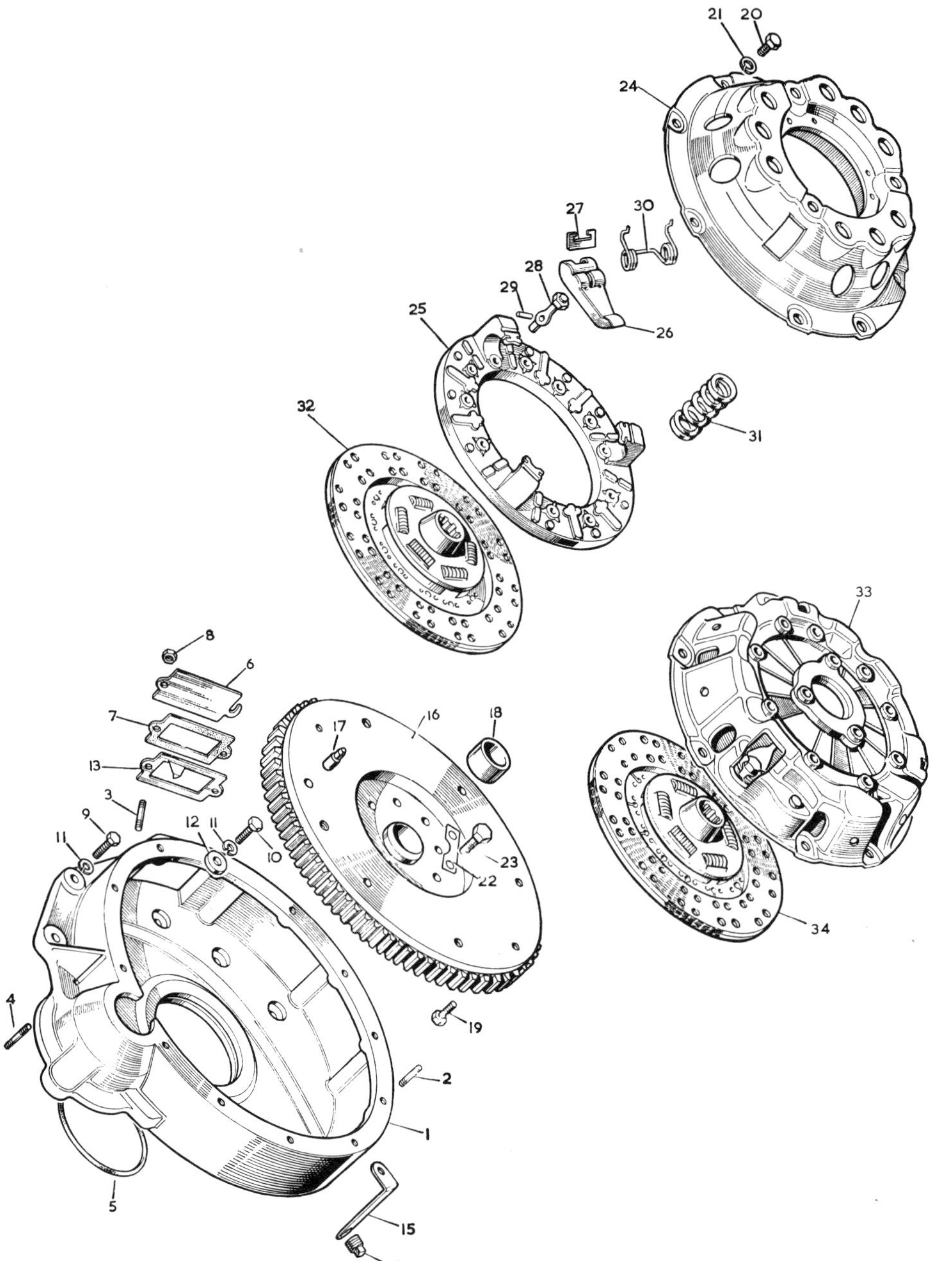

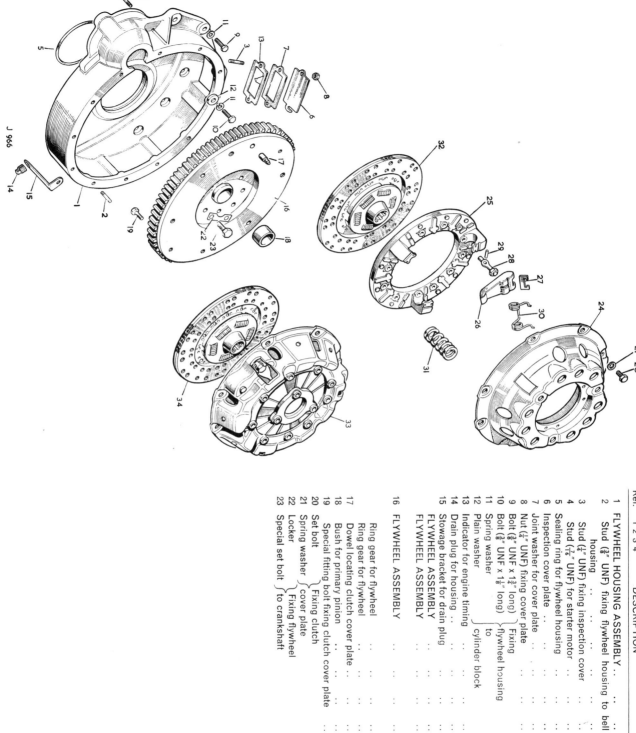

J 966

FLYWHEEL AND CLUTCH, DIESEL ENGINE, $2\frac{1}{4}$ LITRE, Series IIA

Plate Ref.	1 2 3 4	DESCRIPTION	Qty	Part No.	REMARKS
1		FLYWHEEL HOUSING ASSEMBLY	1	515086	
2		Stud ($\frac{3}{8}$" UNF) fixing flywheel housing to bell housing	12	247145	
3		Stud ($\frac{1}{4}$" UNF) fixing inspection cover	2	247146	
4		Stud ($\frac{7}{16}$" UNF) for starter motor	1	277532	
5		Sealing ring for flywheel housing	1	246169	
6		Inspection cover plate	1	56140	
7		Joint washer for cover plate	2	50216	
8		Nut ($\frac{1}{4}$" UNF) fixing cover plate	2	254810	
9		Bolt ($\frac{3}{8}$" UNF x $1\frac{3}{4}$" long) } Fixing flywheel housing to cylinder block	2	256043	
10		Bolt ($\frac{3}{8}$" UNF x $1\frac{1}{8}$" long) } Fixing flywheel housing to cylinder block	6	255248	
11		Spring washer } Fixing flywheel housing to cylinder block	8	3076	
12		Plain washer } Fixing flywheel housing to cylinder block	6	2219	
13		Indicator for engine timing	1	500597	
14		Drain plug for housing	1	3290	
15		Stowage bracket for drain plug	1	276511	
		FLYWHEEL ASSEMBLY	1	546519	For 9" type clutch
		FLYWHEEL ASSEMBLY	1	546518	Steel flywheel } Alternatives } For $9\frac{1}{2}$" diaphragm spring type clutch
16		FLYWHEEL ASSEMBLY	1	566851*	Cast iron flywheel } Alternatives } For $9\frac{1}{2}$" diaphragm spring type clutch
		Ring gear for flywheel	1	510489	For steel flywheel } Check before ordering
		Ring gear for flywheel	1	568431*	For cast iron flywheel } Check before ordering
17		Dowel locating clutch cover plate	2	502116	3 off on vehicles with $9\frac{1}{2}$" clutch
18		Bush for primary pinion	1	08566	
19		Special fitting bolt fixing clutch cover plate ..	6	247166	For steel flywheel } Check before ordering
20		Set bolt } Fixing clutch cover plate	6	255427	For cast iron flywheel } Check before ordering
21		Spring washer } Fixing clutch cover plate	6	3075	For cast iron flywheel } Check before ordering
22		Locker } Fixing flywheel to crankshaft	4	526161	
23		Special set bolt } Fixing flywheel to crankshaft	8	247135	

* Asterisk indicates a new part which has not been used on any previous Rover model

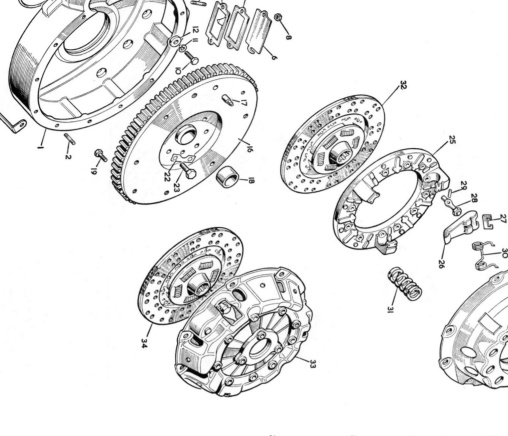

J 966

FLYWHEEL AND CLUTCH, DIESEL ENGINE, 2¼ LITRE, Series IIA

Plate Ref.	1 2 3 4	DESCRIPTION	Qty	Part No.	REMARKS
24		CLUTCH ASSEMBLY, BB 45693/41	1	236684	9" type clutch with black clutch springs
24		Cover plate for clutch, BB 45481 ...	1	231888	
25		Pressure plate for clutch, BB 42652 ...	1	231880	
26		Release lever for clutch, BB 51022 ...	3	534358	For 9" type clutch
27		Strut for release lever, BB 42606 ...	3	231884	
28		Eyebolt and nut for release lever, BB 48508	3	242996	
29		Pin for release lever, BB 42604 ...	3	231885	
30		Anti-rattle spring for release lever, BB 47688	3	231883	
31		Clutch spring (yellow and light green), BB 44780	9	243593	For early type 9" clutch
31		Clutch spring (black), BB 44633	9	275301	Part of 236684. For late type 9" clutch
32		Clutch plate complete, BB 47626/67	1	601294	Check before ordering
		Lining package for clutch plate, BB KL75012	1	517026	Mintex lining / Alterna- / For 9" type clutch
33		CLUTCH ASSEMBLY, BB 75698/28	1	571228	9½" diaphragm spring type clutch. America Dollar Area. Optional equipment for all other areas up to engine suffix 'H' inclusive. Standard equipment from engine suffix 'J' onwards
		Lining package for clutch plate, BB 46627 ...	1	261921	Ferodo lining / tives
34		Clutch plate complete BB 53009/38	1	571712	For 9½" type clutch
		Lining package for clutch, BB SCC75925	1	605997	
		Self-locking nut (5/16" UNF)	6	252211	Fixing clutch cover plate / When steel flywheel is fitted. Check before ordering
		Spacer	6	571227*	

Not part of engine assembly

* Asterisk indicates a new part which has not been used on any previous Rover model

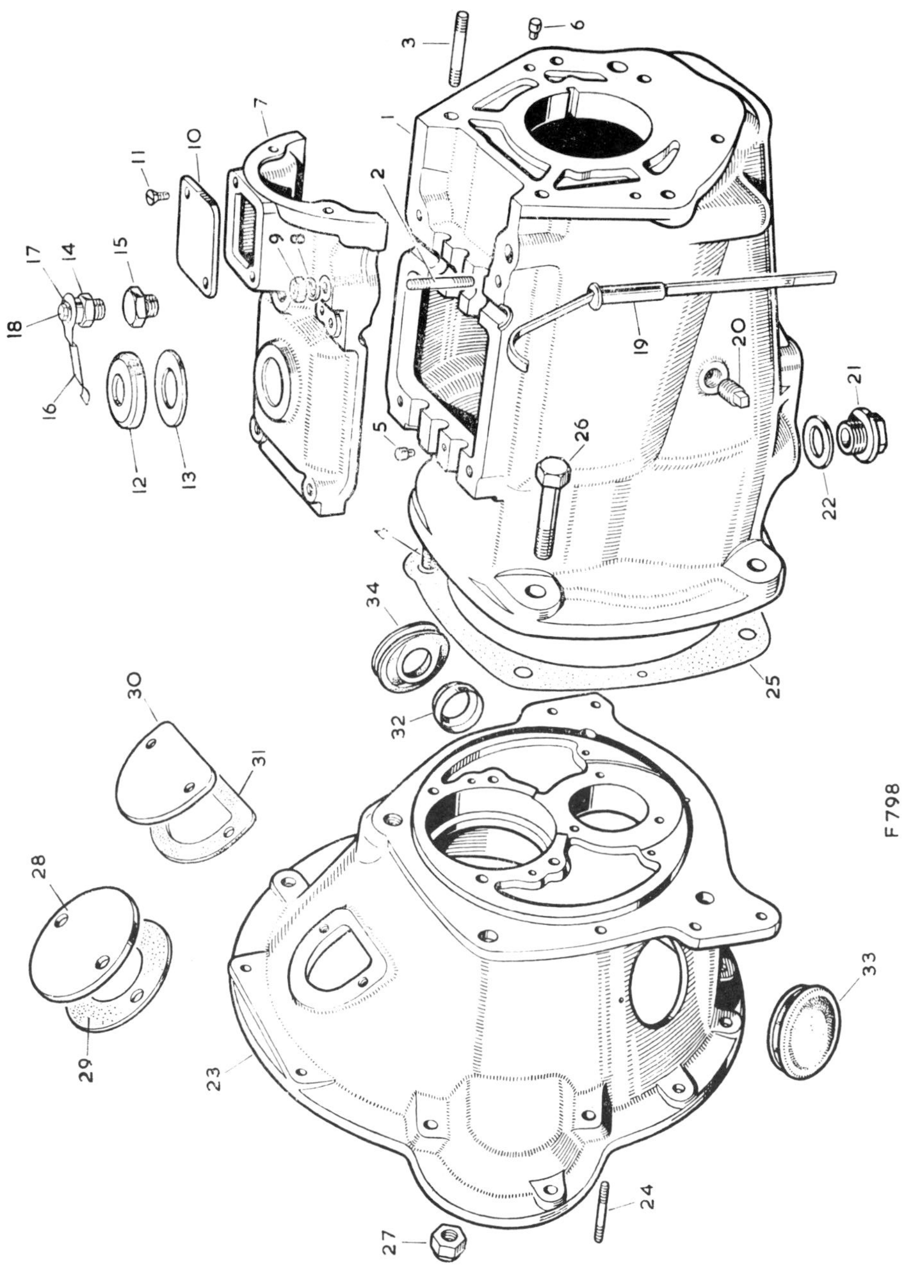

F 798

Plate Ref.	1 2 3 4 DESCRIPTION	Qty	Part No.	REMARKS
	GEARBOX COMPLETE ASSEMBLY	1	248361	2 litre Petrol model
	GEARBOX COMPLETE ASSEMBLY	1	553161†	Diesel and 2¼ litre Petrol models
	GEARBOX COMPLETE ASSEMBLY	1	556461†	2.6 litre Petrol models. Commencing gearbox suffix 'E' onwards
1	GEARBOX CASING ASSEMBLY	1	269960	Up to gearbox suffix 'B' inclusive
	GEARBOX CASING ASSEMBLY	1	605933	From gearbox suffix 'C' onwards
	Stud for top cover, rear	2	3319	
2	Stud } For top cover and gear change plate, front	2	3236	} Alternative fixings
	Set bolt (⅜" Whit x 1²⁹⁄₃₂" long) } For top cover and gear change plate, front	2	561484	} Alternative fixings
	Stud, long } For transfer casing	2	55778	
3	Stud, short } For transfer casing	3	3238	2 off from gearbox suffix 'C' onwards
	Stud } For transfer casing	1	528865	From gearbox suffix 'C' onwards
4	Stud for bell housing	1	231341	
5	Dowel locating top cover	2	07289	
6	Dowel locating transfer casing	2	55636	
7	Top cover for gearbox	1		Not supplied separately
8	Spring washer } Fixing top cover	4	3076	} 2 off when front set bolts are fitted
9	Nut (⅜" BSF) } Fixing top cover	4	2827	} 2 off when front set bolts are fitted
10	Inspection cover plate for selectors	1	05843	
11	Set screw (¼" Whit x 7⁄16" long) fixing cover plate ..	2	52246	
12	Oil filler cap	1	500645	} Up to gearbox suffix 'A' inclusive
13	Joint washer for cap	1	11932	} Up to gearbox suffix 'A' inclusive
14	Plug, retaining cap and selector spring	1	533354	} Up to gearbox suffix 'A' inclusive
15	Plug, retaining selector spring	1	556570	Series IIA. From gearbox suffix 'B' onwards
16	Retaining spring for cap	1	06419	} Series II. Also Series IIA up to gearbox suffix 'A' inclusive
17	Plain washer } Fixing spring to plug	1	2213	} Series II. Also Series IIA up to gearbox suffix 'A' inclusive
18	Set bolt (¼" Whit x 7⁄16" long) } Fixing spring to plug	1	215769	} Series II. Also Series IIA up to gearbox suffix 'A' inclusive
19	Oil level dipstick	1	239192	Series II
	Oil level plug, ⅜" BSP	1	3291	Series IIA up to gearbox suffix 'A' inclusive
20	Oil level and filler plug, ½" BSP	1	3292	From gearbox suffix 'B' onwards
21	Drain plug for gearbox	1	241110	
22	Washer for plug	1	515599	
	BELL HOUSING ASSEMBLY	1	248719	2 litre Petrol model
	BELL HOUSING ASSEMBLY	1	277961	2¼ litre Petrol and Diesel models. Up to gearbox suffix 'A' inclusive
23	BELL HOUSING ASSEMBLY	1	556039	2¼ litre Petrol and Diesel models. From gearbox suffix 'B' onwards
	BELL HOUSING ASSEMBLY	1	556044	2.6 litre Petrol models. Commencing gearbox suffix 'E' onwards
24	Stud for withdrawal race housing	3	269413	
25	Joint washer, bell housing to gearbox	1	07118	
26	Fitting bolt } Fixing gearbox casing to bell housing	3	248720	
27	Self-locking nut (½" BSF) } Fixing gearbox casing to bell housing	4	251324	
28	Top cover for bell housing	1	06846	} 2 litre Petrol models
29	Rubber seal for top cover	1	232606	} 2 litre Petrol models

* Asterisk indicates a new part which has not been used on any previous Rover model
†Gearbox assembly includes all items on pages 229 to 247, 251 to 253, and 315 to 317 inclusive
NOTE—Main gear lever is not included in assembly.

GEARBOX, MAIN CASING

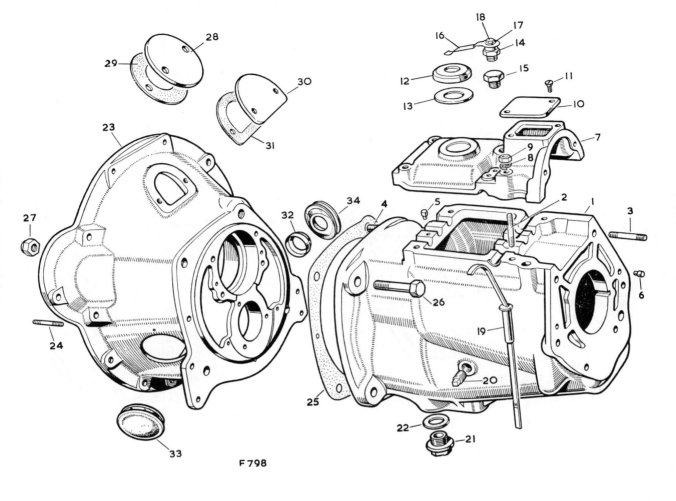

F 798

GEARBOX, MAIN CASING

Plate Ref.	1 2 3 4	DESCRIPTION	Qty	Part No.	REMARKS
30		Top cover for bell housing	1	512237 } 512238 }	2¼ litre Petrol, 2.6 litre Petrol and Diesel models
31		Rubber seal for top cover	1		
32		Centre for dust cover	1	232647	Up to gearbox suffix 'D' Inclusive
33		Grommet for bell housing hole ...	1	232604	Early models
34		Grommet for bell housing shaft ...	1	236281	Up to gearbox suffix 'D' inclusive
		Grommet for bell housing shaft ...	1	553262	From gearbox suffix 'E' onwards
		Plain washer ...	9	2210	
		Plain washer ...	1	2550	
		Nut (3/8" BSF) ... } Fixing gearbox to engine	12	2827 }	2 litre Petrol models
		Nut (5/16" BSF) ... }	1	2828 }	
		Nut (3/8" UNF) ... }	12	254802	Except 2 litre Petrol model
		Gearbox gasket kit	1	600603	

B 504

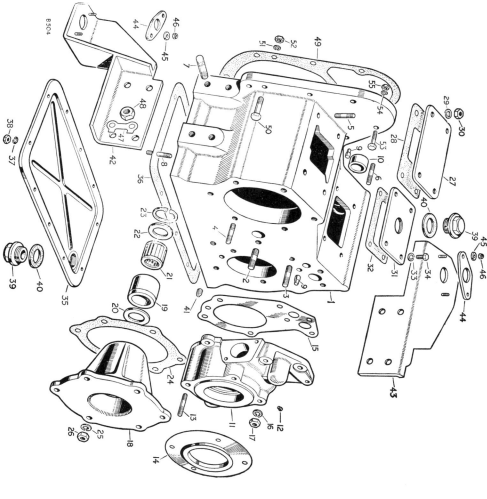

GEARBOX, TRANSFER CASING

Plate Ref.	1 2 3 4	DESCRIPTION	Qty.	Part No.	REMARKS
1		TRANSFER BOX CASING ASSEMBLY	1	533858	Up to gearbox suffix 'A' inclusive
1		TRANSFER BOX CASING ASSEMBLY	1	539787	From gearbox suffix 'B' onwards
2		Stud for intermediate shaft	1	217778	
3		Stud for speedometer housing, short	5	3650	
4		Stud for speedometer housing, long	1	237251	
5		Stud for mainshaft housing	6	3650	
6		Stud for top cover plate	4	217978	
6		Stud long ⎱ For transfer	1	232497	
7		Stud, short ⎰ shaft housing	6	3200	
8		Stud for engine mounting	8	217976	
9		Stud for bottom cover	10	212104	
10		Dowel locating speedometer housing	4	55636	
11		Bush for shaft guide	1	217448	For early type casing with 'oilite' bush
		HOUSING ASSEMBLY FOR SPEEDOMETER PINION			
12		Insert for pinion	1	522318	
13		Stud for transmission brake	1	232846	
14		Mudshield, inner, for housing	1	217973	Not required when transmission brake is fitted
		Shim .040" ⎱	As reqd	522319	
15		Shim .015" ⎰ speedometer	As reqd	217620	
15		Shim .010" ⎱ pinion housing	As reqd	217623	
		Shim .005" ⎰	As reqd	217622	
16		Shim .003"	As reqd	235455	
17		Spring washer ⎱ Fixing housing	6	3076	
17		Nut (⅜" BSF) ⎰ to transfer box	6	2827	
		HOUSING COMPLETE ASSEMBLY, REAR MAINSHAFT			
18		HOUSING ASSEMBLY, REAR MAINSHAFT BEARING	1	230696	
19		Bush for housing	1	533731	Not required when rear power take-off or centre power take-off is fitted
20		Retaining plate, inner	1	217843	
21		Bearing for mainshaft	1	217523	
22		Retaining plate, outer	1	217478	Not required when rear power take-off or centre power take-off is fitted
23		Circlip fixing bearing	1	217523	
24		Joint washer for bearing housing ...	1	217525	
25		Spring washer ⎱ Fixing housing to	6	217680	
26		Nut (⅜" BSF) ⎰ transfer box	6	3076	
27		Cover plate for PTO selector	1	2827	
27				217970	Not required when rear power take-off or centre power take-off is fitted
28		Joint washer for cover plate	1	230140	
29		Spring washer ⎱ Fixing cover plate	4	3075	
30		Nut (⁵⁄₁₆" BSF) ⎰ to transfer box	4	3490	

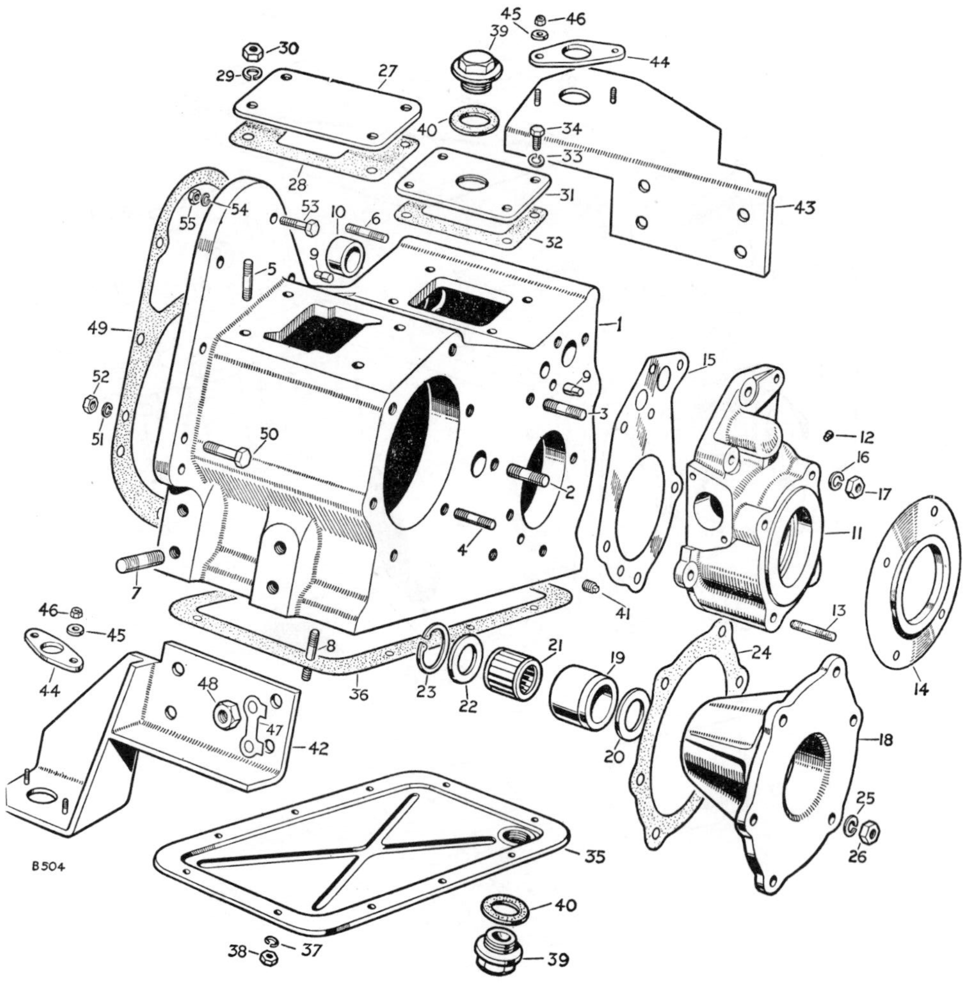

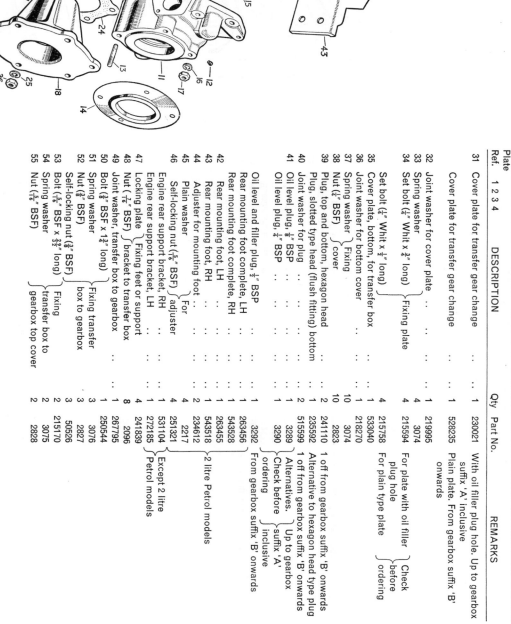

GEARBOX, TRANSFER CASING

Plate Ref.	1 2 3 4 DESCRIPTION	Qty	Part No.	REMARKS
31	Cover plate for transfer gear change	1	230021	With oil filler plug hole. Up to gearbox suffix 'A' inclusive
	Cover plate for transfer gear change	1	528235	Plain plate. From gearbox suffix 'B' onwards
32	Joint washer for cover plate	1	219995	
33	Spring washer } Fixing plate	4	3074	
34	Set bolt ($\frac{1}{4}$" Whit x $\frac{3}{4}$" long) } Fixing plate	4	215594	For plate with oil filler plug hole } Check before ordering
	Set bolt ($\frac{1}{4}$" Whit x $\frac{1}{2}$" long) } Fixing plate	4	215758	For plain type plate } Check before ordering
35	Cover plate, bottom, for transfer box	1	533040	
36	Joint washer for bottom cover	1	218270	
37	Spring washer } Fixing cover	10	3074	
38	Nut ($\frac{1}{4}$" BSF) } Fixing cover	10	2823	
39	Plug, top and bottom, hexagon head	2	241110	1 off from gearbox suffix 'B' onwards
	Plug, slotted type head (flush fitting) bottom ..	1	235592	Alternative to hexagon head type plug
40	Joint washer for plug	2	515599	1 off from gearbox suffix 'B' onwards
41	Oil level plug, $\frac{1}{8}$" BSP	1	3289	Alternatives. Check before ordering } Up to gearbox suffix 'A' inclusive
	Oil level plug, $\frac{1}{4}$" BSP	1	3290	Alternatives. Check before ordering } Up to gearbox suffix 'A' inclusive
	Oil level and filler plug, $\frac{1}{2}$" BSP	1	3292	From gearbox suffix 'B' onwards
	Rear mounting foot complete, LH	1	263456	2 litre Petrol models
	Rear mounting foot complete, RH	1	543528	2 litre Petrol models
42	Rear mounting foot, LH	1	263455	2 litre Petrol models
43	Rear mounting foot, RH	1	543518	2 litre Petrol models
44	Adjuster for mounting foot	2	234612	2 litre Petrol models
45	Plain washer } For adjuster	4	2217	2 litre Petrol models
46	Self-locking nut ($\frac{5}{16}$" BSF) } For adjuster	4	251321	2 litre Petrol models
	Engine rear support bracket, RH	1	531104	Except 2 litre Petrol models
	Engine rear support bracket, LH	1	272185	Except 2 litre Petrol models
47	Locking plate } Fixing feet or support bracket to transfer box	4	241839	
48	Nut ($\frac{7}{16}$" BSF) } Fixing feet or support bracket to transfer box	8	2096	
49	Joint washer, transfer box to gearbox	1	267795	
50	Bolt ($\frac{3}{8}$" BSF x $1\frac{3}{4}$" long) } Fixing transfer box to gearbox	1	250544	
51	Spring washer } Fixing transfer box to gearbox	3	3076	
52	Nut ($\frac{3}{8}$" BSF) } Fixing transfer box to gearbox	3	2827	
	Self-locking nut ($\frac{3}{8}$" BSF) } Fixing transfer box to gearbox	3	50526	
53	Bolt ($\frac{5}{16}$" BSF x $\frac{29}{32}$" long) } Fixing transfer box to gearbox top cover	2	215170	
54	Spring washer } Fixing transfer box to gearbox top cover	2	3075	
55	Nut ($\frac{5}{16}$" BSF) } Fixing transfer box to gearbox top cover	2	2828	

* Asterisk indicates a new part which has not been used on any previous Rover model

GEARBOX, SHAFTS AND GEARS

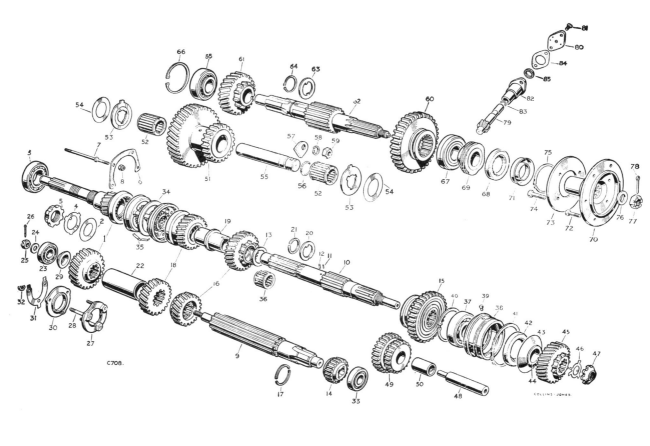

C708.   COLLINS·JONES

GEARBOX, SHAFTS AND GEARS

Plate Ref.	1 2 3 4	DESCRIPTION	Qty	Part No.	REMARKS
1		Primary pinion and constant gear	1	518473	Up to gearbox suffix 'B' inclusive
		Primary pinion and constant gear	1	542231	From gearbox suffix 'C' onwards
2		Shield for primary pinion	1	01017	
3		Ball bearing for primary pinion	1	55714	
4		Lockwasher } Fixing bearing	1	08250	
5		Locknut } to pinion	1	213416	
6		Retaining plate	1	213666	Up to gearbox suffix 'D' inclusive
7		Extension stud	4	214090	
8		Self-locking nut ($\frac{5}{16}$" BSF)	4	251321	
		Retaining plate } Fixing bearing to bell housing	2	556379	From gearbox suffix 'E' onwards
9		Serrated bolt ($\frac{5}{16}$" BSF x 4" long)	4	556147	
		Layshaft	1	09917	Up to gearbox suffix 'A' inclusive
		Layshaft	1	528703	Gearbox suffix 'B' and 'C' only
		Layshaft	1	556040	Gearbox suffix 'B' onwards
10		Mainshaft	1	264250	From gearbox suffix 'D' onwards
11		Peg for 2nd gear thrust washer	1	06405	
12		Peg for mainshaft distance sleeve	1	09561	
		Distance sleeve for mainshaft	1	267572	
13		Thrust washer .125" } For 2nd speed gear		267573	
		Thrust washer .128"		267574	As required
		Thrust washer .130"		267575	
14		1st speed layshaft gear	1	501616	Up to gearbox suffix 'B' inclusive
		1st speed layshaft gear	1	511189	From gearbox suffix 'C' onwards
15		1st speed mainshaft gear	1	501617	Up to gearbox suffix 'B' inclusive
		1st speed mainshaft gear	1	511205	From gearbox suffix 'C' onwards
16		2nd speed layshaft and mainshaft gear	1	245766	Up to gearbox suffix 'C' inclusive
		2nd speed layshaft and mainshaft gear	1	600916	From gearbox suffix 'D' onwards
17		Split ring for 2nd speed layshaft gear	1	239272	Up to gearbox suffix 'C' inclusive
18		3rd speed layshaft and mainshaft gear	1	245767	
19		Distance sleeve for mainshaft	1	239706	
20		Thrust washer .128" } For 3rd speed mainshaft gear		08188	
		Thrust washer .130"		50702	As required
		Thrust washer .135"		231737	
21		Spring ring fixing 2nd and 3rd mainshaft gears	1	06402	
22		Sleeve for layshaft	1	263878	
23		Bearing for layshaft, front	1	08185	
24		Plain washer } Fixing bearing to layshaft	1	09962	
25		Slotted nut ($\frac{1}{2}$" BSF)	1	2980	
26		Split pin	1	09932	
27		BEARING PLATE ASSEMBLY FOR LAYSHAFT	1	213419	
28		Stud for bearing cap	3	213417	
29		Distance piece .312"	1	241651	
		Distance piece .332" For layshaft	1	241650	As required
		Distance piece .352"	1	241649	
30		Retaining plate for layshaft front bearing	1	214792	Up to gearbox suffix 'A' inclusive
31		Lockwasher } Fixing cap and bearing	1	09931	
32		Nut ($\frac{5}{16}$" BSF) } to bell housing	3	2828	

* Asterisk indicates a new part which has not been used on any previous Rover model

GEARBOX, SHAFTS AND GEARS

C708.

COLLINS-JONES

(exploded parts illustration with reference numbers 1–85)

GEARBOX, SHAFTS AND GEARS

Plate Ref. 1 2 3 4	DESCRIPTION	Qty	Part No.	REMARKS
33	Bearing for layshaft, front	1	528701	
	Plain washer ⎱ Fixing bearing	1	528692	
	Slotted nut (¾" UNF) ⎰ to layshaft	1	528691	
	Split pin	1	2766	
34	Bearing plate for layshaft	1	528685	
35	Distance piece .405"	1	528721	From gearbox suffix 'B' onwards
	Distance piece .425" ⎱ For layshaft	1	528720	As required
	Distance piece .445" ⎰	1	528722	
36	Retaining plate for layshaft front bearing	1	528690	
	Lockwasher ⎱ Fixing cap and bearing	1	528683	
	Nut (5/16" UNF) ⎰ to bell housing	1	254861	
37	Bearing for layshaft, rear	1	55715	
38	Housing for mainshaft bearing, rear	1	239160	
39	Ball bearing for mainshaft	1	06395	
40	Roller bearing for mainshaft	1	06397	
41	Distance piece, rear of mainshaft	1	1645	
42	Housing for mainshaft bearing, rear	1	217881	
	Peg, housing to casing	1	09927	Peg fixing ⎱ Alternatives.
	'Loctite' retaining compound, Grade AAV, 10 cc bottle	1	561877*	'Loctite' fixing ⎰ Check before ordering
43	Oil thrower for mainshaft	1	606146	
44	Circlip, housing to casing	1	09960	
	Circlip, bearing to housing	1	09961	
45	Detent spring for clutch	3	236305	
46	Synchronising clutch	1	232415	
47	Mainshaft gear for transfer box	1	218244	
	Lockwasher	1	217476	
	Shim washer ⎱ Fixing gear to mainshaft	1	501501	
	Special nut ⎰	1	217477	
48	Shaft for reverse gear	1	06424 †	Early type
	Shaft for reverse gear	1	561962*†	Late type
49	REVERSE WHEEL ASSEMBLY	1	511203 †	Up to gearbox suffix 'B' inclusive
	REVERSE WHEEL ASSEMBLY	1	217389 †	From gearbox suffix 'C' onwards. For early type shaft and bush
	REVERSE WHEEL ASSEMBLY	1	561960*†	From gearbox suffix 'C' onwards. For late type shaft and bush
50	Bush for reverse wheel	1	56854 †	Early type
	Bush for reverse wheel	1	561954*†	Late type
51	Gear, intermediate	1	533080	From gearbox suffix 'C' onwards
	Gear, intermediate	1	521330	Gearbox suffix 'B' only
	Gear, intermediate	1	219468	Up to gearbox suffix 'A' inclusive
52	Roller bearing for intermediate gear	2	219469	Up to gearbox suffix 'A' inclusive
53	Thrust washer for intermediate gear	2	219466	
54	Shim for intermediate gear	As reqd	234835	
55	Shaft for intermediate gear	1	521329	
56	Sealing ring for intermediate gear	2	521328	
52	Roller bearing for intermediate shaft	2	278025	From gearbox suffix 'B' onwards
53	Thrust washer for intermediate gear	1	267828	
54	Shim for intermediate shaft	As reqd	561197	From gearbox suffix 'B' onwards
55	Shaft for intermediate shaft	1	521326	
56	Sealing ring for intermediate shaft	2	532323	

* Asterisk indicates a new part which has not been used on any previous Rover model
† Interchangeable as a set

GEARBOX, SHAFTS AND GEARS

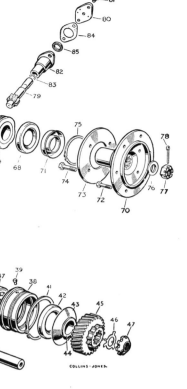

C708.

COLLINS·JONES

GEARBOX, SHAFTS AND GEARS

Plate Ref.	1 2 3 4	DESCRIPTION	Qty	Part No.	REMARKS
57		Retaining plate for shaft	1	217484	
58		Spring washer } Fixing plate to casing		3076	
59		Nut (⅜" BSF) }	1	2827	
60		Low gear wheel	1	235438	Up to gearbox suffix 'B' inclusive
60		Low gear wheel	1	532979	From gearbox suffix 'C' onwards
61		High gear wheel	1	218243	
62		Output shaft, rear drive	1	235985	
63		Thrust washer for high gear wheel	1	217488	
64		Circlip fixing washer to shaft	1	217489	
65		Bearing for output shaft, front	1	217490	
66		Circlip fixing bearing to case	1	217526	
67		Bearing for output shaft, rear	1	217512	
68		Oil seal for output shaft	1	236417	
69		Speedometer worm complete	1	540004	
70		Flange for output shaft, rear drive	1	275238	
71		Mudshield for flange	1	236074	
72		Fixing bolt for brake drum	6	217564	Petrol and 88 Diesel models. Also 109 Diesel from gearbox number 27600880 onwards
73		Retaining flange for brake drum bolts	1	217568	
74		Fitting bolt for propeller shaft	4	217565	
72		Fitting bolt for brake drum	6	275629	} Up to gearbox number 27600879. 109 Diesel models.
73		Dowel for flange	2	55705	
74		Fitting bolt for propeller shaft	4	512701	BSF fixing, alternative to UNF. Check before ordering
75		Circlip retaining bolts and flange	1	217546	UNF fixing, alternative to BSF. Check before ordering
76		Plain washer } Fixing flange to output shaft	1	3300	
77		Slotted nut }	1	3259	
78		Split pin	1	2428	
79		Speedometer pinion	1	237982	
80		Retaining plate for pinion	1	232565	
81		Screw fixing plate to housing	2	2529	
82		Sleeve for pinion	1	268791	
83		Sealing ring for sleeve	1	268828	
84		Joint washer for sleeve	1	267782	
85		Oil seal for pinion	1	211502	

GEARBOX, SHAFTS AND GEARS

* Asterisk indicates a new part which has not been used on any previous Rover model

C 709.

GEARBOX, SELECTORS AND LEVERS

Plate Ref.	1 2 3 4	DESCRIPTION	Qty	Part No.	REMARKS
1		Selector fork, 3rd and 4th speed	1	213637	
2		Shaft for fork, 3rd and 4th speed	1	213636	
3		Selector fork, 1st and 2nd speed	1	06421	
4		SHAFT ASSEMBLY FOR FORK, 1st and 2nd SPEED	1	06422	
5		Interlocking pin	1	55697	
6		Peg fixing interlocking pin	1	55775	
7		Selector fork, reverse	1	217391	
8		Shaft for fork, reverse	1	502201	
9		Set bolt ($\frac{5}{16}$" BSF x 1" long) fixing forks to shafts	3	237160	
10		Stop for 2nd speed	1	210203	
11		Locker	1	210204	
12		Set bolt ($\frac{1}{4}$" BSF x $\frac{5}{8}$" long) ⎫ Fixing stop to	1	250693	
13		Interlocking plunger ⎬ 2nd speed	2	55638	
		⎭ selector shaft			
14		Steel ball for selectors	3	1643	
15		Selector spring, forward	2	03649	
16		Selector spring, reverse	1	56102	
17		Retaining plate, LH	1	05853	
18		Retaining plate, RH ⎬ For selector springs, side	1	05854	
19		Rubber grommet	2	05852	
20		Spring washer	4	3074	
21		Set bolt ($\frac{1}{4}$" Whit x $\frac{1}{2}$" long) ⎬ Fixing	4	215758	
		retaining plates			
22		Sealing ring, forward selector shaft	2	272596	
23		Sealing ring, reverse selector shaft	1	272597	
24		Retaining plate for sealing ring	2	241598	
25		Spring washer ⎫ Fixing	4	3074	
26		Set bolt ($\frac{1}{4}$" Whit x $\frac{7}{16}$" long) ⎬ retaining plate	4	215769	
27		Set bolt (2 BA x $\frac{9}{16}$" long) In cover for ⎫	1	251018	
28		Locknut (2 BA) ⎬ 2nd gear stop	1	2247	
29		Adjustable stop for reverse selector shaft	1	231033	
30		Locknut ($\frac{1}{4}$" BSF) for stop	1	2823	

GEARBOX, SELECTORS AND LEVERS

FRONT OUTPUT SHAFT HOUSING

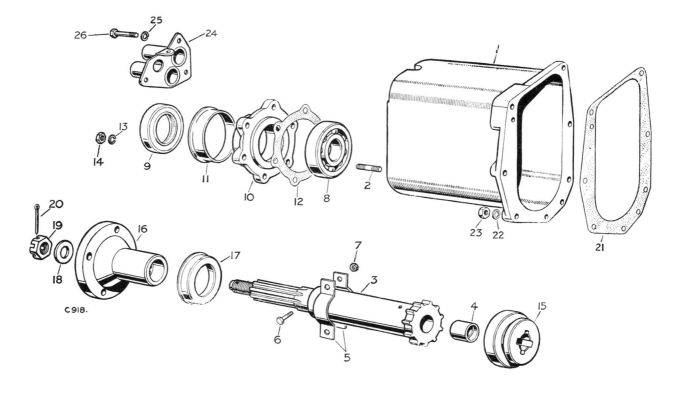

C918.

FRONT OUTPUT SHAFT HOUSING

Plate Ref.	1 2 3 4	DESCRIPTION	Qty	Part No.	REMARKS
1		OUTPUT SHAFT HOUSING ASSEMBLY ...	1	268849	
2		Stud for oil seal retainer ...	6	217978	
3		FRONT OUTPUT SHAFT ASSEMBLY	1	243611	
4		Bush for shaft ...	1	234534	
5		Oil thrower for output shaft ...	2	243873	
6		Bolt (2 BA x $\frac{7}{16}$" long) } Fixing oil thrower to	2	251018	
7		Self-locking nut (2 BA) } front output shaft	2	251335	
8		Bearing for front output shaft ...	1	217325	
9		Oil seal for shaft ...	1	236417	
10		Retainer for oil seal ...	1	236541	
11		Mudshield for retainer ...	1	236548	
12		Joint washer for retainer ...	1	234525	
13		Spring washer } Fixing	6	3075	
14		Nut ($\frac{5}{16}$" BSF) } retainer	6	2828	
15		Locking dog, four-wheel drive ...	1	233241	
16		Flange for transfer shaft ...	1	539993	
17		Mudshield for flange ...	1	236074	
18		Plain washer ...	1	3300	
19		Slotted nut } Fixing flange to transfer shaft	1	3259	
20		Split pin ...	1	2428	
21		Joint washer for transfer housing ...	1	233448	
22		Spring washer } Fixing housing to	7	3075	
23		Nut ($\frac{5}{16}$" BSF) } transfer box	7	2828	
24		Dust cover plate for selector shafts ...	1	266956	
25		Spring washer } Fixing dust cover	3	3074	
26		Bolt ($\frac{1}{4}$" Whit x $\frac{7}{16}$" long) } to housing	3	215769	

* Asterisk indicates a new part which has not been used on any previous Rover model

CLUTCH WITHDRAWAL MECHANISM

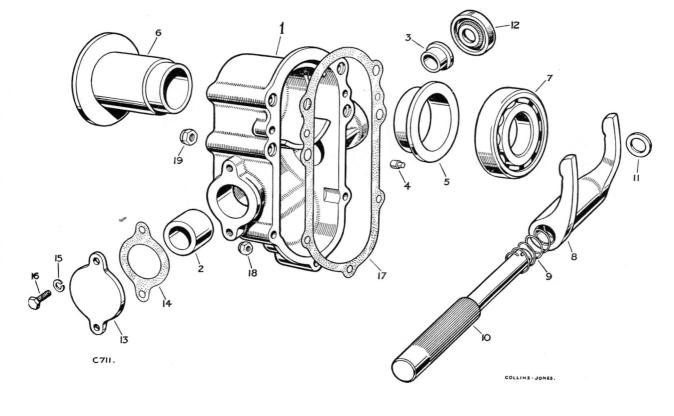

C711.

COLLINS · JONES.

CLUTCH WITHDRAWAL MECHANISM

Plate Ref.	1 2 3 4	DESCRIPTION	Qty	Part No.	REMARKS
1		CLUTCH WITHDRAWAL RACE HOUSING ASSEMBLY	1	231071	Up to gearbox suffix 'A' inclusive
1		CLUTCH WITHDRAWAL RACE HOUSING ASSEMBLY	1	528707	From gearbox suffix 'B' onwards
2		Bush for cross-shaft, large	1	214793	
3		Bush for cross-shaft, small	2	214794	
4		Dowel locating housing	2	213700	
5		Bush for withdrawal race sleeve ...	1	231075	
6		Clutch withdrawal sleeve	1	231074	
7		Withdrawal race thrust bearing	1	214797	
8		Operating fork for clutch	1	264807	
9		Spring for operating fork	1	264806	
10		Cross-shaft for clutch operation ...	1	231943	
11		Thrust washer for cross-shaft ...	1	213660	
12		Oil seal for cross-shaft	1	214787	
13		Cover plate for cross-shaft ...	1	213661	
14		Joint washer for cover plate or housing ...	1	213662	
15		Spring washer	2	3074	
16		Set bolt (¼" Whit x 2¼" long) } Fixing cover plate or oil seal housing	2	215593	
17		Joint washer for clutch withdrawal housing ...	1	213663	Up to gearbox suffix 'A' inclusive
17		Joint washer for clutch withdrawal housing ...	1	526697	From gearbox suffix 'B' onwards
18		Self-locking nut (¼" BSF) } Fixing withdrawal housing to bell housing	3	251320	
19		Self-locking nut (⁵⁄₁₆" BSF) } Fixing withdrawal housing to bell housing	4	251321	

MAIN GEAR LEVER

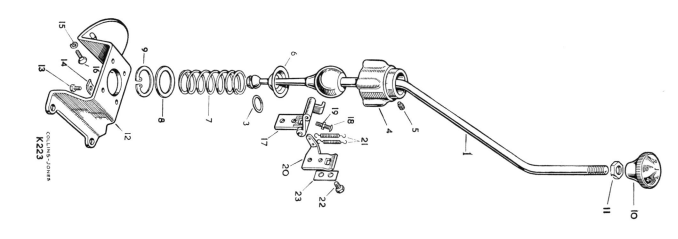

COLLINS-JONES
K223

MAIN GEAR LEVER

Plate Ref. 1 2 3 4	DESCRIPTION	Qty	Part No.	REMARKS
	MAIN GEAR-CHANGE LEVER ASSEMBLY ...	1	544917	RHStg } 4-cylinder models
	MAIN GEAR-CHANGE LEVER ASSEMBLY ...	1	544918	LHStg } models
	MAIN GEAR-CHANGE LEVER ASSEMBLY ...	1	561182	RHStg } 6-cylinder models
	MAIN GEAR-CHANGE LEVER ASSEMBLY ...	1	556462	LHStg } models
1	Gear-change lever	1	544827	RHStg } 4-cylinder
2	Gear-change lever	1	544828	LHStg } models
2	Gear-change lever	1	561181	RHStg } 6-cylinder models
1	Gear-change lever	1	556463	LHStg } models
3	'O' ring for gear-change lever ...	1	540354	
4	Housing for lever	1	219714	
5	Locating pin for lever ball ...	1	507447	
6	Spherical seat for gear lever ...	1	219721	
7	Retaining spring for lever ...	1	219723	
8	Retaining plate for spring ...	1	219722	
9	Circlip fixing retaining plate ...	1	219797	
10	Knob for lever	1	217735	
11	Locknut (¼" BSF) for knob ...	1	3905	
12	Mounting plate for gear change ...	1	232608	
13	Set bolt (5/16" Whit x 1¾" long) } Fixing housing to mounting plate	4	215647	
14	Locker	4	2499	
15	Spring washer	2	3074	
16	Set bolt (¼" Whit x 2¼" long) } Fixing mounting plate to bell housing	2	215593	
17	Reverse stop hinge complete ...	1	502202	
18	Adjusting screw } For hinge	1	76653	
19	Locknut (2 BA) } hinge	1	2247	
20	Bracket for reverse stop spring ...	1	502205	
21	Spring for reverse stop	1	514650	
22	Set bolt (¼" BSF x ½" long) } Fixing hinge and bracket to reverse selector shaft	2	237139	
23	Locker	1	3421	

* Asterisk indicates a new part which has not been used on any previous Rover model

FOUR-WHEEL DRIVE SELECTORS AND TRANSFER GEAR LEVER

C712.

COLLINS-JONES.

FOUR-WHEEL DRIVE SELECTORS AND TRANSFER GEAR LEVER

Plate Ref.	1 2 3 4	DESCRIPTION	Qty	Part No.	REMARKS
1		Selector shaft, four-wheel drive	1	233416	
2		Selector fork complete, four-wheel drive	1	233397	
3		Bush for selector fork	2	217521	
4		Spring for selector fork	2	233449	
5		Block for selector shaft	1	233406	
6		Special screw	1	233409	Fixing block to selector shaft
7		Castle nut ($\frac{5}{16}$" BSF)	1	251601	
8		Split pin	1	2422	
9		Selector shaft, transfer gear change	1	233398	
10		Sealing ring for transfer gear change shaft	1	515572	
11		Selector fork, transfer gear change	1	217584	
12		Set bolt ($\frac{5}{16}$" BSF x 1" long) fixing fork	1	237160	
13		Distance tube for transfer selector shaft	1	233438	
14		Locating bush for selector shaft	1	233437	
15		Spring for gear-change selector shaft	1	217445	
16		Connector, gear change to pivot shaft	1	538536	
17		Block for selector shaft	1	233406	
18		Special screw	1	233409	Fixing block to selector shaft
19		Castle nut ($\frac{5}{16}$" BSF)	1	251601	
20		Split pin	1	2422	
21		Pivot shaft for selector shafts	1	549168	
22		Coupling, selector shafts to pivot	1	233407	
23		Special screw	2	233407	Fixing coupling to pivot shaft
24		Castle nut ($\frac{5}{16}$" BSF)	1	234252	
25		Split pin	1	2827	
26		Shakeproof washer	1	70823	
27		Nut ($\frac{3}{8}$" BSF)	1	233416	
28		Plunger for transfer selector shaft	1	56102	
29		Spring for plunger	1	233441	
30		Plug retaining plunger	1	549169	
31		Link for selector shaft	1	251601	Fixing link to selector shaft
32		Special nut	1	77626	
33		Shakeproof washer	1	238329	
34		LEVER ASSEMBLY, FOUR-WHEEL DRIVE	1	3764	
35		Bush for lever	1	268847	
36		Special bolt, hexagon head, lever to housing	1	230086	
37		Locking pin, four-wheel drive lever	1	540842	
38		Sealing ring for four-wheel drive locking pin	1	232464	
39		Plain washer	1	266992	Fixing locking pin to lever
40		Split pin	1	2876	
41		Selector rod, four-wheel drive	1	2389	
42		Clevis complete for rod	1	571855	
43		Split pin for clevis	1	215808	
44		Spring for selector rod	1	2389	
45		Special bush for spring	1	561221	
46		Control knob for rod	1	234658	
47		Locknut ($\frac{1}{4}$" BSF) for knob and clevis	2	232813	
48		Transfer gear-change lever	1	2823	
49		Spring for transfer gear-change lever	1	576210	
50		Knob for gear-change lever	1	243714	
51		Locknut for knob	1	219521	
				3764	

Overall length $1\frac{5}{32}$" Alternatives.
Overall length $1\frac{3}{8}$" Check before ordering

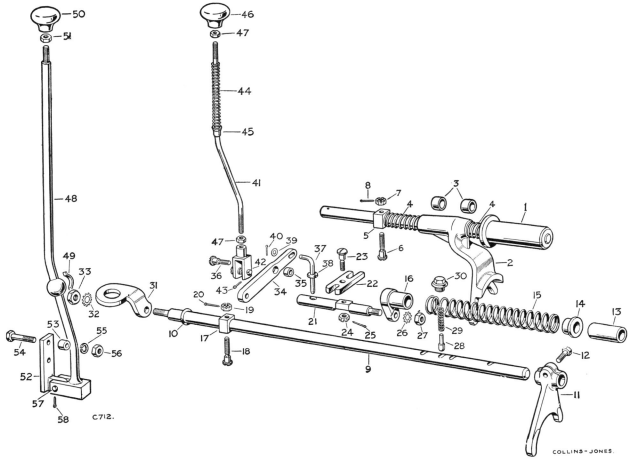

C712.

COLLINS-JONES.

FOUR-WHEEL DRIVE SELECTORS AND TRANSFER GEAR LEVER

Plate Ref.	1 2 3 4	DESCRIPTION	Qty	Part No.	REMARKS
52		Bracket for gear-change lever	1	219709	
53		Distance piece for bracket, $\frac{3}{8}$" long	2	266955	
54		Bolt ($\frac{5}{16}$" BSF x $1\frac{19}{32}$" long)	2	215170	
55		Spring washer	2	3075	Alternative to UNF fixings
56		Nut ($\frac{5}{16}$" BSF)	2	2828	
		Bolt ($\frac{5}{16}$" UNF x $1\frac{3}{8}$" long)	2	256221	Fixing bracket to bell housing
		Spring washer	2	3075	Alternative to BSF fixings
		Nut ($\frac{5}{16}$" UNF)	2	254831	
57		Clevis pin ⎱ Fixing transfer gear	1	216421	
58		Split pin ⎰ lever to bracket	1	2392	Alternative fixings. Check before ordering
		Bolt ($\frac{5}{16}$" UNF x $1\frac{3}{8}$" long) ⎱ Fixing transfer gear	1	256221	
		Nut ($\frac{5}{16}$" UNF) ⎰ lever to bracket	1	252161	

REAR AXLE AND PROPELLER SHAFT

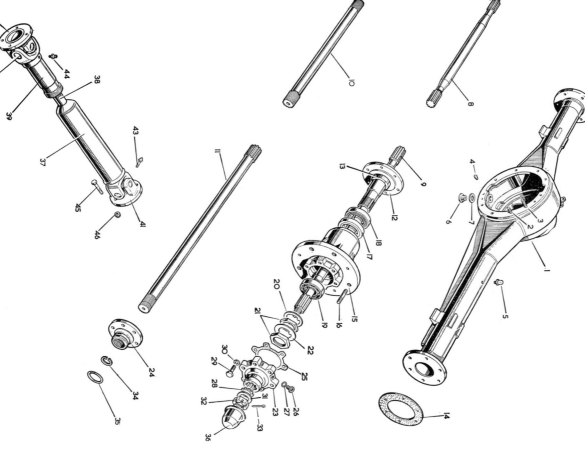

K130

REAR AXLE AND PROPELLER SHAFT

Plate Ref.	1 2 3 4	DESCRIPTION	Qty	Part No.	REMARKS
1		REAR AXLE ASSEMBLY, 4.7 RATIO	1	540220†	88. Rover type
		REAR AXLE ASSEMBLY, 4.7 RATIO	1	540223†	109. Rover type
		REAR AXLE ASSEMBLY, 4.7 RATIO, ENV A90084	1	549160†	109. ENV type. Optional
		Rear axle casing complete	1	549021	88 } Rover
		Rear axle casing complete	1	549025	109 } type
		Rear axle casing complete, ENV A16286	1	533567	109. ENV type. Optional
2		Special bolt, short	Fixing	4	01565	Bolt type
		Special bolt, long	differential	6	01564	fixing } 109. ENV type. Optional
3		Special stud, short	to	4	561195	Stud type } Check before
		Special stud, long	housing	6	561196	fixing } ordering
		Set bolt (⅜" UNF x 1⅛" long)	axle	10	255248	ENV type axle.
		Spring washer	casing	10	3076	109. Optional
4		Dowel locating differential housing		2	55705	
5		Breather complete for rear axle		1	515845	Early type axles
6		Drain plug for rear axle casing		1	235592	
7		Joint washer for drain plug		1	515599	
8		Rear axle shaft, RH		1	549495	Rover type
9		Rear axle shaft, LH		1	549496	axles
10		Rear axle shaft, RH, ENV A16228		1	533579	ENV type axle.
11		Rear axle shaft, LH, ENV A16229		1	533580	109. Optional
12		Rear hub bearing sleeve		2	273886	
13		Distance piece for bearing sleeve		2	217351	
14		Joint washer, bearing sleeve to axle casing		2	500978	
15		REAR HUB ASSEMBLY		2	576475	Grease-packed type
16		Stud for road wheel		10	561590	
17		Hub bearing, inner		2	217269	
18		Oil seal for inner bearing		2	239422	
19		Hub bearing, outer		2	217270	
20		Key washer		2	217352	
21		Special nut	Fixing hub bearing	2	217355	
22		Locker		4	217353	
23		Driving member for rear hub		2	571235	Grease-packed type, Rover axles
24		Driving member for rear hub		2	571711	Grease-packed type. ENV axle. 109. Optional
25		Joint washer for driving member		2	231505	
26		Filler plug for driving member		2	556204	Early models with
27		Joint washer for filler plug		2	232038	oil filled hub
28		Oil seal for rear axle shaft		2	501412	Rover type axles
29		Set bolt (⅜" BSF x 1¼" long) } Fixing driving		12	215331	
30		Spring washer } member to rear hub		12	3076	
31		Plain washer		12	571922	
32		Slotted nut		2	3259	Rover type axles
33		Split pin		2	2428	
34		Circlip, ENV X14344 } Fixing		2	549473	ENV type axle.
35		"O" ring, ENV X14220 } axle shaft to		2	553263	109. Optional
36		Hub cap, rear	driving member	2	219098	

* Asterisk indicates a new part which has not been used on any previous Rover model
† Rear axle assemblies include differential and brakes

REAR AXLE AND PROPELLER SHAFT

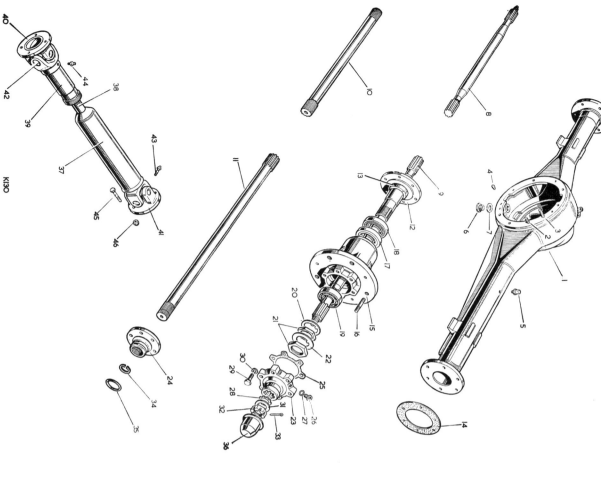

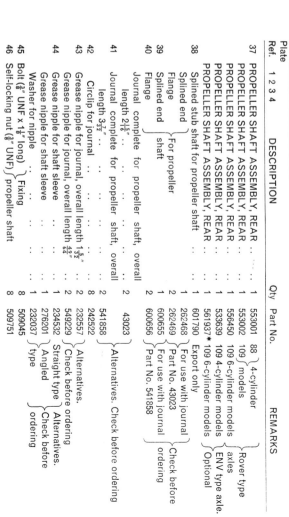

REAR AXLE AND PROPELLER SHAFT

Plate Ref.	1 2 3 4	DESCRIPTION	Qty	Part No.	REMARKS
37		PROPELLER SHAFT ASSEMBLY, REAR ...	1	553001	88 ⎫ 4-cylinder
		PROPELLER SHAFT ASSEMBLY, REAR ...	1	553002	109 ⎬ models ⎫ Rover type axles
		PROPELLER SHAFT ASSEMBLY, REAR ...	1	556450	109 6-cylinder models
		PROPELLER SHAFT ASSEMBLY, REAR ...	1	533639	109 4-cylinder models ⎫ ENV type axle.
		PROPELLER SHAFT ASSEMBLY, REAR ...	1	561937*	109 6-cylinder models ⎭ Optional
38		Splined stub shaft for propeller shaft ...	1	601790	Export only
		Splined end ⎫	1	262468	For use with journal Part No. 43023 ⎫ Check before ordering
		Flange ⎬ For propeller	2	262469	
39		Splined end ⎬ shaft	1	600655	For use with journal Part No. 541858 ⎫ Check before ordering
40		Flange ⎭	2	600656	
		Journal complete for propeller shaft, overall length $2\frac{1}{16}''$ ...	2	43023	
41		Journal complete for propeller shaft, overall length $3\frac{7}{32}''$ ...	2	541858	Alternatives. Check before ordering
42		Circlip for journal ...	8	242522	
43		Grease nipple for journal, overall length $1\frac{5}{32}''$	2	232557	Alternatives. Check before ordering
44		Grease nipple for journal, overall length $1\frac{49}{64}''$	2	549229	
		Grease nipple for shaft sleeve ...	1	234532	Straight type ⎫ Alternatives.
		Grease nipple for shaft sleeve ...	1	276201	Angled type ⎬ Check before ordering
		Washer for nipple ...	1	232037	
45		Bolt ($\frac{3}{8}''$ UNF x $1\frac{1}{8}''$ long) ⎫ Fixing	8	509045	
46		Self-locking nut ($\frac{3}{8}''$ UNF) ⎬ propeller shaft	8	509751	

* Asterisk indicates a new part which has not been used on any previous Rover model

FRONT AXLE AND PROPELLER SHAFT

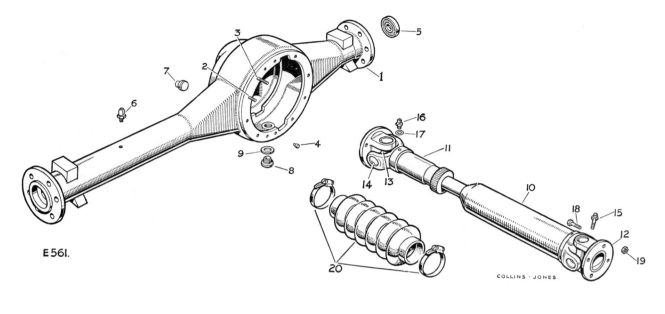

E 561.   COLLINS · JONES.

FRONT AXLE AND PROPELLER SHAFT

Plate Ref. 1 2 3 4	DESCRIPTION	Qty	Part No.	REMARKS
1	FRONT AXLE ASSEMBLY, 4.7 RATIO	1	540588 †	RHStg 88, reinforced
1	FRONT AXLE ASSEMBLY, 4.7 RATIO	1	540589 †	LHStg type
1	FRONT AXLE ASSEMBLY, 4.7 RATIO	1	540850 †	RHStg 109, reinforced — 4-cylinder models
1	FRONT AXLE ASSEMBLY, 4.7 RATIO	1	540851 †	LHStg type — 4-cylinder models
1	FRONT AXLE ASSEMBLY, 4.7 RATIO	1	561311 †	RHStg 109 standard
1	FRONT AXLE ASSEMBLY, 4.7 RATIO	1	564471 †	LHStg type
1	FRONT AXLE ASSEMBLY, 4.7 RATIO	1	565076 *†	RHStg 109 reinforced — 6-cylinder models
1	FRONT AXLE ASSEMBLY, 4.7 RATIO	1	565038 *†	LHStg type — 6-cylinder models
2	Front axle casing complete	1	549027	Reinforced type
3	Special bolt, short	4	01565	Alternative to UNF stud type fixing. Check before ordering
3	Special bolt, long	6	01564	} Fixing differential housing to axle casing
3	Special stud, short	4	561196	Alternative to BSF. Check before ordering
3	Special stud, long	6	561195	
4	Dowel locating differential housing	2	55705	
5	Oil seal in axle casing	1	217400	Fitted to early type axles
6	Breather complete for front axle	1	515845	
7	Oil filler plug, front, for axle casing	1	3294	
8	Drain plug for front axle casing	1	235592	
9	Joint washer for drain plug	1	540870	
10	PROPELLER SHAFT ASSEMBLY, FRONT	1	515599	Round head type
10	PROPELLER SHAFT ASSEMBLY, FRONT	1	553000	Hexagon head type } Alternatives
	Splined stub shaft for propeller shaft	1	556449	With spline grommet, 4-cylinder models
	Splined stub shaft for propeller shaft	1	601790	With spline grommet, 6-cylinder models
11	Splined shaft	1	262468	Export only
11	Flange } For propeller shaft	1	262469	For use with journal Part No. 43023 } Check before ordering
11	Splined end	1	600655	
12	Flange	1	600656	For use with journal Part No. 541858 } Check before ordering
13	Journal complete for propeller shaft, overall length 2 15/16"	2	43023	Alternatives. Check before ordering
13	Journal complete for propeller shaft, overall length 3 7/32"	2	541858	Alternatives. Check before ordering
14	Circlip for journal	8	242522	
15	Grease nipple for journal, overall length 1 5/32"	2	232557	Alternatives. Check before ordering
15	Grease nipple for journal, overall length 49/64"	2	549229	
16	Grease nipple for shaft sleeve	1	276201	Angled type
17	Washer for nipple	1	232037	
18	Grease nipple for shaft sleeve	1	234532	Straight type nipple } Alternatives
19	Bolt (3/8" UNF x 1 1/8" long)	8	509045	} Fixing propeller shaft
19	Bolt (3/8" UNF x 1 7/32" long)	4	509046	
20	Self-locking nut (3/8" UNF)	8	509751	
20	Rubber grommet set for propeller shaft splines	1	276484	Part of front propeller shaft

* Asterisk indicates a new part which has not been used on any previous Rover model
† Front axle assemblies include differential, brakes, universal joints, front hubs and steering track rod

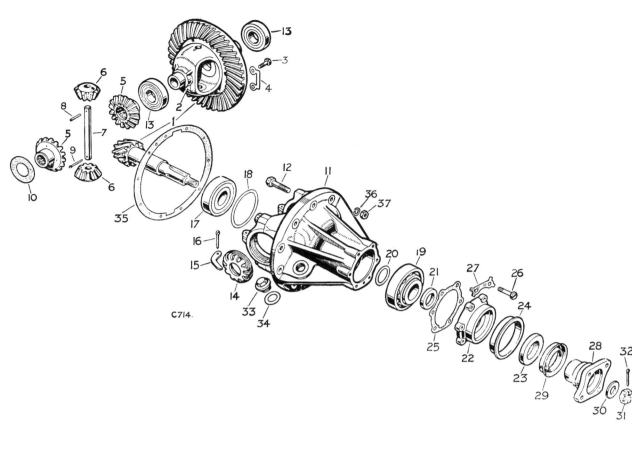

C714.

DIFFERENTIAL, FRONT AND REAR, ROVER TYPE AXLES

Plate Ref.	1 2 3 4	DESCRIPTION	Qty	Part No.	REMARKS
		DIFFERENTIAL ASSEMBLY, 4.7 RATIO	2	539732	
1		Crown wheel and bevel pinion	2	542341	
2		Differential casing	2	273441	
3		Set bolt (⅜" BSF x 1" long) ⎫ Fixing	16	237339	
4		Special fitting bolt ⎬ crown wheel	4	272934	
5		Locker, double type ⎭ to differential casing	10	272922	
4		Differential wheel	4	533794	
5		Differential pinion	4	533777	
6		Spindle for pinion	4	539703	
7		Plain pin ⎫ For	2	11379	
8		Split pin ⎬ spindle	2	2396	
9		Thrust washer .040" ⎫ For	4	533786	
10		Thrust washer .045" ⎬ differential	4	533787	As required
		Thrust washer .050" ⎭ wheels	4	533788	
11		Bevel pinion housing complete ...	2	539744	
		Bevel pinion housing complete ...	2	528257	Up to axle suffix 'A' inclusive
		Bevel pinion housing complete ...	2	539744	From axle suffix 'B' onwards
12		Special set bolt fixing bearing cap ...	8	40742	
13		Taper roller bearing for differential ...	4	41045	
14		Serrated nut ⎫ For	4	40756	
15		Lock tab ⎬ bearing adjustment	4	40758	
16		Split pin fixing lock tab ...	2	2766	
17		Bearing for bevel pinion, pinion end	2	219544	Up to axle suffix 'A' inclusive
		Bearing for bevel pinion, pinion end	2	539706	From axle suffix 'B' onwards
		Shim .003" ⎫	4	230438	
18		Shim .005" ⎬	4	230439	
		Shim .010" ⎬ For bearing adjustment,		230440	
		Shim .020" ⎭ pinion end		233678	
		Shim .022" ⎫		539711	
		Shim .024" ⎬		539713	
		Shim .030" ⎭		539715	As reqd
19		Bearing for bevel pinion, flange end	2	539717	
		Bearing for bevel pinion, flange end	2	219550	Up to axle suffix 'A' inclusive
		End washer for bearing shim ...	2	539707	From axle suffix 'B' onwards
		Shim .003" ⎫	4	50224	†† Up to axle suffix 'A' inclusive
20		Shim .005" ⎬		219547	
		Shim .010" ⎬ For bearing adjustment,		219648	
		Shim .020" ⎭ flange end		219549	
		Shim .072" ⎫		233677	As reqd
		Shim .074" ⎬		539718	As reqd
		Shim .076" ⎬		539720	As reqd
		Shim .081" ⎭		539722	As reqd
				539724	As reqd

* Asterisk indicates a new part which has not been used on any previous Rover model
†† From axles numbered: Front: 1419041190, 144902375, 151963127, 154901273 onwards
Rear: 141907744, 151905087 onwards

DIFFERENTIAL, FRONT AND REAR, ROVER TYPE AXLES

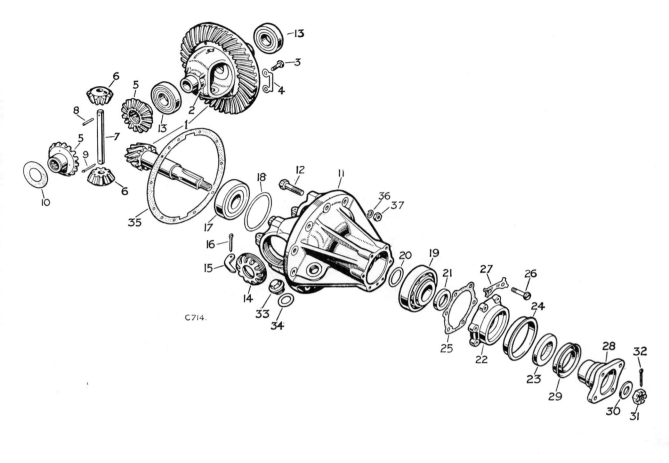

C714.

Plate Ref. 1 2 3 4	DESCRIPTION	Qty	Part No.	REMARKS
21	Washer for pinion bearing ...	2	231242	Up to axle suffix 'A' inclusive
22	Washer for pinion bearing ...	2	539745	From axle suffix 'B' onwards
23	Retainer for oil seal ...	2	236547	
24	Oil seal for pinion ...	2	217507	
25	Mudshield for retainer ...	2	236546	
26	Joint washer for oil seal retainer ...	2	553412	
27	Special bolt } Fixing oil	12	571916*	
	Locker } seal retainer	6	41049	} Alternatives
28	Spring washer ...	6	3075	
29	Driving flange for bevel pinion ...	2	236632	
30	Mudshield for driving flange ...	2	236072	
31	Plain washer ...	2	504433‡	
32	Special nut } Fixing driving flange	2	3259	
	Plain washer } to bevel pinion	2	513454††	
33	Oil filler plug ...	2	2428	
34	Joint washer for plug ...	2	533358	
35	Split pin ...	2	06009	
36	Joint washer, differential to axle casing ...	20	07316	
37	Spring washer ...	20	2827	} UNF fixing
	Nut (⅜" BSF) } Fixing differential to	20	3076	} Alternative to
	Self-locking nut (⅜" UNF) } axle casing	20	252162	} BSF fixing
	'Hylomar' sealing compound, 4 oz tube ..	As reqd	534244	For use with joint washer 553412
	'POWR-LOK' DIFFERENTIAL			
	Differential casing complete ...	1	600875	'POWR-LOK' type optional equipment.
	Special set bolt fixing casing halves together	8	600876	Primarily intended for America Dollar
	Differential wheel ...	8	605024	Area. Replaces normal differential at
	Gear ring for differential wheel ...	2	600877	**rear** only
	Differential bevel pinion ...	2	600878	
	Cross-shaft for pinion ...	4	600879	
	Friction plate ...	2	600880	
	Friction disc ...	4	600881	
	Belleville plate ...	2	600882	
	Belleville disc ...	2	600883	
	Belleville disc ...	2	600884	

* Asterisk indicates a new part which has not been used on any previous Rover model

‡ Up to axles numbered: Front: 141904189, 151903126, 144902374, 151905086
   Rear: 141907743, 151905086

†† From axles numbered: Front: 141904190, 151903127, 144902375, 154901273 onwards
   Rear: 141907744, 151905087 onwards

264

H625

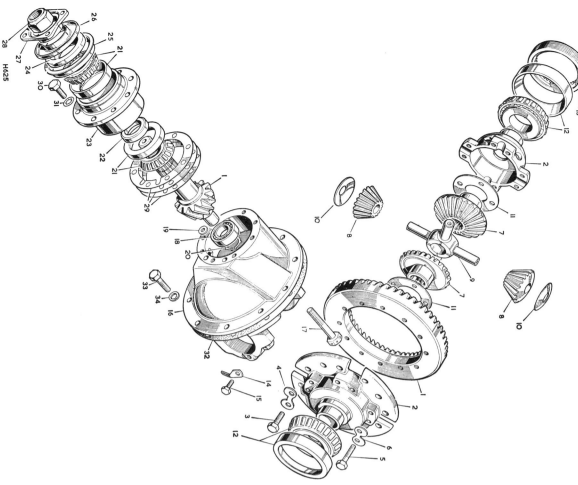

Plate Ref.	1 2 3 4	DESCRIPTION	Qty	Part No.	REMARKS
		DIFFERENTIAL ASSEMBLY, 4.7 RATIO, ENV D90205 ...	1		
1		Crown wheel and bevel pinion, ENV C16157 ...	1	533568	
2		Differential casing, ENV D80210 ...	1	549493	
3		Set bolt (5/16" UNF x 1" long), ENV X10332	12	600901	Fixing crown-wheel to differential casing
4		Locking plate, ENV D16138	6	255066	
5		Set bolt (5/16" UNF x 1 3/4" long), ENV X10350	8	549460	Fixing differential casing halves together
6		Locking plate, double type, ENV D16145	4	256062	
7		Differential wheel, ENV D16142 ...	2	549464	
8		Differential pinion, ENV D16143 ...	4	549461	
9		Spindle for pinion, ENV D16144 ...	1	549462	
10		Spherical differential pinion washer, ENV D16141 ...	4	549463	
11		Differential wheel washer, ENV D16140 ...	2	549466	
12		Taper roller bearing for differential, ENV X13268 ...	2	549465	
13		Differential bearing adjuster, ENV D16102 ...	2	549457	
14		Locking plate, ENV D16135 } For bearing adjuster	2	549411	
15		Set bolt (1/4" UNC x 1/2" long), ENV D80777	2	549447	
16		Nose piece and bearing cap complete,	1	253004	
17		Special set bolt fixing bearing cap, ENV D16139	4	600902	
18		Bearing for bevel pinion, nose end, ENV X13623	1	549409	
19		Retaining washer, ENV D16107 } Fixing bearing to nose piece	1	549417	
19		Alum. rivet (3/8" dia. x 1 1/4" long), ENV X14672	4	549418	
20		Circlip, ENV X14338	1	4560	
21		Bearing for bevel pinion, ENV X13265 ...	2	549419	
22		Spacer .370", ENV D16112	As reqd	549420	
		Spacer .373", ENV D16114	As reqd	549425	
		Spacer .376", ENV D16116	As reqd	549427	
		Spacer .379", ENV D16118	As reqd	549429	
		Spacer .382", ENV D16120	As reqd	549431	For pinion shaft bearing adjustment
		Spacer .385", ENV D16122	As reqd	549433	
		Spacer .388", ENV D16124	As reqd	549435	
		Spacer .391", ENV D16126	As reqd	549437	
		Spacer .394", ENV D16128	As reqd	549439	
		Spacer .397", ENV D16130	As reqd	549441	
		Spacer .400", ENV D16132	As reqd	549443	
				549445	

* Asterisk indicates a new part which has not been used on any previous Rover model

H625

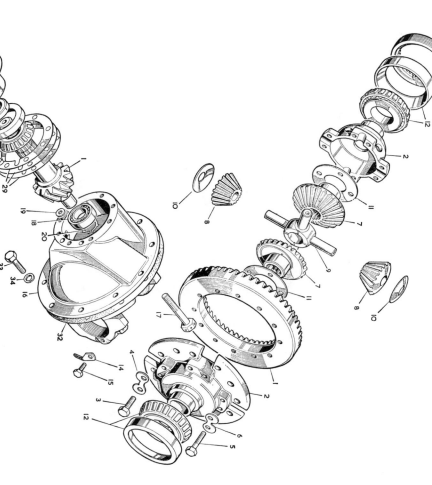

DIFFERENTIAL, REAR, ENV TYPE AXLE, 109 OPTIONAL

Plate Ref.	1 2 3 4	DESCRIPTION	Qty	Part No.	REMARKS
23		Bevel pinion housing complete, ENV D16103 ...	1	549413	
24		Oil seal for pinion, ENV X13022	1	549412	
25		Mudshield for bevel pinion bearing housing, ENV D17603	1	549454	
26		Mudshield for driving flange, ENV D17602 ...	1	549453	
27		Driving flange for bevel pinion, ENV D16108 ...	1	549421	
28		Special nut ($\frac{7}{8}$" UNF), ENV X12403, fixing driving flange to bevel pinion	1	549448	
29		Shim .002", ENV D16104 ⎫ For differential	As reqd	549414	
		Shim .003", ENV D16105 ⎬ bearing hous-	As reqd	549415	
		Shim .010", ENV D16106 ⎭ ing adjustment	As reqd	549416	
30		Set bolt ($\frac{3}{8}$" UNF x 1$\frac{1}{8}$" ⎫ Fixing pinion			
		long) ⎬ housing to	8	255248	
31		Shakeproof washer ⎭ nose piece	8	70823	
32		Joint washer, differential to axle casing, ENV A16429	1	533645	
33		Set bolt ($\frac{3}{8}$" UNF x 1$\frac{1}{8}$" ⎫ Fixing			
		long) ⎬ differential to	10	255248	
34		Spring washer ⎭ axle casing	10	3076	

UNIVERSAL JOINTS AND FRONT HUBS

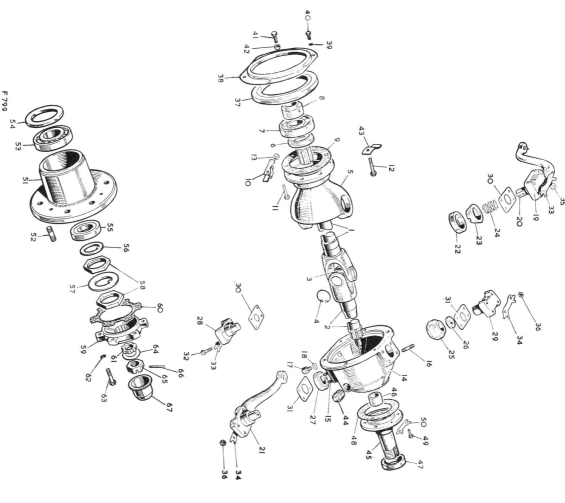

F 799

Plate Ref.	1 2 3 4	DESCRIPTION	Qty	Part No.	REMARKS
		Halfshaft complete for front axle, RH	1	269265	
		Halfshaft complete for front axle, LH	1	269266	
1		Halfshaft only, RH	1	276720	
1		Halfshaft only, LH	1	276719	
2		Stub shaft	1	242520	
3		Journal assembly	2	242521	
4		Circlip for journal	8	242522	
5		Housing for swivel pin bearing ...	2	234789	
6		Distance piece for bearing ...	2	244151	
7		Bearing for halfshaft	2	244150	
8		Retaining collar for bearing ...	2	217398	
9		Joint washer for housing ...	2	232413	
10		Location stop for jack	1	519206	
11		Bolt ($\frac{3}{8}$" BSF x 1$\frac{1}{4}$" long) } Fixing stop	10	237340	With standard axle casing
12		Bolt ($\frac{3}{8}$" BSF x 1$\frac{1}{8}$" long) } and housing	2	237343	10 off with reinforced axle casing
12		Bolt ($\frac{3}{8}$" BSF x 1$\frac{1}{2}$" long) } to front axle	2	237342	With reinforced axle casing
13		Self-locking nut ($\frac{3}{8}$" BSF) ...	12	50526	
14		HOUSING ASSEMBLY FOR SWIVEL PIN, LH } casing	1	508153	8 off for axles with non-pendant type ball joints and $\frac{3}{8}$" studs
14		HOUSING ASSEMBLY FOR SWIVEL PIN, RH	1	531005	8 off for axles with non-pendant type $\frac{7}{16}$" steering lever studs
		HOUSING ASSEMBLY FOR SWIVEL PIN, LH	1	531004	For axles with
		HOUSING ASSEMBLY FOR SWIVEL PIN, RH } $\frac{3}{8}$" steering lever studs	1	524875	
		Special stud, $\frac{3}{8}$", for steering lever and bracket ...	2	524874	For axles with
15		Special stud, $\frac{7}{16}$" for steering lever and bracket	4	531043	8 off for axles with non-pendant type ball joints and $\frac{3}{8}$" studs
16		Stud, $\frac{7}{16}$", steering lever	4	531494	joints and $\frac{7}{16}$" studs
17		Set bolt ($\frac{7}{16}$" BSF x 1$\frac{3}{8}$" long) fixing steering lever	8	237357	Alternative fixing to two lines above
18		Drain plug for housing	8	236070	Not part of
19		Joint washer for drain plug ...	2	230511	assemblies 531004/5
19		Swivel pin and steering lever complete, RH	2	274145	RHStg
		Swivel pin and steering lever complete, LH	1	274147	RHStg
		Swivel pin and steering lever complete, RH	1	274146	LHStg
20		Swivel pin and steering lever complete, LH	1	274148	LHStg } ball joints
		Swivel pin only	1	239017	For axles with pendant type
		Grooved pin fixing swivel pin	2	50453	
		Swivel pin and steering lever complete, RH	1	502710	RHStg
		Swivel pin and steering lever complete, LH	1	502711	RHStg } For axles with non-pendant type ball joints and $\frac{7}{16}$" studs
		Swivel pin and steering lever complete, RH	1	502713	LHStg
		Swivel pin and steering lever complete, LH	1	502712	LHStg
		Swivel pin and steering lever complete, RH	1	530988	RHStg
21		Swivel pin and steering lever complete, LH	1	530989	RHStg } For axles with non-pendant type ball joints and $\frac{3}{8}$" studs
		Swivel pin and steering lever complete, RH	1	530990	LHStg
		Swivel pin and steering lever complete, LH	1	530991	LHStg
		'O' ring for steering levers ...	2	531433	For axles with non-pendant type ball joints

* Asterisk indicates a new part which has not been used on any previous Rover model

UNIVERSAL JOINTS AND FRONT HUBS

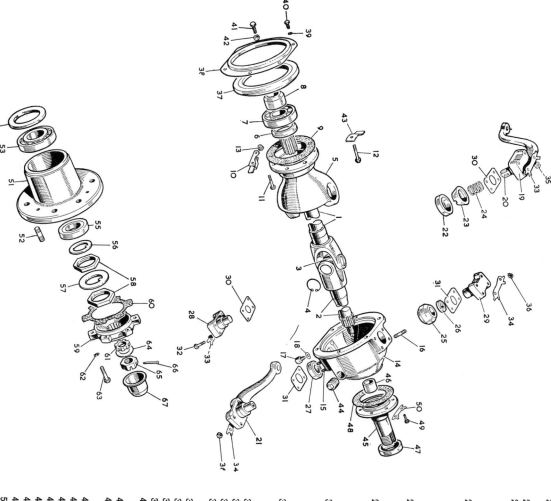

F 799

UNIVERSAL JOINTS AND FRONT HUBS

Plate Ref. 1 2 3 4	DESCRIPTION	Qty	Part No.	REMARKS
22	Cone seat for swivel pin, top ...	2	230858	Cone and spring type steering damping. Up to axles numbered:
23	Cone bearing for swivel pin, top ...	2	238853	24109240 88 RHStg 25107785 109 RHStg
24	Spring washer for cone bearing ...	2	242742	24405088 88 LHStg 25404199 109 LHStg
26	Thrust washer for swivel pin ...	2	528702	Bush and thrust washer type steering damping. From axles numbered:
25	Railko bush and housing ...	2	539742	24109241 88 RHStg 25107786 109 RHStg
				24405089 88 LHStg 25404200 109 LHStg onwards
27	Bearing for swivel pin, bottom ...	2	217268	
28	Swivel pin and bracket complete ...	2	530992	For axles with non-pendant type ball joints and $\frac{7}{16}$" studs steering damping
	Swivel pin and bracket complete ...	2	217421	For axles with $\frac{7}{16}$" studs
	Swivel pin and bracket complete ...	2	502714	For axles with $\frac{3}{8}$" studs
29	Swivel pin and bracket complete ...	2	532963	For axles with pendant type ball joints
	Swivel pin and bracket complete ...	2	530987	For axles with non-pendant type ball joints and $\frac{3}{8}$" studs. Bush and thrust washer type steering damping
	Swivel pin and bracket complete ...	2	530986	For axles with $\frac{7}{16}$" studs
	Swivel pin and bracket complete ...	2	530985	For axles with $\frac{7}{16}$" studs
	Swivel pin and bracket complete ...	2	530984	For axles with $\frac{3}{8}$" studs
30	Shim .003"	As reqd	217453	For swivel pin bearing
	Shim .005"	As reqd	217454	
	Shim .010"	As reqd	217455	
	Shim .030"	As reqd	230007	
31	Shim .003"	As reqd		For swivel pin bearing
	Shim .005"	As reqd		
	Shim .010"	As reqd		
	Shim .030"	As reqd		
32	Set bolt ($\frac{3}{8}$" BSF x 1$\frac{3}{8}$" long)	8	250542	Fixing swivel pin to swivel pin housing
33	Locker	8	270287	
34	Locker	8	531001	
35	Nut ($\frac{3}{8}$" BSF)	8	2827	
36	Nut ($\frac{7}{16}$" BSF)	16	2096	
37	Oil seal for swivel pin bearing housing	2	502406	
38	Retainer for oil seal	2	235968	
39	Spring washer	10	3074	Fixing retainer to swivel pin housing
40	Set bolt ($\frac{1}{4}$" BSF x $\frac{1}{2}$" long)	10	237139	
41	Plain washer	10	3840	
42	Locknut ($\frac{1}{4}$" BSF) for stop bolt	2	2823	
43	Adjustable lock stop bolt	2	250696	
44	Lock stop plate	2	508175	
45	Oil filler plug for swivel pin housing	2	3292	
	STUB AXLE ASSEMBLY ...	2	556373	
46	Bush for driving shaft	2	217354	For early type stub axle
47	Distance piece for inner bearing	2	217351	
48	Joint washer, stub axle to swivel pin housing	2	237289	
49	Set bolt ($\frac{3}{8}$" BSF x 1" long)	12	237339	Fixing stub axle to swivel pin housing. 4 off on 109 and when 7.00 x 16, or 8.20 x 15 tyres are fitted
50	Locker	6	277311	

* Asterisk indicates a new part which has not been used on any previous Rover model

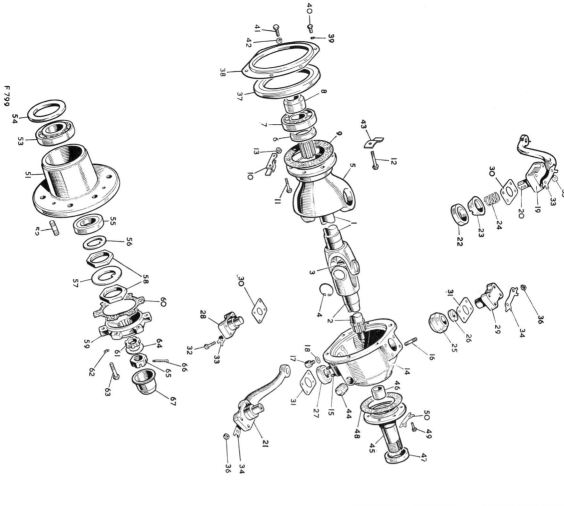

F 799

UNIVERSAL JOINTS AND FRONT HUBS

Plate Ref. 1 2 3 4	DESCRIPTION	Qty	Part No.	REMARKS
51	FRONT HUB ASSEMBLY	2	576475	Grease-packed type
52	Stud for road wheel	10	561590	
53	Bearing for front hub, inner	2	217269	
54	Oil seal for inner bearing	2	239422	
55	Bearing for front hub, outer	2	217270	
56	Key washer	2	217352	
57	Locker ⎱ Fixing front hub bearing	2	217353	
58	Special nut ⎰	4	217355	
59	Driving member complete for front hub	2	571235	Grease-packed type
60	Joint washer ⎱ For driving	2	232038	Early models
	Oil filler plug ⎰ member	2	556204 ⎱ with oil filled hub
61	Joint washer for driving member	2	231505 ⎰
62	Oil seal for stub shaft	2	501412	
63	Spring washer ⎱ Fixing driving	12	3076	
	Set bolt (⅜" BSF x 1³⁄₃₂" long) ⎰ member to front hub	12	215331	
64	Plain washer ⎱ Fixing	12	571922	
65	Slotted nut ⎰ driving member	2	3259	
66	Split pin ⎰ to driving shaft	2	2428	
67	Hub cap, front	2	219098	
	Swivel pin stud conversion kit	1	532329	To convert ⅜" stud to ⁷⁄₁₆" stud on non-pendant type steering levers only
	Swivel pin conversion kit	1	532268	To convert cone and spring steering damping to bush and thrust washer type

STEERING COLUMN

H753

STEERING COLUMN

Plate Ref. 1 2 3 4	DESCRIPTION	Qty	Part No.	REMARKS
1 2 3 4	STEERING UNIT ASSEMBLY	1	509449	RHStg With centre horn push. For 5/16" diameter support bracket bolts. Up to vehicle suffix 'A'
	STEERING UNIT ASSEMBLY	1	509448	LHStg With centre horn push. For 5/16" diameter support bracket bolts. Up to vehicle suffix 'A'
	STEERING UNIT ASSEMBLY	1	537541	RHStg With centre horn push. For 3/8" diameter support bracket bolts. Vehicle suffix 'B' only
	STEERING UNIT ASSEMBLY	1	537542	LHStg With centre horn push. For 3/8" diameter support bracket bolts. Vehicle suffix 'B' only
	STEERING UNIT ASSEMBLY	1	551702	RHStg With centre horn push. For 3/8" diameter support bracket bolts. From vehicle suffix 'C' onwards
	STEERING UNIT ASSEMBLY	1	551703	LHStg With centre horn push. For 3/8" diameter support bracket bolts. From vehicle suffix 'C' onwards
1	STEERING BOX ASSEMBLY	1	515508	RHStg Dowel type. For 5/16" diameter support bracket bolts. Up to vehicle suffix 'A'
	STEERING BOX ASSEMBLY	1	515509	LHStg Dowel type. For 5/16" diameter support bracket bolts. Up to vehicle suffix 'A'
	STEERING BOX ASSEMBLY	1	541899	RHStg Dowel type. For 3/8" diameter support bracket bolts. From vehicle suffix 'B' onwards
	STEERING BOX ASSEMBLY	1	541900	LHStg Dowel type. For 3/8" diameter support bracket bolts. From vehicle suffix 'B' onwards
2	Bush for rocker shaft	1	261850	Models with centre horn push, From vehicle suffix 'C' onwards
3	Outer column, overall length 23½"	1	271364	Models with side horn push. Check before ordering
	Outer column, overall length 24½"	1	600468	Models with centre horn push. Up to vehicle suffix 'B' inclusive
	Outer column	1	600984	Models with centre horn push, From vehicle suffix 'C' onwards
4	Joint washer, steel	As reqd	261858	
5	Joint washer, paper	As reqd	261857	
	Joint washer, steel	As reqd	271379	
	Bolt (5/16" UNC x 2 9/32" long) } Fixing outer column	4	271380	
	Spring washer }	4	3075	
6	Inner column	1	271367	RHStg Models with side horn push.
	Inner column	1	271368	LHStg Models with side horn push.
	Inner column	1	518676	RHStg Models with centre horn push. Up to vehicle suffix 'B' inclusive
	Inner column	1	518677	LHStg Models with centre horn push. Up to vehicle suffix 'B' inclusive
7	Inner column	1	600985	RHStg Models with centre horn push. From vehicle suffix 'C' onwards
	Inner column	1	600986	LHStg Models with centre horn push. From vehicle suffix 'C' onwards
8	Bush for inner column	1	501245	Part of outer column.
9	Spring ring for inner column bush	1	501246	Up to vehicle suffix 'B' inclusive
10	Ball bearing for inner column	1	279072	From vehicle suffix 'C' onwards
11	Dust shield for inner column	1	267016	Models with side horn push.
12	MAIN NUT ASSEMBLY	1	261862	RHStg
	MAIN NUT ASSEMBLY	1	261865	LHStg
	Set bolt (2 BA x ½" long) } Fixing retainer	2	234603	
	Lockwasher for bolt }	2	261868	
13	Steel ball (⅜") for main nut	12	261869	
14	Roller for main nut	2	1643	
15	Adjustable ball race	1	271384	
16	Steel balls (.280") for adjustable race	20	260823	

* Asterisk indicates a new part which has not been used on any previous Rover model

STEERING COLUMN

H753

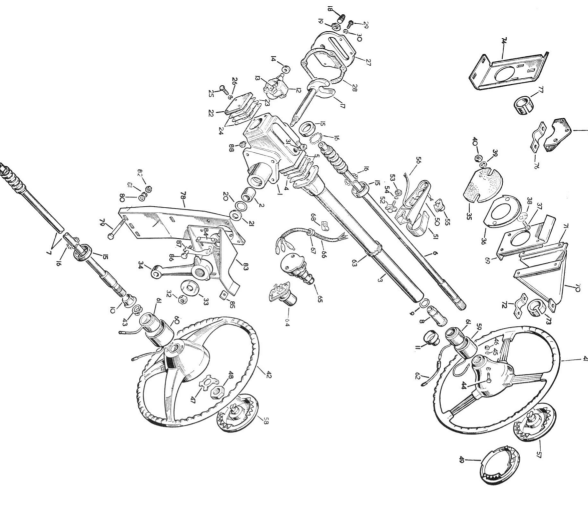

STEERING COLUMN

Plate Ref.	1 2 3 4	DESCRIPTION	Qty	Part No.	REMARKS
17		Rocker shaft	1	271372	
18		Adjuster screw for rocker shaft	1	261873	
19		Locknut for adjuster screw	1	261874	
20		Oil seal for rocker shaft	1	261851	
21		Washer for rocker shaft oil seal	1	271377	
22		End plate	1	271378	
23		Joint washer, steel	As reqd	261858	
24		Joint washer, paper	As reqd	261857	
25		Joint washer, paper	As reqd	271379	
26		Bolt ($\frac{5}{16}$" UNC x $\frac{3}{32}$" long) ⎱ Fixing end plate	4	271380	
		Spring washer ⎰	4	3075	
27		Side cover plate	1	272181	Plain type ⎱ Alternatives.
			1	261880	Dowel type ⎰ Check before ordering
28		Joint washer for side cover plate	1	515848	
29		Bolt ($\frac{5}{16}$" UNC x $1\frac{1}{16}$" long) ⎱ Fixing side	4	515849	
30		Spring washer ⎰ cover plate	4	271382	
31		Oil filler plug	1	3075	
32		Special nut ⎱ Fixing	1	303996	
33		Lockwasher ⎰ drop arm	1	77861	
34		Steering drop arm	1	303997	Up to vehicle suffix 'B' inclusive
35		Rubber seal for steering column	1	303995	From vehicle suffix 'C' onwards
36		Cover for steering column seal	1	4115	
37		Screw (2 BA x $\frac{3}{4}$" long) ⎱ Fixing cover and	2	517507	
38		Special washer ⎰ seal to dash	2	254877	
39		Spring washer	2	271386	
40		Nut (2 BA)	2	3073	
41		Steering wheel	1	2247	
42		Steering wheel	1	512322	Up to vehicle suffix 'B' inclusive
43		Special spring washer on inner column for wheel	1	551985	From vehicle suffix 'C' onwards
44		Bolt ($\frac{5}{16}$" UNF x 2" long) ⎱ Fixing steering wheel	1	551984	From vehicle suffix 'C' onwards
45		Plain washer ⎰	1	256226	
46		Nut ($\frac{5}{16}$" UNF)	1	2550	
47		Tag washer ⎱ Fixing steering	1	254431	Up to vehicle suffix 'B' inclusive
48		Special nut ⎰ wheel	1	552572	From vehicle suffix 'C' onwards
49		Steering wheel centre cover	1	552804	
50		Horn push bracket	1	268284	
51		Clip for horn push bracket	1	271496	
52		Yoke assembly for horn push bracket	1	277560	
53		Nut ($\frac{1}{4}$" BSF) ⎱ Fixing horn push	2	270724	
54		Shakeproof washer ⎰ bracket	2	2823	
55		Horn push	1	78114	Models with side horn push.
		Special screw, LU 153564 ⎱ Fixing	2	217279	
		Spring washer ⎰ horn push	2	3279	
		Nut (5 BA) to bracket	2	3072	
				4011	
56		Lead, horn push to junction box	1	270961	Check before ordering

STEERING COLUMN

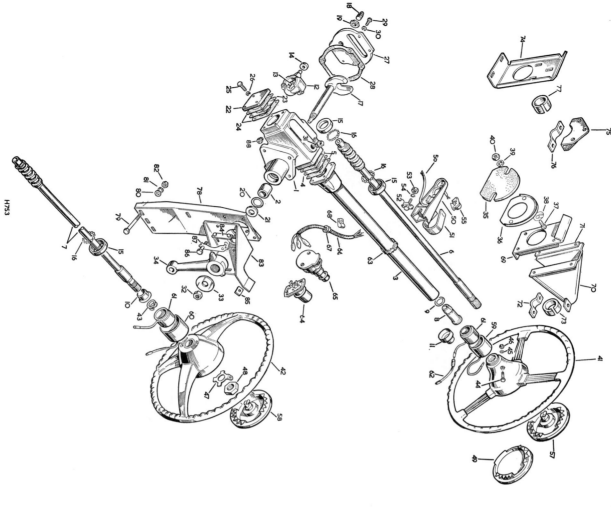

H753

STEERING COLUMN

Plate Ref.	1 2 3 4	DESCRIPTION	Qty	Part No.	REMARKS
57		Horn push and centre cover for steering wheel ...	1	512352	Up to vehicle suffix 'B' ⎱ Models with centre horn push. Check before ordering
				551979	From vehicle suffix 'C' onwards ⎰
58		Horn push and centre cover for steering wheel	1	512359	Up to vehicle suffix 'B' ⎱ Models with centre horn push. Check before ordering
				67184	From vehicle suffix 'C' onwards ⎰
59		Dust cover and horn contact	1	552575	From vehicle suffix 'C' onwards. Check before ordering
60		Dust cover and horn contact	1	519753	Up to vehicle suffix 'B'
				519755	Series II only
61		Slip ring complete for horn contact	1	3902	
62		Lead, slip ring to junction box	1	240431	
63		Cable cleat on steering column	1	232026	
64		Dip switch	1	77869	1958-60. Except 2¼ litre Petrol LHStg models when toe box heat shields are fitted
		Screw (2 BA x ¼" long)	2	3073	
		Spring washer	2	3823	
		Nut (2 BA)	2	332199	
		Heat protection cover for dip switch	1	237121	LHStg. 1958-60
		Bolt (2 BA x ⅝" long)	2	3851	
		Plain washer	2	257011	1958-60 2¼ litre Petrol LHStg models when toe box heat shields are fitted
		Spring washer	2	3073	
		Nut (No. 10 UNF)	2	257011	
		Dip switch, LU 34790	1	514827	
		Spring washer	2	236389	
		Nut (2 BA)	2	8885	
65		Mounting plate for dip switch	1	502087	Alternatives
		Screw (No. 10 UNF x 1" long)	2	541602	
		Spring washer	2	78316	
		Nut (10 UNF)	2	3073	Alternatives. 1961 models
		Bolt (No. 10 UNF x ½" long)	2	257011	UNF ⎱ fixing Check before ordering
		Spring washer	2	78173	2 BA ⎰
		Nut (No. 10 UNF)	2	3823	
		Dip switch, LU type 103SA	1	503428	1958-60 models
		Dip switch, LU type 21SA	1	519638	RHStg 1961
		Cable clip, lead to mounting plate	1	514827	RHStg Series II models
		Grommet for lead in toe box floor	1	514840	LHStg Series IIA
		Lead, dip switch to junction box	1	531501	RHStg 4-cylinder models
66		Screw (2 BA x ¼" long) ⎱ Fixing dip switch	2	560533	LHStg Series IIA
		Spring washer ⎰	2	560534	RHStg 6-cylinder models
		Nut (2 BA)	2	560968	LHStg Series IIA
		Lead	1	531501	LHStg 4 and 6 cylinder models

H753

Plate Ref. 1 2 3 4	DESCRIPTION		Qty	Part No.	REMARKS
67	Grommet for lead in toe box floor	...	1	236389	1961 models onwards
68	Clip fixing clip switch lead to floor	...	2	50639	
	Screw		2	77941	5 off on LHStg
	Spring washer	} Fixing clip	2	3073	
	Nut (2 BA)		2	2247	
69	Support bracket on dash	...	1	302986	
	Bolt ($\frac{5}{16}$" UNF x $\frac{5}{8}$" long)	} Fixing	4	255025	
	Plain washer, small	support	2	3868	
	Spring washer	bracket	2	3662	
	Plain washer, large	to dash	4	3075	
	Nut ($\frac{5}{16}$" UNF)		4	254831	
70	Support bracket for steering column	...	1	303701	
	Bolt ($\frac{5}{16}$" UNF x $\frac{3}{4}$" long)	} Fixing support	4	255226	
	Plain washer, large	bracket to	2	3830	
	Spring washer	dash bracket	4	3662	
	Nut ($\frac{5}{16}$" UNF)		4	254831	
71	Packing piece for steering column support bracket	...	1	332729	
72	Clip for steering column	...	1	300715	
73	Rubber strip for clip	...	2	300716	Up to vehicle suffix 'C' inclusive
	Bolt ($\frac{1}{4}$" UNF x $\frac{3}{4}$" long)	} Fixing clip	2	255207	
	Spring washer	to support	2	3074	
	Nut ($\frac{1}{4}$" UNF)	bracket	2	254810	
74	Support bracket on dash	...	1	348743	
	Bolt ($\frac{5}{16}$" UNF x $\frac{3}{4}$" long)	} Fixing	4	255226	From vehicle suffix 'D' onwards
	Plain washer, small	support	2	3830	
	Spring washer	bracket	4	3662	
	Plain washer, large	to dash	2	3075	
	Nut ($\frac{5}{16}$" UNF)		4	254831	
75	Clamp, upper, for steering column	...	1	348744	
	Bolt ($\frac{5}{16}$" UNF x $\frac{3}{4}$" long)	} Fixing	3	255226	
	Plain washer, small	support	3	3830	
	Spring washer	bracket	3	3075	
	Nut ($\frac{5}{16}$" UNF)		4	254831	
76	Clamp, lower, for steering column	...	1	348745	
	Spring washer	} Fixing upper and lower clamps to steering column	2	254831	
	Nut ($\frac{1}{4}$" UNF)		2	348747	
77	Rubber strip for clamp	...	1	255208	
	Bolt ($\frac{1}{4}$" UNF x $\frac{7}{8}$" long)	} Fixing upper clamp to support bracket	1	3074	
	Spring washer		3	254810	
	Nut ($\frac{1}{4}$" UNF)		3	254831	
	Plain washer		3	3830	
	Plain washer, small		6	3075	
78	Support bracket, RH	} For steering box on chassis	1	277294	For $\frac{5}{16}$" diameter steering box fixing bolts. Up to vehicle suffix 'A'
	Support bracket, LH		1	277295	
	Support bracket, RH		1	537535	For $\frac{3}{8}$" diameter steering box fixing bolts. From vehicle suffix 'B' onwards
	Support bracket, LH		1	537533	
79	Packing piece for support bracket	...	1	256233	
	Bolt ($\frac{5}{16}$" UNF x $3\frac{1}{4}$" long)	} Fixing brackets to chassis frame	6	569522*	LHStg
80	Plain washer, thin		6	256233	
81	Plain washer, thick		6	3830	
	Spring washer		6	3898	
82	Nut ($\frac{5}{16}$" UNF)		6	3075	
			6	254831	

STEERING COLUMN

H753

STEERING COLUMN

Plate Ref.	1 2 3 4	DESCRIPTION		Qty	Part No.	REMARKS
83		Stiffener bracket		1	504276	RHStg } For 5/16" diameter steering box
		Stiffener bracket		1	504272	LHStg } fixing bolts. Up to vehicle suffix 'A'
		Stiffener bracket		1	537537	RHStg } For 3/8" diameter steering box
		Stiffener bracket	For steering box	1	537539	LHStg } fixing bolts. From vehicle suffix 'B' onwards
		Bolt (1/4" UNF x 9/16" long)	Fixing stiffener bracket to front face toe box	4	255005	Alternatives to line below
		Plain washer		4	3840	
				4	3074	
84		Bolt plate		2	330908	Alternative to 2 lines above
		Bolt (1/4" UNF x 1" long)		4	254810	
		Nut (1/4" UNF)		4	254810	
		Spring washer		1	255009	RHStg
		Plain washer		1	3840	3 off on LHStg
		Nut (1/4" UNF)	Fixing stiffener bracket to top face of toe box	1	3074	
				1	254810	
85		Shim washer		As reqd	504279	RHStg
		Shim washer		As reqd	504275	LHStg
		Bolt (5/16" UNF x 1 3/4" long)		2	256021	
		Set bolt (5/16" UNC x 3/4" long)	Fixing steering box	1	253026	Except 109 LHStg
		Set bolt (5/16" UNC x 7/8" long)		1	253027	109 LHStg
		Locking plate		1	517877	
86		Set bolt (3/8" UNC x 7/8" long)	Fixing steering box support bracket	1	253046	5/16" fixings. Up to vehicle suffix 'A' inclusive
87		Locking plate		2	517878	
88		Self-locking nut (5/16" UNF)		2	252211	
		Bolt (3/8" UNF x 1 1/8" long)		2	252211	
		Set bolt (3/8" UNC x 3/4" long)	Fixing steering box support bracket	2	253045	
		Set bolt (3/8" UNC x 7/8" long)		2	253046	
		Self-locking nut (3/8" UNF)		2	252050	3/8" fixings. From vehicle suffix 'B' onwards
		Set bolt (3/8" UNC x 7/8" long)		1	253046	
		Locking plate	Fixing steering box to chassis	1	537543	Except 109 LHStg
		Set bolt (3/8" UNC x 7/8" long)		1	253046	109 LHStg
		Locking plate		1	537544	
		Self-locking nut (3/8" UNF)	support bracket	2	252212	

* Asterisk indicates a new part which has not been used on any previous Rover model

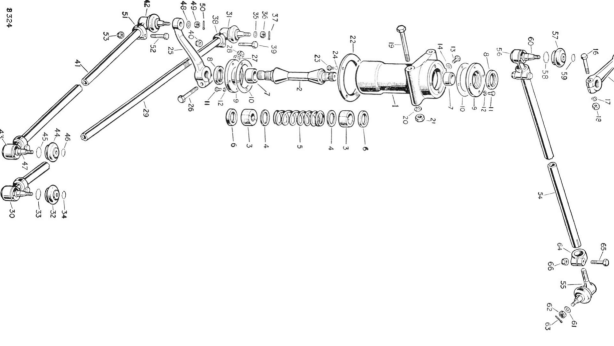

B 324

STEERING RELAY AND LINKAGE

Plate Ref. 1 2 3 4	DESCRIPTION	Qty	Part No.	REMARKS
	STEERING RELAY COMPLETE ASSEMBLY	1	543351	
1	Housing for relay shaft	1	562874	Housing with plug oil hole
1	Housing for relay shaft	1	543972	Housing without oil plug hole } Alternatives
2	Shaft for steering relay levers	1	562875	With oil plug hole
			537877	Without oil plug hole } Alternatives
3	Split bush for housing, halves	4	230760	
4	Washer for spring	1	230759	
5	Spring for housing	1	241388	
6	Thrust washer for shaft	2	230184	For early type shaft only
7	Distance piece for shaft	2	213340	
8	Oil seal for shaft	2	230294	
9	Retainer for oil seal	2	230295	
10	Joint washer for retainer	2	544337	} Fixing retainer
11	Special bolt	8	3074	} to housing
12	Spring washer	8	3052	
13	Plug for oil hole	2	237138	} For housing with oil plug hole
14	Joint washer for plug	2	531040	
15	Relay lever, upper	1	256465	
16	Bolt ($\frac{7}{16}$" UNF x 2$\frac{1}{4}$" long)	1	252163	} Fixing lever to shaft
18	Self-locking nut ($\frac{7}{16}$" UNF)	1	256432	
19	Bolt ($\frac{5}{16}$" UNF x 3$\frac{1}{2}$" long)	2	254825*	} Fixing housing to chassis frame
20	Spring washer	2	254821	
21	Nut ($\frac{5}{16}$" UNF)	2	217694	
22	Flange plate for relay mounting	1	255206	} Fixing flange plate to chassis frame
23	Set bolt ($\frac{1}{4}$" UNF x $\frac{5}{8}$" long)	4	3074	
24	Spring washer	4	3052	
25	Relay lever, lower	1	535286	} Fixing lever to shaft
26	Bolt ($\frac{7}{16}$" UNF x 2$\frac{1}{4}$" long)	1	256465	
28	Self-locking nut ($\frac{7}{16}$" UNF)	1	252163	
29	STEERING TRACK ROD ASSEMBLY (length between centres of ball joints 46$\frac{1}{4}$")	1	269267	} For pendant type ball joints. Check before ordering
	STEERING TRACK ROD ASSEMBLY (length between centres of ball joints 45$\frac{5}{8}$")	1	269269	
	STEERING TRACK ROD ASSEMBLY (length between centres of ball joints 43$\frac{3}{4}$")	1	526993	} For non-pendant type ball joints.
	Steering track rod only, overall length 43$\frac{3}{8}$"	1	526994	} Check before ordering
30	BALL JOINT ASSEMBLY, RH THREAD	1	233183	
31	BALL JOINT ASSEMBLY, LH THREAD	1	231184	
32	Rubber cover for ball joint	2	214649	
33	Spring ring, cover to body	2	214685	
34	Spring ring } Cover	2	214684	
	Retainer } to ball	2	214662	
35	Plain washer	2	3680	
36	Castle nut } Fixing ball joints to lever	2	2822	
37	Split pin	2	2393	

B 324

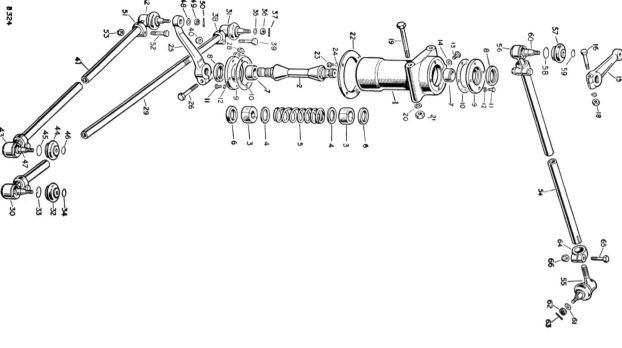

Plate Ref. 1 2 3 4	DESCRIPTION	Qty	Part No.	REMARKS
38	Clip for ball joint	2	552038	
39	Bolt (¼" BSF x 1½" long)	2	250517	BSF type, short bolt
39	Self-locking nut (¼" BSF)	2	251320	
40	Bolt (¼" UNF x 1⅞" long)	2	256006	UNF type, long bolt
40	Plain washer	2	3911	} Fixing ball joint clips
40	Self-locking nut (¼" UNF)	2	252210	Alternative fixings. Check before ordering
41	STEERING DRAG LINK ASSEMBLY	1	535070	
42	Steering drag link only	1	269270	
43	BALL JOINT ASSEMBLY, RH THREAD	1	231183	
44	BALL JOINT ASSEMBLY, LH THREAD	1	231184	
45	Rubber cover for ball joint	2	214649	
46	Spring ring, cover to body	2	214685	
47	Spring ring	2	214684	} to ball
48	Retainer	2	214662	
49	Plain washer	2	3680	
50	Castle nut	2	2822	} Fixing ball joints to levers
51	Split pin	2	2393	
52	Clip for ball joint	2	552038	
53	Bolt (¼" BSF x 1½" long)	2	250517	BSF type, short bolt
53	Self-locking nut (¼" BSF)	2	251320	
53	Bolt (¼" UNF x 1⅞" long)	2	256006	UNF type, long bolt
53	Plain washer	2	3911	} Fixing ball joint clips
53	Self-locking nut (¼" UNF)	2	255210	Alternative fixings. Check before ordering
54	LONGITUDINAL STEERING TUBE ASSEMBLY	1	276784	
55	Longitudinal steering tube only	1	276785	
56	BALL JOINT ASSEMBLY, RH THREAD	1	231183	
57	BALL JOINT ASSEMBLY, LH THREAD	1	231184	
58	Rubber cover for ball joint	2	214649	
59	Spring ring, cover to body	2	214685	
60	Spring ring	2	214684	} to ball
61	Retainer	2	214662	
62	Plain washer	2	3680	
63	Castle nut	2	2822	} Fixing ball joints to levers
64	Split pin	2	2393	
65	Clip for ball joint	2	552038	
65	Bolt (¼" BSF x 1½" long)	2	250517	BSF type, short bolt
65	Self-locking nut (¼" BSF)	2	251320	
66	Bolt (¼" UNF x 1⅞" long)	2	256006	UNF type, long bolt
66	Plain washer	2	3911	} Fixing ball joint clips
66	Self-locking nut (¼" UNF)	2	252210	Alternative fixings. Check before ordering

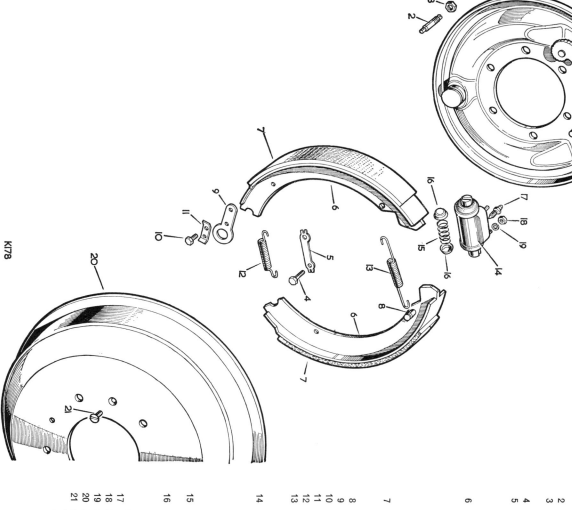

K/78

Plate Ref.	1 2 3 4	DESCRIPTION	Qty	Part No.	REMARKS
1		BRAKE ANCHOR PLATE ASSEMBLY, LH, GI 64270176	2	515406	
		BRAKE ANCHOR PLATE ASSEMBLY, RH, GI 64270177	2	515405	
2		Shoe, steady post, G, GB 37991	8	219620	
3		Locknut for steady post, GI 30-BS-32	8	212835	Up to **front** axles numbered 141001222 and up to **rear** axles numbered 141002007
4		Set bolt (⅜" BSF x 1" long) } Fixing front anchor	12	237339	
5		Locker } plate to axle case	6	232416	
5		Bolt (⅜" BSF x 1⅜" long) } Fixing	12	250642	
		Spring washer } rear anchor plate	12	3076	
		Nut (⅜" BSF) } to axle case	12	2827	
6		BRAKE SHOE ASSEMBLY, LH, BOXED PAIRS, FRONT AND REAR	2	505675	Up to early type shoe with riveted linings
		BRAKE SHOE ASSEMBLY, RH, BOXED PAIRS, FRONT AND REAR	2	505676	Bonded type linings
		Lining complete with rivets, for brake shoe, GI GB 35484 BG	8	241090	For early type shoe with riveted linings. Check before ordering
7		Lining complete with rivets, for brake shoe, GI GB 35484 CQ	8	504577	Riveted type lining for replacement of bonded originals on Export model
					Part of brake shoe
8		Spring post for brake shoe, GI GB 37769	8	232074	
9		Anchor for brake shoe, GI GB 42600	4	236993	
10		Special set screw fixing anchor, GI 25-BS-72	8	238542	
11		Locking plate for bolt, GI GB 42620	4	236995	
12		Pull-off spring for brake shoe, GI GB 36741	4	219883	
13		Pull-off spring for leading shoe, GI GB 16337	4	503981	
14		WHEEL CYLINDER ASSEMBLY, RH FRONT, GI 390073 W	1	243297	
		WHEEL CYLINDER ASSEMBLY, LH FRONT GI 390072 W	1	242296	
		WHEEL CYLINDER ASSEMBLY, RH REAR, GI 390015 W	1	243303	
		WHEEL CYLINDER ASSEMBLY, LH REAR, GI 390014 W	1	243302	
15		Spring for piston, front, GI 378868	2	212919	
		Spring for piston, rear, GI 378860	2	212943	
16		Washer for spring, front, GI 378109	2	232107	
		Washer for spring, rear, GI 378108	2	242117	
17		Bleed screw, GI 377685	4	248509	Ball seated
		Steel ball for bleed screw, GI 27240	4	3746	} type
18		Bleed screw, GI 64470444	4	556508	Cone seated } Alternatives
19		Special nut } Fixing	8	243301	type
20		Spring washer } wheel cylinder	8	219025	
20		Brake drum	4	218149	
21		Set screw fixing brake drum	12	1510	
		Wheel cylinder overhaul kit, front, GI SP 2051	1	275744	
		Wheel cylinder overhaul kit, rear, GI SP 2060	1	266687	

* Asterisk indicates a new part which has not been used on any previous Rover model

290

BRAKES, FOOT, FRONT, 11", 109 4-CYLINDER MODELS

291

BRAKES, FOOT, FRONT, 11", 109 4-CYLINDER MODELS

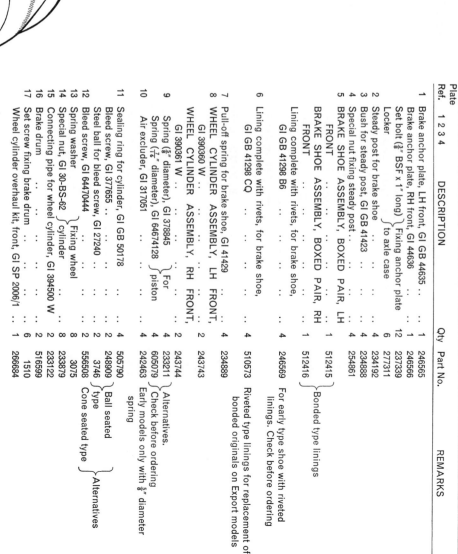

K179

Plate Ref.	1 2 3 4	DESCRIPTION	Qty	Part No.	REMARKS
1		Brake anchor plate, LH front, GI GB 44635	1	246565	
1		Brake anchor plate, RH front, GI 44636	1	246566	
2		Set bolt (⅜" BSF x 1" long) } Fixing anchor plate to axle case	12	237339	
		Locker	6	277311	
3		Steady post for brake shoe	4	234192	
4		Bush for steady post, GI GB 41423	4	234888	
4		Special nut fixing steady post	4	254861	
5		BRAKE SHOE ASSEMBLY, BOXED PAIR, LH FRONT	1	512415 } Bonded type linings	
5		BRAKE SHOE ASSEMBLY, BOXED PAIR, RH FRONT	1	512416	
6		Lining complete with rivets, for brake shoe, GI GB 41298 B6	4	246569	For early type shoe with riveted linings. Check before ordering
6		Lining complete with rivets, for brake shoe, GI GB 41298 CQ	4	510573	Riveted type linings for replacement of bonded originals on Export models
7		Pull-off spring for brake shoe, GI 41429 ...	4	234889	
8		WHEEL CYLINDER ASSEMBLY, LH FRONT, GI 390360 W	2	243743	
8		WHEEL CYLINDER ASSEMBLY, RH FRONT, GI 390361 W	2	243744	
9		Spring (⅝" diameter), GI 378845 } For	4	233211	Alternatives.
9		Spring (₇⁄₁₆" diameter), GI 64674128 } piston	4	605079	Check before ordering
10		Air excluder, GI 317051	4	242463	Early models only with ⅝" diameter spring
11		Sealing ring for cylinder, GI GB 50178 ...	4	505790	
12		Bleed screw, GI 377655 ... ... } cylinder	2	248909	Ball seated
12		Steel ball for bleed screw, GI 27240 ...	2	3746	type
13		Bleed screw, GI 64470444 ... ... } Fixing wheel	2	556508	Cone seated type } Alternatives
13		Spring washer	8	3075	
14		Special nut, GI 30-BS-62	8	233879	
15		Connecting pipe for wheel cylinder, GI 394500 W	2	233122	
16		Brake drum	2	516599	
17		Set screw fixing brake drum	6	1510	
		Wheel cylinder overhaul kit, front, GI SP 2006/1	1	266684	

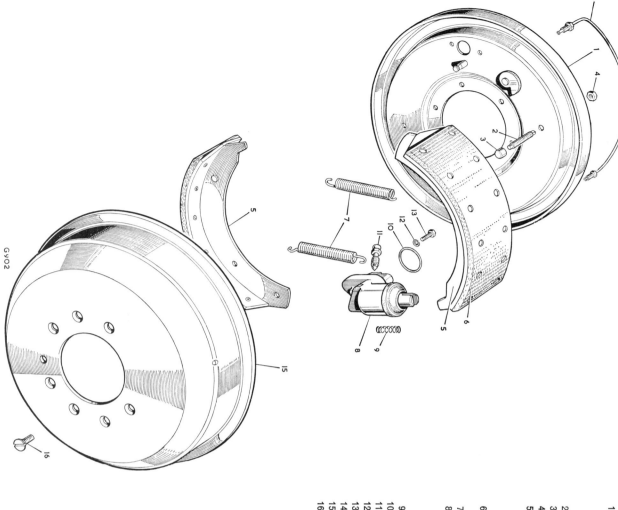

GyO2

BRAKES, FOOT, FRONT, 11″, 109 6-CYLINDER MODELS

Plate Ref.	1 2 3 4	DESCRIPTION	Qty	Part No.	REMARKS
1		Brake anchor plate, LH front, GI 64271324	1	600202	
1		Brake anchor plate, RH front, GI 64271325	1	600203	
2		Set bolt (⅜″ BSF x 1″ long) ⎱ Fixing anchor plate	12	237339	
		Locker ⎰ to axle case	6	277311	
2		Steady post for brake shoe, GI GB 40038	4	234192	
3		Bush for steady post, GI GB 41423	4	234888	
4		Special nut fixing steady post	4	254861	
5		BRAKE SHOE ASSEMBLY, BOXED PAIR, LH FRONT	1	600204	
		BRAKE SHOE ASSEMBLY, BOXED PAIR, RH FRONT	1	600205	
6		Lining complete with rivets, boxed pair, for brake shoe, GI SP 1445	2	600211	
7		Pull-off spring for brake shoe, GI 41429	4	234889	
8		WHEEL CYLINDER ASSEMBLY, LH FRONT, GI 6467107	2	600200	
		WHEEL CYLINDER ASSEMBLY, RH FRONT, GI 6467108	2	600201	
9		Spring, GI 6467128	4	600212	
10		Sealing ring for cylinder, GI GB 50178	4	505790	
11		Bleed screw, GI 6470444	2	556608	
12		Spring washer, GI 40S32 ⎱ Fixing wheel	8	600208	
13		Special set bolt, GI 6410339 ⎰ cylinder	8	600207	
14		Connecting pipe for wheel cylinder, GI 64474364	2	600206	
15		Brake drum	2	522593	
16		Set screw fixing brake drum	6	1510	
		Brake wheel cylinder overhaul kit, front, GI SP 2189	1	600210	

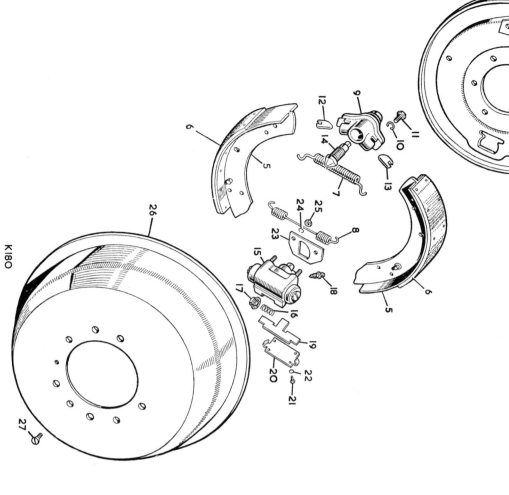

K180

Plate Ref.	1 2 3 4	DESCRIPTION		Qty	Part No.	REMARKS
1		Brake anchor plate, LH rear, GI 42841	...	1	237710	
1		Brake anchor plate, RH rear, GI 42842	...	1	237711	
2		Bolt (⅜" BSF × 1¾" long)	} Fixing rear anchor plate	12	250542	
		Spring washer		12	3076	
		Nut (⅜" BSF)	} to axle case	12	2827	
2		Steady post for brake shoe, GI GB 40038	...	4	234192	
3		Bush for steady post, GI GB 41423	...	4	234888	
4		Special nut for steady post	...	4	254861	
5		Lining complete with rivets, for brake shoe, GI GB 47095	...	4	268763	For early type shoe with riveted linings. Check before ordering. Bonded type linings
6		Lining complete with rivets, for brake shoe	...	4	510575	Riveted type linings for replacement of bonded originals on Export models
5		BRAKE SHOE ASSEMBLY, REAR, BOXED PAIRS		2	532044	
7		Spring, adjuster end, GI GB 2453		2	42318	
8		Spring, wheel cylinder end, GI GB 39757	} For brake shoe	2	242905	
9		Adjuster housing, GI GB 41321		2	234901	
10		Spring washer		4	3076	
11		Special set bolt, GI GB 41442	} Fixing adjuster housing	2	234908	
12		Plunger, LH, GI 41322		2	234902	
13		Plunger, RH, GI 41323		2	234903	
14		Cone for adjuster, GI GB 3D		2	04965	
15		WHEEL CYLINDER ASSEMBLY, LH, GI 64673536I		1	521543	
15		WHEEL CYLINDER ASSEMBLY, RH, GI 64673537		1	521544	
16		Spring, GI 378845		2	232211	
17		Air excluder, GI 317261		4	248240	
18		Bleed screw, GI 377685		2	248909	} Ball seated type
18		Steel ball for bleed screw, GI 27240		2	3746	} Cone seated type Alternatives
		Bleed screw, GI 64470444		2	556508	
19		Brake shoe abutment plate, GI 378133		2	248248	
20		Retainer for brake shoe abutment plate, GI 487143		2	234960	
21		Screw, GI 13112	} Fixing retainer and	8	234962	
22		Shakeproof washer, GI 12437	} abutment plate	8	234963	
23		Dust cover plate for brake wheel cylinder, GI GB 42106	...	2	248247	
24		Spring washer	} Fixing wheel	6	2284	
25		Self-locking nut, GI GB 5232	} cylinder	6	235222	
26		Brake drum	...	2	516599	
27		Set screw fixing brake drum	...	6	1510	
26		Brake wheel cylinder overhaul kit, rear, SP 2004	...	1	266683	With steel dust cover
		Brake wheel cylinder overhaul kit, rear, GI SP 2103	1	523164	With rubber dust cover } Alternatives. Check before ordering

* Asterisk indicates a new part which has not been used on any previous Rover model

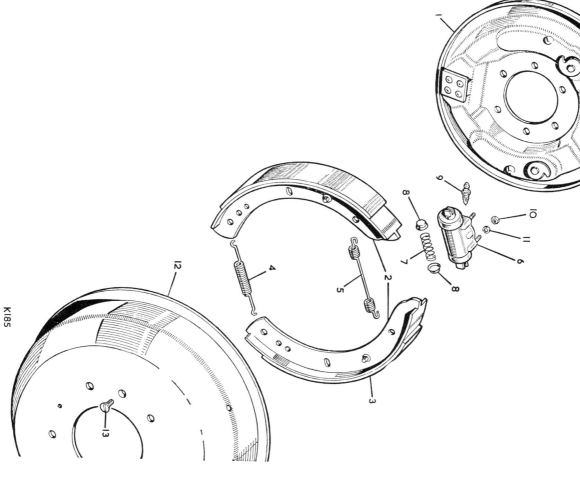

K185

Plate Ref.	1 2 3 4	DESCRIPTION	Qty	Part No.	REMARKS
1		Brake anchor plate, LH rear, GI 64270811	1	531888	
		Brake anchor plate, RH rear, GI 64270812	1	531889	
		Bolt (⅜" BSF x 1⅜" long) ⎱ Fixing rear	12	250542	
		Spring washer ⎰ anchor plate	12	3076	
		Nut ⅜" BSF ⎰ to axle case	12	2827	
2		BRAKE SHOE ASSEMBLY, REAR, BOXED PAIRS	2	532044	Bonded type linings
			2	510575	Riveted type linings for replacement of bonded originals on Export models
3		Lining complete with rivets, for brake shoe ...	4		
4		Spring, abutment end, GI 6437119 ⎱ For brake	2	531893	
5		Spring, wheel cylinder end, GI 64378393 ⎰ shoe	2	548169	
6		WHEEL CYLINDER ASSEMBLY, LH, GI 390072 W	1	243296	
		WHEEL CYLINDER ASSEMBLY, RH, GI 390073 W	1	243297	
7		Spring for piston, GI 378868	2	212919	
8		Washer for spring, GI 378108	4	242117	
		Bleed screw, GI 377685	2	248909 ⎱ Ball seated	
		Steel ball for bleed screw, GI 27240	2	3746 ⎰ type	Alternatives
9		Bleed screw, GI 64470444	4	556608 ⎰ Cone seated type	
10		Special nut ⎱ Fixing wheel	4	243301	
11		Spring washer ⎰ cylinder	4	219025	
12		Brake drum	2	516699	
13		Set screw fixing brake drum	6	1510	
		Brake wheel cylinder overhaul kit, rear, GI SP 2051	1	275744	

BRAKE PIPES, 4-CYLINDER MODELS

K260

BRAKE PIPES, 4 CYLINDER MODELS

Plate Ref. 1 2 3 4	DESCRIPTION	Qty	Part No.	REMARKS
1	Bracket for junction piece	1	241676	
2	Drive screw fixing bracket	2	72626	
3	5-way junction piece for brake pipes	1	279412	
4	4-way junction piece for brake pipes	1	241690	
5	Bolt (¼" UNF x 1¼" long) ⎫ Fixing junction piece	1	256001	
6	Spring washer ⎬	1	3074	
7	Nut (¼" UNF) ⎭	1	254810	
8	Stop lamp switch, LU 34545	1	560075	Hydraulic type. Up to vehicle suffix 'E' inclusive. See page 333 for mechanical type
9	Brake pipe, GI 3306610W ⎫	1	508148	RHStg 88. Up to vehicles numbered
	Brake pipe, GI 3310610W ⎬ Master cylinder to 5-way junction piece	1	512838	LHStg 88. Up to vehicle suffix 'E' inclusive
			271086693D	Diesel
	Brake pipe ⎫	1	564786*	RHStg 88. From vehicles numbered
	Brake pipe ⎬	1	564783*	LHStg 88. From vehicle suffix
	Brake pipe ⎬ Master cylinder to 4-way junction piece	1	569223*	RHStg 'F' onwards
	Brake pipe ⎬	1	569148*	LHStg 109. From vehicle suffix
	Brake pipe ⎬	1	569237*	LHStg 'F' onwards
	Brake pipe ⎬	1	24131337D	Petrol
	Brake pipe ⎭	1	24131336D	Petrol
			27108693D	Diesel
			27108694D	Diesel Up to vehicle suffix 'E' inclusive
10	Brake pipe, junction piece to LH front	1	277922	109. Up to vehicle suffix
11	Brake pipe, junction piece to RH front	1	277923	109. From vehicle suffix
12	Bracket for LH front brake pipe	1	508566	
13	Clip for LH front brake pipe	1	255226	
	Bolt (5/16" UNF x ¾" long) ⎫ Fixing bracket to chassis frame	1	508565	
	Plain washer ⎬	1	70822	
	Shakeproof washer ⎬	1	254831	
	Nut (5/16" UNF) ⎭	1	3830	
	Bolt (2 BA x ½" long) ⎫ Fixing clip	1	2226	
	Plain washer ⎬	1	2247	
	Nut (2 BA) ⎭	1	2226	
14	Hose complete for front wheels, GI 3700628W	2	268341	2¼ litre Petrol, 2 litre and 2¼ litre Diesel models
15	Hose complete to rear axle, GI 3700628W	1	268341	88
	Hose complete to rear axle, GI 64047389	1	235208	109
16	Joint washer for hoses, GI 378711	3	233220	
17	Shakeproof washer, GI 76-BS-27 ⎫ Fixing hose to bracket	3	233305	
18	Special nut ⎭	3	254852	
19	'T' piece on rear axle, GI 35201W	1	234928	Rover type axles
20	'T' piece on rear axle, GI 64474341	1	537761	ENV type axle, 109 optional
21	Nut (5/16" UNF)	1	254831	
22	Spring washer	1	3075	
23	Bolt (5/16" UNF x ¾" long) ⎫ Fixing 'T' piece	1	255226	
24	Bracket for 'T' piece ⎭	1	255225	ENV type axle, 109 optional
	Bolt (5/16" UNF x 5/8" long) ⎫ Fixing bracket to rear axle	1	234928	
	Spring washer ⎬	1	3075	
	Nut (5/16" UNF) ⎭	1	254831	

* Asterisk indicates a new part which has not been used on any previous Rover model

K260

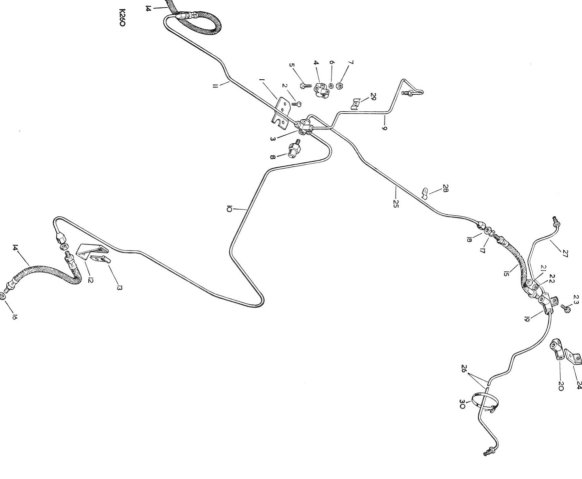

BRAKE PIPES, 4 CYLINDER MODELS

Plate Ref. 1 2 3 4	DESCRIPTION		Qty	Part No.	REMARKS
25	Brake pipe to rear hose, GI 3311230W	... ...	1	279418	88
	Brake pipe to rear hose	... ...	1	279452	109
26	Brake pipe, LH rear to 'T' piece, GI 3307510W	...	1	531905	88. Also 109 with
27	Brake pipe, RH rear to 'T' piece, GI 3304410W	...	1	531906	Rover type axle
	Brake pipe, LH rear to 'T' piece, GI 6474132	...	1	551651	ENV type axle
	Brake pipe, RH rear to 'T' piece, GI 6475774	...	1	551652	optional
28	Clip, brake pipes to chassis frame	... ...	As reqd	41379	
29	Clip, brake and clutch pipes to dash	... As reqd		508945	LHStg
	Clip fixing pipe to steering box stiffener bracket ...		1	513940	RHStg
	Drive screw		As reqd	72626	
	Screw (2 BA x ⅝" long) ⎫ Fixing clips		As reqd	75940	
	Shakeproof washer ⎬		As reqd	71082	LHStg
	Nut (2 BA) ⎭		As reqd	2247	
30	Clip on rear axle for LH pipe	... ...	1	11820	
	Clip		4	56666	
	Rubber grommet		4	06660	
	Bolt (2 BA x ½" long) ⎫ Fixing LH brake pipe		4	234603	
	Plain washer ⎬ to chassis frame		4	3851	
	Spring washer ⎪		4	3073	
	Nut (2 BA) ⎭		4	2247	

* Asterisk indicates a new part which has not been used on any previous Rover model

H044

BRAKE SERVO UNIT AND BRAKE PIPES, 6 CYLINDER MODELS

Plate Ref.	1 2 3 4	DESCRIPTION	Qty	Part No.	REMARKS
1		BRAKE SERVO UNIT	1	562678	
2		Support bracket for brake servo	1	562675	
3		Set bolt ($\frac{5}{16}$" UNC x $\frac{5}{8}$" long) Fixing servo	2	253025	
4		Plain washer to support	2	3830	
5		Spring washer bracket	2	3075	
6		Bolt ($\frac{5}{16}$" UNF x $\frac{3}{4}$" long)	1	255226	
7		Plain washer Fixing support bracket	1	3830	
8		Spring washer to chassis frame	1	3075	
9		Nut ($\frac{5}{16}$" UNF)	1	254831	
10		Spring washer Fixing servo to	2	3075	
11		Nut ($\frac{5}{16}$" UNF) air cleaner support	2	254831	
12		Banjo for servo	1	538614	
13		Banjo bolt	1	538615	
14		Gasket Fixing banjo to servo	2	538616	
15		Gasket	1	538612	
16		Pipe complete, inlet manifold to hose	1	562666	
17		Adaptor Fixing servo pipe to	1	562677	
18		Gasket inlet manifold	1	538616	
19		Clip, fixing servo pipe to water outlet pipe	1	50642	
20		Rubber hose connecting manifold pipe to servo	1	565476	
21		Clip, fixing rubber hose to pipe and brake servo	2	50302	
22		Adaptor for servo pipe	1	538613	
23		Gasket for adaptor	1	538612	
24		Pipe complete, master cylinder to union	1	562904	RHStg
			1	562906	LHStg
25		Pipe complete, master cylinder to union	2	562906	
26		Union for pipe, GI 374102	2	504765	
27		Pipe complete, union to servo	1	562905	
28		Pipe complete, brake servo to junction piece	1	562908	
29		Bracket for junction piece	1	241676	
30		Drive screw fixing bracket	2	72626	
31		Five-way junction piece for brake pipes	1	279412	Up to vehicle suffix 'E' inclusive
			1	241690	From vehicle suffix 'F' onwards
32		4-way junction piece for brake pipes	1	256001	
33		Bolt ($\frac{1}{4}$" UNF x $1\frac{1}{4}$" long) Fixing junction	1	3074	
		piece to			
34		Spring washer support bracket	1	254810	
35		Nut ($\frac{1}{4}$" UNF)	1	560775	Hydraulic type. Up to vehicle suffix 'E' inclusive. See page 333 for mechanical type
36		Stop lamp switch			
35		Brake pipe, junction piece to LH front	1	562711	
36		Brake pipe, junction piece to RH front	1	277923	
37		Hose complete for front brakes	2	266341	
38		Hose complete to rear axle, GI 3700628W	1	235208	
39		Joint washer for hoses, GI 3700631W	3	233220	
40		Shakeproof washer, GI 74330	3	233305	
41		Special nut	3	254852	
42		'T' piece on rear axle, GI 353201W	1	224928	
43		Bolt ($\frac{5}{16}$" UNF x $\frac{3}{4}$" long) Fixing hose	1	255226	
44		Spring washer to bracket	1	3075	
45		Nut ($\frac{5}{16}$" UNF) Fixing 'T' piece	1	254831	
46		Brake pipe to rear hose	1	279452	
47		Brake pipe, LH rear to 'T' piece	1	531905	
48		Brake pipe, RH rear to 'T' piece	1	531906	

* Asterisk indicates a new part which has not been used on any previous Rover model

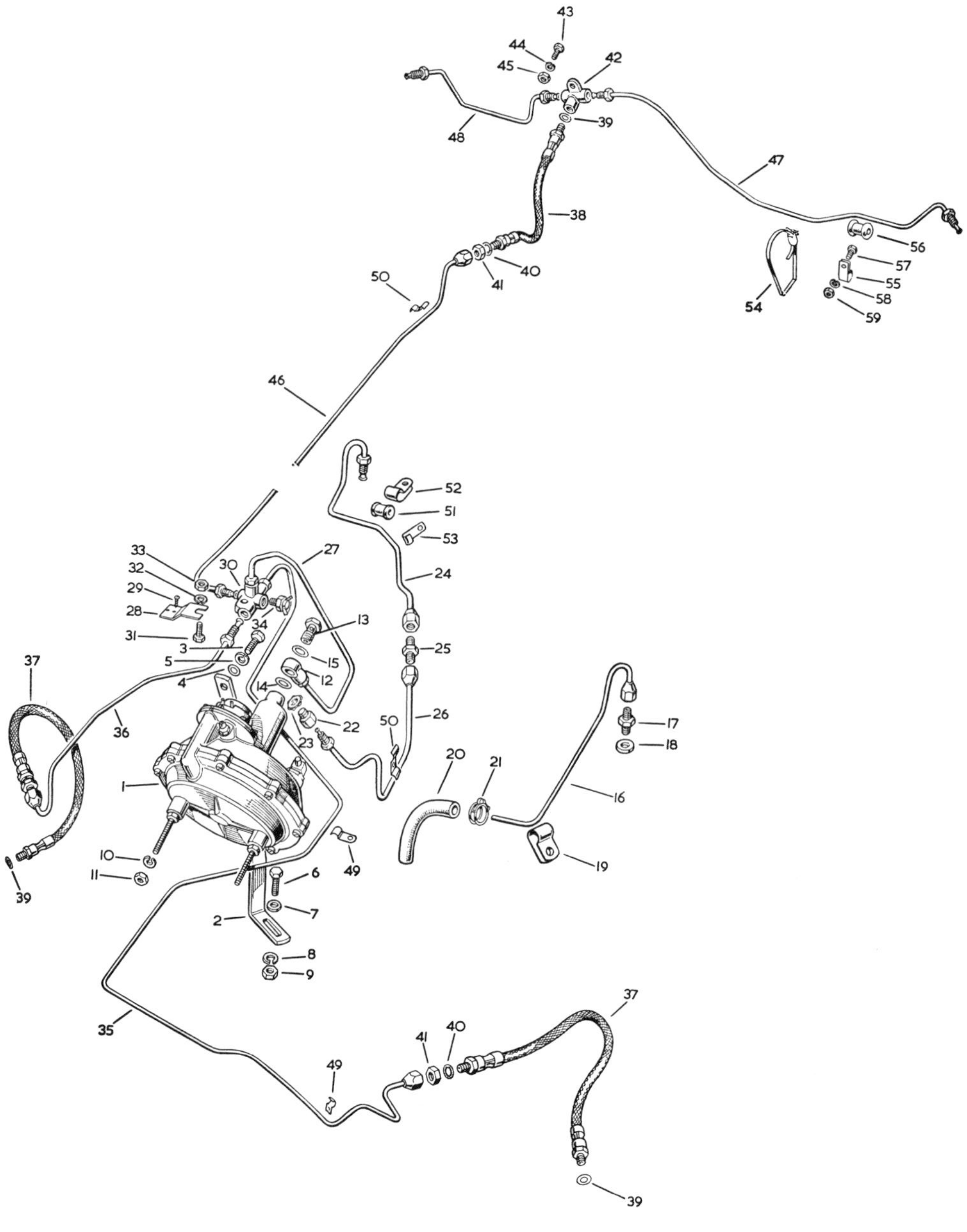

H944

Plate Ref.	1 2 3 4	DESCRIPTION	Qty	Part No.	REMARKS
49		Single clip, brake pipes to chassis frame ...	As reqd	41379	
50		Double clip, brake pipes to chassis frame ...	As reqd	233274	
		Drive screw fixing clips	As reqd	72626	
51		Grommet ⎱ Fixing brake servo pipe	1	06860	
52		Clip ⎰ to pedal bracket top cover	1	56666	
53		Clip, fixing brake pipe to steering box bracket ...	1	513940	RHStg
		Clip, fixing brake pipe to dash	3	338035	LHStg
54		Clip on rear axle for LH pipe	1	11820	
55		Clip for bush	4	56666	
56		Rubber bush ...	4	06860	
57		Bolt (2 BA x ½″ long) ⎱ Fixing LH axle pipe to chassis frame	4	234603	
58		Spring washer ⎰	4	3073	
59		Nut (2 BA)	4	2247	
		Brake servo major overhaul kit, Clayton Dewandre 701688 ...	1	601980	
		Brake servo minor overhaul kit, Clayton Dewandre 701687 ...	1	601981	
		Brake servo poppet valve overhaul kit, Clayton Dewandre 700986 ...	1	600434	Export only
		Brake servo non-return valve overhaul kit, Clayton Dewandre 700985 ...	1	600435	

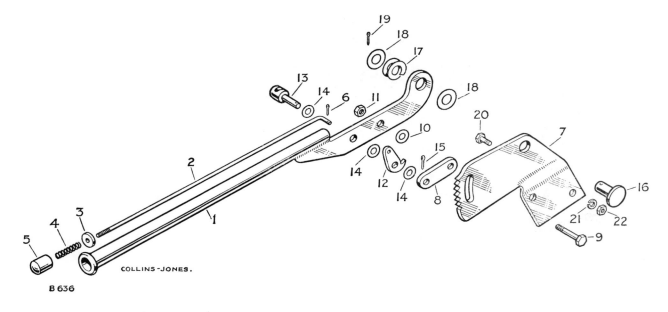

COLLINS-JONES.

B 636

HAND BRAKE LEVER, RHStg. Up to vehicle suffix 'C' inclusive

Plate Ref.	1 2 3 4	DESCRIPTION	Qty	Part No.	REMARKS
		HAND BRAKE LEVER ASSEMBLY ...	1	508581	
1		Hand brake lever and rod complete	1	529338	
		Handbrake lever	1	503468	}Cranked type Alternative types.
			1	503467	}Straight type Check before ordering
2		Plunger rod	1	500252	
3		Washer for plunger rod ...	1	242109	
4		Spring for plunger rod ...	1	58702	
5		Plunger	1	242108	
6		Split pin fixing rod to catch ...	1	2388	
7		Ratchet for hand brake ...	1	218382	
8		Locating plate	1	01384	
9		Bolt (³⁄₈" BSF x 1⅛" long)	1	237179	}Fixing
10		Plain washer	1	2210	}locating plate and
11		Nut (³⁄₈" BSF)	1	2827	}lever to ratchet
12		Brake catch	1	50235	
13		Pin	1	07164	
14		Plain washer	3	2210	}Fixing catch to lever
15		Split pin	1	2392	
16		Fulcrum pin for hand brake ...	1	231318	
17		Spring washer	1	2290	}Fixing
18		Plain washer	1	2218	}pin to ratchet
19		Split pin	1	2974	}and lever
20		Bolt (³⁄₈" UNF x ⁷⁄₈" long)	2	255046	}Fixing
21		Spring washer	2	3076	}hand brake
22		Nut (³⁄₈" UNF)	2	254812	}lever

* Asterisk indicates a new part which has not been used on any previous Rover model

H656

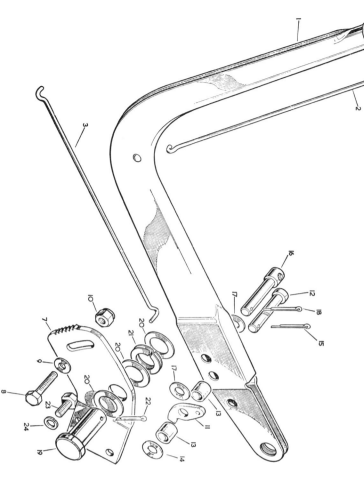

Plate Ref.	1 2 3 4	DESCRIPTION	Qty	Part No.	REMARKS
		HAND BRAKE LEVER ASSEMBLY			
1		Hand brake lever	1	543303	
1		Hand brake lever	1	543265	
2		Plunger rod, upper	1	543292	
3		Plunger rod, lower	1	543291	
4		Washer for plunger spring ...	1	552857	
5		Spring for plunger rod	1	58702	
6		Plunger	1	552856	
7		Ratchet for hand brake	1	543546	
8		Bolt (⅜" UNF x 1¼" long) ...	1	255249	} Fixing lever to ratchet
9		Plain washer	2	2210	
10		Self-locking nut (⅜" UNF) ...	1	252162	
11		Brake catch	1	50235	
12		Pin	1	559563	
13		Distance piece } Fixing catch	2	543275	
14		Plain washer } to lever	1	4574	
15		Split pin	1	2392	
16		Pin for hand brake adjuster rod ...	1	543281	
17		Plain washer } Fixing pin to	2	4574	
18		Split pin } hand brake lever	1	2392	
19		Fulcrum pin for hand brake lever ...	1	231318	
20		Plain washer	3	2218	
21		Spring washer } Fixing pin to ratchet and lever	1	2290	
22		Split pin	1	2974	
23		Bolt (⅜" UNF x ⅞" long) } Fixing hand brake	2	255046	
24		Spring washer } lever to	2	3076	
		Nut (⅜" UNF) } chassis frame	2	254812	

HAND BRAKE LEVER, LHStg. Up to vehicle suffix 'C' inclusive

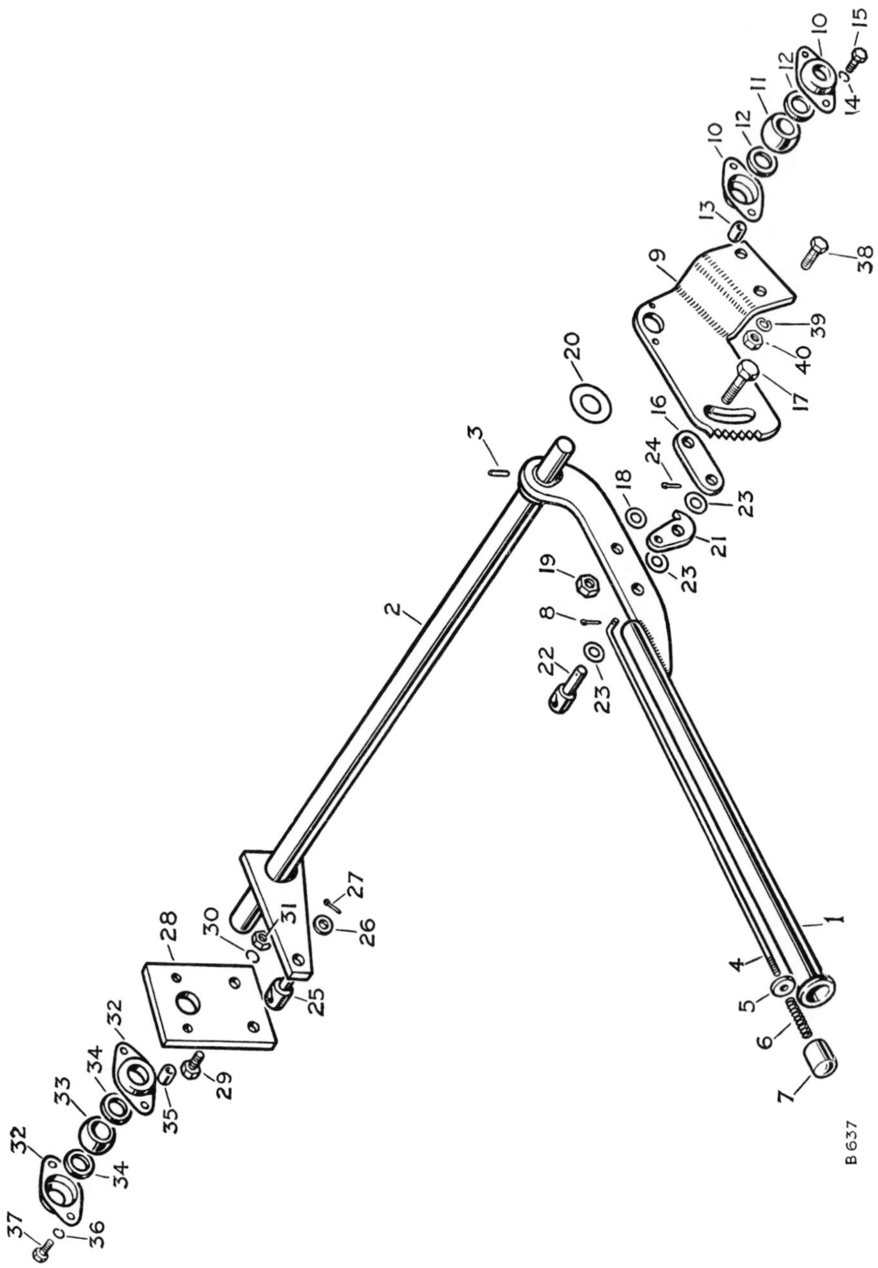

HAND BRAKE LEVER, LHStg. Up to vehicle suffix 'C' inclusive

Plate Ref.	1 2 3 4 DESCRIPTION	Qty	Part No.	REMARKS
	HAND BRAKE LEVER ASSEMBLY	1	508582	
	Hand brake lever, cross shaft and rod complete	1	529337	Cranked type
	Hand brake lever and cross shaft	1	503471	
	Hand brake lever	1	503469	Cranked type } Alternative types.
1	Hand brake lever	1	500254	Straight type } Check before ordering
2	Cross shaft for hand brake	1	218402	
3	Pin fixing lever to shaft	1	50472	
	Plunger rod	1	503467	Cranked type } Alternative types.
4	Plunger rod	1	500252	Straight type } Check before ordering
5	Washer for plunger spring	1	242109	
6	Spring for plunger rod	1	58702	
7	Plunger	1	242108	
8	Split pin fixing rod to catch	1	2388	
9	Ratchet for hand brake	1	218407	
10	Housing for cross shaft bearing	2	217983	
11	Spherical bearing for cross shaft	1	217984	
12	Felt ring for bearing	2	217985	
13	Distance piece } Fixing	2	218410	
14	Spring washer } bearing and	2	3075	
15	Set bolt ($\frac{5}{16}$" BSF x $\frac{7}{8}$" long) } housing to ratchet	2	237161	
16	Locating plate	1	01384	
17	Bolt ($\frac{3}{8}$" BSF x $1\frac{1}{8}$" long) } Fixing	1	237179	
18	Plain washer } locating plate and	1	2210	
19	Nut ($\frac{3}{8}$" BSF) } lever to ratchet	1	2827	
20	Plain washer between lever and ratchet	1	2218	
21	Brake catch	1	50235	
22	Pin }	1	07164	
23	Plain washer } Fixing catch to lever	3	2210	
24	Split pin }	1	2392	
25	Pin for hand brake adjuster rod	1	218408	
26	Plain washer } Fixing pin to	1	2208	
27	Split pin } cross shaft lever	1	2392	
28	Support plate for hand brake bearing housing ..	1	218409	
29	Bolt ($\frac{3}{8}$" BSF x $\frac{7}{8}$" long) } Fixing	2	237178	
30	Spring washer } support plate	2	3076	
31	Nut ($\frac{3}{8}$" BSF) } to chassis frame	2	2827	
32	Housing for cross shaft bearing	2	217983	
33	Spherical bearing for cross shaft	1	217984	
34	Felt ring for bearing	2	217985	
35	Distance piece } Fixing housing	2	218410	
36	Spring washer } and bearing to	2	3075	
37	Set bolt ($\frac{5}{16}$" BSF x $\frac{7}{8}$" long) } support plate	2	237161	
38	Bolt ($\frac{3}{8}$" UNF x $\frac{7}{8}$" long) } Fixing	2	255046	
39	Spring washer } hand brake	2	3076	
40	Nut ($\frac{3}{8}$" UNF) } lever	2	254812	

* Asterisk indicates a new part which has not been used on any previous Rover model

H657

HAND BRAKE LEVER, LHStg. From vehicle suffix 'D' onwards

Plate Ref.	1 2 3 4	DESCRIPTION	Qty	Part No.	REMARKS
		HAND BRAKE LEVER ASSEMBLY ...	1	552577	
1		Hand brake lever	1	543265	
2		Cross shaft for hand brake ...	1	552578	
3		Plunger rod, upper	1	543292	
4		Plunger rod, lower	1	543291	
5		Washer for plunger rod ...	1	552857	
6		Spring for plunger spring ...	1	58702	
7		Plunger	1	552856	
8		Ratchet for hand brake ...	1	552583	
9		Housing for cross shaft bearing	1	217983	
10		Spherical bearing for cross shaft	1	217984	
11		Felt ring for bearing ...	2	217985	
12		Distance piece } Fixing bearing	2	552805	
13		Spring washer } and housing	2	3075	
14		Set bolt ($\frac{5}{16}$" UNF x 1$\frac{1}{2}$" long) } Fixing	2	255029	
15		Bolt ($\frac{3}{8}$" UNF x 1$\frac{1}{4}$" long) } lever to	1	255053	
16		Plain washer } ratchet	2	2210	
17		Self-locking nut ($\frac{3}{8}$" UNF)	1	252162	
18		Plain washer between lever and ratchet ...	1	2218	
19		Brake catch	1	50235	
20		Pin	1	559563	
21		Distance piece } Fixing catch	2	543275	
22		Plain washer } to lever	1	4574	
23		Split pin ...	1	2392	
24		Pin for hand brake adjuster rod ...	1	218408	
25		Plain washer } Fixing pin to	1	2208	
26		Split pin } cross shaft lever	1	2392	
27		Support plate for hand brake bearing housing	1	552858	
28		Bolt ($\frac{3}{8}$" UNF x $\frac{7}{8}$" long) } Fixing support	2	255046	
29		Spring washer } plate to	2	3076	
30		Nut ($\frac{3}{8}$" UNF) } chassis frame	2	254812	
30		Housing for cross shaft bearing ...	2	217983	
31		Spherical bearing for cross shaft	1	217984	
32		Felt ring for bearing ...	2	217985	
33		Distance piece	2	218410	
34		Spring washer } Fixing housing	2	3075	
35		Set bolt ($\frac{5}{16}$" UNF x $\frac{7}{8}$" long) } and bearing to	2	255227	
36		Bolt ($\frac{3}{8}$" UNF x $\frac{7}{8}$" long) } support plate	2	255046	
37		Spring washer } Fixing hand brake	2	3076	
		Nut ($\frac{3}{8}$" UNF) } lever to chassis frame	2	254812	

HAND BRAKE LEVER, LHStg. From vehicle suffix 'D' onwards

* Asterisk indicates a new part which has not been used on any previous Rover model

HAND BRAKE, TRANSMISSION

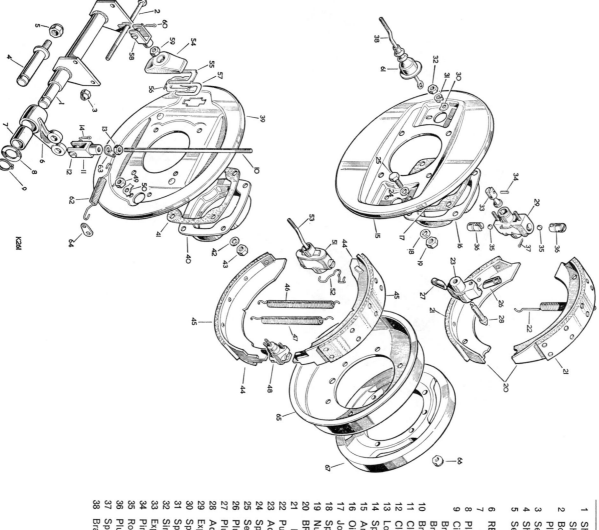

K264

HAND BRAKE, TRANSMISSION

Plate Ref. 1 2 3 4	DESCRIPTION	Qty	Part No.	REMARKS
1	Shaft for hand brake relay lever	1	240829	
2	Shaft for hand brake relay lever	2	267193	88 } RHStg } Up to vehicle suffix 'C' inclusive
2	Shaft for hand brake relay lever	2	256253	109 } LHStg
	Bolt (⅜" UNF x 4" long) } Fixing shaft to chassis frame	2	2210	
3	Plain washer	2	252162	} From vehicle suffix 'D' onwards
4	Self-locking nut (⅜"UNF) } Fixing shaft to chassis frame	2	252162	
5	Self-locking nut (⅜" UNF) fixing shaft to chassis frame	1	552746	From vehicle suffix 'D' onwards
6	RELAY LEVER ASSEMBLY FOR HAND BRAKE	1	275199	
7	Bush for relay lever	1	216421	
8	Plain washer } Fixing lever to spindle	2	254831	
9	Circlip	2	2392	
10	Brake rod, relay to hand brake lever	1	219002	
10	Brake rod, relay to hand brake lever	1	238386	
11	Clevis fork end } Fixing brake rod	1	3299	
12	Clevis pin complete } to relay and hand brake lever	1	09203	
13	Locknut ($\frac{5}{16}$" UNF)	3	277921	
14	Split pin	3	278020	
15	Anchor plate, transmission brake	1	552407	
16	Oil catcher for transmission brake	1	561856*	
17	Joint washer for oil catcher	1	561368*	
18	Spring washer } Fixing anchor plate and oil catcher to speedometer housing	4	3076	
19	Nut (⅜" BSF)	4	2827	
20	BRAKE SHOE ASSEMBLY, BOXED PAIR	1	264374	
21	Lining complete with rivets, for brake shoe, GI 34920	4	219007	
22	Pull-off spring for brake shoe, GI GB 2478	2	219023	
23	Adjuster housing, GI GB 1398	2	219010	
24	Spring washer } Fixing adjuster housing	2	219025	
25	Set bolt	2	219024	
26	Plunger RH, GI GB 1399	2	219011	Up to gearboxes numbered: 146000565
27	Plunger LH, GI GB 1399A	2	219012	156000430
28	Adjuster cone, GI GB 522	1	219013	151005187
29	Expander housing	1	230741	
30	Special washer	2	219022	
31	Spring washer } Fixing expander housing	2	219027	
32	Simmonds nut	2	232361	
33	Expander cone, GI GB 1404	1	232361	
34	Pin fixing cone to brake rod, GI GB 67	2	219021	
35	Roller for expander, GI GB 535	2	219017	
36	Plunger for expander, GI GB 1403	2	219018	
37	Split pin fixing plunger	2	219019	
38	Brake rod, expander to relay lever	1	219020	

* Asterisk indicates a new part which has not been used on any previous Rover model

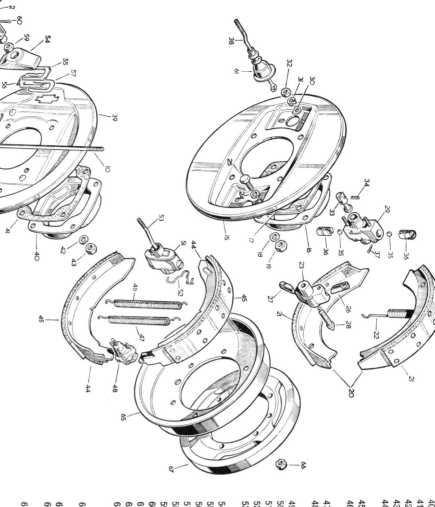

K26I

HAND BRAKE, TRANSMISSION

Plate Ref. 1 2 3 4	DESCRIPTION	Qty	Part No.	REMARKS
39	Anchor plate, transmission brake, GI 64270168 ...	1	515365	
40	Oil catcher for transmission brake ...	1	561368*	
41	Joint washer for oil catcher ...	1	561856*	
42	Spring washer for oil catcher } Fixing anchor plate and oil	4	3076	
43	Nut (3/8" BSF) } catcher to speedometer housing	4	2827	
44	BRAKE SHOE ASSEMBLY, BOXED PAIR, GI GB 47989DV ...	1	516031	1 off from gearboxes numbered 146000566, 156000431, 151005188
45	Lining complete for shoe, boxed pair, GI SP 1444	1	541994	From gearboxes numbered: 146000566 156000431 151005188 onwards
46	Pull-off spring, expander end, GI GB 40655 ...	1	515465	
47	Return spring, adjuster end, GI GB 40655 ...	1	515465	
48	ADJUSTER UNIT ASSEMBLY, GI GB 49001	1	515925	
49	Repair kit for adjuster unit GI SP 1856 ...	1	515924	For brake shoe
50	Nut (1/4" UNF) } Fixing	2	254810	
51	Tab washer, GI 64271598 } adjuster unit	1	542515	
52	Clip retaining tappets, GI GB 46277	1	515366	
53	EXPANDER UNIT ASSEMBLY, GI GB 64270170	1	515927	
	Brake rod, expander to relay lever, GI 64270172 ...	1	515926	
	Repair kit for expander unit, GI SP 1807 ...	1	515923	
54	Dust cover for expander unit, GI GB 46278	1	515466	
55	Packing plate ...	1	515467	
56	Locking plate ...	1	515470	
57	Retaining spring } Fixing expander unit	1	515468	
58	Clevis complete }	1	515809	
59	Locknut (5/16" BSF) } relay lever	1	3490	
60	Split pin } brake rod to	1	2392	
61	Dust cover for brake rod, GI GB 2501	1	219028	
62	Return spring for brake rod ...	1	59663	
63	Anchor for spring ...	2	240708	
64	Anchor for spring, on transfer box ...	1	267412	From gearboxes numbered 146000566, onwards
65	Brake drum ...	1	274423	From gearboxes numbered 146000431, 151005188 onwards
66	Self-locking nut (5/16" BSF) fixing brake drum and damper	6	251321	
67	Transmission damper at rear end of gearbox ...	1	275239	109 Diesel. Up to gearbox number 27600879

CHASSIS FRAME, ROAD SPRINGS AND SHOCK ABSORBERS

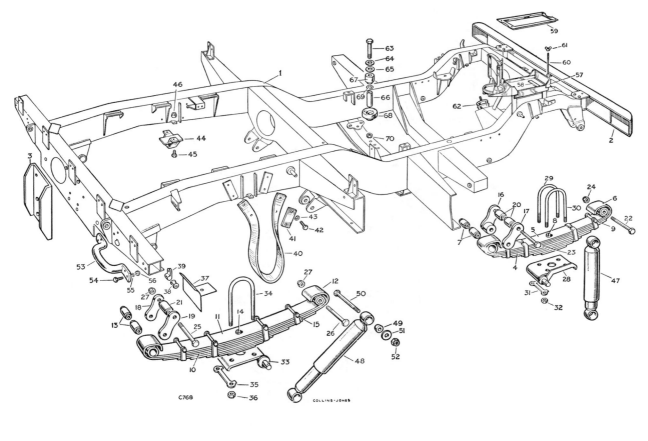

C768   COLLINS-JONES

CHASSIS FRAME, ROAD SPRINGS AND SHOCK ABSORBERS

Plate Ref. 1 2 3 4	DESCRIPTION	Qty	Part No.	REMARKS
1	Chassis frame	1	562502	88. See footnote †
	Cross-member No. 3 } For chassis frame	1	240841	88
	Cross-member No. 4 }	1	218453	88
	Chassis frame	1	562532	109. 4 cylinder models. See footnote †
	Chassis frame	1	562574	109. 6 cylinder models
	Chassis frame	1	278665	109 Station Wagon, 4 cylinder models.
	Chassis frame	1	562533	109 Station Wagon 6 cylinder models. Up to vehicle suffix 'C' inclusive
	Chassis frame	1	564704	From vehicle suffix 'D' onwards
	Cross-member No. 3 } For chassis frame	1	552398	See footnote †
	Cross-member No. 4 }	1	241130	109 and 109 Station Wagon 4 cylinder models
	Cross-member No. 3 For chassis frame	1	543946	
	Cross-member No. 4 frame	1	241130	109 and 109 Station Wagon 6 cylinder models
2	Front bumper	1	256253	
3	Bolt (3/8" UNF x 4" long) } Fixing front bumper to chassis frame	8		
	Plain washer }	8	3822	
	Self-locking nut (3/8" UNF) }	4	252162	
4	Road spring complete, front, Driver's	1	238207	88 Petrol models, 2 litre
	Road spring complete, front, Passenger's	1	238208	88 Petrol models, 2 litre
	Road spring complete, front, Driver's	1	241283	88 Petrol models, 2¼ litre
	Road spring complete, front, Passenger's	1	242863	88 Petrol models, 2¼ litre
	Road spring complete, front, Driver's	1	265459	109 Petrol and Diesel
	Road spring complete, front, Passenger's	1	265627	109 Petrol and Diesel
	Road spring complete, front, Driver's	1	264563	88 Diesel models
	Road spring complete, front, Passenger's	1	264565	109 Petrol and Diesel
	Road spring complete, front, Driver's	1	276034	88 Diesel models
	Road spring complete, front, Passenger's	1	256627	109 Diesel models See page 330 for heavy duty road springs
5	Main leaf } For front spring	2	243121	88 Petrol models, 2 litre
	2nd leaf }	2	243122	2 litre
6	Main leaf For front spring	2	243125	88 Petrol models, 2¼ litre
	2nd leaf spring	2	243126	2¼ litre
7	Bush for front spring	4	265460	109 Petrol and Diesel and 88 Diesel models
		4	242825	
8	Dowel and nut for front spring	2	243129	88 Petrol models, 2 litre
	Dowel and nut for front spring	2	243131	88 Petrol models, 2¼ litre
	Dowel and nut for front spring	2	265461	109 Petrol and Diesel and 88 Diesel models
9	Bolt (3/8" BSF x 3½" long) } For spring clip	2	250551	
	Nut (3/8" BSF) }	2	2827	

* Asterisk indicates a new part which has not been used on any previous Rover model
NOTE: Part numbers of road springs are stamped on underside of 3rd or bottom leaf

† Footnote. When supplying this chassis for vehicles prior to vehicle suffix 'D', the following must be fitted:
Shaft for handbrake relay, 1 off   Part No. 552746
Self-locking nut (3/8" UNF), 1 off   Part No. 252162

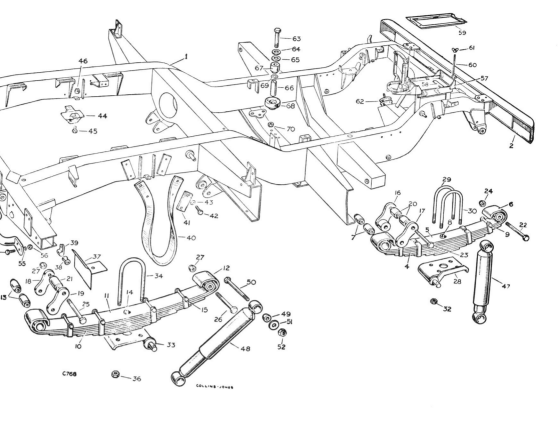

C768   COLLINS-JONES

CHASSIS FRAME, ROAD SPRINGS AND SHOCK ABSORBERS

Plate Ref.	1 2 3 4	DESCRIPTION	Qty	Part No.	REMARKS
10		Road spring complete, rear, Driver's	1	241445‡	} See page 330 for heavy duty road springs
		Road spring complete, rear, Passenger's	1	241446‡	} 88 models
		Road spring complete, rear, Driver's	1	517588†‡	
		Road spring complete, rear, Passenger's	1	517589†‡	} 109 models
		Road spring complete, rear, Driver's	1	279678	
		Road spring complete, rear, Passenger's	1	279679	} 109 models
11		Main leaf } For rear spring	2	243123	} 88 models
12		2nd leaf }	2	243124	
		Main leaf } For rear spring	2	501390	} 109 models
		2nd leaf }	2	501391	
13		Bush for rear spring	4	242825	
14		Dowel for rear spring	2	243130	88 models
		Dowel for rear spring	2	501389	109 models
15		Bolt ($\frac{3}{8}$" BSF x $3\frac{1}{2}$" long) } For spring clip	4	250551	
		Nut ($\frac{3}{8}$" BSF) }	4	2827	
16		Shackle plate, tapped } For rear springs	2	270521	$\frac{1}{2}$" BSF type. Up to vehicle suffix 'A'
		Shackle plate, tapped }	2	537687	$\frac{9}{16}$" UNF type. From vehicle suffix 'B' onwards
17		Shackle plate, plain	2	270520	
18		Shackle plate, tapped } For rear springs	2	244163	$\frac{1}{2}$" BSF type. Up to vehicle suffix 'A'
		Shackle plate, tapped }	2	537686	$\frac{9}{16}$" UNF type. From vehicle suffix 'B' onwards
19		Shackle plate, plain } For rear springs	2	244162	
		Shackle plate, tapped }	2	279969	$\frac{1}{2}$" BSF type. Up to vehicle suffix 'A'
		Shackle plate, tapped }	2	537685	$\frac{9}{16}$" UNF type. From vehicle suffix 'B' onwards
20		Shackle plate, plain, front spring	2	279970	
21		Bush in chassis frame, front spring	4	263354	
22		Bush in chassis frame, rear	2	242825	
23		Shackle pin, front end of front spring	2	270741	
24		Self-locking nut ($\frac{1}{2}$" BSF), front	6	236625	$\frac{1}{2}$" BSF type. Up to vehicle suffix 'A'
25		Shackle pin, rear end of front spring	2	251324	
26		Shackle pin, front end of rear spring	2	236969	
27		Self-locking nut ($\frac{1}{2}$" BSF), rear	6	270741	$\frac{1}{2}$" BSF type. Up to vehicle suffix 'A'
		Shackle pin, rear end of rear spring	2	251324	
		Shackle pin, front end of front spring	4	251324	
		Self-locking nut ($\frac{9}{16}$" UNF), front	6	537742	$\frac{9}{16}$" UNF type. From vehicle suffix 'B'
		Shackle pin, rear end of front spring	2	537741	
		Self-locking nut ($\frac{9}{16}$" UNF), front	4	252165	
		Shackle pin, front end of rear spring	4	537742	
		Self-locking nut ($\frac{9}{16}$" UNF), rear	2	252165	From vehicle suffix 'B'
		Shackle pin, rear end of rear spring	4	537741	
		Shackle pin, front end of rear spring	6	537740	onwards
		Self-locking nut ($\frac{7}{16}$" UNF), rear	6	252165	

*   Asterisk indicates a new part which has not been used on any previous Rover model
†   Up to vehicles numbered:
    141100189   141100190
    142100095   142100096
    144100147   144100148
    149100018   149100019 onwards
‡‡ From vehicles numbered:
    146100079   146100080
    147100006   147100007
    144100148   149100019 onwards

NOTE: Part numbers of road springs are stamped on underside of 3rd or bottom leaf

CHASSIS FRAME, ROAD SPRINGS AND SHOCK ABSORBERS

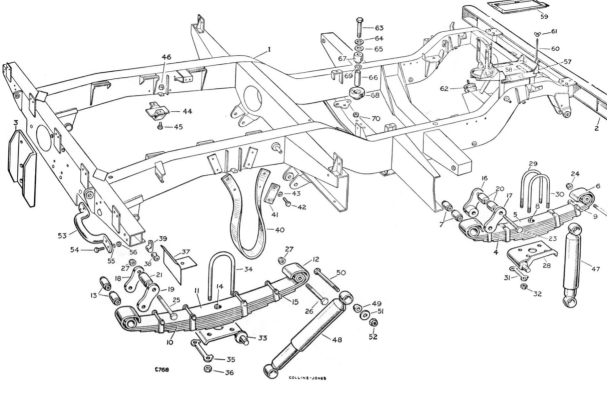

C768   COLLINS-JONES

CHASSIS FRAME, ROAD SPRINGS AND SHOCK ABSORBERS

Plate Ref. 1 2 3 4	DESCRIPTION	Qty	Part No.	REMARKS
28	Bottom plate for front spring, LH	1	264022	88 Petrol ⎱ 88 models
28	Bottom plate for front spring, RH	1	264023	⎰
28	Bottom plate for front spring, LH	1	264466	109 Petrol and Diesel ⎱ and 88 Diesel models
28	Bottom plate for front spring, RH	1	264467	⎰
29	'U' bolt	3	217259	88 Petrol ⎱ 88 models
30	'U' bolt	1	03459	⎰
29	'U' bolt	3	543108	88 Diesel ⎱ models
30	'U' bolt	1	569028	⎰
30	'U' bolt	3	543108	1 off with reinforced axle ⎱ 109 Petrol and Diesel models
29	'U' bolt	1	569028	3 off with reinforced axle ⎰
32	Self-locking nut (7/16" BSF) — Fixing front springs to axle	8	251323	
28	Bottom plate for rear spring	1	264020	⎱ 88 models
33	Bottom plate for rear spring, LH	1	264021	⎰
33	Bottom plate for rear spring, RH	1	265642	109. Rover type axle
34	'U' bolt	2	550786	ENV type axle, 109, optional
34	'U' bolt	1	242127	Rover type axle
36	Self-locking nut (7/16" BSF) — Fixing rear springs to axle	8	251323	
36	Self-locking nut (7/16" BSF)	8	562734*	ENV type axle, 109, optional
37	Shield for brake pipe, RH ⎱ springs to axle	1	251323	109 optional
37	Shield for brake pipe, LH ⎰	1	243038	
38	Self-locking nut (7/16" BSF)	1	243039	
39	Rubber grommet for brake pipe	2	56666	
40	Clip for grommet — Fixing clip to shield	2	06860	
40	Bolt (2 BA x 1/2" long)	2	234603	
41	Spring washer	8	3073	
42	Nut (2 BA)	8	2247	
40	Check strap for rear axle, RH	1	237100	88 ⎱ models
40	Check strap for rear axle, LH	1	274469	⎰
40	Check strap for rear axle, RH	1	278699	1958-60 models
40	Check strap for rear axle, LH	1	278698	2 off on 1961 models onwards
			562735*	ENV type axle, 109. Optional
41	Check strap for rear axle	2	219574	Alternative
41	Plate ⎱ For check strap	4	2851	fixings
42	Plain washer ⎰	4	255245	Alternative fixings
42	Set bolt (3/8" UNF x 7/8" long)	8	3076	
43	Spring washer	8	255247	Alternative to nut type fixing
43	Bolt (3/8" UNF x 1" long) ⎱ Fixing check strap to chassis frame	8	254812	Alternative to bolt and set bolt fixing
44	Nut (3/8" UNF) ⎰	4	241380	
44	Rubber buffer for axles, front and rear	8	255207	
45	Bolt (1/4" UNF x 3/4" long) ⎱ Fixing buffers	8	255210	
46	Self-locking nut (1/4" UNF) ⎰ to chassis frame	8	501444	
47	Shock absorber, front	2	501445	
48	Shock absorber, front	2	502532	See page 330 for heavy
48	Shock absorber, rear	2	276937	109 ⎱ duty shock absorbers
48	Shock absorber, rear	2		⎰
49	Rubber bush for shock absorbers	16	243057	12 off on 109

* Asterisk indicates a new part which has not been used on any previous Rover model

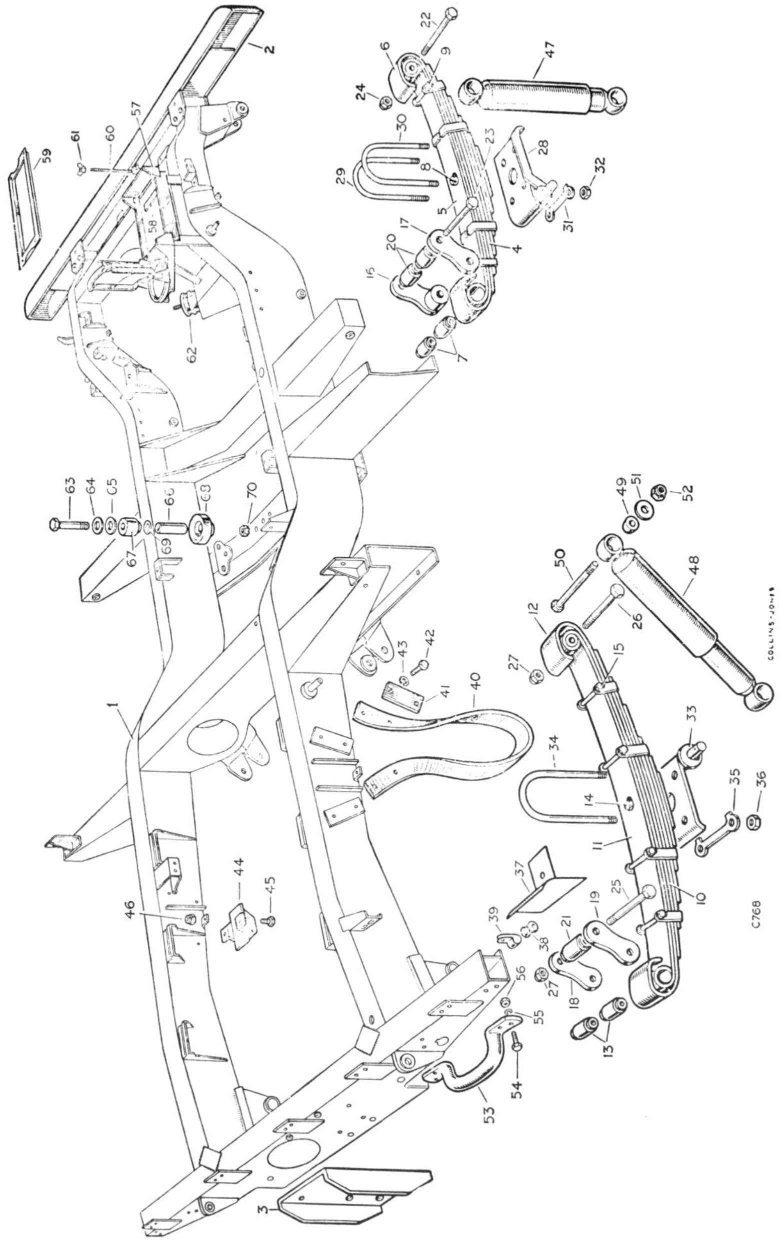

Plate Ref.	1 2 3 4 DESCRIPTION	Qty	Part No.	REMARKS
50	Bolt ($\frac{7}{16}$" UNF x $6\frac{1}{2}$" long) } Fixing shock absorbers	4	253817	2 off on 109
	Bolt ($\frac{7}{16}$" UNF x $7\frac{1}{4}$" long) } Fixing shock absorbers	2	253826	109
51	Plain washer } Fixing shock absorbers	4	243022	
52	Self-locking nut ($\frac{7}{16}$" UNF) } Fixing shock absorbers	4	252163	
	Plain washer } Fixing shock absorber at bottom	4	264024	2 off on 109
	Split pin } Fixing shock absorber at bottom	4	4063	2 off on 109
	Guide washer } Fixing rear shock absorber at bottom	4	508895	109
	Retainer for washer } Fixing rear shock absorber at bottom	4	500746	109
	Rubber pad } Fixing rear shock absorber at bottom	4	552818	109
	Self-locking nut ($\frac{1}{2}$" UNF) } Fixing rear shock absorber at bottom	2	252164	109
53	Lifting handle, rear	2	300816	
54	Bolt ($\frac{5}{16}$" UNF x $\frac{3}{4}$" long) } Fixing handles to chassis frame	8	255226	
55	Spring washer } Fixing handles to chassis frame	8	3075	
56	Nut ($\frac{5}{16}$" UNF) } Fixing handles to chassis frame	8	254831	
57	Battery casing and air cleaner support	1	562928	4 cylinder models
	Bridge plate and air cleaner support	1	562663	6 cylinder models
	Battery carrier, LH	1	273598	1958 2 litre Diesel 88
	Battery carrier, LH	1	500999	1959–61 2 litre Diesel 88
	Battery carrier, LH	1	501006	2 litre Diesel 109
58	Rubber strip for battery	2	334242	4 off on Diesel models
	Bolt ($\frac{1}{4}$" UNF x $\frac{5}{8}$" long) } Fixing battery casing and support to chassis frame	5	255206	4 cylinder models
	Fan disc washer } Fixing battery casing and support to chassis frame	2	510170	4 cylinder models
	Spring washer } Fixing battery casing and support to chassis frame	4	3074	4 cylinder models
	Nut ($\frac{1}{4}$" UNF) } Fixing battery casing and support to chassis frame	5	254810	4 cylinder models
	Bolt ($\frac{1}{4}$" UNF x $\frac{5}{8}$" long) } Fixing bridge plate and support to chassis frame	3	255206	6 cylinder models
	Spring washer } Fixing bridge plate and support to chassis frame	3	3074	6 cylinder models
	Nut ($\frac{1}{4}$" UNF) } Fixing bridge plate and support to chassis frame	3	254810	6 cylinder models
	Bolt ($\frac{5}{16}$" UNF x $\frac{3}{4}$" long) } Fixing LH battery carrier to chassis frame	3	255226	2 litre Diesel models
	Plain washer } Fixing LH battery carrier to chassis frame	3	3830	2 litre Diesel models
	Shakeproof washer } Fixing LH battery carrier to chassis frame	4	70822	2 litre Diesel models
	Nut ($\frac{5}{16}$" UNF) } Fixing LH battery carrier to chassis frame	3	254831	2 litre Diesel models
	Valance for battery support, LH	1	271155	88 } 2 litre Diesel models
	Valance for battery support, LH	1	279800	109 } 2 litre Diesel models
	Bolt (2 BA x $\frac{1}{2}$" long) } Fixing valance to battery support	4	234603	2 litre Diesel models
	Plain washer } Fixing valance to battery support	4	2226	2 litre Diesel models
	Nut (2 BA) } Fixing valance to battery support	4	2247	2 litre Diesel models
59	Battery cover at frame or seat base	1	236986	2 off on $2\frac{1}{4}$ litre Diesel models
	Wing nut for air cleaner support	1	250431	
60	Battery fixing rod	2	270420	4 cylinder models
61	Wing nut } Fixing battery	2	250431	
	Plain washer } Fixing battery	2	2876	
	Battery fixing rod, seat base	2	332524	$2\frac{1}{4}$ litre Diesel and 2.6 litre Petrol models
	Battery cover, LH of frame	1	271163	2 litre Diesel models
	Battery fixing rod, LH of frame	2	271151	2 litre Diesel models
	Nut ($\frac{1}{4}$" BSF) } Fixing battery, LH of frame or seat base	2	2823	Diesel models
	Wing nuts } Fixing battery, LH of frame or seat base	2	250431	Diesel models
	Plain washer } Fixing battery, LH of frame or seat base	2	2876	Diesel models

* Asterisk indicates a new part which has not been used on any previous Rover model

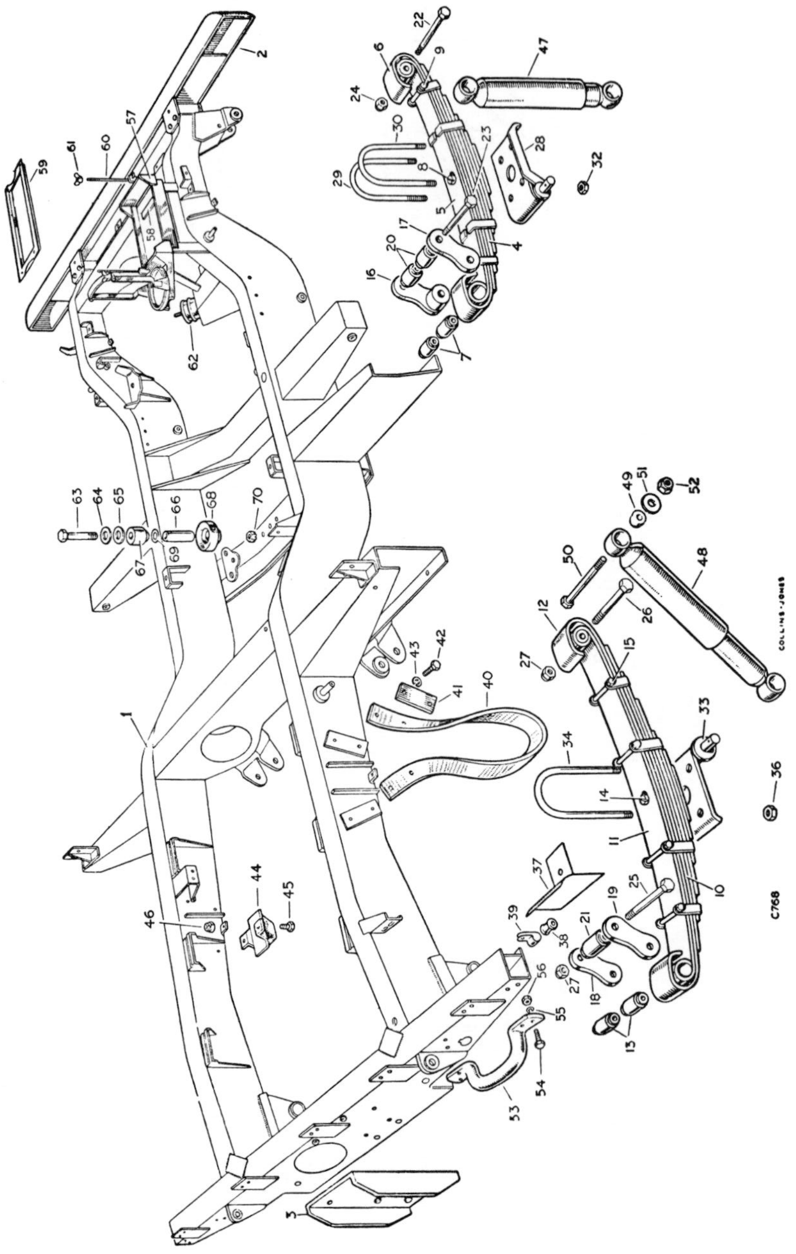
C768
COLLINS-JONES

CHASSIS FRAME, ROAD SPRINGS AND SHOCK ABSORBERS

Plate Ref.	1 2 3 4 DESCRIPTION	Qty	Part No.	REMARKS
	Engine mounting, front, RH	1	272506	Diesel and $2\frac{1}{4}$ litre Petrol models
	Engine mounting, rear, LH	1	272498	
	Engine mounting, rear, RH	1	272501	
	Bolt ($\frac{3}{8}$" UNF x 1" long) — Fixing front mounting to chassis frame	2	255247	
	Plain washer	2	2251	
	Spring washer	2	3076	
	Nut ($\frac{3}{8}$" UNF)	2	254812	
	Bolt ($\frac{3}{8}$" UNF x 1" long) — Fixing rear mounting to chassis frame	4	255247	
	Plain washer	4	2251	
	Spring washer	4	3076	
	Nut ($\frac{3}{8}$" UNF)	4	254812	
	Suspension rubber for engine, front	2	562688	Diesel models
	Suspension rubber for engine, front	2	231489	2 litre Petrol models
62	Suspension rubber for engine, front	2	526777	$2\frac{1}{4}$ litre Petrol models
	Plain washer, large — Fixing suspension rubber to engine and chassis	4	2851	1 off on Diesel and $2\frac{1}{4}$ litre Petrol models
	Plain washer, small	3	2251	Diesel and $2\frac{1}{4}$ litre Petrol models
	Spring washer	4	3076	
	Nut $\frac{3}{8}$" BSF)	4	2827	
	Susp(ension rubber for engine, rear	2	272191	$2\frac{1}{4}$ litre and 2.6 litre Petrol models
	Suspension rubber for engine, rear	2	562688*	Diesel models
	Self-locking nut ($\frac{3}{8}$" BSF) fixing rear suspension rubber	4	251322	
63	Bolt ($\frac{7}{16}$" BSF x $3\frac{1}{4}$" long) rear — Fixing engine unit to chassis frame at front and rear	2	274997	2 litre Petrol models
64	Plain washer, top	2	04799	
65	Rubber washer	2	50052	
66	Distance tube	2	274996	
67	Top rubber	2	07091	
68	Bottom rubber	2	04788	
69	Shim	As reqd	04800	
70	Special nut	2	240418	
	Split pin	2	2981	
	Road wheel, well base rim type	5	231601	88. Except America Dollar Area
	Tyre 6.00" x 16", Dunlop RK3 or Avon TM	5	239742	
	Inner tube 6.00" x 16"	5	232122	
	Road wheel, 6L x 15"	5	526753	88 models. America Dollar Area
	Tyre, 7.10" x 15", Goodyear 'Ultragrip'	5	548232	
	Inner tube, 7.10" x 15"	5	548285	
	Road wheel, well base rim type, 5.50" x 16"	5	272309	† 109. Offset from rim centre line $1\frac{13}{16}$" — Alternatives. Check before ordering
	Road wheel, well base rim type, 5.50" x 16"	5	568966	† 109. Offset from rim centre line $1\frac{5}{16}$"
	Tyre, 7.50" x 16" Dunlop RK3 or Avon TM	5	278434	109
	Inner tube, 7.50" x 16"	5	270324	
	Road wheel, detachable rim type	5	217267	Optional on 88 models
	Special bolt and nut for detachable rim type wheel	8	243471	
	Rim band for detachable rim type wheel	5	232053	

* Asterisk indicates a new part which has not been used on any previous Rover model
† Optional on 88 models when Dunlop RK3 or Avon TM 7.50" x 16" tyres are fitted

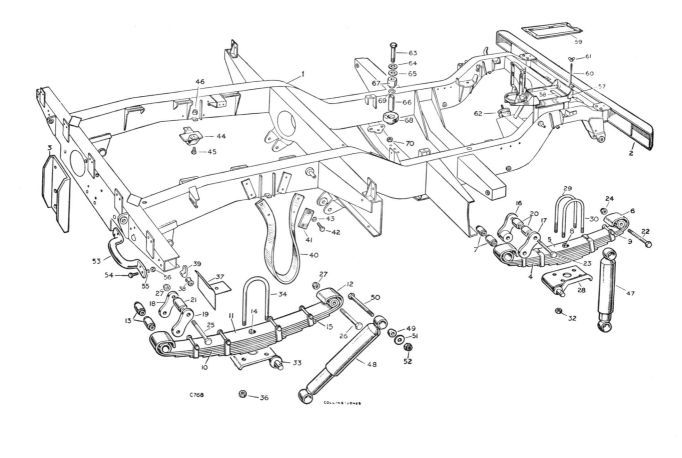

C768
COLLINS-JONES

Plate Ref.	1 2 3 4	DESCRIPTION	Qty	Part No.	REMARKS
		Tyre 6.00" x 16", Dunlop T28 tread ...	5	232121	
		Tyre 6.50" x 16", Dunlop RK3 tread ...	5	542865	
		Tyre 6.50" x 16", Dunlop RK3 tread ...	5	232559	
		Tyre 7.00" x 16", Dunlop 'Crosscountry', T29A tread ...	5	242861	
		Tyre 7.00" x 16", Dunlop 'Crosscountry', T29A tread ...	5	232123	Optional on 88 models
		Tyre 7.00" x 16", Dunlop curved bar, tractor tread ...	5	248806	
		Tyre 7.00" x 16", Dunlop RK3 or Avon TM ...	5	233041	
		Tyre 7.00" x 16", Dunlop 'Fort' ...	5	244298	
		Sand tyre 7.00" x 16" ...	5	232560	
		Inner tube 6.50" x 16" ...	5	235560	
		Inner tube 6.50" x 16" ...	5	233042	
		Tyre 7.00" x 16", Dunlop 'Fort' ...	5	278434 ††	
		Tyre 7.50" x 16" Dunlop RK3, or Avon TM ...	5	545547 ††	Optional on 109 models
		Tyre 7.50" x 16" Avon TM 8PR ...	5	270324 ††	
		Inner tube 7.50" x 16" ...	5	511878	Optional on 109 models
		Tyre 7.50" x 16", Michelin XY ...	5	511879	
		Inner tube	5	272310	
		Tyre 7.50" x 16", Dunlop 'Roadtrack Major' ...	5	270323	
		Tyre 7.50" x 16", Dunlop T29A tread ...	5	502526	Optional on 109 models
		Sand tyre 7.50" x 16", Dunlop Gold Seal Heavy Duty	5	552231	
		Inner tube 7.50" x 16", for Michelin X 'Sahara'.	5	552232	
		Inner tube 7.50" x 16", for Michelin sand tyre	5	264643	
		Sand tyre 8.20" x 15" ...	5	264644	Optional
		Inner tube 8.20" x 15" } For	5	526753	
		Road wheel } sand tyre	5	542863	Optional
		Sand tyre 9.00" x 15", Dunlop block tread } For	5	542864	Optional
		Inner tube 9.00" x 15" } well base	5	526753	
		Road wheel }	5	263142	
		Balance weight for road wheel } For road wheel	As reqd	3850	
		Special screw fixing balance weight } rim type	As reqd	217361	Alternatives.
		Nut for road wheel, 1" long ...	20	561254	Check before ordering
		Nut for road wheel, 11/16" long ...	20		

* Asterisk indicates a new part which has not been used on any previous Rover model
†† Optional on 88 models when road wheels 568966 are fitted

HEAVY DUTY SUSPENSION. OPTIONAL

Plate Ref. 1 2 3 4	DESCRIPTION	Qty	Part No.	REMARKS
	Front spring, passenger's side	1	242863	⎫
	Front spring, driver's side	1	241283	88
	Main leaf for front spring	2	243125	⎬ 2 litre
	Second leaf for front spring	2	243126	Petrol
	Dowel for front spring	2	243131	⎭
	Rear spring, passenger's side	1	265989	
	Rear spring, driver's side	1	241285	88
	Main leaf for rear spring	2	243127	2 litre
	Second leaf for rear spring	2	537965	and
	Dowel for rear spring	2	243132	2½ litre
	Shock absorber, front	2	512102	Petrol
	Shock absorber, rear	2	512086	⎫ Heavy duty road springs and
	Rear spring, driver's side	1	241285	⎬ shock absorbers.
	Rear spring, passenger's side	1	265989	⎭ Must be fitted in complete sets
	Main leaf for rear spring	2	243127	when replacing standard
	Second leaf for rear spring	2	537965	equipment
	Dowel for rear spring	2	243132	88
	Shock absorber, front	2	512102	Diesel
	Shock absorber, rear	2	512086	
	Rear spring, driver's side	1	272967	
	Rear spring, passenger's side	1	272968	
	Main leaf for rear spring	2	537964	⎫
	Second leaf for rear spring	2	537966	⎬ 109
	Dowel for rear spring	2	279762	
	Shock absorber, front	2	508033	
	Shock absorber, rear	2	508034	⎭

* Asterisk indicates a new part which has not been used on any previous Rover model

CLUTCH AND BRAKE PEDALS AND MASTER CYLINDERS

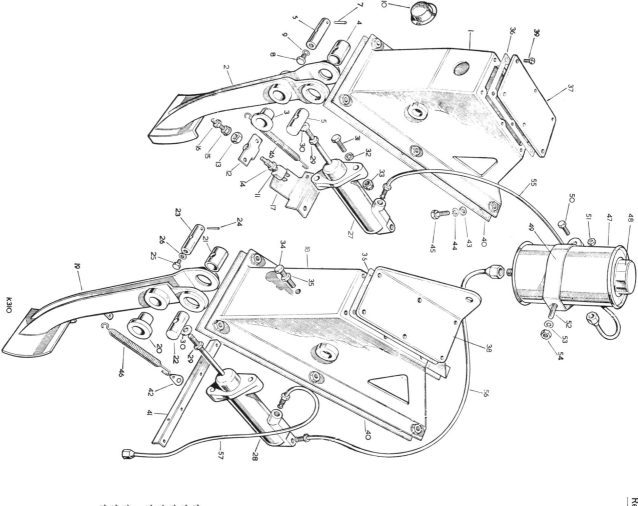

K3IO

CLUTCH AND BRAKE PEDALS AND MASTER CYLINDERS

Plate Ref.	1 2 3 4	DESCRIPTION	Qty	Part No.	REMARKS
		BRAKE PEDAL AND BRACKET ASSEMBLY ..	1	568894	88. Up to vehicles numbered: 141102873, 142101568, 144103476 Petrol models 146100972, 147100215, 149100343 Diesel models
		BRAKE PEDAL AND BRACKET ASSEMBLY	1	523916	88 From vehicles numbered: 141102874, 142101569, 144103477 Petrol models 146100973, 147100216, 149100344 Diesel models
		BRAKE PEDAL AND BRACKET ASSEMBLY	1	569054*	88 From vehicles numbered: 141102874, 142101569, 144103477 Petrol models 146100973, 147100216, 149100344 Diesel models } Up to vehicle suffix 'E' inclusive
		BRAKE PEDAL AND BRACKET ASSEMBLY	1	568894	109 Up to vehicle suffix 'E' inclusive
		BRAKE PEDAL AND BRACKET ASSEMBLY	1	569084*	109 From vehicle suffix 'F' onwards
		Bracket for brake pedal	1	272632	88 Up to vehicles numbered: 141102873, 142101568, 144103476 Petrol models
		Brake pedal and bushes	1	568896	88 Up to vehicle suffix
		Bracket for brake pedal	1	523695	88 From vehicles numbered: 141102874, 142101569, 144103477 } Up to
		Brake pedal and bushes	1	523696	88 }vehicle suffix
1		Bracket for brake pedal	1	569055*	88 From vehicle suffix
2		Brake pedal and bushes	1	569057*	88 'F' onwards
1		Bracket for brake pedal	1	272632	109 Up to vehicle suffix
2		Brake pedal and bushes	1	568896	109 'E' inclusive
1		Bracket for brake pedal	1	569085*	109 from vehicle suffix
2		Brake pedal and bushes	1	569086*	'F' onwards
3		Bush for pedal	2	272714	
4		Distance piece for pedal trunnion ..	1	269783	
5		Trunnion for pedal ..	1	568883	
6		Shaft for pedal ..	1	272712	
7		Pin locating pedal shaft	1	50446	
8		Oil plug ..	1	255202	
9		Joint washer for oil plug	2	3052	
10		Grommet for brake pedal bracket ..	1	509970	88
11		Stop lamp switch, mechanical, LU 35914	1	560864*	
12		Mounting bracket for stop lamp switch	1	560058*	
13		End stop, LU 54325966 } Fixing switch	1	569117*	
14		Special locknut, LU 54132410 } to mounting bracket	1	569116*	
15		Special bolt } Brake pedal	1	560223	
16		Locknut (5/16" UNF) } stop in lever	1	254881	
17		Switch protector plate and spring anchor ..	1	569201*	

* Asterisk indicates a new part which has not been used on any previous Rover model

CLUTCH AND BRAKE PEDALS AND MASTER CYLINDERS

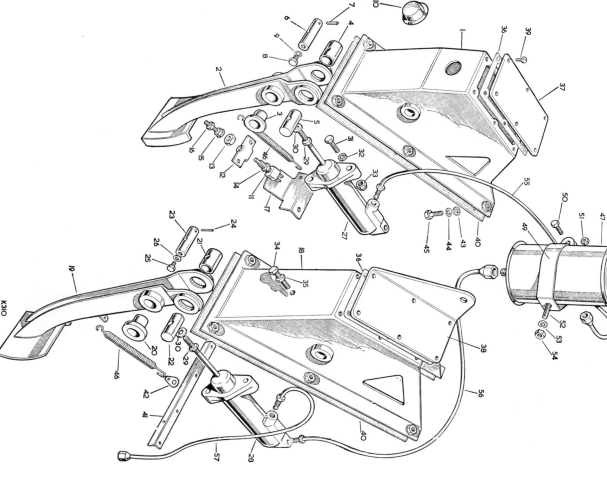

K310

CLUTCH AND BRAKE PEDALS AND MASTER CYLINDERS

Plate Ref. 1 2 3 4	DESCRIPTION	Qty	Part No.	REMARKS
	CLUTCH PEDAL AND BRACKET ASSEMBLY	1	568893	RHStg
	CLUTCH PEDAL AND BRACKET ASSEMBLY	1	568894	LHStg
18	Bracket for clutch pedal	1	272632	
19	Clutch pedal and bushes	1	568895	RHStg
19	Clutch pedal and bushes	1	568896	LHStg
20	Bush for pedal	2	272714	
21	Distance piece for pedal trunnion	1	269783	
22	Trunnion for pedal	1	568883	
23	Shaft for pedal	1	568883	
24	Pin locating pedal shaft	1	272712	
25	Oil plug	1	50446	
26	Joint washer for oil plug	1	255202	
			3052	
27	Brake master cylinder, GI 64068893, CV type	1	520849	88 Up to vehicles numbered: 24131336D, 24431621D Petrol models; 27108693D, 27405072D Diesel models
27	Brake master cylinder, GI 64067720, CB type	1	569126*	88 From vehicles numbered: 24131337D, 24431622D Petrol models; 27108694D, 27405073D onwards Diesel models
	Brake master cylinder, GI 64068750, CB type	1	569128*	109 4 cylinder models. Up to vehicle suffix 'E' inclusive
	Brake master cylinder, GI 64068750, CV type	1	564944*	109 4 cylinder models. From vehicle suffix 'F' onwards
	Brake master cylinder, GI 64068830, CV type	1	564706	109 6 cylinder models
28	End cap and spacer for brake master cylinder	1	516410	109 4 cylinder models
	Packing piece for brake master cylinder	1	523084	109 } For CB type
	Clutch master cylinder, GI 64068893	1	569126	109 } master cylinder
29	Nut for master cylinder push rod	1	254831	
30	Plain washer for master cylinder push rod	1	4148	
31	Bolt (5/16" UNF x 1 5/8" long)	4	255029	Fixing master cylinder to pedal bracket
	Bolt (5/16" UNF x 1 1/2" long)	2	256222	
32	Plain washer	4	3830	
33	Self-locking nut (5/16" UNF)	2	252211	
34	Bolt (1/4" UNF x 1" long)	4	255009	In bracket for pedal stop
35	Nut (1/4" UNF)	2	254810	1 off when mechanical stop lamp switch is fitted
36	Gasket for pedal bracket top cover	2	272819	2 off on 109 4 cylinder models
37	Top cover for pedal bracket	1	272713	109 4 cylinder models
38	Pedal bracket top cover and reservoir tank support	1	277840	All 109, All 88 RHStg. Up to vehicles numbered on 88 LHStg: 144103476 Petrol models; 149100343 Diesel models
38	Pedal bracket top cover and reservoir tank support	1	517907	88 LHStg. From vehicles numbered: 144103477 onwards Petrol models; 149100344 onwards Diesel models

* Asterisk indicates a new part which has not been used on any previous Rover model

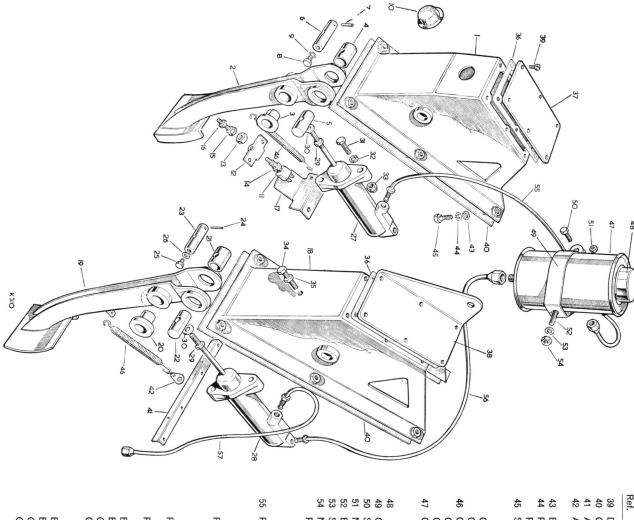

K310

CLUTCH AND BRAKE PEDALS AND MASTER CYLINDERS

Plate Ref. 1 2 3 4	DESCRIPTION	Qty.	Part No.	REMARKS
39	Drive screw fixing top cover to bracket	12	78227	
40	Gasket for pendant pedal brackets	2	562940	
41	Anchor for pedal springs	1	272730	Early type
42	Anchor for pedal springs	2	240708	Late type. 1 off when mechanical stop lamp is fitted
43	Bolt ($\frac{5}{16}$" UNF x $\frac{7}{8}$" long)	12	255227	Fixing pedal bracket and anchor to dash
44	Plain washer	10	2223	
45	Spring washer	3	3522	
		12	3075	
46	Clutch and brake pedal return spring, $3\frac{1}{2}$" long	2	568866*	Alternatives. Check before ordering
46	Clutch and brake pedal return spring, $3\frac{1}{2}$" long	2	272729	
47	Clutch and brake reservoir tank and pipe	1	569701*	RHStg
47	Clutch and brake reservoir tank and pipe	1	504135**	RHStg
47	Clutch and brake reservoir tank, GI 64047276	1	504136**	LHStg
	Filter for clutch and brake reservoir tank, GI 64473410	1	504105††	
	Sealing washer for filter, GI 362920	1	518682††	
48	Filler cap for supply tank, GI 378434	1	264767††	
49	Clip for reservoir tank	1	500201††	
50	Screw (2 BA x $\frac{7}{8}$" long)	1	217636	Fixing clip to tank
51	Nut (2 BA)	1	78214	
52	Bolt ($\frac{1}{4}$" UNF x $\frac{3}{8}$" long)	2	2247	Fixing tank clip to mounting bracket
53	Spring washer	2	255206	
54	Nut ($\frac{1}{4}$" UNF)	2	3074	
	Pipe, reservoir tank to brake master cylinder	1	254810	
55	Pipe, reservoir tank to brake master cylinder	1	504106†	All 88 LHStg. Up to vehicles numbered on 88 RHStg
55	Pipe, reservoir tank to brake master cylinder	1	564785*	88 RHStg. From vehicles numbered: 24131336D Petrol models 27108693D Diesel models
		1	504106	All 109 6 cylinder models
		1	569149*	88 Up to vehicles numbered: 24131337D Petrol models 27108694D onwards Diesel models
	Pipe complete, reservoir tank to brake master cylinder	1	569147*	109 4 cylinder models. From vehicle suffix 'F' onwards
	Pipe complete, reservoir tank to brake master cylinder	1	504104†	Up to vehicle suffix 'E' inclusive on 109 4 cylinder models
	Pipe complete, reservoir tank to brake master cylinder	1	216909†	88 Up to vehicles numbered: 24431621D Petrol models 27405072D Diesel models
	Pipe complete, reservoir tank to brake master cylinder	1	267601†	Up to vehicle suffix 'E' inclusive on 109 4 cylinder models;
	Banjo, GI 352201W	1	216909	
	Banjo bolt, GI 376207W	1	267601	}LHStg
	Gasket, small, GI 378700	1	216914	Fixing pipe to brake master cylinder
	Gasket, large, GI 1378731	1	504104	
	Banjo, GI 352201W	1	216909†	
	Banjo bolt, GI 376207W	1	267601†	}LHStg
	Gasket, small, GI 378700	1	216914†	Fixing pipe to brake master cylinder
	Gasket, large, GI 1378731	1	504104†	}109 6 cylinder models

* Asterisk indicates a new part which has not been used on any previous Rover model

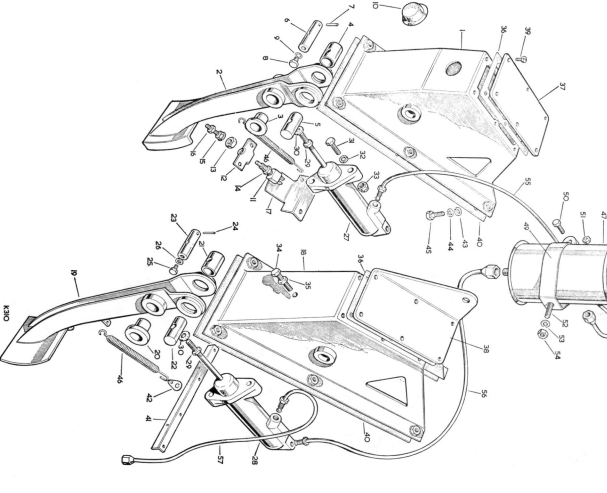

K3IO

Plate Ref.	1	2	3	4	DESCRIPTION	Qty	Part No.	REMARKS
56					Pipe, reservoir tank to clutch master cylinder	1	277929	RHStg
					Pipe, reservoir tank to clutch master cylinder	1	562978	LHStg
57					Pipe, 3/16", clutch master cylinder to hose	1	277930	RHStg
					Pipe, 3/16", clutch master cylinder to hose	1	512839	LHStg, Early models. Alternative to 1/4" pipe. Check before ordering
					Pipe, 1/4", clutch master cylinder to hose	1	552026	LHStg 88 Up to vehicles numbered: 24431621D Petrol models, 27405072D Diesel models
					Pipe, 1/4", clutch master cylinder to hose	1	569705	LHStg 88 From vehicles numbered: 24431622D Petrol models, 27405073D Diesel models onwards
					Pipe, 1/4", clutch master cylinder to hose	1	569705	LHStg 109 4 cylinder models
					Pipe, 3/16", clutch master cylinder to hose	1	562932	RHStg 109 6 cylinder models } Check before ordering
					Pipe, 3/16", clutch master cylinder to slave cylinder	1	562933	LHStg 109 6 cylinder models
					Pipe, 1/4", clutch master cylinder to slave cylinder	1	568947*	Early type, Nylon pipe / Late type, metal pipe } Check before ordering
					Pipe, 1/4", clutch master cylinder to hose	1	533822	109 2¼ litre Petrol models. Optional equipment
					BRAKE SERVO CONVERSION KIT	1	502333	88. For CB type master cylinder, GI 64067720 } Check before ordering
					Brake master cylinder overhaul kit, GI SP 1980/1	1	601611	88. For CV type master cylinder, GI 64068893
					Brake master cylinder overhaul kit, GI SP 1967/4	1	503754	109 4 cylinder models. For CB type master cylinder, GI 64067722
					Brake master cylinder overhaul kit, GI SP 1989	1	601611	109 6 cylinder models. For CB type master cylinder, GI 64068750 } Check before ordering
					Brake master cylinder overhaul kit, GI SP 2385	1	605127	109 4 cylinder models. For CV type master cylinder, GI 64068830
					Brake master cylinder overhaul kit, GI SP 2385	1	606023*	109 6 cylinder models
					Brake master cylinder overhaul kit, GI SP 2472	1	601611	
					Clutch master cylinder overhaul kit, GI SP 1967/4	1	601611	

* Asterisk indicates a new part which has not been used on any previous Rover model

** Up to vehicles numbered:

141800001	144800001	151800001	156800001
141800271	144800059	151800036	156800007
142800375	147800125	152800114	157800018

† From vehicles numbered:

141800002	144800002	151800002	156800002
141800272	144800060	151800037	156800008
142800376	147800126	152800115	157800019

‡‡ From vehicles numbered:

141800002	144800002	151800002	156800002
141800272	144800060	151800037	156800008
142800376	147800126	152800115	157800019

For vehicles prior to these supply part number 504135 for RHStg models and part number 504136 for LHStg models

K.497

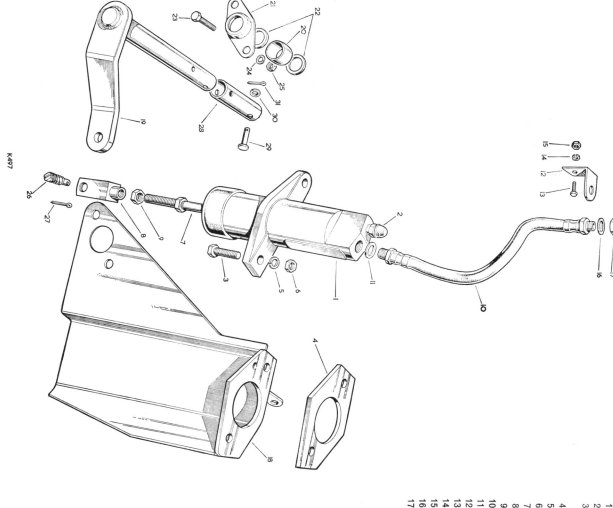

CLUTCH SLAVE CYLINDER

Plate Ref. 1 2 3 4	DESCRIPTION	Qty	Part No.	REMARKS
1	Clutch slave cylinder, GI 64067749	1	266694	
2	Bleed screw for clutch slave cylinder, GI 64470444	1	556508	
3	Bolt ($\frac{5}{16}$" UNF x 1" long) ... } Fixing slave	2	255228	When 9" type clutch is fitted
3	Bolt ($\frac{5}{16}$" UNF x 1$\frac{1}{4}$" long) } cylinder to	2	255030	When 9$\frac{1}{2}$" diaphragm
4	Packing piece } support bracket	1	544686	spring clutch is fitted
5	Spring washer	2	3075	
6	Nut ($\frac{5}{16}$" UNF)	2	254831	
7	Push rod for clutch slave cylinder, GI 64674324	1	537601	
8	Clevis for push rod	1	275199	
9	Locknut ($\frac{5}{16}$" UNF) for push rod	1	254831	
10	Hose, pipe to clutch slave cylinder, GI 3700628W ...	1	268341	
11	Gasket for hose at slave cylinder, GI 378711	1	233220	
12	Mounting bracket for hose	1	569462	All RHStg.
13	Bolt (2 BA x $\frac{1}{2}$" long) } Fixing	2	234603	Also LHStg.
14	Spring washer } bracket	2	3073	with $\frac{3}{16}$" pipe
15	Nut (2 BA) } to dash	2	2247	
16	Shakeproof washer } Fixing hose	1	233305	
17	Special nut, GI 10233 } to bracket	1	254852	
	Hose, pipe to clutch slave cylinder, GI 3703925W ...	1	552057	LHStg with $\frac{1}{4}$" pipe
	Adaptor, pipe to clutch master cylinder, GI 3746662W	1	512650	
	Gasket, adaptor to master cylinder, GI 378711 ...	1	233220	
	Banjo, hose to slave cylinder, GI 35206W	1	538068	
	Gasket, banjo to slave cylinder, GI 378711 ...	1	233220	4 cylinder models
	Banjo bolt, GI 376102 } Fixing hose	1	512235	
	Gasket, GI 378700 } to banjo	2	216914	
	Mounting bracket for hose	1	569462	4 cylinder models
	Shakeproof washer } Fixing hose	1	512651	
	Special nut, GI 10236 } to bracket	1	216912	
	Bolt (2 BA x $\frac{1}{2}$" long) } Fixing bracket to dash	2	234603	
	Spring washer	2	3073	
	Nut (2 BA)	2	2247	
	Clip fixing clutch and brake pipes	2	50639	

* Asterisk indicates a new part which has not been used on any previous Rover model

K.497

CLUTCH SLAVE CYLINDER

CLUTCH SLAVE CYLINDER

Plate Ref. 1 2 3 4	DESCRIPTION	Qty	Part No.	REMARKS
	Adaptor, GI 374662W ⎫ Fixing pipe to	1	512650	For early type nylon pipe. Check before ordering
	Gasket, GI 378711 ⎬ clutch slave cylinder	1	233220	
	Adaptor, GI 64473617 ⎫ Fixing pipe to	1	562943	Check before ordering
	Gasket, GI 378711 ⎬ clutch master cylinder	1	233220	
	Clip for pipe, clutch master cylinder to slave cylinder	1	562947	
	Drive screw, fixing clip to dash	1	72626	
	Adaptor, GI 374662W ⎫ Fixing hose to clutch	1	512650	
	Gasket, GI 378711 ⎬ slave cylinder	1	233220	
	Hose, adaptor to pipe, GI 370392SW	1	552057	LHStg 109 6 cylinder models
	Mounting bracket for hose	1	569462*	
	Shakeproof washer	1	512651	
	Special nut, GI 10236 ⎫ Fixing hose	1	216912	For late type metal pipe. Check before ordering
	Bolt (2 BA x ½" long) ⎬ to bracket	2	234603	
	Spring washer ⎫ Fixing	2	3073	
	Nut (2 BA) ⎬ bracket	2	2247	
	Adaptor ⎫ Fixing pipe to clutch	1	512650	
	Gasket ⎬ master cylinder	1	233220	
	Clip ⎫ Fixing pipe to	2	3621	
	Clip, fixing brake pipe to toebox	1	50639	
	Screw (2 BA x ½" long) ⎫ clutch pipe	2	78318	2 litre Petrol models
	Nut (2 BA) ⎬ to dash	2	3823	
18	Support bracket for clutch slave cylinder	1	272487	2 litre Petrol models
	Support bracket for clutch slave cylinder	1	509856	Except 2 litre Petrol models. Up to engine suffix 'A'
	Support bracket for clutch slave cylinder	1	561242	2¼ litre Petrol models and 2¼ litre Diesel models. From engine suffix 'B' onwards
	Support bracket for clutch slave cylinder	1	561762*	2.6 litre Petrol models.
	Shaft and operating lever for clutch	1	273061	2 litre Petrol models
	Shaft and operating lever for clutch	1	273077	Except 2 litre Petrol models. Up to engine suffix 'A'
	Shaft and operating lever for clutch	1	537603	From engine suffix 'A'. 2¼ litre Petrol and 2¼ litre Diesel models. From engine suffix 'B' onwards
19	Shaft and operating lever for clutch	1	544980	2.6 litre Petrol models
20	Spherical bearing	1	217984	
21	Housing for spherical bearing	1	217983	
22	Felt ring for spherical bearing	2	217985	
23	Bolt (5/16" UNF x ¾" long) ⎫ Fixing	2	255226	4 cylinder models
	Bolt (5/16" UNF x 1" long) ⎬ spherical	2	255228	6 cylinder models
24	Spring washer ⎬ bearing to	2	3075	
25	Nut (5/16" UNF) ⎬ support bracket	2	254831	
26	Clevis pin, lever to fork end	1	216421	
27	Split pin for clevis pin	1	2392	
28	Connecting tube for clutch cross-shaft	2	561661	
29	Pin ⎫ Fixing tube to	2	536803	
30	Plain washer ⎬ clutch cross and	2	10882	
31	Split pin ⎬ operating shafts	2	2422	
	Return spring for clutch operating lever	1	278690	Up to engine suffix 'A'
	Anchor plate for return spring at slave cylinder support bracket	1	272920	
	Clutch slave cylinder overhaul kit, GI SP 2029	1	502335	Except 2 litre Petrol models. Up to engine suffix 'A'

* Asterisk indicates a new part which has not been used on any previous Rover model

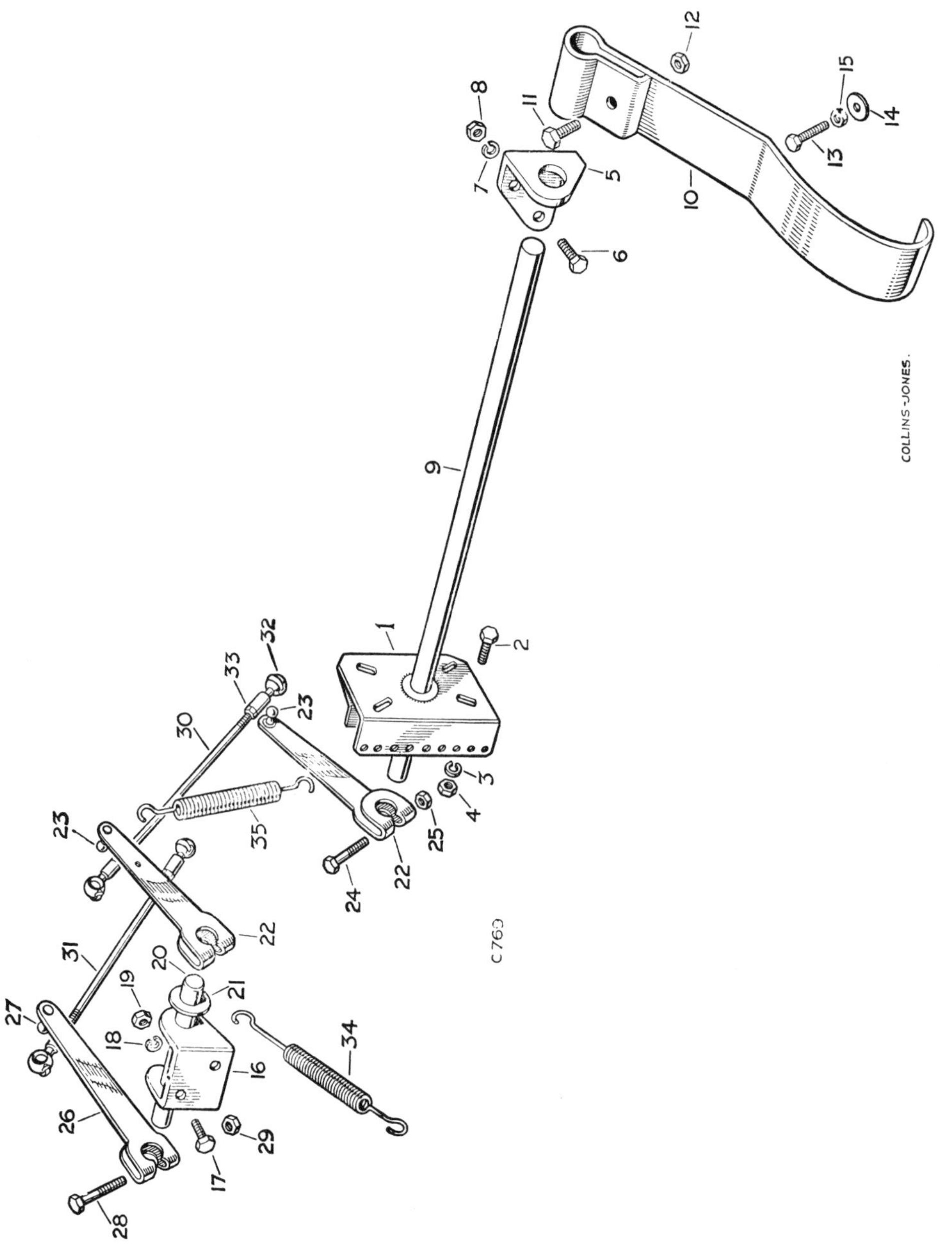
C 763
COLLINS-JONES.

ACCELERATOR LEVER AND RODS, 2 LITRE PETROL

Plate Ref.	1 2 3 4	DESCRIPTION		Qty	Part No.	REMARKS
1		Housing for accelerator shaft and pedal stop ..		1	277103	
2		Bolt ($\frac{1}{4}$" UNF x $\frac{1}{2}$" long)	Fixing housing and pedal stop to dash	2	255204	
3		Spring washer		2	3074	
4		Nut ($\frac{1}{4}$" UNF)		2	254810	
5		Bracket for accelerator pedal shaft		1	272804	
6		Bolt ($\frac{1}{4}$" UNF x $\frac{5}{8}$" long)	Fixing bracket to dash	2	255206	RHStg
		Bolt ($\frac{1}{4}$" UNF x $\frac{3}{4}$" long)		2	255207	LHStg
7		Spring washer		2	3074	
8		Nut ($\frac{1}{4}$" UNF)		2	254810	
9		Shaft for accelerator pedal		1	236658	
10		Accelerator pedal		1	500257	
		Pad for accelerator pedal		1	509463	LHStg
11		Bolt ($\frac{5}{16}$" UNF x $\frac{7}{8}$" long)	Fixing pedal to shaft	1	255027	
12		Nut ($\frac{5}{16}$" UNF)		1	254831	
13		Bolt ($\frac{5}{16}$" UNF x $2\frac{1}{2}$" long)	Pedal stop in floor	1	255038	
14		Plain washer		2	2258	
15		Nut ($\frac{5}{16}$" UNF)		2	254831	
16		Bracket for accelerator cross shaft		1	236665	RHStg
		Bracket for accelerator cross shaft		2	236998	LHStg
17		Bolt ($\frac{1}{4}$" UNF x $\frac{5}{8}$" long)	Fixing bracket to dash	2	255206	4 off on LHStg
18		Spring washer		2	3074	4 off on LHStg
19		Nut ($\frac{1}{4}$" UNF)		2	254810	4 off on LHStg
20		Cross shaft for accelerator		1	277153	RHStg
		Cross shaft for accelerator		1	277154	LHStg
		Stop clip for cross shaft		2	269132	LHStg
21		Distance washer for lever		2	3680	
22		LEVER ASSEMBLY FOR ACCELERATOR ..		2	277105	
23		Ball end for lever		2	1481	
24		Bolt ($\frac{1}{4}$" UNF x $1\frac{1}{8}$" long)	Fixing levers and stop clip to shaft	2	256200	4 off on LHStg
25		Nut ($\frac{1}{4}$" UNF)		2	254810	4 off on LHStg
26		LEVER ASSEMBLY FOR CROSS SHAFT		1	278079	
27		Ball end for lever		1	1481	
28		Bolt ($\frac{1}{4}$" UNF x $1\frac{1}{8}$" long)	Fixing lever to cross shaft	1	256200	
29		Nut ($\frac{1}{4}$" UNF)		1	254810	
30		Control rod pedal shaft to cross shaft		1	509556	
31		Control rod, cross shaft to engine		1	277983	
32		Ball joint socket for rods		4	531324	
33		Locknut for socket		4	2247	
34		Return spring for throttle		1	272729	
35		Return spring for pedal		1	277455	

* Asterisk indicates a new part which has not been used on any previous Rover model

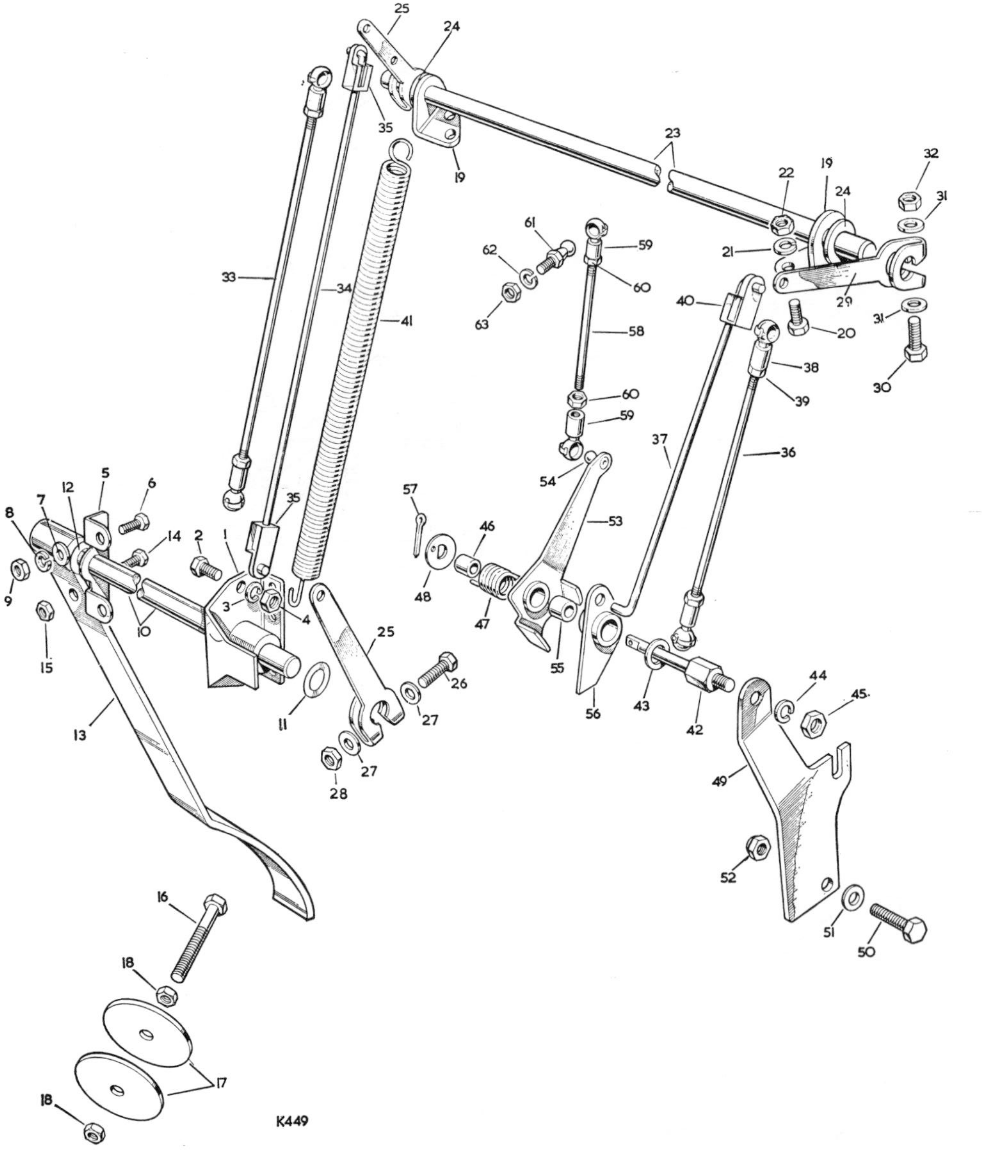

ACCELERATOR LEVER AND RODS, 2¼ LITRE PETROL

Plate Ref.	1 2 3 4 DESCRIPTION		Qty	Part No.	REMARKS
1	Housing for accelerator shaft and pedal stop		1	277103	
2	Bolt ($\frac{1}{4}''$ UNF x $\frac{1}{2}''$ long)	Fixing housing and pedal stop to dash	2	255204	
3	Spring washer		2	3074	
4	Nut ($\frac{1}{4}''$ UNF)		2	254810	
5	Bracket for accelerator pedal shaft		1	272804	
6	Bolt ($\frac{1}{4}''$ UNF x $\frac{5}{8}''$ long)	Fixing bracket to dash	1	255206	
7	Plain washer		2	3821	
8	Spring washer		2	3074	
9	Nut ($\frac{1}{4}''$ UNF)		2	254810	
10	Shaft for accelerator pedal		1	236658	RHStg
	Shaft for accelerator pedal		1	277153	LHStg
11	Special washer	On accelerator shaft	1	508962	
12	Plain washer		1	3843	
13	Accelerator pedal		1	500257	
	ACCELERATOR PEDAL RESTRICTOR KIT		1	516057	88 models. From vehicles numbered: 141001495 142000991 144002270 onwards Up to vehicle suffix 'C' inclusive
	Accelerator restrictor lever complete		1	509407	
	Spring, accelerator restrictor		1	264362	
	Control rod, pedal shaft to cross shaft		1	509556	
	Control rod, relay shaft to engine		1	509558	
	Stop clip for relay shaft		1	269132	RHStg
	Bolt ($\frac{1}{4}''$ UNF x $1\frac{1}{8}''$ long)	Fixing clip to shaft	1	256200	RHStg
	Nut ($\frac{1}{4}''$ UNF)		1	254810	RHStg
	Anchor plate for restrictor spring		1	240708	
	Bolt ($\frac{5}{16}''$ UNF x $\frac{1}{2}''$ long)	Fixing anchor plate to toe board	1	255223	RHStg
	Spring washer		1	3075	RHStg
	Nut ($\frac{5}{16}''$ UNF)		1	254831	RHStg
	Bolt ($\frac{1}{4}''$ UNF x $1\frac{1}{4}''$ long)	Lever stop on toe board	1	255211	
	Nut ($\frac{1}{4}''$ UNF)		2	254810	
14	Bolt ($\frac{5}{16}''$ UNF x $\frac{7}{8}''$ long)	Fixing pedal to shaft	1	255027	
15	Nut ($\frac{5}{16}''$ UNF)		1	254831	
16	Bolt ($\frac{5}{16}''$ UNF x $1\frac{1}{8}''$ long)	Pedal stop in floor	1	255029	
17	Plain washer		2	2258	
18	Nut ($\frac{5}{16}''$ UNF)		2	245831	
19	Bracket for accelerator cross shaft, 'L' shaped		2	236998	Alternatives. Check before ordering
	Bracket for accelerator cross shaft, 'U' shaped		2	236665	
20	Bolt ($\frac{1}{4}''$ UNF x $\frac{5}{8}''$ long)	Fixing bracket to dash	4	255206	2 off on LHStg
21	Spring washer		4	3074	2 off on LHStg
22	Nut ($\frac{1}{4}''$ UNF)		4	254810	2 off on LHStg
23	Cross shaft for accelerator		1	277154	RHStg
	Cross shaft for accelerator		1	277153	LHStg
24	Distance washer for lever		2	3680	
	LEVER ASSEMBLY FOR ACCELERATOR		2	277105	Ball joint type. Alternatives. Check before ordering
	Ball end for lever		2	1841	
25	Lever for accelerator		2	277475	Linkage clip type. Alternatives. Check before ordering
26	Bolt ($\frac{1}{4}''$ UNF x $1\frac{1}{4}''$ long)	Fixing levers to shaft	2	256001	
27	Plain washer		4	3911	
28	Nut ($\frac{1}{4}''$ UNF)		2	254810	

* Asterisk indicates a new part which has not been used on any previous Rover model

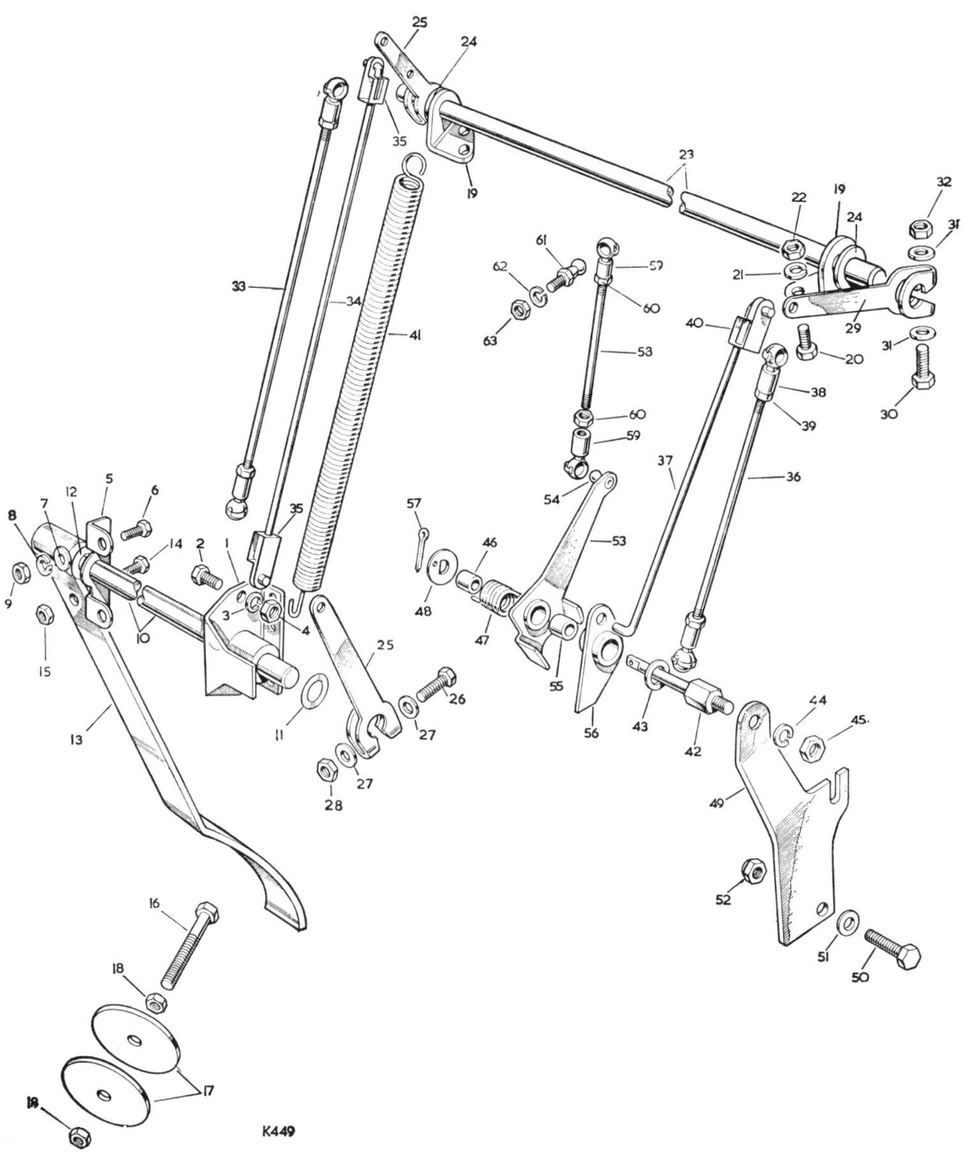

K449

ACCELERATOR LEVER AND RODS, 2¼ LITRE PETROL

Plate Ref. 1 2 3 4	DESCRIPTION	Qty	Part No.	REMARKS
29	LEVER ASSEMBLY FOR CROSS SHAFT	1	277106	Up to vehicle suffix 'A'
	Ball end for lever	1	1481	
30	Lever for cross shaft	1	531395	From vehicle suffix 'B' onwards
31	Plain washer	1	256001	
32	Nut (¼" UNF)	2	3911	
33	Bolt (¼" UNF x 1¼" long)	1	254810	} Fixing lever to cross shaft
34	Control rod, pedal shaft to cross shaft	1	509556	Ball joint type } Alternatives. Check before ordering
35	Linkage clip for control rod	2	531390	Linkage clip type }
36	Control rod, cross shaft to engine	1	531394	Ball joint type
36	Control rod, cross shaft to engine	1	503434	Ball joint type 1958-60
36	Control rod, cross shaft to engine	1	509558	Linkage clip type 1961. Also Series IIA up to vehicle suffix 'A'
36	Control rod, cross shaft to engine	1	531389	Linkage clip type. From vehicle suffix 'B' to suffix 'C' inclusive
36	Control rod, cross shaft to engine	1	531388	Linkage clip type. 'D' onwards
37	Linkage clip for control rod, cross shaft to engine	2	531324	2 off (cross shaft to engine) on linkage clip type. Up to vehicle suffix 'A'
37	Linkage clip for control rod, cross shaft to engine	4	2247	Linkage clip type. From vehicle suffix 'B' onwards
38	Ball joint socket for rods	1	531394	Ball joint type
39	Locknut for socket	1	277455	
40	Torsion spring for bell crank	1	502900	
41	Return spring for pedal	1	2765	
42	Spindle for carburetter bell crank	1	3076	
43	Plain washer	1	254812	
44	Spring washer	1	502899	} Fixing spindle
45	Nut (⅜" UNF)	1	502982	Up to engine suffix 'H' inclusive
45	Nut (⅜" UNF)	1	552223	From engine suffix 'J' onwards
46	Spacer for spindle	1	502808	
47	Torsion spring for bell crank	1	503430	For 5/16" diameter bolt fixing. Up to engine suffix 'C'
48	Special washer for torsion spring	1	537707	For ⅜" diameter bolt fixing. From engine suffix 'D' onwards
49	Bracket for accelerator controls	1	255228	RHStg.
50	Bolt (5/16" UNF x 1" long)	2	3830	RHStg.
51	Plain washer	2	252211	suffix 'C'
52	Self-locking nut (5/16" UNF)	2	255247	RHStg. } Fixing bracket to steering column support
50	Bolt (⅜" UNF x 1" long)	2	3036	From engine suffix 'D' onwards
51	Plain washer	2	252212	From engine suffix 'D' onwards
52	Self-locking nut (⅜" UNF)	2	504619	Up to engine suffix 'C'
53	CARBURETTER BELL CRANK LEVER ASSEMBLY	1	531267	From engine suffix 'D' to suffix 'H' inclusive
53	CARBURETTER BELL CRANK LEVER ASSEMBLY	1	1481	to suffix 'H' inclusive
53	CARBURETTER BELL CRANK LEVER ASSEMBLY	1	277120	From engine suffix 'J' onwards
54	Ball end for lever	1	552097	
55	Bush for bell crank	1	277120	
54	Ball end for lever	1	1481	From engine suffix 'J' onwards
55	Bush for bell crank	1	277120	

* Asterisk indicates a new part which has not been used on any previous Rover model

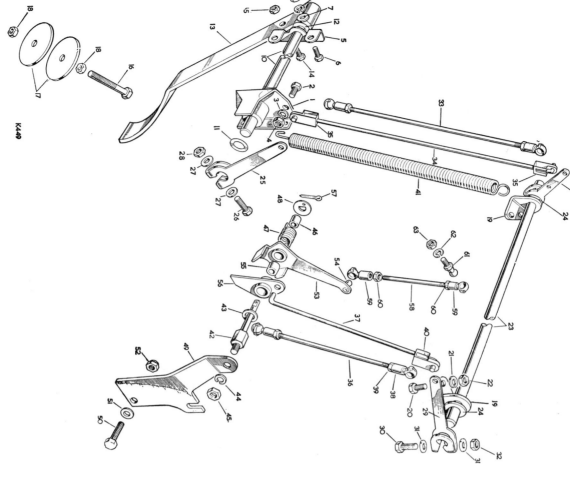

K449

ACCELERATOR LEVER AND RODS, 2¼ LITRE PETROL

Plate Ref. 1 2 3 4	DESCRIPTION	Qty	Part No.	REMARKS
	CARBURETTER RELAY LEVER ASSEMBLY	1	504621	} Up to engine
56	Ball end for lever	1	1481	} suffix 'C'
56	Carburetter relay lever	1	504620	From engine suffix 'D' onwards
57	Split pin fixing levers to spindle	1	2392	
	ROD ASSEMBLY, BELL CRANK TO CARBURETTER	1	506210	
	Ball joint for rod	2	277633	
	Locknut for ball joint (¼" UNF)	2	254810	} Up to engine suffix 'C'
58	Spring washer } Fixing rod to bell	2	3074	
58	Nut (¼" UNF) } crank and carburetter	2	254810	
58	Control rod, bell crank to carburetter	1	531439	} From engine suffix 'D' onwards
59	Ball joint } For	2	531324	
60	Locknut (2 BA) } control rod	2	2247	
61	Ball end for carburetter lever	1	535168	
62	Spring washer } Fixing ball end to	1	3074	
63	Nut (¼" UNF) } carburetter lever	1	254810	

* Asterisk indicates a new part which has not been used on any previous Rover model

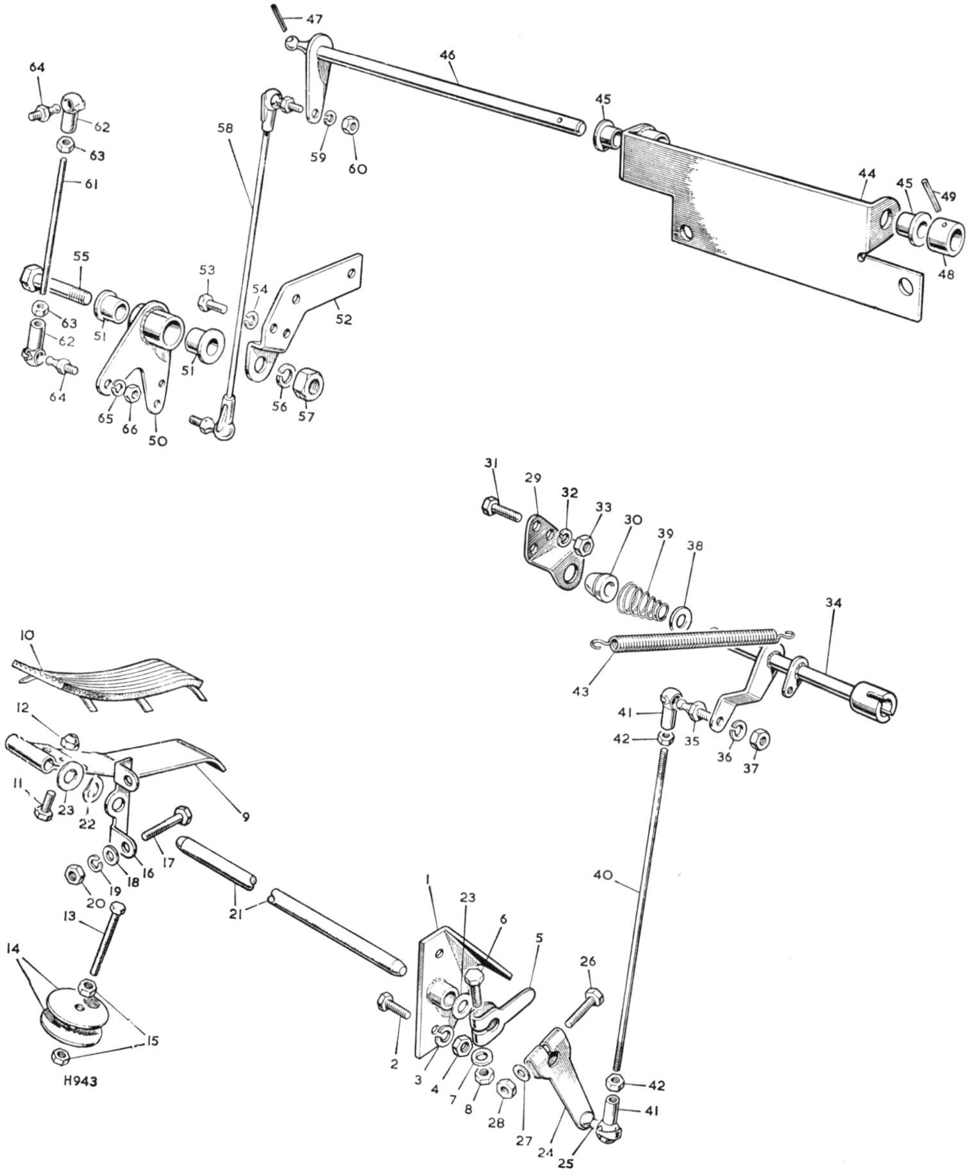

ACCELERATOR LEVERS AND RODS, 2.6 LITRE PETROL

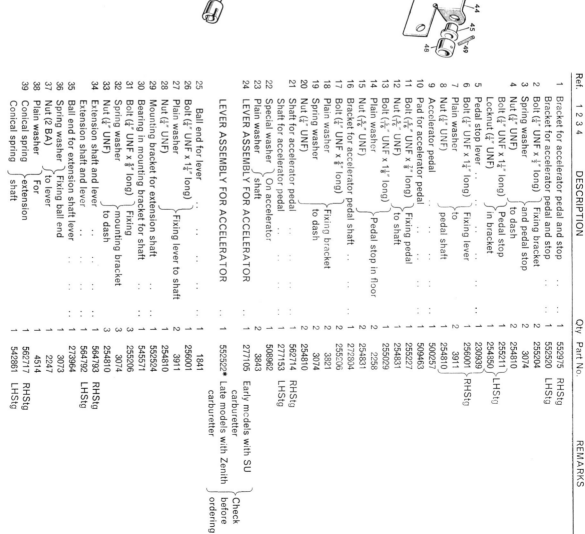

H943

ACCELERATOR LEVERS AND RODS, 2.6 LITRE PETROL

Plate Ref. 1 2 3 4	DESCRIPTION	Qty.	Part No.	REMARKS
1	Bracket for accelerator pedal and stop ...	1	552975	RHStg
1	Bracket for accelerator pedal and stop ...	1	552520	LHStg
2	Bolt (¼" UNF x ½" long) ⎫ Fixing bracket	2	255204	
3	Spring washer ⎬ and pedal stop	2	3074	
4	Nut (¼" UNF) ⎭ to dash	2	254810	
5	Pedal stop lever ... ⎫ Pedal stop	1	255211	RHStg
	⎬ in bracket	1	230939	LHStg
	Locknut (¼" UNF) ⎭	1	254350	
6	Bolt (¼" UNF x 1¼" long) ⎫ Fixing lever	1	256001	
7	Plain washer ⎬ to	1	2258	
8	Nut (¼" UNF) ⎭ pedal shaft	1	254831	
9	Accelerator pedal	1	255029	
10	Pad for accelerator pedal	1	509463	
11	Bolt (5⁄16" UNF x 7⁄8" long) ⎫ Fixing pedal	1	500257	
12	Nut (5⁄16" UNF) ⎭ to shaft	2	254810	
13	Bolt (5⁄16" UNF x 1⅛" long) ⎫	1	3911	
14	Plain washer ⎬ Pedal shaft	2	255206	
15	Nut (5⁄16" UNF) ⎭	2	3821	
16	Bracket for accelerator pedal shaft	1	3074	
17	Bolt (¼" UNF x 5⁄8" long) ⎫ Fixing bracket	2	272804	
18	Plain washer ⎬ to dash	2	254831	
19	Spring washer ⎬	2	255227	
20	Nut (¼" UNF) ⎭	2	254810	
21	Shaft for accelerator pedal	1	277105	
22	Special washer ⎫ On accelerator	2	3843	
23	Plain washer ⎭ shaft	2	508962	
			277153	LHStg
			562714	RHStg
24	LEVER ASSEMBLY FOR ACCELERATOR	1	552522*	Early models with SU carburetter ⎫ Check before ordering. Late models with Zenith carburetter ⎭
25	Ball end for lever	1	1841	
26	Bolt (¼" UNF x 1¼" long) ⎫ Fixing lever to shaft	1	256001	
27	Plain washer ⎬	2	3911	
28	Nut (¼" UNF) ⎭	1	254810	
29	Mounting bracket for extension shaft	1	552524	
30	Bearing in mounting bracket for shaft	1	254571	
31	Bolt (¼" UNF x 5⁄8" long) ⎫ Fixing	3	255206	
32	Spring washer ⎬ mounting bracket	3	3074	
33	Nut (¼" UNF) ⎭ to dash	3	254810	
34	Extension shaft and lever ...	1	564793	RHStg
34	Extension shaft and lever ...	1	564792	LHStg
35	Ball end for extension shaft lever	1	273964	
36	Spring washer ⎫ Fixing ball end	1	3073	
37	Nut (2 BA) ⎭ to lever	1	2247	
38	Plain washer ⎫ For extension	1	4514	
39	Conical spring ⎬ shaft	1	562717	RHStg
39	Conical spring ⎭	1	542861	LHStg

* Asterisk indicates a new part which has not been used on any previous Rover model

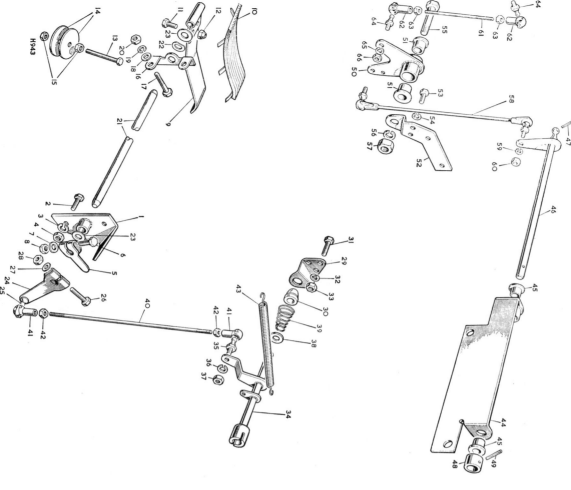

H943

ACCELERATOR LEVERS AND RODS, 2.6 LITRE PETROL

Plate Ref. 1 2 3 4	DESCRIPTION	Qty	Part No.	REMARKS
40	Control rod, pedal shaft to extension shaft ...	1	564797	RHStg
	Control rod, pedal shaft to extension shaft	1	552526	LHStg
41	Ball joint socket for control rod	1	531324	
42	Locknut (2 BA) fixing socket to control rod ...	2	2247	
43	Return spring, bell crank to extension shaft	1	277455	
44	BRACKET ASSEMBLY FOR ACCELERATOR			
	CROSS SHAFT			
45	Bearing for cross shaft	1	552784	
46	Accelerator cross shaft and lever	2	511127	
47	Accelerator cross shaft and lever ...	1	552980	RHStg
	Accelerator cross shaft and lever	1	552989	LHStg
47	Spiral pin for cross shaft	1	542783	
48	Boss for cross shaft	1	542762	
49	Spring dowel fixing boss to shaft ...	1	534021	
50	BELL CRANK LEVER AND BEARINGS			
	ASSEMBLY	1	552984	
51	Bearing for bell crank	2	238793	
52	Support bracket for bell crank	1	562670	
53	Set bolt (¼" UNC × ⁷⁄₁₆" long) Fixing support	3	253205	
	bracket to cylinder head			
54	Spring washer	3	3074	
55	Centre pin Fixing bell crank	1	542776	
56	Spring washer to	1	3075	
57	Nut (⁵⁄₁₆" UNC) support bracket	1	256801	
58	Control rod, cross shaft to bell crank ...	1	564794	
59	Spring washer Fixing control rod	2	3073	
60	Nut (10 UNF) to levers	2	257011	
61	Control rod, bell crank to carburetter	1	531324	
62	Ball joint For	2	531324	
63	Locknut (2 BA) control rod	2	2247	
64	Ball end Fixing control rod	2	273964	
65	Spring washer to bell crank lever	2	3073	1 off with Zenith type carburetter
66	Nut (2 BA) and carburetter	2	2247	

* Asterisk indicates a new part which has not been used on any previous Rover model

ACCELERATOR LEVER AND RODS, DIESEL

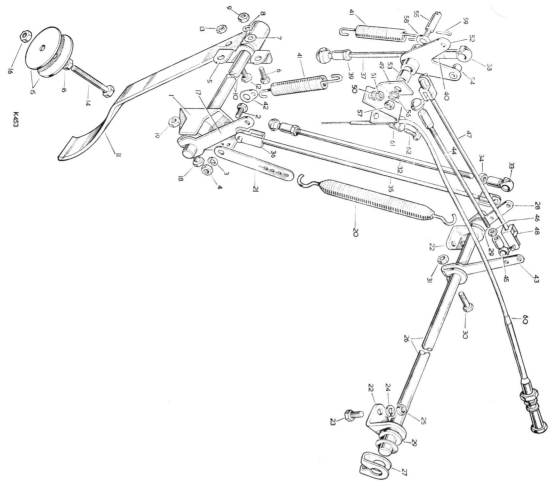

K453

ACCELERATOR LEVER AND RODS, DIESEL

Plate Ref.	1	2	3	4	DESCRIPTION	Qty	Part No.	REMARKS
1					Housing for accelerator shaft and pedal stop	1	277103	
2					Bolt (¼" UNF × ½" long) } Fixing housing	2	255204	
3					Spring washer } and pedal stop	2	3074	
4					Nut (¼" UNF) } to dash	2	254810	
5					Bracket for accelerator pedal shaft	1	272804	
6					Bolt (¼" UNF × ⅝" long) } Fixing bracket	2	255206	
7					Plain washer } to dash	2	2224	
8					Spring washer }	2	3074	
9					Nut (¼" UNF) }	2	254810	
10					Shaft for accelerator pedal	1	236658 / 277153	RHStg / LHStg
11					Accelerator pedal	1	500257	
12					Bolt (5/16" UNF × ⅞" long) } Fixing pedal	1	255227	
13					Nut (5/16" UNF) } to shaft	1	254831	
14					Bolt (5/16" UNF × 2½" long) } Pedal	1	255038	
15					Plain washer } stop in	2	2258	
16					Nut (5/16" UNF) } floor	2	254831	
					LEVER ASSEMBLY FOR ACCELERATOR ON PEDAL SHAFT			
17					Lever for accelerator on pedal shaft	1	277105 / 1481	Ball joint type / type — Alternatives. Check before ordering
18					Bolt (¼" UNF × 1¼" long) } Fixing	1	256001	
19					Plain washer } lever to	2	3911	
20					Nut (¼" UNF) } shaft	1	254810	
21					Return spring for pedal	1	231393	
22					Anchor for return spring	1	543498	
23					Bracket for accelerator cross shaft	2	236998	
24					Bolt (¼" UNF × ⅝" long) } Fixing	2	255206	
25					Spring washer } brackets	4	3074	
26					Nut (¼" UNF) } to dash	4	254810	
27					Accelerator cross shaft	1	277154	
28					Stop clip for cross shaft	2	269132 / 277472 / 277473	2 off on LHStg — Ball joint type / Linkage clip type before ordering
29					Accelerator lever on cross shaft from pedal	1	256001	4 off on LHStg
30					Distance washer for cross shaft	3	3680	8 off on LHStg
31					Bolt (¼" UNF × 1¼" long) } Fixing levers and	3	256001	4 off on LHStg
					Plain washer } stop clip to		3911	
32					Nut (¼" UNF) } cross shaft	6	254810	
33					Control rod, pedal shaft to cross shaft	2	531324	Ball joint type — Alternatives.
34					Ball joint socket } For	2	2247	Check
35					Locknut (2 BA) } rod	2	503556	before ordering
36					Control rod, pedal shaft to cross shaft	1	531390	RHStg — Linkage clip type before ordering
37					Linkage clip for control rod	2	537792	LHStg
					Control rod, pedal shaft to cross shaft	2	531394	
					Control rod, bell crank to accelerator lever	1	277779	Overall length 3"
					Control rod, bell crank to accelerator lever	1	537791	Overall length 3⅜" — Alternatives. Check before ordering
38					Ball socket }	2	531324	
39					Nut (2 BA) } For bell crank control rod	2	2247	
40					Adjuster nut	1	277778	

* Asterisk indicates a new part which has not been used on any previous Rover model

ACCELERATOR LEVER AND RODS, DIESEL

K453

ACCELERATOR LEVER AND RODS, DIESEL

Plate Ref.	1 2 3 4	DESCRIPTION	Qty	Part No.	REMARKS
41		Return spring for accelerator and stop levers on distributor pump	2	277502	
42		Anchor for return spring	1	277565	
43		Accelerator lever on cross shaft to engine	1	277612	Ball joint type } Alternatives. Check before ordering
		Ball end } Fixing ball end to accelerator lever	1	273964	
		Spring washer		3073	
		Nut (2 BA)		2247	
43		Accelerator lever on cross shaft to engine	1	531395	Linkage clip type
44		Control rod, cross shaft to bell crank	1	504699	
		Ball joint		531324	Ball joint type } Alternatives. Check before ordering
45		Ball joint } For control rod	2	2247	
46		Locknut (2 BA)	2	531391	Linkage clip type
47		Control rod, cross shaft to bell crank	2	531394	
48		Linkage clip for control rod	2	2247	Linkage clip type
				3073	
		Accelerator control lever on distributor pump		257023	type bracket
		Stop lever complete on distributor pump		277450 DPA 3240095	Use with distributor pump } Series II 2 litre Diesel
		Accelerator control lever on distributor pump		277619 DPA 3240095	Use with distributor pump
		Stop lever complete on distributor pump		277553	Use with distributor pump
		Accelerator control lever on distributor pump		509009	Use with distributor pump
		Stop lever complete on distributor pump		509011 DPA 3240099	Use with distributor pump
		Swivel clamp for stop lever		219582	
49		Bracket for bell crank on distributor pump		544451	Up to engine suffix 'J' inclusive
50		Bracket for bell crank on distributor pump		564911*	From engine suffix 'K' onwards
		Spring washer } Fixing bracket to distributor pump	4	3073	2 off on late type bracket
51		Nut (10 UNF)	4	257023	
52		Bell crank complete on distributor pump	1	2208	
53		Bell crank complete on distributor pump	1	2423	
54		Bush for bell crank	2	78114	
55		Ball end for bell crank		254810	
				1481	1 off on linkage clip type
56		Pin for bell crank		277454	
		Shakeproof washer } Fixing pin to bell crank bracket		277453	
57		Nut (¼" UNF)		531392	Ball joint type
58		Plain washer } Fixing bell crank lever to pin		277450	Linkage clip type } Alternatives. Check before ordering
		Split pin		277478	From engine suffix 'H' to suffix 'J' inclusive
59		'Engine stop' control		552688	Up to engine suffix 'G' inclusive
60		'Engine stop' control	1	552688*	From engine suffix 'J' inclusive
61		Clip } Fixing control	1	239673	From engine suffix 'K' onwards
62		Screw (2 BA x ⅝" long) } outer cable to abutment bracket on distributor pump	1	75940	Alternatives to 3 lines below. Check before ordering
		Bolt (2 BA x ⅞" long)	1	250962	Alternatives to line above. Check before ordering
		Spring washer	1	3073	
		Nut (2 BA)	1	2247	before ordering

* Asterisk indicates a new part which has not been used on any previous Rover model

H617

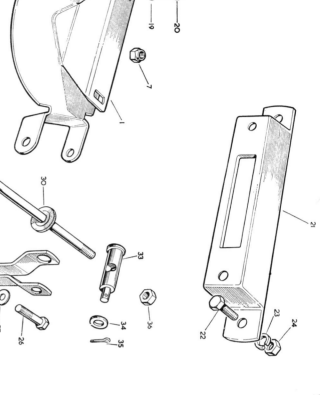

Plate Ref.	1 2 3 4	DESCRIPTION	Qty	Part No.	REMARKS
1		Housing for control quadrant	1	275706	
2		Lever and ball end for control	1	275709	
3		Bush for lever	1	275714	
4		Washer for lever	1	275715	
5		Bolt ($\frac{5}{16}$" UNF x 1" long)	1	255228	Fixing
6		Plain washer	1	2223	control lever
7		Self-locking nut ($\frac{5}{16}$" UNF)	1	252211	to housing
8		Knob for lever	1	552703	
9		Nylon spacer for knob	1	552555	
10		Special screw	1	219673	Fixing knob
11		Nut (2 BA)	to lever	1	2247	to lever
12		Quadrant plate	1	275713	
13		Special screw		2	278386	Fixing quadrant plate
14		Plain washer	to housing	2	3840	to housing
15		Spring washer		2	3074	
16		Nut ($\frac{1}{4}$" UNF)		2	254810	
17		Bolt (2 BA x $\frac{1}{2}$" long)		2	234603	
18		Plain washer	Fixing control	2	3685	
19		Spring washer	to dash, upper	2	3073	Up to vehicle suffix 'C' inclusive
20		Nut (2 BA)		2	2247	
21		Mounting box for hand speed control ...	1	255206	
22		Bolt ($\frac{1}{4}$" UNF x $\frac{1}{2}$" long)	Fixing mounting box	4	3074	
23		Spring washer	and speed control	4	3074	
24		Nut ($\frac{1}{4}$" UNF)	to dash, lower	4	254810	
25		Operating lever for hand engine speed control	1	276013	
26		Bolt ($\frac{1}{4}$" UNF x 1$\frac{1}{8}$" long)	Fixing control	1	256001	From vehicle suffix 'D' to vehicle suffix 'E' inclusive
27		Plain washer	lever to accelerator	1	3911	
28		Nut ($\frac{1}{4}$" UNF)	cross shaft	1	254810	
29		Control rod for engine speed control	1	276014	
30		Grommet in dash for control rod ...	1	312937	
31		Nut (2 BA)	For	1	2247	
32		Ball socket	control rod	1	276015	
33		Joint pin		1	531324	
34		Plain washer	For engine speed control rod	1	2208	
35		Split pin		1	3958	
36		Nut (2 BA) fixing control rod to joint pin ...	2	2247	

HAND CONTROL, ENGINE SPEED, DIESEL. From vehicle suffix 'F' onwards

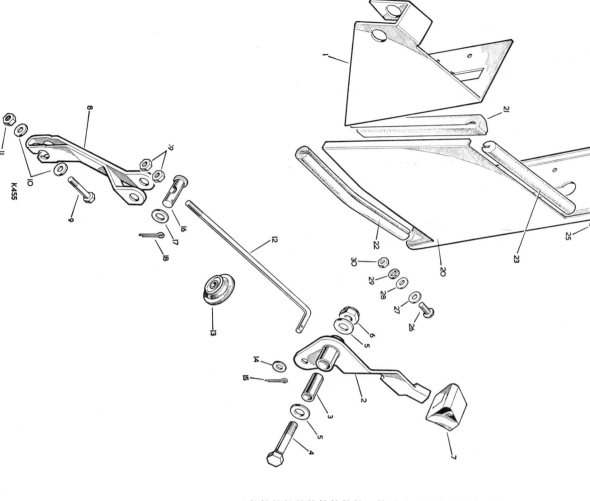

HAND CONTROL, ENGINE SPEED, DIESEL. From vehicle suffix 'F' onwards

Plate Ref.	1 2 3 4	DESCRIPTION		Qty	Part No.	REMARKS
1		Housing and quadrant for control	… …	1	569243*	
2		Lever for control	… …	1	569245*	
3		Distance tube for housing	…	1	569247*	
4		Bolt (¼" UNF x 1½" long)	Fixing control	1	256003	
5		Plain washer	lever and tube	2	3665	
6		Self-locking nut (¼" UNF)	to housing	1	252160	
7		Knob for lever	… …	1	569249*	
8		Operating lever for cross shaft		1	569250*	
9		Bolt (¼" UNF x 1⅛" long)	Fixing operating	1	256001	
10		Plain washer	lever to accelerator	2	3911	
11		Nut (¼" UNF)	cross shaft	1	254810	
12		Control rod for engine speed control		1	569251*	
13		Grommet in dash for control rod	…	1	312937	
14		Plain washer	Fixing control rod	1	3902	
15		Split pin	to control lever	1	2392	
16		Joint pin	Fixing	1	276015	
17		Plain washer	control rod to	1	2208	
18		Split pin	operating lever	1	3958	
19		Nut (2 BA) fixing control rod to joint pin	…	1	2247	
20		Mounting panel, RH	For control	1	569653*	RHStg
		Mounting panel, LH	housing	1	569654*	LHStg
21		Rubber seal, rear		1	348537	
22		Rubber finisher, bottom	For mounting panel	1	348541	
23		Rubber finisher, top		1	348535	
24		Drive screw	Fixing mounting panel	6	78174	
25		Plain washer	to control housing	6	4428	
26		Screw (2 BA x ½" long)		1	77869	
27		Plain washer	Fixing mounting	1	4030	
28		Packing washer	panel to	1	4172	
29		Spring washer	dash, lower	1	3073	
30		Nut (2 BA)		1	3823	

* Asterisk indicates a new part which has not been used on any previous Rover model

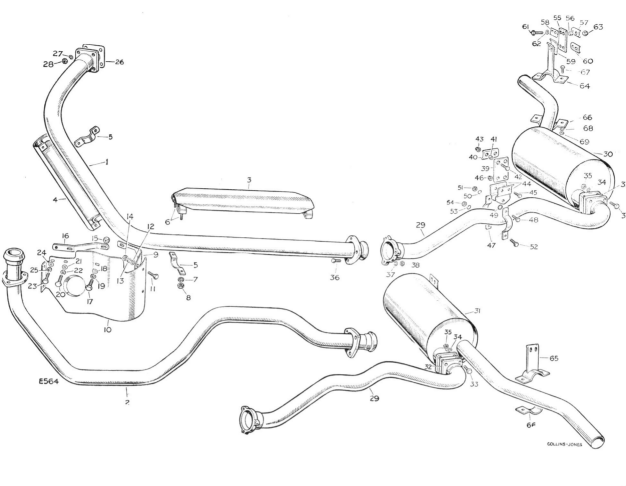

E564

COLLINS-JONES

EXHAUST SYSTEM, 4 CYLINDER MODELS

Plate Ref. 1 2 3 4	DESCRIPTION	Qty	Part No.	REMARKS
1	Front exhaust pipe complete	1	562858	2 litre Petrol and 88 Diesel
1	Front exhaust pipe complete	1	505855	109 Diesel
2	Front exhaust pipe complete	1	505853	88 2¼ litre Petrol 1958-60
2	Front exhaust pipe complete	1	517469	88 2¼ litre Petrol 1961 onwards
2	Front exhaust pipe complete	1	505826	109 2¼ litre Petrol 1958-60
2	Front exhaust pipe complete	1	517632	109 2¼ litre Petrol 1961 onwards
3	Exhaust pipe heat shield, long	1	503310	Except LHStg when toe box heat shields are fitted — 1958-60 Petrol models
4	Exhaust pipe heat shield, short	1	503309	Diesel models up to vehicle suffix 'C' inclusive
5	Pipe clamp	4	503307	
6	Bolt (¼" UNF x ⅞" long) — Fixing heat shield to front exhaust pipe	8	255208	
7	Spring washer	8	3074	
8	Nut (¼" UNF)	8	254810	
	HEAT SHIELD ASSEMBLY (TOE BOX)	1	516496	RHStg — Up to engine suffix 'H' inclusive
	HEAT SHIELD ASSEMBLY (TOE BOX)	1	516945†	LHStg — Up to engine suffix 'H' inclusive
9	Plain washer on rear manifold stud	1	515506	
	Bracket, exhaust manifold to heat shield stud	1	515597	Up to engine suffix 'H' inclusive
	Plain washer for manifold stud	1	2920	
10	Shield, exhaust manifold	1	557871	Not part of 516496 and 516945 — From engine suffix 'J' onwards
	Distance washer for heat shield	1	515331	
15		1	515505	
16	Steady strip for heat shield	1	254831	
17	Bolt (5/16" UNC x 1⅛" long) — Fixing steady strip, heat shield and distance washer	1	3075	
	Plain washer	1	2550	
11	Bolt (5/16" UNF x ¾" long) — Fixing bracket to heat shield	1	255226	
	Spring washer	1	3075	
12	Plain washer	1	2550	
13	Spring washer	1	3075	
14	Nut (¼" UNF)	1	253029	
	Support strip, heat shield to starter motor — Fixing support strip to heat shield	1	255226	1958-60 models
	Bolt (5/16" UNF x ¾" long)	1	255226	
	Spring washer	1	3075	
	Plain washer	1	2550	
	Nut (5/16" UNF)	1	253027	
18	Plain washer — Fixing heatshield and steady strip to manifold	1	2550	Up to engine suffix 'H' inclusive
19	Spring washer	1	3075	
20	Bolt (5/16" UNC x ⅞" long)	1	253027	
21	Spring washer — Fixing heatshield and steady strip to manifold	1	3075	
	Plain washer	1	2550	
22	Spring washer	1	3075	
23	Bolt (¼" UNF x 7/16" long)	1	255003	
24	Plain washer	1	3840	
25	Spring washer	1	3074	
	Distance washer	1	515505	
	Plain washer	1	253027	
	Spring washer	1	2550	
	Bolt (5/16" UNC x ⅞" long) — Fixing heat shield to inlet manifold	1	3075	From engine suffix 'J' onwards
	Plain washer	1	253025	
	Spring washer	1	2550	
		1	3075	

2¼ litre petrol models

* Asterisk indicates a new part which has not been used on any previous Rover model
† Also includes dipswitch lead, mounting plate and fixings

EXHAUST SYSTEM, 4 CYLINDER MODELS

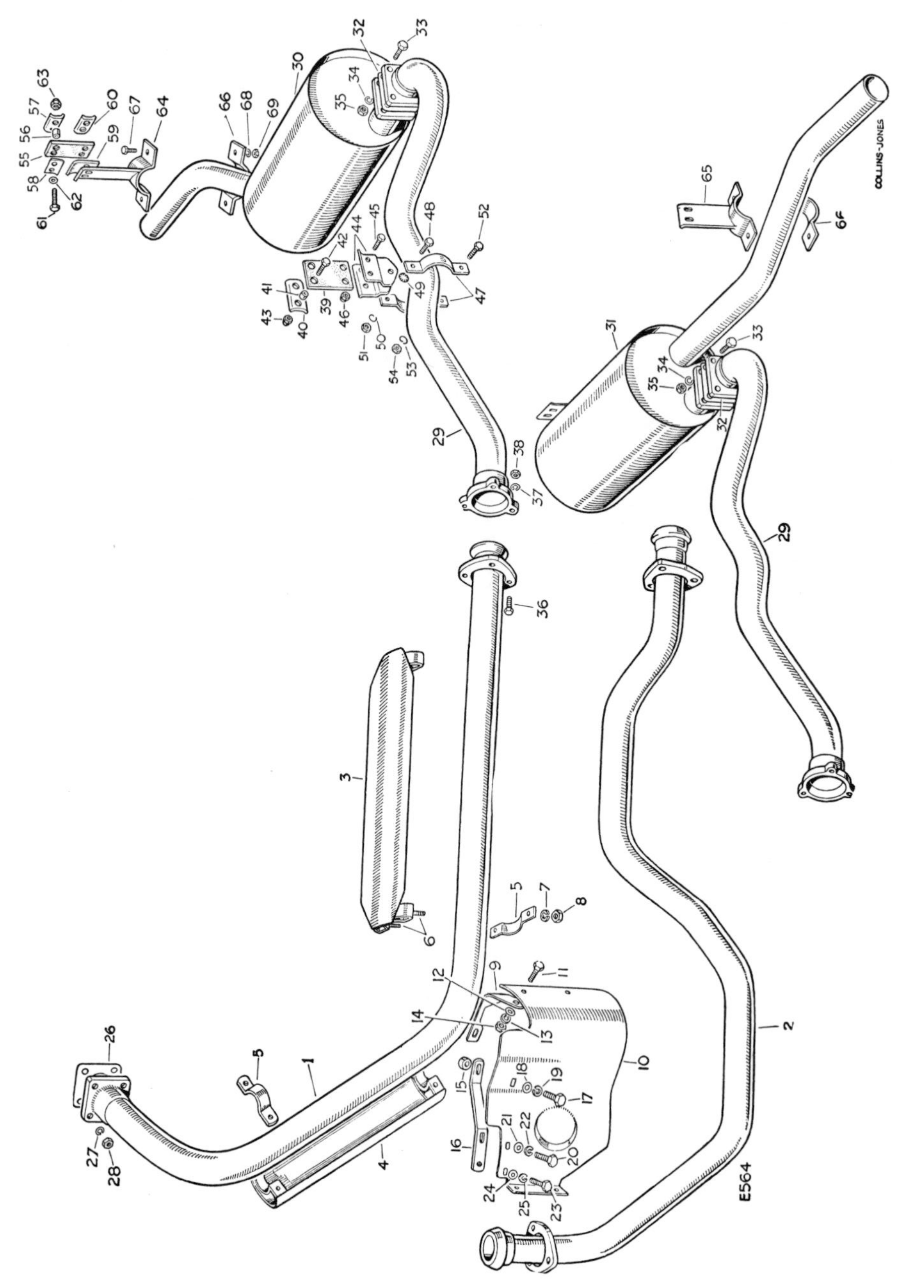
E564
COLLINS-JONES

EXHAUST SYSTEM, 4 CYLINDER MODELS

Plate Ref.	1 2 3 4 DESCRIPTION	Qty	Part No.	REMARKS
	Heat shield for exhaust pipe	1	515598	1958–60 models; 2¼ litre Petrol models
	Bolt (¼" UNF x ⅝" long) – Fixing exhaust pipe shield to manifold shield	2	255206	1958–60 models; 2¼ litre Petrol models
	Plain washer – Fixing exhaust pipe shield to manifold shield	2	3840	1958–60 models; 2¼ litre Petrol models
	Spring washer – Fixing exhaust pipe shield to manifold shield	2	3074	1958–60 models; 2¼ litre Petrol models
	Nut (¼" UNF) – Fixing exhaust pipe shield to manifold shield	2	254810	1958–60 models; 2¼ litre Petrol models
	Heat shield between exhaust pipe and toe box	1	509582	1958–60 LHStg; 2¼ litre Petrol models
	Pipe clamp, heat shield to pipe	2	503307	1958–60 LHStg; 2¼ litre Petrol models
26	Joint washer for exhaust pipe	1	213358	Diesel and 2 litre Petrol models
27	Spring washer – Fixing exhaust pipe to manifold	4	3075	2 litre Petrol models
28	Nut (5/16" BSF) – Fixing exhaust pipe to manifold	4	3771	2 litre Petrol models
	Spring washer – Fixing exhaust pipe to manifold	3	3075	4 off on Diesel models
	Nut (5/16" UNF) – Fixing exhaust pipe to manifold	3	253809	4 off on Diesel models
	Intermediate exhaust pipe	1	244449	88 Petrol models 1958–60. Diesel models 1958 onwards
29	Intermediate exhaust pipe	1	264195	88 2¼ litre Petrol models. 1961 onwards
	Intermediate exhaust pipe	1	276141	109
	Intermediate exhaust pipe	1	500289	109 Station Wagon
30	Silencer complete	1	552482†	RHStg
31	Silencer complete	1	526779	LHStg
	Silencer complete	1	500290	109 Station Wagon
	Silencer complete	1	523655	109 Station Wagon. When hydraulic front winch is fitted
32	Joint washer – Fixing silencer to intermediate pipe	1	213358	
33	Bolt (5/16" UNF x 1" long) – Fixing silencer to intermediate pipe	4	255228	
34	Spring washer – Fixing silencer to intermediate pipe	4	3075	
35	Nut (5/16" UNF) – Fixing silencer to intermediate pipe	4	254831	
36	Bolt (5/16" UNF x 1½" long) – Fixing intermediate pipe to front exhaust pipe	3	256222	
37	Spring washer – Fixing intermediate pipe to front exhaust pipe	3	3075	
38	Nut (5/16" UNF) – Fixing intermediate pipe to front exhaust pipe	3	254831	
39	Flexible mounting, intermediate exhaust pipe	2	244316	88
40	Clamp plate – Fixing flexible mounting to chassis	1	239715	88
41	Distance piece – Fixing flexible mounting to chassis	4	244009	88
42	Bolt (¼" UNF x 1" long) – Fixing flexible mounting to chassis	2	255209	88
43	Self-locking nut (¼" UNF) – Fixing flexible mounting to chassis	2	252220	88
44	Plate for flexible mounting	2	239710	88
45	Bolt (¼" UNF x 1" long) – Fixing plate to flexible mounting	2	255209	88
46	Self-locking nut ("¼ UNF) – Fixing plate to flexible mounting	2	252220	88
47	Pipe clamp	2	239711	88
48	Bolt (5/16" UNF x 1" long) – Fixing pipe clamp to flexible mounting plate	1	255228	88
49	Shakeproof washer – Fixing pipe clamp to flexible mounting plate	2	72614	88
50	Spring washer – Fixing pipe clamp to flexible mounting plate	1	3075	88
51	Nut (5/16" UNF) – Fixing pipe clamp to flexible mounting plate	1	254831	88
52	Bolt (¼" UNF x 1" long) – Fixing pipe clamp to exhaust pipe	1	255209	88
53	Spring washer – Fixing pipe clamp to exhaust pipe	1	3074	88
54	Nut (¼" UNF) – Fixing pipe clamp to exhaust pipe	1	254810	88

* Asterisk indicates a new part which has not been used on any previous Rover model
† Alternative designs, round or oval, interchangeable

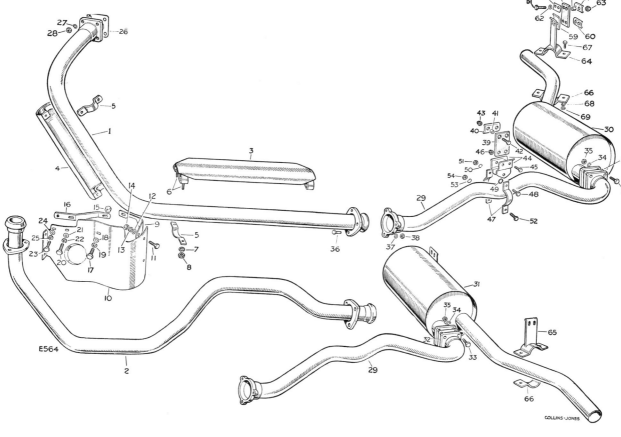

E564
COLLINS-JONES

EXHAUST SYSTEM, 4 CYLINDER MODELS

Plate Ref. 1 2 3 4	DESCRIPTION		Qty	Part No.	REMARKS
	Flexible mounting for silencer		2	244316	
	Distance piece		4	244009	
	Packing plate, upper		1	245926	
	Packing plate, lower		1	274870	
	Clamp plate	Fixing flexible mounting to chassis and silencer	2	239715	
	Bolt (¼" UNF x 1⅛" long)		4	255210	LHStg
	Plain washer		4	3840	
	Self-locking nut (¼" UNF)		4	252220	
55	Flexible mounting for tail pipe		2	244316	
56	Distance piece		4	244009	
57	Clamp plate, upper		1	245926	
58	Packing plate, upper	Fixing flexible mounting to chassis frame and clamp bracket	1	239715	
59	Packing plate, lower		1	274870	
60	Clamp plate, lower		1	239715	
61	Bolt (¼" UNF x 1⅛" long)		4	255210	
62	Plain washer		4	3840	
63	Self-locking nut (¼" UNF)		4	252220	
64	Pipe clamp bracket		1	239712	RHStg
				239711	LHStg
65	Pipe clamp bracket		1	239716	LHStg
				239715	RHStg
66	Saddle for clamp bracket	Fixing exhaust pipe to clamp bracket	1	240087	
67	Bolt (5/16" UNF x 1" long)		2	255228	
68	Spring washer		2	3075	
69	Nut (5/16" UNF)		2	254831	
	Flexible mounting, intermediate pipe, front		2	244316	
	Distance piece		4	244009	
	Packing plate, lower		1	274870	
	Clamp plate	Fixing flexible mounting to chassis	2	239715	
	Bolt (¼" UNF x 1" long), upper		2	255209	
	Bolt (¼" UNF x 1⅛" long), lower		2	255210	
	Plain washer		4	3840	
	Self-locking nut (¼" UNF)		4	252220	
	Nut (5/16" UNF)		2	254831	
	Spring washer	Fixing saddle and clamp bracket to intermediate exhaust pipe	2	3075	
	Bolt (5/16" UNF x 1" long)		2	255228	
	Saddle for pipe clamp bracket		1	264886	
	Clamp bracket		1	264889	
	Self-locking nut (¼" UNF)		4	252220	
	Plain washer		4	3840	
	Bolt (¼" UNF x 1⅛" long), lower	Fixing clamp plate and clamp bracket	2	255210	
	Clamp plate, lower		2	239715	
	Packing plate, lower		2	274870	
	Packing plate, upper		2	255209	
	Clamp plate, upper		2	239715	
	Distance piece		4	244009	
	Flexible mounting, intermediate pipe, rear		2	244316	
	Nut (5/16" UNF)		2	254831	
	Spring washer	Fixing bracket to intermediate exhaust pipe	2	3075	
	Bolt (5/16" UNF x 1" long)		2	255228	
	Plate for flexible mounting	Fixing flexible mounting to mounting plate and chassis	2	239710	
	Clamp plate, upper		2	239715	
	Distance piece		4	244009	
	Pipe clamp		2	239711	RHStg
	Bolt (5/16" UNF x 1" long)	Fixing pipe clamp to mounting plate	1	255228	
	Shakeproof washer		2	72614	
	Spring washer		1	3075	
	Nut (5/16" UNF)		1	254831	
	Bolt (¼" UNF x 1" long)	Fixing pipe clamp to exhaust pipe	1	255209	
	Spring washer		1	3074	
	Nut (¼" UNF)		1	254810	

88, 109 (assembly reference brackets)

* Asterisk indicates a new part which has not been used on any previous Rover model

EXHAUST SYSTEM, 4 CYLINDER MODELS

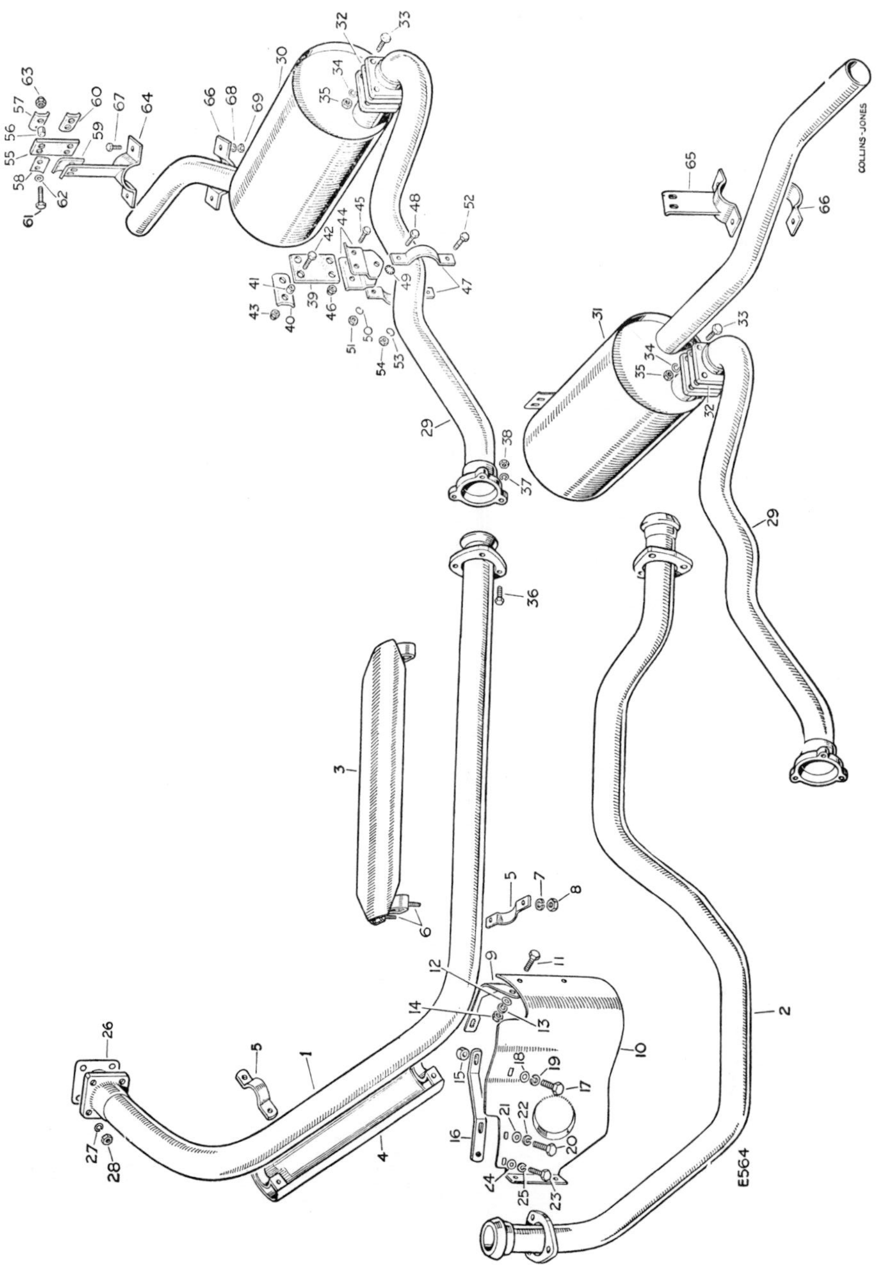

EXHAUST SYSTEM, 4 CYLINDER MODELS

Plate Ref. 1 2 3 4	DESCRIPTION		Qty	Part No.	REMARKS
	Flexible mounting for silencer		2	244316	LHStg; 109
	Distance piece	Fixing flexible mounting to chassis and silencer	4	244009	LHStg; 109
	Clamp plate	Fixing flexible mounting to chassis and silencer	2	239715	LHStg; 109
	Packing plate, lower	Fixing flexible mounting to chassis and silencer	1	274870	LHStg; 109
	Packing plate, upper	Fixing flexible mounting to chassis and silencer	1	245926	LHStg; 109
	Bolt ($\frac{1}{4}$" UNF x $1\frac{1}{8}$" long)	Fixing flexible mounting to chassis and silencer	2	255210	LHStg; 109
	Plain washer	Fixing flexible mounting to chassis and silencer	4	3840	LHStg; 109
	Nut ($\frac{1}{4}$" UNF)	Fixing flexible mounting to chassis and silencer	2	252220	LHStg; 109
	Flexible mounting for tail pipe		2	244316	109
	Distance piece	Fixing flexible mounting to chassis frame and mounting plate	4	244009	109
	Clamp plate, upper	Fixing flexible mounting to chassis frame and mounting plate	1	239715	109
	Bolt ($\frac{1}{4}$" UNF x 1" long)	Fixing flexible mounting to chassis frame and mounting plate	4	255209	109
	Plain washer	Fixing flexible mounting to chassis frame and mounting plate	4	3840	109
	Self-locking nut ($\frac{1}{4}$" UNF)	Fixing flexible mounting to chassis frame and mounting plate	4	252220	109
	Plate for flexible mounting		2	239710	109
	Pipe clamp		1	264865	1958–60; 109
	Pipe clamp		2	517567	1961 onwards; 109
	Bolt ($\frac{1}{4}$" UNF x $\frac{7}{8}$" long)	Lower fixing for clamp to tail pipe	1	255208	1961 onwards; 109
	Spring washer	Lower fixing for clamp to tail pipe	1	3074	1961 onwards; 109
	Nut ($\frac{1}{4}$" UNF)	Lower fixing for clamp to tail pipe	1	254810	1961 onwards; 109
	Bolt ($\frac{5}{16}$" UNF x 1" long)	Fixing clamp to tail pipe and mounting plate	1	255228	109
	Spring washer	Fixing clamp to tail pipe and mounting plate	1	3075	109
	Nut ($\frac{5}{16}$" UNF)	Fixing clamp to tail pipe and mounting plate	1	254831	109

* Asterisk indicates a new part which has not been used on any previous Rover model

S6OH

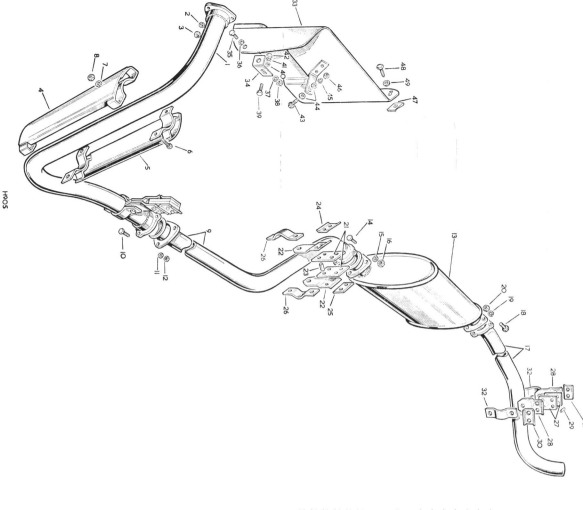

EXHAUST SYSTEM, 6 CYLINDER MODELS

Plate Ref.	1	2	3	4	DESCRIPTION	Qty	Part No.	REMARKS
1					Front exhaust pipe complete	1	562885	
2					Spring washer ⎫ Fixing front exhaust	3	3075	
3					Nut (5/16" UNF) ⎬ pipe to manifold	3	253809	
4					Heat shield ⎱ For front	1	552720	
5					Heat shield ⎰ exhaust pipe	1	562781	
6					Bolt (1/4" UNF x 7/8" long) ⎫ Fixing heat shields	4	255208	
7					Spring washer ⎬ together on	4	3074	
8					Nut (1/4" UNF) ⎭ front exhaust pipe	4	254810	
9					Intermediate exhaust pipe complete	1	562785	
10					Bolt (5/16" UNF x 1 1/2" long) ⎫ Fixing front and	3	256222	
11					Spring washer ⎬ intermediate exhaust	3	3075	
12					Nut (5/16" UNF) ⎭ pipes together	3	254831	
13					Exhaust silencer	1	562737	
14					Bolt (5/16" UNF x 1 1/2" long) ⎫ Fixing	3	256222	
15					Spring washer ⎬ intermediate pipe	3	3075	
16					Nut (5/16" UNF) ⎭ to silencer	3	254831	
17					Tail pipe complete	1	562787	109 Long
						1	569214*	109 Station Wagon
18					Bolt (5/16" UNF x 1 1/2" long) ⎫ Fixing silencer	3	256222	
19					Spring washer ⎬ to	3	3075	
20					Nut (5/16" UNF) ⎭ tail pipe	3	254831	
21					Plate for flexible mounting	4	239710	
22					Flexible mounting for front and intermediate pipes	4	244316	
23					Distance piece	8	239710	
24					Clamp plate ... ... ⎫ Fixing	2	244009	
25					Packing plate ⎬ flexible	2	239715	
26					Bolt (1/4" UNF x 1" long) ⎱ mounting to	8	245926	
					Self-locking nut (1/4" UNF) ⎰ mounting plate	8	252220	
					Pipe clamp, front and intermediate exhaust	4	562823	
					Bolt (5/16" UNF x 1" long) ⎱ Fixing	4	255228	
26					Spring washer ⎬ pipe clamp	4	3075	
27					Nut (5/16" UNF) ⎭	4	254831	
28					Flexible mounting for tail pipe	2	239710	
29					Plate for flexible mounting	2	244009	
30					Distance piece ⎫ Fixing flexible	4	239710	
31					Clamp plate ⎬ mounting to	1	239715	
					Packing plate ⎱ mounting plate	4	245926	
					Bolt (1/4" UNF x 1" long) ⎬ and	4	245209	
					Self-locking nut (1/4" UNF) ⎭ chassis frame	4	252220	
32					Clamp for tail pipe	2	562823	
					Bolt (5/16" UNF x 1" long) ⎱ Fixing	2	255228	
					Plain washer ⎬ tail	2	3899	
					Spring washer ⎱ pipe clamp	2	3075	
					Nut (5/16" UNF) ⎰	2	254831	

* Asterisk indicates a new part which has not been used on any previous Rover model

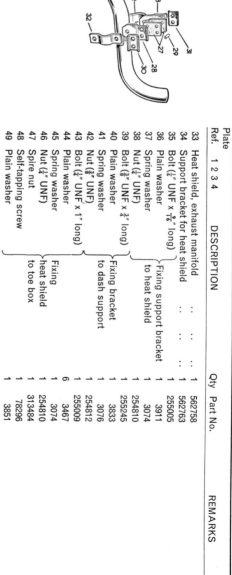

H905

EXHAUST SYSTEM, 6 CYLINDER MODELS

Plate Ref.	1 2 3 4	DESCRIPTION			Qty	Part No.	REMARKS
33		Heat shield, exhaust manifold			1	562758	
34		Support bracket for heat shield			1	562763	
35		Bolt ($\frac{1}{4}$" UNF x $\frac{9}{16}$" long)		..	1	255005	
36		Plain washer	..	..	1	3911	
37		Spring washer			1	3074	
38		Nut ($\frac{1}{4}$" UNF)	} Fixing support bracket		1	254810	
39		Bolt ($\frac{3}{8}$" UNF x $\frac{3}{4}$" long)	} to heat shield		1	255245	
40		Plain washer			1	3076	
41		Spring washer	} Fixing bracket		1	3833	
42		Nut ($\frac{3}{8}$" UNF)	} to dash support		1	254812	
43		Bolt ($\frac{1}{4}$" UNF x 1" long)			1	255009	
44		Plain washer			6	3467	
45		Spring washer	} Fixing		1	3074	
46		Nut ($\frac{1}{4}$" UNF)	} heat shield		1	254810	
47		Spire nut	} to toe box		1	313484	
48		Self-tapping screw			1	78296	
49		Plain washer			1	3851	

* Asterisk indicates a new part which has not been used on any previous Rover model

RADIATOR AND FITTINGS

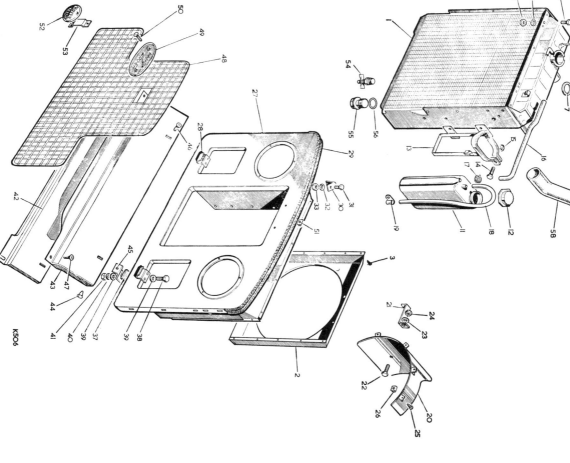

K506

RADIATOR AND FITTINGS

Plate Ref.	1 2 3 4	DESCRIPTION	Qty.	Part No.	REMARKS
1		RADIATOR BLOCK ASSEMBLY	1	279848	2 litre Petrol models
		RADIATOR BLOCK ASSEMBLY	1	548073	2¼ litre Series II Petrol models
		RADIATOR BLOCK ASSEMBLY } With straight bottom connection.	1	568842*	2¼ litre Series IIA Petrol models
		RADIATOR BLOCK ASSEMBLY } With angled bottom connection.	1	568843*	2¼ litre Series IIA Petrol models when Lucas 2AC type 12 volt A.C./D.C. generator is fitted
		RADIATOR BLOCK ASSEMBLY } With overflow bottle provision.	1	568845*	2¼ litre Diesel models
		RADIATOR BLOCK ASSEMBLY	1	562922	With overflow bottle provision. 2 litre Petrol models
2		RADIATOR BLOCK ASSEMBLY	1	269953	2 litre Petrol models
		RADIATOR BLOCK ASSEMBLY	1	509217	2¼ litre Series II Petrol models
		RADIATOR BLOCK ASSEMBLY	1	544848	2¼ litre Series IIA Petrol models
		RADIATOR BLOCK ASSEMBLY	1	562944	2.6 litre Petrol models
		RADIATOR BLOCK ASSEMBLY } Without overflow bottle provision.	1	279846	2.6 litre Diesel models
		RADIATOR BLOCK ASSEMBLY	1	526772	2¼ litre Diesel models
		RADIATOR BLOCK ASSEMBLY	1	279863	2¼ litre Diesel models
3		RADIATOR BLOCK ASSEMBLY } With overflow bottle provision.	1	559579	2¼ litre Diesel models
		Drive screw fixing cowl to radiator block	10	77789	4-cylinder models
		Self-tapping screw } Fixing cowl to	9	562979	6-cylinder models
4		Spring washer, double coil } radiator block	9	564741	models
		Filler cap for radiator, 4 lb pressure	1	217437	2 litre Petrol models
		Filler cap for radiator, 10 lb pressure	1	242399	Diesel } Alternatives. Check before ordering
		Filler cap for radiator, 9 lb pressure	1	509767	2¼ litre Petrol models }
5		Chain for filler cap	1	230328	For 4 lb or 10 lb pressure caps. Check before ordering
		Chain for filler cap	1	509769	For 9 lb pressure cap. Check before ordering
6		Retainer for chain	1	230329	For 4 lb or 10 lb pressure caps. Check before ordering
7		Joint washer for filler cap	1	516914	} Without overflow bottle provision

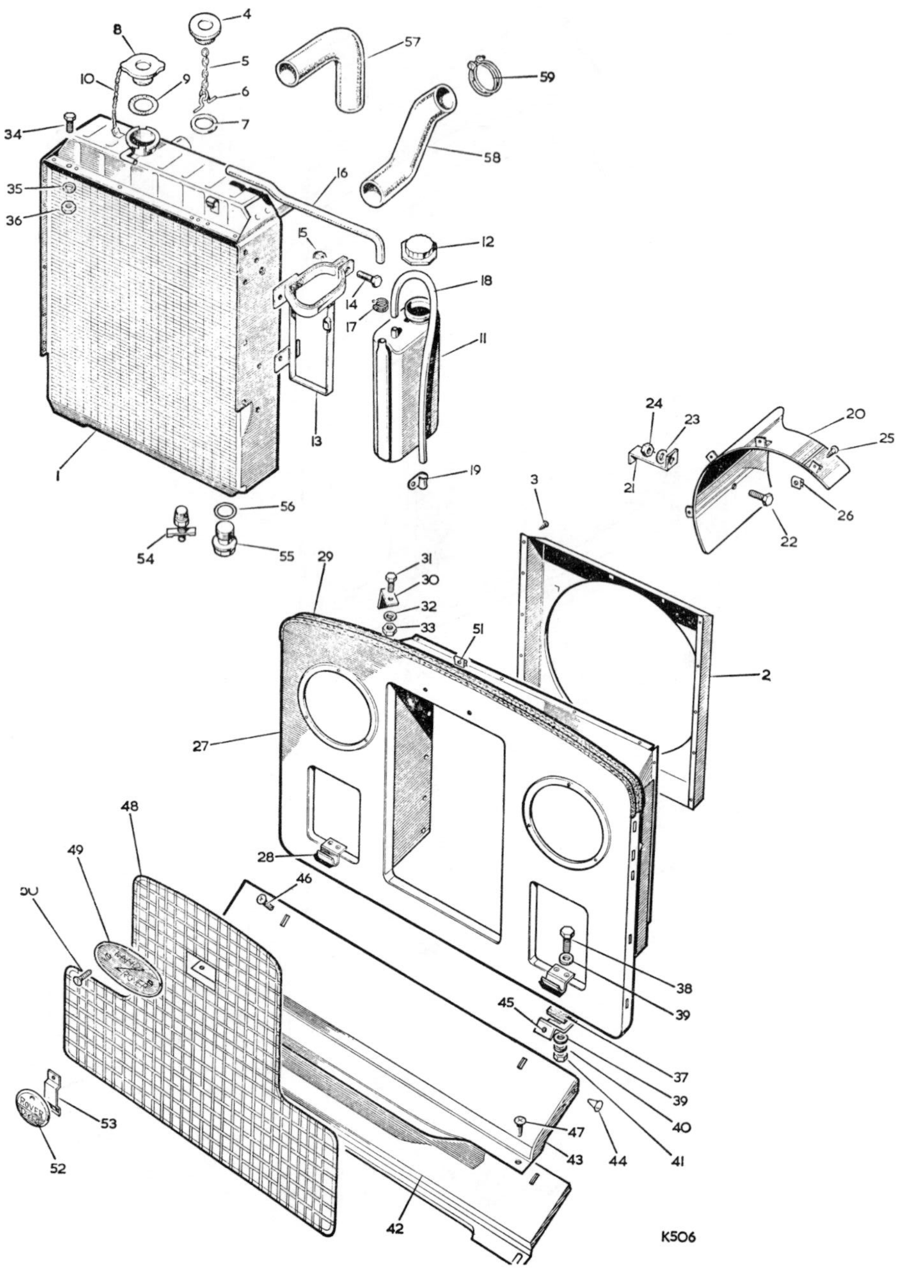

Plate Ref.	1 2 3 4	DESCRIPTION	Qty	Part No.	REMARKS
8		Filler cap for radiator, 9 lb pressure	1	564713	With overflow bottle provision. $2\frac{1}{4}$ litre Petrol, 2.6 litre Petrol and $2\frac{1}{4}$ litre Diesel models
9		Joint washer for filler cap	1	564999*	
10		Chain for filler cap	1	509769	
11		Overflow bottle for radiator	1	564718	For radiator block with overflow provision. $2\frac{1}{4}$ litre Petrol, 2.6 litre Petrol and $2\frac{1}{4}$ litre Diesel models
12		Cap for overflow bottle	1	564719	
13		Carrier bracket for overflow bottle	1	564717	
14		Bolt ($\frac{1}{4}''$ UNF x $\frac{7}{8}''$ long) } Clamping bottle	1	255208	
15		Self-locking nut ($\frac{1}{4}''$ UNF) } to carrier	1	252160	
16		Hose, radiator to overflow bottle	1	564720	
17		Clip, fixing hose	2	50300	
18		Flexible pipe, overflow bottle outlet	1	564724	
19		Clip, fixing outlet pipe	1	219677	
20		Shroud for fan cowl	1	551714	$2\frac{1}{4}$ litre Petrol models. Except Series IIA fitted with Lucas 2AC type 12 volt AC/DC generator
21		Steady strip for shroud	1	531332	
22		Bolt ($\frac{1}{4}''$ UNF x $\frac{5}{8}''$ long) } Fixing steady strip to shroud	1	255206	
23		Spring washer	1	3074	
24		Nut ($\frac{1}{4}''$ UNF)	1	254810	
		Bolt (2 BA x $\frac{1}{2}''$ long) } Fixing shroud	4	234603	Alternative fixings
		Nut (2 BA)	4	251335	
25		Drive screw	4	78436	
26		Spire nut	4	78237	
		Shroud for fan cowl	1	559580	$2\frac{1}{4}$ litre Diesel. Except models fitted with Lucas 2AC type 12 volt AC/DC generator
		Drive screw } Fixing shroud	3	78436	
		Spire nut	3	78237	
		Bolt (2 BA x $\frac{3}{8}''$ long)	1	237119	
		Spring washer	1	3073	
		Nut (2 BA)	1	2247	
		Shroud for fan cowl	1	569212*	2.6 litre Petrol models
		Drive screw, fixing shroud to fan cowl	1	78436	
		Radiator grille panel complete	1	330950	All America Dollar Area. Up to vehicle suffix 'A' for other areas
27		Radiator grille panel complete	1	336466	From vehicle suffix 'B' onwards, except America Dollar Area
28		Support clip for grille mesh	2	330150	
		Pop rivet fixing clip to grille panel	4	300789	
29		Bonnet rest strip, 35″ long	1	300824	
		Bifurcated rivet fixing strip	10	68087	Alternative fixings. Check before ordering
		Spring clip, fixing strip	10	338380	
30		Protection plate for headlamp	1	348182	From vehicle suffix 'B' onwards
31		Bolt ($\frac{1}{4}''$ UNF x $\frac{5}{8}''$ long) } Fixing plate to grille panel	1	255206	
32		Spring washer	1	3074	
33		Nut ($\frac{1}{4}''$ UNF)	1	254810	
34		Bolt ($\frac{1}{4}''$ UNF x $\frac{5}{8}''$ long) } Fixing radiator block to grille panel	13	255206	11 off on 2 litre Petrol and 2 litre Diesel models
35		Spring washer	13	3074	
36		Nut ($\frac{1}{4}''$ UNF)	13	254810	
37		Rubber buffer } Fixing grille panel and front apron bracket to chassis frame	As reqd	306465	
38		Bolt ($\frac{5}{16}''$ UNF x $1\frac{1}{2}''$ long)	3	256222	
39		Plain washer	6	3830	
40		Spring washer	3	3075	
41		Nut ($\frac{5}{16}''$ UNF)	3	252161	

* Asterisk indicates a new part which has not been used on any previous Rover model

RADIATOR AND FITTINGS

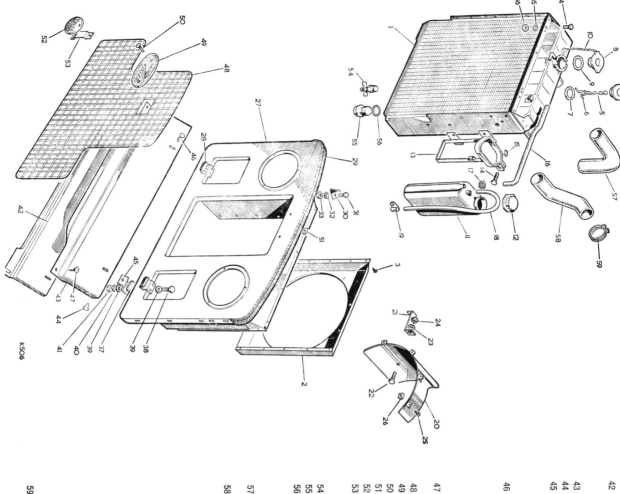

K.506

RADIATOR AND FITTINGS

Plate Ref. 1 2 3 4	DESCRIPTION	Qty	Part No.	REMARKS
42	Front apron panel	1	332640	Flat type panel.
			332656	Alternative to shaped type.
	Canvas strip for front apron panel	1	78226	Check before ordering
43	Rivet fixing strip	8	—	
43	Front apron panel	1	336786	Shaped type panel. Alternative to flat type. Check before ordering
44	Rubber buffer for front apron panel	4	310877	
45	Securing bracket for panel	2	345192	
45	Special set bolt	2	78208	Fixing securing brackets to apron panel
	Plain washer	2	3900	
	Spring washer	2	3074	
46	Special thread-forming screw	2	78796	Late type fixing }
			3075	Early type fixing } Check before ordering
			3900	
47	Special thread-forming screw (1/4" UNC x 3/8" long)	2	255226	For flat type front apron panel
			254831	
			72628	Early type fixing }
			78795	Late type fixing } For shaped type front apron panel
	Drive screw	2	330149	
48	Grille for radiator	1	332670	
49	'Land-Rover' nameplate, 7 3/16" overall	1	78438	Late type fixing
	Bolt (1/4" UNC x 1/2" long)	2	78237	Fixing nameplate and grille to grille panel
	Spring washer	2	320276	
	Nut (5/16" UNF)	2	320333	Diesel models
50	Drive screw	1	78255	Diesel models
51	Spire nut	1	78256	
52	'Diesel' badge	1	538608	Alternative to drain plug
53	Fixing bracket for badge	1	542232	Alternative to drain tap
	Rivet	2	213959	Fixing bracket and badge to grille
	Lockwasher	2	268037	
	Fixing bracket	1	268038	
54	Drain tap for radiator	1	538608	Alternative to drain plug
55	Drain plug for radiator	1	542232	Alternative to drain tap
56	Joint washer for plug	1	213959	
57	Hose for radiator, top	1	268037	2 litre } Petrol models
	Hose for radiator, bottom	1	268038	1958-60
	Hose for radiator, top	1	247872	1961
	Hose for radiator, bottom	1	517613	1961 onwards } 2 1/4 litre Petrol models
	Hose for radiator, top	1	278539	Series II
	Hose for radiator, bottom	1	534130	Series IIA
58	Hose for radiator, bottom	1	530585	For radiator with straight bottom connection
	Hose for radiator, bottom	1	543629	For radiator with angled bottom connection when Lucas 2AC type 12 volt AC/DC generator is fitted
	Hose for radiator, top	1	559876	2.6 litre
	Hose for radiator, top	1	552516	2.6 litre Petrol models
	Hose for radiator, bottom	1	508060	1958-60
	Hose for radiator, top	1	517613	1961 } 2 litre Petrol models
	Hose for radiator, bottom	1	273529	
	Hose for radiator, top	1	517613	2 litre Series II Diesel models
	Hose for radiator, bottom	1	534130	
59	Clip for radiator hoses	4	50316	2 1/4 litre
	Clip, top	1	50316	Diesel models
	Clip, bottom	1	50316	2 litre Petrol models
	Clip, top	1	50316	Diesel models
	Clip, bottom	1	50319	2 litre Petrol models
	Clip, top and bottom	2	50320	Diesel models
	Clip, top	1	50316	and 2 1/4 litre Petrol models
	Clip, top and bottom, for radiator hoses	4	50319	2.6 litre Petrol models

* Asterisk indicates a new part which has not been used on any previous Rover model

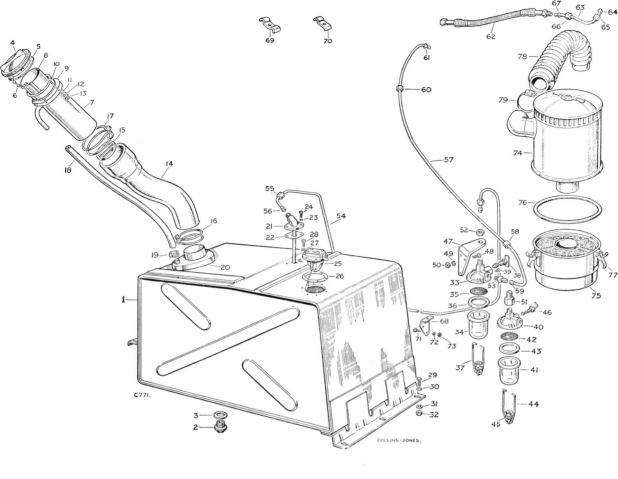

C771.

COLLINS-JONES.

FUEL TANK, FILTER AND PIPES, 4-CYLINDER MODELS

Plate Ref.	1	2	3	4	DESCRIPTION	Qty	Part No.	REMARKS
1					Fuel tank complete	1	552174	Petrol models
1					Fuel tank complete	1	543164†	Petrol models, 109 Station Wagon
1					Fuel tank complete	1	552175	Diesel models
1					Fuel tank complete	1	543165†	Diesel models, 109 Station Wagon
2					Drain plug for petrol tank	1	235592	
3					Joint washer for drain plug	1	51599	
4					Telescopic filler tube complete	1	277259	
5					Filler cap	1	277260	With ball valve in filler cap.
6					Joint washer for cap	1	505244	
7					Chain for filler cap	1	231190	Alternatives to 5 lines below
8					Filler tube	1	277261	
9					Extension tube for filler	1	277262	Diesel models, 109 Station Wagon
5					Filler cap	1	504655	With felt pad in filler cap.
6					Joint washer for cap	1	505244	
7					Chain for filler cap	1	231190	Alternatives to 5 lines above.
8					Filler tube	1	504657	
9					Extension tube for filler	1	504656	
10					Grommet for fuel tank filler	1	500710	
11					Screw (2 BA x 3/8" long)	4	77941	Fixing filler to body side
12					Plain washer	4	4030	Fixing filler to body side
13					Spring washer	4	3073	Fixing filler to body side
					Special nut	4	313385	Fixing filler to body side
					Nut (2 BA)	3	3823	Fixing filler to body side
14					Hose, tank to filler tube	1	543764	
15					Hose, tank to filler tube	1	543782	109 Station Wagon
16					Anti-theft grid for filler	1	219687	
					Clip for hose, bottom	1	50028	
17					Clip for hose, top	1	50343	
18					Breather hose for fuel tank	1	543765	
					Breather hose for fuel tank	1	543767	109 Station Wagon
19					Clip for breather hose	2	50302	
					Clip for breather hose	2	543766	109 Station Wagon
20					Clip for air balance hose	2	50302	109 Station Wagon
					Air balance hose for fuel tank	1	504233	109 Station Wagon
					Rubber seal for filler pipe	1	504673	2 off on Diesel models
					Joint washer for fuel tank filler seal	1		
21					Outlet elbow complete for tank	1	503492	Petrol models except 109 Station Wagon
					Outlet elbow complete for tank	1	500432	Petrol models, 109 Station Wagon
					Outlet elbow complete for tank	1	544658	Diesel models except 109 Station Wagon
					Outlet elbow complete for tank	1	544674	Diesel models, 109 Station Wagon
					Return elbow complete for tank	1	271872	Diesel models except 109 Station Wagon
					Return elbow complete for tank	1	500818	Diesel models, 109 Station Wagon
22					Joint washer for outlet and return elbow		267837	2 off on Diesel models
23					Spring washer	2	3101	Fixing elbow to tank — 4 off on Diesel models
24					Screw (3 BA x 7/16" long)	2	3972	Fixing elbow to tank — Diesel models

* Asterisk indicates a new part which has not been used on any previous Rover model
† See 109 Station Wagon section for fuel tank fixings

FUEL TANK, FILTER AND PIPES, 4-CYLINDER MODELS

C771.

COLLINS-JONES.

FUEL TANK, FILTER AND PIPES, 4-CYLINDER MODELS

Plate Ref.	1 2 3 4	DESCRIPTION	Qty	Part No.	REMARKS
25		Gauge unit for fuel tank, SM Y 79976	1	236017	Series II Petrol models
		Gauge unit for fuel tank, SM FT 5300/89	1	501620	Series II Petrol models, 109 Station Wagon
		Gauge unit for fuel tank, SM FT 5347/07	1	519838	Series IIA Petrol models } Up to vehicle suffix 'C' inclusive
		Gauge unit for fuel tank, SM FT 5301/18	1	519875	Series IIA Petrol models, 109 Station Wagon }
		Gauge unit for fuel tank, SM TB 1114/001	1	555844	Series II Diesel models
		Gauge unit for fuel tank, SM TB 1114/000	1	555846	Series II Diesel models, 109 Station Wagon
		Gauge unit for fuel tank, SM FT 5387/00	1	529969	Series IIA Diesel models } Up to vehicle suffix 'C' inclusive
		Gauge unit for fuel tank, SM FT 5340/06	1	529970	Series IIA Diesel models, 109 Station Wagon }
		Gauge unit for fuel tank, SM FT 5387/00	1	271220	Series IIA Petrol models } From vehicle suffix 'D' onwards
		Gauge unit for fuel tank, SM FT 5342/00	1	501621	Series IIA Petrol models, 109 Station Wagon }
		Gauge unit for fuel tank, SM TB 1214/001	1	555845	Series IIA Diesel models
		Gauge unit for fuel tank, SM TB 1214/000	1	555847	Series IIA Diesel 109 Station Wagon models
26		Joint washer for gauge unit	1	546488	
27		Spring washer	6	3101	Fixing gauge
28		Screw (3 BA x 9/16" long)	6	3972	unit to tank
29		Bolt (5/16" UNF x 3/4" long)	6	255026	Fixing
30		Plain washer	12	3830	fuel
31		Spring washer	6	3075	tank to
32		Nut (5/16" UNF)	6	254831	chassis frame
		Bolt (5/16" UNF x 3/4" long)	3	255226	Fixing front
		Plain washer	6	3830	of fuel tank
		Spring washer	3	3075	to chassis
		Nut (5/16" UNF)	3	254831	frame
		Bolt (5/16" UNF x 1½" long)	1	2765	Fixing rear
		Plain washer	1	543803	of fuel tank
		Rubber bush	2	508545	to chassis
		Stiffener plate	1	543808	frame
		Spring washer	1	3075	
		Nut (5/16" UNF)	1	254831	
33		Petrol sediment bowl complete, AC 1524895	1	274323	Alternative to ten lines below
			1	274381	Alternative to four lines above
34		Body only, AC 7950351	1	236891	
35		Bowl only, AC 1523620	1	274380	2 litre Petrol models.
36		Gauze for bowl, AC 1524524	1	241225	Check carefully before ordering
37		Joint washer for bowl, AC 5593321	1	268797	
38		Retainer for bowl, AC 5592050	1	241224	
39		Tap and gland complete, AC 1524897	1	241226	
		Special screw for tap, AC 132900	1	235682	
40		Body only	1	262278	
41		Bowl only	1	236896	2 litre Petrol models.
42		Gauze for bowl	1	236894	Alternative to the 8 lines above.
43		Joint washer for bowl	1	236895	Check carefully before ordering
44		Retainer for bowl	1	240234	
45		Screw cap for retainer	1	240235	
46		Tap and gland complete	1	240236	

* Asterisk indicates a new part which has not been used on any previous Rover model

FUEL TANK, FILTER AND PIPES, 4-CYLINDER MODELS

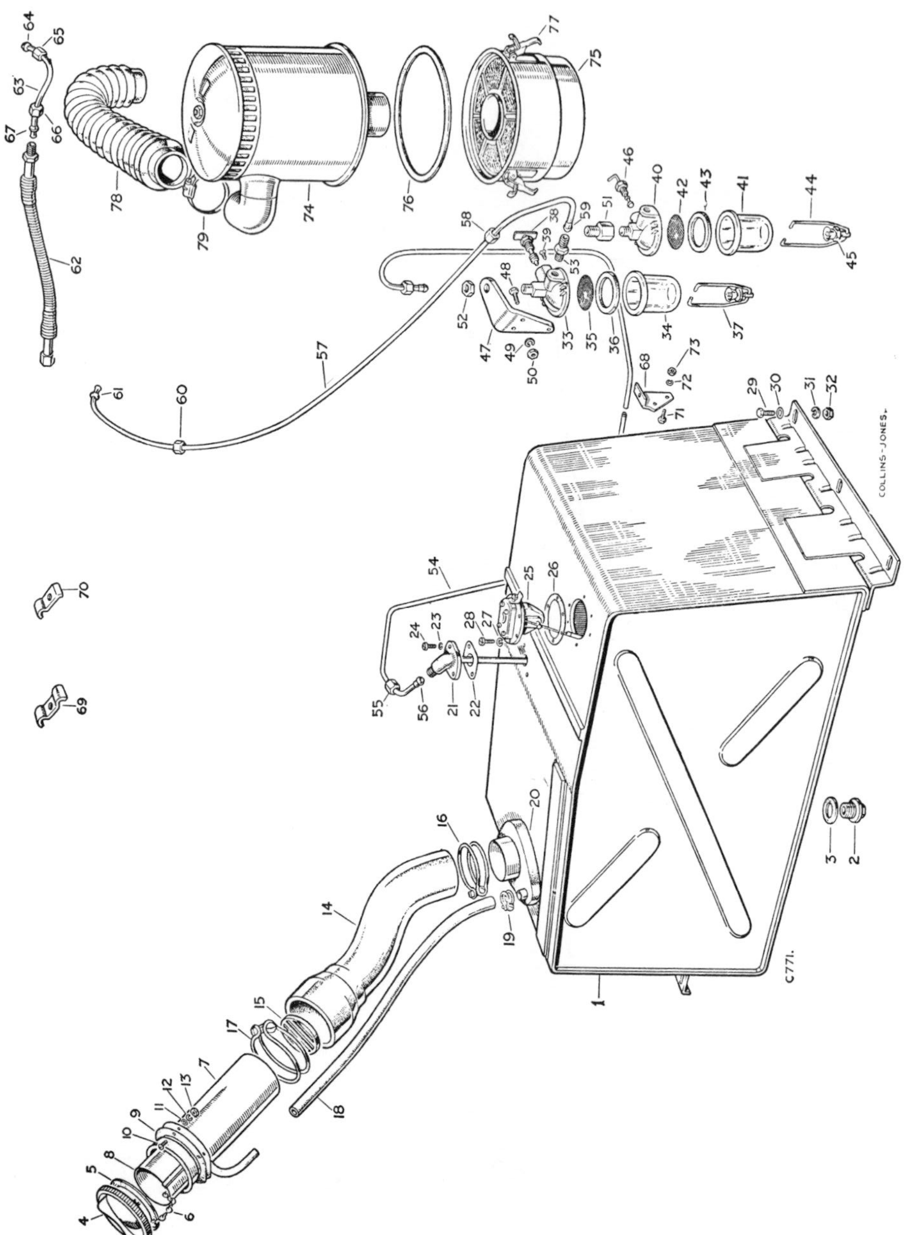

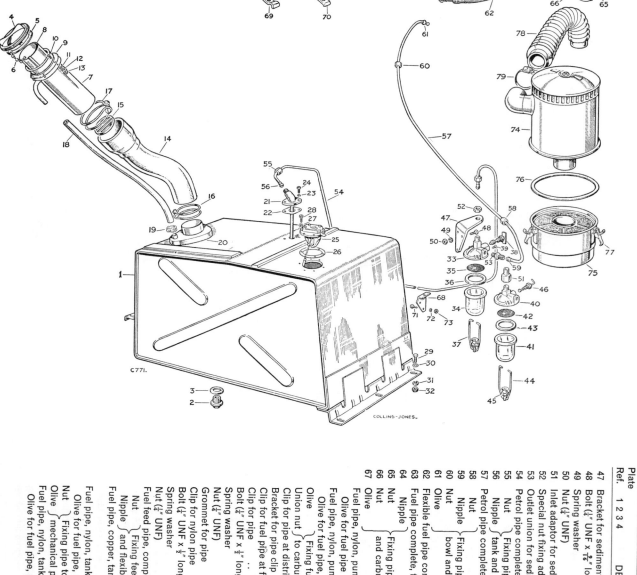

C771.
COLLINS-JONES.

FUEL TANK, FILTER AND PIPES, 4-CYLINDER MODELS

Plate Ref. 1 2 3 4	DESCRIPTION	Qty	Part No.	REMARKS
47	Bracket for sediment bowl	1	218468	
48	Bolt (¼" UNF x 9/16" long) } Fixing bracket to dash	3	255005	
49	Spring washer	3	3074	
50	Nut (¼" UNF)	3	254810	
51	Inlet adaptor for sediment bowl	1	218471	
52	Special nut fixing adaptor and bowl to bracket ..	1	218470	
53	Outlet union for sediment bowl	1	218469	
54	Petrol pipe complete, tank to bowl	1	508985	2 litre Petrol models
55	Nut } Fixing pipe to tank and bowl	2	2729	
56	Nipple	2	2728	
57	Petrol pipe complete, bowl to pump	1	240207	
58	Nut } Fixing pipe to bowl and pump	1	2729	
59	Nipple	1	2728	
60	Nut	1	50421	
61	Olive	1	3136	
62	Flexible fuel pipe complete	1	276267	1958–60 Petrol models
63	Fuel pipe complete, flex to carburetter	1	218955	
64	Nipple } Fixing pipe to flex and carburetter	1	2728	
65	Nut	1	2729	2 litre Petrol models
66	Nut	1	50421	
67	Olive	1	3136	
	Fuel pipe, nylon, pump to carburetter	1	557432	Up to engine suffix 'H' inclusive } 2¼ litre Petrol models
	Olive for fuel pipe	1	557531	
	Fuel pipe, nylon, pump to carburetter	1	550471	From engine suffix 'J' onwards
	Olive for fuel pipe, pump end	1	557531	
	Olive } Fixing fuel pipe to carburetter	1	568575	
	Union nut	1	537297	
	Clip for pipe at distributor adaptor stud	1	219676	
	Bracket for pipe clip	1	501428	Alternative fixings
	Clip for fuel pipe at front cover	1	219676	
	Clip for pipe	1	232425	
	Bolt (¼" UNF x ½" long) } Fixing clip	1	255204	
	Spring washer	1	3074	
	Nut (¼" UNF)	1	254810	
	Grommet for pipe	1	06860	
	Clip for nylon pipe	1	50637	
	Bolt (¼" UNF x ½" long) } Fixing clip to oil filler pipe support bracket	1	255204	Series II models
	Spring washer	1	3074	
	Nut (¼" UNF)	1	254810	
	Fuel feed pipe, complete, copper	1	277894	2 litre Diesel and 2¼ litre Petrol models,
	Nut } Fixing feed pipe to tank and flexible pipe	2	270141	
	Nipple	2	270133	
	Fuel pipe, copper, tank to flexible pipe	1	500438	1958–60 Diesel and Petrol models. 109 Station Wagon
	Fuel pipe, nylon, tank to pump	1	552435	88 and 109 } 2¼ litre Petrol models, 1961 onwards. Diesel models. Up to engine suffix 'J' inclusive
	Olive for fuel pipe, pump end	1	557531	
	Nut } Fixing pipe to mechanical pump	1	270115	Early type fixing only
	Olive	1	270105	
	Fuel pipe, nylon, tank to pump	1	552436	109 Station Wagon
	Olive for fuel pipe, pump end	1	557531	

* Asterisk indicates a new part which has not been used on any previous Rover model

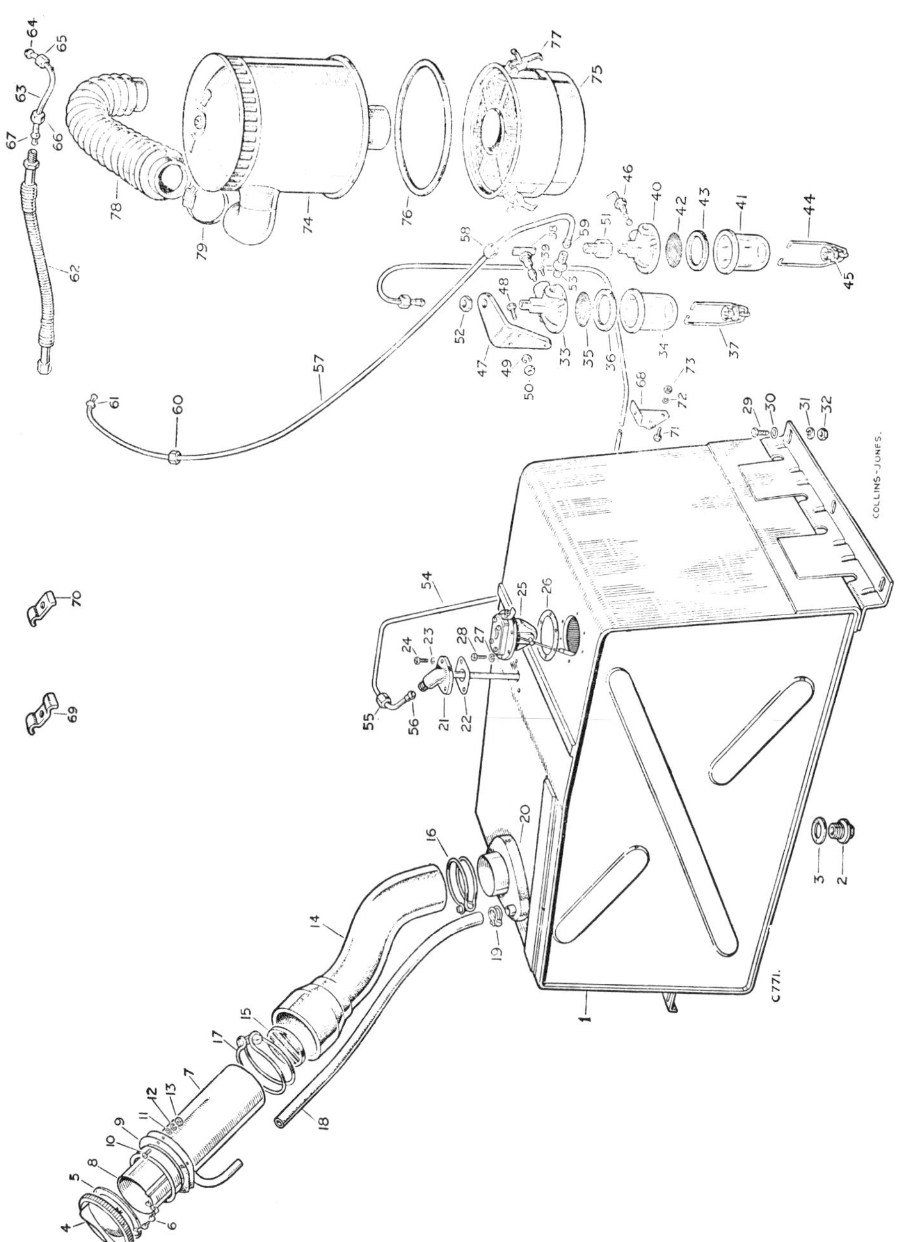

C 771.

COLLINS-JONES.

FUEL TANK, FILTER AND PIPES, 4-CYLINDER MODELS

Plate Ref.	DESCRIPTION	Qty	Part No.	REMARKS
	Fuel pipe, nylon, tank to pump	1	564898 *	88 and 103 Except Diesel models
	Olive for fuel pipe, pump end	1	557531	88 and 103 models
	Fuel pipe, nylon, tank to pump	1	552436	109 Station Wagon
	Olive for fuel pipe, pump end	1	557531	109 Station Wagon
	Fuel pipe, tank to sedimentor	1	564936 *	88 and 109 Models with sedimentor
	Olive for fuel pipe, sedimentor end	1	557531	88 and 109 Models with sedimentor
	Fuel pipe, tank to sedimentor	1	564963 *	109 Station Wagon with sedimentor
	Olive for fuel pipe, sedimentor end	1	557531	109 Station Wagon with sedimentor
	Fuel return pipe	2	277918	Diesel models, From engine suffix 'K' onwards
	Nut } Fixing return pipe to	2	270141	Diesel models, 1958-60
	Nipple } tank and flexible pipe	2	270133	Diesel models, 1958-60
	Fuel pipe, flex to tank return	1	500821	Diesel models, 1958-60, 109 Station Wagon
	Fuel pipe, spill return to tank, nylon	1	509909	Diesel models
	Bolt (2 BA x ¾" long) } Fixing spill return	1	250961	
	Spring washer } pipe to bracket	1	3073	
	Nut (2 BA) } on injector stud	1	2247	
	Grommet for spill return pipe	1	273370	
68	Fuel pipe retaining bracket	1	270297	Petrol models, 1958-60
69	Clip for fuel pipes	2	509412	Petrol models, 1958-60
70	Clip for fuel pipes	1	50183	1961 onwards
	Clip for fuel pipes	2	216708	Petrol models, 1961 onwards
	Drive screw fixing clip and pipe to chassis	1	72626	
71	Bolt (2 BA x ½" long) } Fixing clip and	3	234603	
72	Spring washer } bracket to chassis	3	3073	All Diesel and 1958-60 Petrol models
73	Nut (2 BA) } crossmember	3	2247	
74	Air cleaner, AC 7222906	1	263148	2 litre Petrol models
	Air cleaner, AC 7950·0	1	279652	2¼ litre Petrol and 2¼ litre Diesel models
	Air cleaner, AC 7223457	1	269139	2 litre, Series II Diesel models
	Element for air cleaner, AC 7222911	1	600613	2 litre, Series II Diesel models
75	Oil container, AC 7955536	1	264777	2 litre Petrol models
	Oil container, AC 7364533	1	600400	2¼ litre Petrol and 2¼ litre Diesel models
76	Washer for container	1	261414	2¼ litre Petrol and 2¼ litre Diesel models
77	Toggle	3	262068	
78	Connection, air cleaner to carburetter	1	262149	2 litre Petrol models
	Connection, air cleaner to manifold	1	275250	2¼ litre Series II Diesel models
	Connection, air cleaner to manifold	2	509730	2¼ litre Series IIA Diesel models
	Clip fixing connection	2	500225	2 litre Petrol models
	Clip fixing connection	2	50340	Diesel models
79	Clip, air cleaner connection to rocker cover	1	517502	2¼ litre Series IIA
	Plain washer for clip	1	2219	
	Carburetter elbow	1	279651	Up to engine suffix 'H' inclusive
	Carburetter elbow	1	554383	From engine suffix 'J' onwards
	Clip fixing elbow to connection	1	279653	Up to engine suffix 'H' inclusive
	Rubber seal, carburetter to elbow	1	276426	From engine suffix 'J' onwards
	Rubber seal, carburetter to elbow	1	554418	From engine suffix 'J' onwards
	Clip fixing connection	2	271870	2¼ litre Petrol models
	Connection, elbow to air cleaner	1	543580	2¼ litre Petrol models
	Clip fixing connection	1	50340	2¼ litre Petrol models

* Asterisk indicates a new part which has not been used on any previous Rover model

K749

FUEL TANK, FILTER AND PIPES, 6-CYLINDER MODELS

Plate Ref.	1 2 3 4	DESCRIPTION	Qty	Part No.	REMARKS
1		Fuel tank complete	1	552971	109
1		Fuel tank complete	1	543164	109 Station Wagon
2		Drain plug	1	235592	
3		Joint washer	For	1	515599	
		Telescopic filler tube complete	fuel tank	1	277259	
		Filler cap	1	277260	Alternatives to 5 lines below
4		Joint washer for cap	1	277262	
		Chain for filler cap	1	231190	in filler cap.
		Filler tube	1	277261	
		Filler cap	1	504655	With ball valve
4		Joint washer for cap	1	505244	in filler cap.
5		Chain for filler cap	1	231190	With felt pad
6		Filler tube	1	504657	Alternatives to 5 lines above
7		Extension tube for filler	1	504656	
8		Filler tube	1	504657	
9		Grommet for fuel tank filler	1	500710	
		Screw (2 BA x 3/8" long) Fixing filler to	4	77941	
		Plain washer body side	4	4030	
		Spring washer	4	3073	
		Special nut	1	313385	
		Nut (2 BA)	3	3823	
10		Hose, tank to filler tube	1	543782	
11		Clip for hose, bottom	1	50328	
12		Clip for hose, top	1	50343	
13		Breather hose for fuel tank	1	543767	
14		Clip for breather hose	2	50300	
15		Air balance hose for fuel tank	1	543766	
16		Clip for air balance hose	2	50302	
17		Outlet elbow complete for tank	1	500432	
18		Joint washer for outlet elbow	1	267837	
19		Spring washer Fixing elbow	2	3101	
20		Screw (3 BA x 7/16" long) to tank	2	3972	
21		Gauge unit for fuel tank	1	560612 *	
22		Gauge unit for fuel tank	1	605183 *	109
		Joint washer for gauge unit, SM TB 1114/000	1	555846	109
23		Joint washer for gauge unit	1	546488	Station Wagon
23		Spring washer	6	3101	
24		Set screw (3 BA x 3/8" long) Fixing gauge unit	6	3890	
25		Support for fuel tank to tank	1	568802	
25		Bolt (1/4" UNF x 3/4" long) Fixing support	4	255207	
		Plain washer to tank	2	3840	
		Spring washer	4	3074	Alternative to 3/8" UNF fixing
		Locking plate	4	277491	
26		Square nut (1/4" UNF)	2	277490	
		Nut (1/4" UNF)	2	254810	
27		Bolt (3/8" UNF x 3/4" long)	2	255245	
28		Plain washer	4	2251	Alternative to 1/4" UNF fixing
29		Spring washer	2	3076	
29		Nut (3/8" UNF)	2	254812	

* Asterisk indicates a new part which has not been used on any previous Rover model

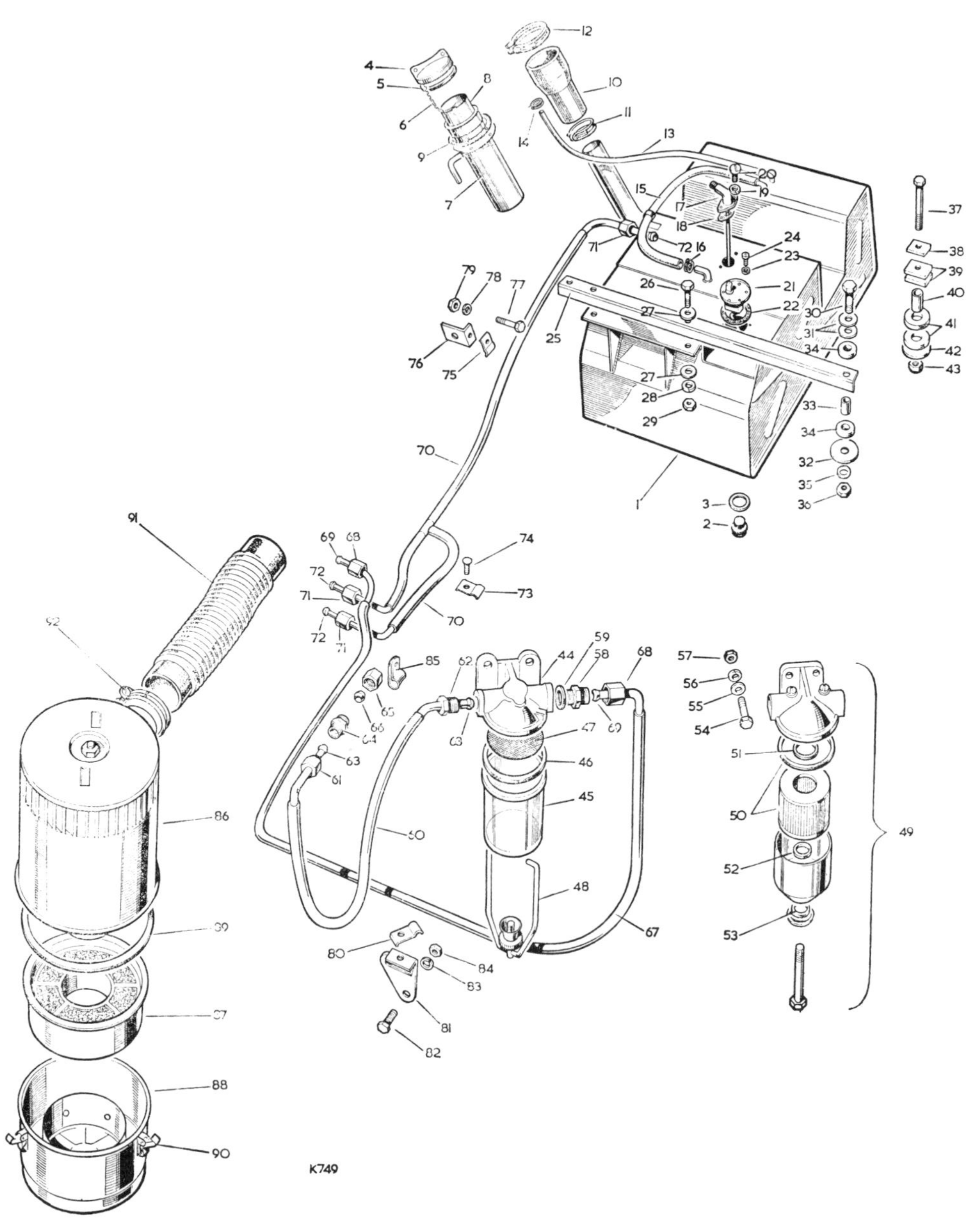

Plate Ref.	1 2 3 4 DESCRIPTION		Qty	Part No.	REMARKS	
30	Bolt ($\frac{3}{8}$" UNF x $1\frac{3}{8}$" long)	Fixing front of fuel tank to support and chassis frame	2	256040	109	
31	Plain washer, small		4	2251		
32	Plain washer, large		2	3109		
33	Distance tube		2	508544		
34	Rubber bush		4	508545		
35	Spring washer		2	3076		
36	Nut ($\frac{3}{8}$" UNF)		2	254812		
37	Bolt ($\frac{3}{8}$" UNF x $2\frac{1}{2}$" long)	Fixing rear of fuel tank to support and chassis frame	2	256047		
38	Distance piece		2	501070		
39	Distance piece		4	500588		
40	Distance tube		2	500446		
41	Mounting rubber		4	500447		
42	Plain washer		2	3933		
43	Self-locking nut ($\frac{3}{8}$" UNF)		2	252162		
	Bolt ($\frac{3}{8}$" UNF x $1\frac{3}{8}$" long)	Fixing fuel tank to frame at front	2	256040	109 Station Wagon	
	Plain washer		2	3109		
	Distance piece		2	508544		
	Mounting rubber		2	508545		
	Spring washer		2	3076		
	Nut ($\frac{3}{8}$" UNF)		2	254812		
	Bolt ($\frac{3}{8}$" UNF x $2\frac{1}{2}$" long)	Fixing fuel tank to frame at rear	2	256047		
	Distance piece, top		2	501070		
	Plain washer		4	3933		
	Spring washer		1	2265		
	Distance piece		2	500446		
	Mounting rubber		4	500447		
	Distance piece, bottom		2	500588		
	Self-locking nut ($\frac{3}{8}$" UNF)		2	252162		
	FUEL SEDIMENT BOWL COMPLETE		1	267494		
44	Body only		1	268793	AC Delco type	Alternatives. Check before ordering
45	Bowl only		1	236891		
46	Joint washer for bowl..		1	241225		
47	Gauze for bowl		1	241223		
48	Retainer for bowl		1	268797		
	Body only		1	514269	Wipac type	
	Bowl only		1	236896		
	Joint washer for bowl..		1	236895		
	Gauze for bowl		1	514270		
	Retainer for bowl		1	240234		
	Screw cap for retainer		1	240235		
49	FUEL FILTER COMPLETE		1	577064*		
50	Element and seal for filter, AC 60		1	606204*		
51	Centre seal, upper, AC 7961152		1	606205*		
52	Centre seal, lower, AC 7961102		1	606206*		
53	Seal for centre bolt, AC 7961099		1	606207*		

* Asterisk indicates a new part which has not been used on any previous Rover model

K749

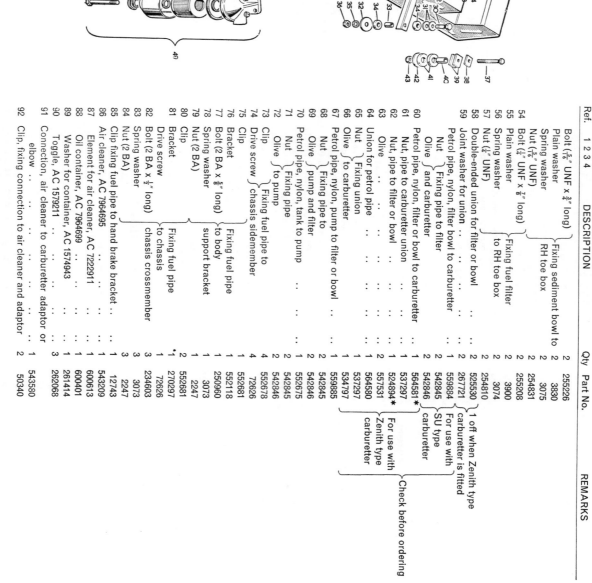

FUEL TANK, FILTER AND PIPES, 6-CYLINDER MODELS

Plate Ref. 1 2 3 4	DESCRIPTION	Qty	Part No.	REMARKS
54	Bolt (5/16" UNF x 3/4" long)	2	255226	⎱ Fixing sediment bowl to
55	Plain washer	2	3830	⎰ RH toe box
56	Spring washer	2	3075	
57	Nut (5/16" UNF)	2	254831	
54	Bolt (1/4" UNF x 7/8" long)	2	255208	⎱ Fixing fuel filter
55	Plain washer	2	3900	⎰ to RH toe box
56	Spring washer	2	3074	
57	Nut (1/4" UNF)	2	254810	
58	Double-ended union for filter or bowl	2	525530	
59	Joint washer for union	2	267721	
60	Petrol pipe, nylon, filter bowl to carburetter	1	524894 *	1 off when Zenith type carburetter is fitted
61	Olive	2	542845	
62	Nut, pipe to carburetter union	1	542846	⎫ Fixing pipe to filter or bowl and carburetter
63	Olive	2	542846	
64	Nut	1	564581 *	For use with Zenith type carburetter
65	Nut, pipe to carburetter union	1	557531	
66	Nut, pipe to filter or bowl	1	564580	For use with SU type carburetter
67	Olive	1	537297	
68	Union for petrol pipe	1	534797	
69	Olive	1	559885	⎱ Fixing union
70	Nut	2	542845	⎰
71	Olive	2	542846	
72	Petrol pipe, nylon, pump to filter or bowl	2	552675	
73	Olive	2	542846	⎱ Fixing pipe to
74	Nut	2	542846	⎰ pump and filter
75	Olive	2	552678	
76	Petrol pipe, nylon, tank to pump	1	72626	
77	Olive	2	552681	⎱ Fixing pipe
78	Nut	2	270297	⎰ to pump
79	Olive	2	72626	
80	Clip	4	234603	⎱ Fixing fuel pipe to chassis sidemember
81	Drive screw	4	3073	⎰
82	Clip	4	2247	Fixing fuel pipe to chassis crossmember
83	Bracket	3	3073	⎱
84	Nut (2 BA)	3	250960	⎬ to body support bracket
85	Bolt (2 BA x 5/8" long)	3	552118	⎭
86	Spring washer	1	552681	
87	Clip fixing fuel pipe to hand brake bracket	1	72626	
86	Air cleaner, AC 7964695	1	543209	⎫ Check before ordering
87	Element for air cleaner, AC 7222911	1	600613	
88	Oil container, AC 7964699	1	600401	
89	Washer for container, AC 1574943	1	261414	
90	Toggle, AC 1579211	3	262068	
91	Connection, air cleaner to carburetter adaptor or elbow	1	543580	
92	Clip, fixing connection to air cleaner and adaptor	2	50340	

* Asterisk indicates a new part which has not been used on any previous Rover model

PETROL PUMP, 4-CYLINDER MODELS

C 919.

COLLINS-JONES

PETROL PUMP, 4-CYLINDER MODELS

Plate Ref.	1 2 3 4	DESCRIPTION	Qty.	Part No.	REMARKS
1		Petrol pump complete, SU AUA 79, 'L' type	1	268374	
2		Coil complete, SU AUA 6001	1	262286	
3		Spring for armature, SU AUA 1449	1	262304	
4		Diaphragm complete, SU AUA 6011	1	260550	
5		Roller for diaphragm, SU AUA 1433	11	260583	
6		Plate body, SU AUA 4081/1	1	260299	
7		Joint washer for plate body, SU AUA 4081/1	1	262300	
8		Body, SU AUA 4080/1	1	262287	
9		Screw fixing coil housing to body, SU AUA 4080/1	5	262301	
10		Special spring washer, SU AUA 1863 ⎱ For earth	5	262303	
11		Special screw, SU AUA 4850 ⎰ terminal	1	262829	
12		Valve cage, SU AUA 1416	1	262288	
13		Disc for valve, SU AUA 839	2	262297	
14		Spring clip retaining valve disc, SU AUA 840	1	262289	
15		Washer for valve cage, SU AUA 1479	1	262290	
16		Outlet union, SU AUA 1422	1	262294	
17		Washer for outlet union, SU AUA 1442	1	262293	
18		Inlet union, SU AUA 887	1	262296	
19		Washer for inlet union, SU AUA 1405	1	262298	
20		Filter, SU AUA 1464	1	262292	⎱ 2 litre Petrol models
21		Plug for filter, SU AUA 1421	1	262295	⎰
22		Washer for filter plug, SU AUA 1442	1	262293	
23		Contact set complete, twin points, SU AUA 6021	1	268674	
24		Special screw for contact blade, SU AUA 847	2	262310	
25		Moulding for end plate, SU AUA 1467	1	260585	
26		Screw fixing moulding, SU AUA 1459	2	262308	
27		Terminal screw, SU AUA 1468	1	262307	
28		Cover for end plate, SU AUA 1466	1	234309	
29		Terminal nut, SU AUA 1661	2	262306	
30		Tag for terminal, SU AUA 1662	1	262305	
31		Spring washer	1	3073	
32		Nut (2 BA)	2	212556	
33		Elbow for pump, SU AUA 1478	1	549761	1 off on Series IIA
33		Olive ⎱ Fixing elbow	1	236891	
34		Nut ⎰ to pump	1	268797	
35		Mounting plate	1	506796	
36		Rubber bush ⎱ Fixing electric petrol pump	2	241225	
37		Plain washer ⎰	2	275565	
38		Special set bolt	2	252211	
39		Mechanical petrol pump complete, AC 7950446	1	247929	⎱ 2¼ litre Petrol models.
40		Joint washer, pump to cylinder block	1	243967	Not part
41		Self-locking nut (⁵⁄₁₆" UNF) fixing fuel pump	2	50499	of engine
42		Inlet and outlet union for pump	2	270115	⎱ assembly
43		Joint washer for inlet and outlet union	2	270105	⎰
44		Elbow for inlet union	2	272069	⎱ 2¼ litre
45		Nut ⎱ Fixing elbow to	2	524684	⎰ Petrol models
46		Olive ⎰ fuel pump	2		
45		Repair kit for petrol pump, AC 1524507			Early type
46		Overhaul kit for petrol pump, AC 7950463	1		fixing only

* Asterisk indicates a new part which has not been used on any previous Rover model

PETROL PUMP, 6-CYLINDER MODELS

G944

PETROL PUMP, 6-CYLINDER MODELS

Plate Ref.	1 2 3 4	DESCRIPTION	Qty	Part No.	REMARKS
		FUEL PUMP, DOUBLE ENTRY TYPE, SU AUF 503			
1		Coil complete, SU AUB 6007	1	545306	
2		Spring for armature, SU AUB 759	2	530884	
3		Diaphragm complete, SU AUB 6040	2	530885	
4		Roller for diaphragm, SU AUB 1433	2	542320	
5		Joint washer, SU AUB 593, diaphragm to body, outer	22	26083	
5		Joint washer, SU AUB 809, diaphragm to body, inner	2	538499	
6		Screw, SU AJD 3204, fixing coil housings to body	12	605932*	
7		Body, SU AUB 650	1	538501	
8		Special screw, SU AUB 699	1	538485	
9		Spring washer, SU AUA 585 } For earth terminal	1	538502	
10		'Lucar' connector, SU AUA 692	1	538503	
11		Valve assembly, SU AUB 6062	4	538510	
12		Sealing washer for valve assembly, SU AUB 676	4	600723	
13		Screw for valve assembly, SU AUB 597	4	538487	
14		Filter, SU AUB 617	2	538489	
15		Unions, inlet and outlets, SU AUB 655	2	538488	
16		Sealing ring for unions, SU AUB 654	3	538498	
17		Contact set complete, SU AUB 6022	2	538497	
18		Dished washer, SU AUA 566	2	268674	
19		Screw, SU AUA 565	2	268874	
20		Pedestal assembly, SU AUA 6034	2	538507	
21		Screw, SU AU A 1459 } Fixing pedestal	4	538506	
22		Spring washer, SU AUA 1863 } assembly	4	538505	
23		Joint washer for dished cover, SU AUA 573	1	262308	
24		Dished washer, SU AUA 575 } Fixing dished cover	1	538490	
25		Spring washer, SU LWZ 303 } to body	1	538491	
26		Bolt, SU AUC 2694	1	538493	
27		Condenser, SU AUA 5060	1	605322	
28		Diaphragm for air bottle, SU AUB 656	1	245382	
29		Joint washer, diaphragm to housing, SU AUB 795	4	538494	
30		Screw, SU AUC 2588, fixing air bottle cover	4	542319	
31		Sealing ring, SU AUB 657, for air bottle cover	1	538496	
32		Clip for condenser, SU AUA 5059	1	538495	
33		Terminal screw, SU AUA 1468	1	245384	
34		Spring washer for terminal, SU AUA 1863	2	262307	
35		Terminal nut, SU AUA 1661	2	262303	
36		Lead washer, SU AUA 1662	2	262306	
37		Washer for terminal screw, SU AUB 609	2	262305	
38		Cover, black, SU AUB 626	2	538509	
39		Shakeproof washer, SU LWN 403	2	538508	
40		'Lucar' connector, SU AUA 692	2	538510	

* Asterisk indicates a new part which has not been used on any previous Rover model

PETROL PUMP, 6-CYLINDER MODELS

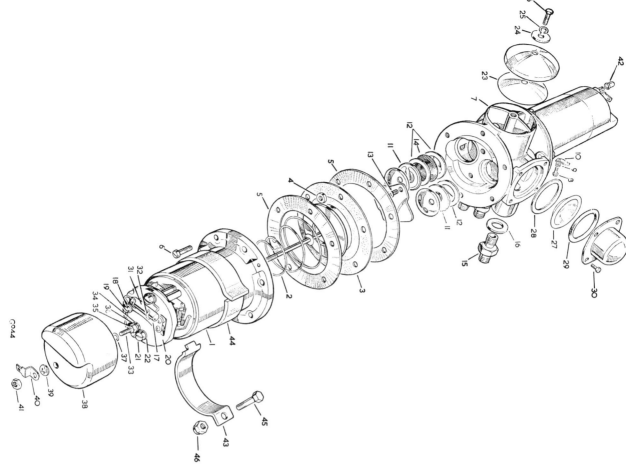

G044

PETROL PUMP, 6-CYLINDER MODELS

Plate Ref.	1 2 3 4	DESCRIPTION	Qty	Part No.	REMARKS
41		Terminal nut, SU AUA 878, fixing 'Lucar' connector	2	262309	
42		Insulating sleeve, SU AUB 611, for terminal	2	533511	
43		Clamp plate	2	529141	
44		Rubber strip	2	534200	
45		Bolt ($\frac{1}{4}$" UNF x $\frac{7}{8}$" long)	2	255208	Fixing petrol pump to support bracket
46		Self-locking nut ($\frac{1}{4}$" UNF)	2	252160	
		Support bracket for fuel pump	1	562719	
		Bolt ($\frac{1}{4}$" UNF x $\frac{5}{8}$" long)	4	255206	Fixing support bracket to chassis frame
		Spring washer	4	3074	
		Nut ($\frac{1}{4}$" UNF)	4	254810	

* Asterisk indicates a new part which has not been used on any previous Rover model

INSTRUMENTS, POSITIVE EARTH SYSTEM
Series II and IIA models. Up to vehicle suffix 'C' inclusive

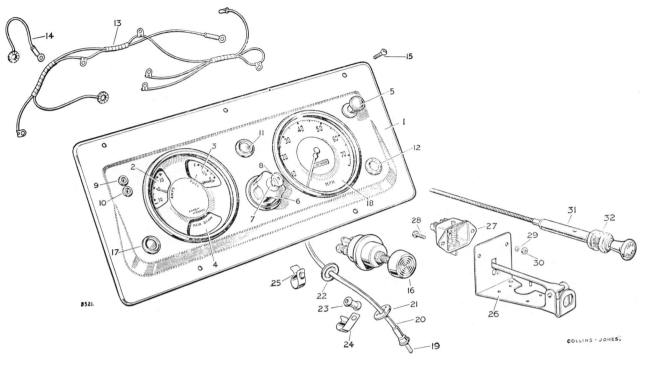

B521.

COLLINS - JONES.

INSTRUMENTS, POSITIVE EARTH SYSTEM
Series II and IIA models. Up to vehicle suffix 'C' inclusive

Plate Ref. 1 2 3 4	DESCRIPTION	Qty	Part No.	REMARKS
1	Instrument panel	1	302958	1958
	Instrument panel	1	332466	1959 2¼ litre Petrol models onwards
	Instrument panel	1	332467	1959 Diesel models onwards
2	Gauge and ammeter panel complete, JA PA 133	1	239566	⎫
3	Ammeter, SM AM 6200/13	1	239567	⎪
4	Fuel gauge, SM FG 6237/13	1	239568	⎬ Series II models
	Warning light for headlamp beam, SM 53334/2	1	239569	⎪
	Bulb for warning light, LU 987	1	232590	⎭
	Gauge and ammeter panel complete	1	519827	Germany only
	Gauge and ammeter panel complete	1	519840	⎫
	Ammeter	1	519841	⎪ Series IIA models
	Fuel gauge	1	519873	⎬ except 109 Station
	Warning light for headlamp beam, red	1	519874	⎪ Wagon
	Warning light for headlamp beam, blue	1	545417	Germany only
	Bulb for warning light, LU 987	1	545405	Germany only
	Gauge and ammeter panel complete	1	545408	Germany only
	Gauge and ammeter panel complete	1	528950	⎫ Series IIA models,
		1	232590	⎭ 109 Station Wagon
5	Switch for panel lights	1	240908	
6	Switch for lamps, LU 54033253	1	239570	Series II models
	Switch for lamps, LU 031412	1	240714	America. Dollar area. Series II models ⎫ Alternatives.
	Switch for lamps, locking type	1	519775	Series IIA models ⎬ Check before
	Switch for lamps, Lucas 54033283, non-locking type	1	536917	Late Series IIA Diesel models ⎭ ordering
	Switch for lamps, locking type	1	531530	America Dollar Area. Series IIA Petrol models
	Switch for lamps, non-locking type	1	547102	America Dollar Area. Series IIA Diesel models
7	Knob for ignition and lamp switch, LU 316436	1	537284	
8	Barrel lock for ignition or electrical services	1	59895	For locking type switch
9	Key for lock	1	29979	State key number when ordering
10	Socket for inspection lamp, black, LU 304989	1	273937	
11	Socket for inspection lamp, red, LU 305021	1	273936	
12	Warning light, charging, LU 38084A	1	238018	1958
	Warning light, charging	1	519743	1959-60
	Warning light, oil	1	519740†	
	Warning light, mixture or heater plug, LU 38086A	1	238020	1958
	Bulb for warning light, LU 987	2	232590	
	Warning light, mixture	1	519744	1959 2¼ litre Petrol models onwards
	Warning light, heater plug	1	501563	Diesel models
	Bulb for heater plug warning light	1	507129	Diesel models

* Asterisk indicates a new part which has not been used on any previous Rover model

† NOTE: Positions of warning lights, oil and charging are reversed on 1961 models onwards. Word 'Oil' is printed on lens face of oil warning light.

INSTRUMENTS, POSITIVE EARTH SYSTEM
Series II and IIA models. Up to vehicle suffix 'C' inclusive

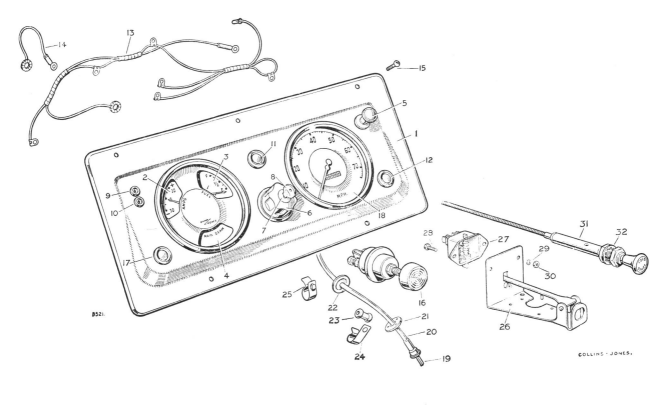

B521.

COLLINS · JONES.

INSTRUMENTS, POSITIVE EARTH SYSTEM
Series II and IIA models. Up to vehicle suffix 'C' inclusive

Plate Ref. 1 2 3 4	DESCRIPTION	Qty	Part No.	REMARKS
13	Panel harness	1	238723	1958 Petrol models
	Panel harness, LU 864385	1	272743	1958 Diesel models
	Panel harness	1	279400	1959-61 Series II models
14	Lead for panel harness, LU 864386	1	272850	} Series II models
	Lead, ammeter to inspection socket, LU 825887	1	243815	}
15	Bulb for panel light, LU 987	2	232590	
	Set screw (2 BA x 3/8" long)	5	77941	
	Plain washer	5	3885	} Fixing instrument panel
	Spring washer	5	3073	
	Nut (2 BA)	2	2247	
16	Starter switch	1	530034	Petrol models
	Knob for starter switch, LU 764505	1	605280*	Petrol models
	Starter heater plug switch, LU 34510A	1	530071	Non-barrel lock type } Petrol models. Alternatives. Check before ordering
	Starter heater plug switch, Lucas 35577	1	551565	Barrel lock type }
	Key for starter heater plug switch, LU 312305	1	505798	For switch 530071, Non-barrel type } Diesel models
	Barrel lock for starter heater switch	1	536913	For switch 277615, Barrel lock type }
	Key for starter switch, LU 54330767	1	542091	For switch 530071, LU 34510A type
	Key for starter heater switch	1	29979	Barrel lock type. Diesel models
17	Warning light, fuel	1	519742	Diesel models
	Warning light, oil, LU WL 3/1 38085A	1	238019	1958
	Warning light, oil	1	519740	1959-60
	Warning light, charging	1	519743†	1961 models onwards
	Bulb for warning light, LU 987	2	232590	
18	Speedometer, miles, SM SN 3391/01	1	239564	
	Speedometer, kilometres, SM SN 3391/02	1	239565	} 88
	Speedometer, miles, SM SN 3391/03	1	279340	
	Speedometer, kilometres, SM SN 3391/04	1	279341	} 109. Also optional on 88 when 7.50" x 16" tyres are fitted
	Cable complete for speedometer, SM 1105/00	1	270716	
19	Cable, inner	1	270718	
20	Cable, outer	1	270717	
21	Retaining plate for cable	1	232566	
	Spring washer	3	3073	} Fixing retaining plate
	Set bolt (2 BA x ¼" long)	3	250957	
22	Felt washer for speedometer	1	241387	
	Rubber grommet for speedometer	1	233243	} For speedometer cable
	Rubber grommet, in dash	3	06860	
23	Rubber grommet, on cable			
	Rubber grommet for speedometer cable at fuel pump	1	273370	Diesel and 2¼ litre Petrol models

* Asterisk indicates a new part which has not been used on any previous Rover model

† NOTE: Positions of warning lights, oil and charging are reversed on 1961 models onwards. Word 'Oil' is printed on lens face of oil warning light.

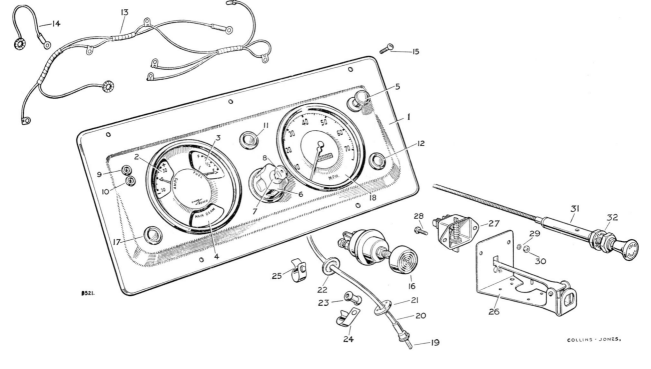

Ð521.

COLLINS · JONES.

INSTRUMENTS, POSITIVE EARTH SYSTEM
Series II and IIA models. Up to vehicle suffix 'C' inclusive

Plate Ref.	1 2 3 4	DESCRIPTION	Qty	Part No.	REMARKS
24		Clip	1	239600	
25		Clip ⎫ For speedometer cable	1	240407	
		Clip ⎬	2	278010	
		Clip ⎭	2	240417	
26		Bracket for mixture control	1	216380	
27		Switch for mixture warning light, LU 31109A	1	214223	
28		Bolt (2 BA x ½" long) ⎫ Fixing	1	234603	
29		Spring washer ⎬ switch	2	3073	
30		Nut (2 BA) ⎭ to bracket	2	2247	
31		Mixture control complete	1	500571	Must be used with carburetter type **without** starter heater element. Check before ordering ⎫ Petrol models
		Mixture control	1	523637	Must be used with carburetter type **with** starter heater element. Check before ordering ⎬
32		Shakeproof washer for control	1	233017	

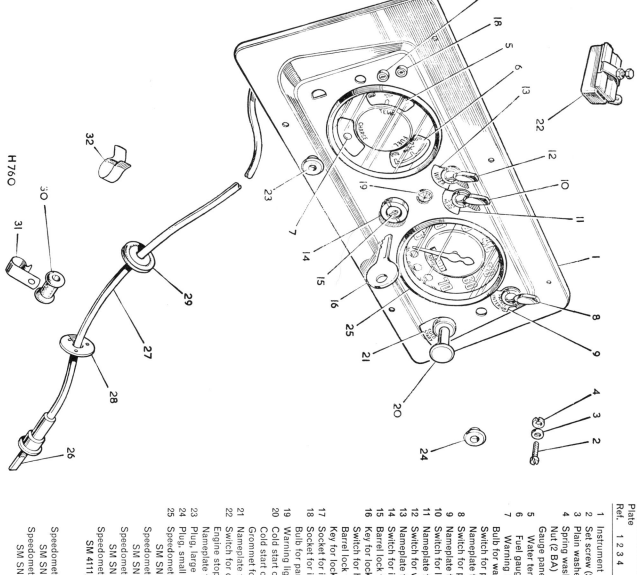

H760

INSTRUMENTS, NEGATIVE EARTH SYSTEM
Series IIA models. From vehicle suffix 'D' onwards

Plate Ref.	1 2 3 4	DESCRIPTION	Qty	Part No.	REMARKS
1		Instrument panel ...	1	345052	
2		Set screw (2 BA x ⅜" long)	5	77941	Fixing instrument panel
3		Plain washer ...	5	3885	
4		Spring washer ...	5	3073	
		Nut (2 BA) ...	2	2247	
5		Gauge panel complete ...	1	560744	
5		Water temperature gauge ...	1	560746	
6		Fuel gauge ...	1	555835	
7		Warning light, charging, red ...	1	555837	
7		Bulb for warning light, LU 987	2	232590	
		Switch for panel lights ...	1	536173	
8		Nameplate for switch 'PANEL'	1	560404	Except Station Wagon
9		Switch for panel and interior lights, LU 34734	1	555877	Station Wagon
9		Nameplate for switch 'PANEL/INTERIOR'	1	560405	
10		Switch for lamps, LU 31956B	1	555779	
11		Nameplate for switch 'SIDE/HEAD'	1	560407	
12		Switch for wiper motor, LU 34475	1	555778	
13		Nameplate for switch 'WIPER'	1	560410	
14		Switch for ignition and starter, LU 35598	1	551508	
15		Barrel lock for ignition switch ...	1	59895	2¼ litre and 2.6 litre Petrol models
16		Switch for heater plug and starter	1	29979	2¼ litre Diesel models
16		Barrel lock for starter heater switch	1	575081	2¼ litre Diesel models
		Key for lock ...	1	536913	
17		Socket for inspection lamp, black	1	29979	
18		Socket for inspection lamp, red	1	555661	
19		Bulb for panel lights, LU 987	2	555662	
20		Warning light, fuel (blue)	1	232590	
20		Cold start control complete ...	1	560756	2¼ litre Diesel models
21		Cold start control complete ...	1	552682	2¼ litre Petrol models
		Grommet for cold start cable	1	552683	2.6 litre Petrol models
21		Nameplate for control 'COLD START'	1	235113	2¼ litre and 2.6 litre Petrol models
22		Switch for cold start warning light	1	552682	
		Engine stop control complete	1	560411	2¼ litre Diesel models
		Nameplate for control 'ENGINE STOP'	1	563318	
23		Plug, large ⎫ For redundant	1	552688	
24		Plug, small ⎬ holes in panel	3	560412	
		⎭		338016	
25		Speedometer and warning lights, miles, SM 4111/05	1	338013	
		Speedometer and warning lights, miles, SM SN 4111/02	1	540118	7.50" x 16" tyres are fitted, on 88 models when
		Speedometer and warning lights, kilometres, SM SN 4111/03	1	540117	103 models. Also optional ⎱ 2¼ litre Petrol and 2¼ litre Diesel models
		Speedometer and warning lights, miles, SM SN 4111/04	1	540115	88 models
		Speedometer and warning lights, kilometres, SM SN 4111/10	1	540114	88 models
		Speedometer and warning lights, miles, SM SN 4111/11	1	559159	109 ⎱ 2¼ litre Petrol models, 2.6 litre
		Speedometer and warning lights, kilometres,	1	559160	Petrol models ⎱ 2¼ litre Diesel models

* Asterisk indicates a new part which has not been used on any previous Rover model

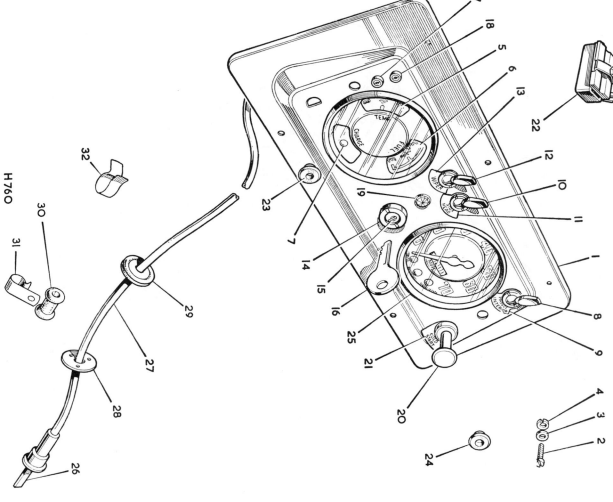

H760

INSTRUMENTS, NEGATIVE EARTH SYSTEM
Series IIA models. From vehicle suffix 'D' onwards

Plate Ref.	1 2 3 4	DESCRIPTION	Qty	Part No.	REMARKS
		Bulb for warning lights in speedometer, LU 987	3	232590	
		Cable complete for speedometer, SM 1105/00	1	270716	} 2¼ litre Petrol
26		Cable, inner	1	270718	and 2¼ litre Diesel
27		Cable, outer	1	270717	} models
		Cable complete for speedometer	1	560751	}
		Cable, inner	1	560752	2.6 litre Petrol models
		Cable, outer	1	560753	}
28		Retaining plate for cable	1	232566	
		Spring washer	3	3073	} Fixing
		Set bolt (2 BA x ¼" long)	3	250957	} retaining plate
		Felt washer for speedometer	1	241387	
29		Rubber grommet, in dash	1	233243	} For speedometer
30		Rubber grommet, on cable	4	06660	} cable
		Rubber grommet for speedometer cable at fuel pump	1	273370	5 off on 2.6 litre Petrol models
31		Clip	1	239600	3 off on 2.6 litre Petrol models
32		Clip	1	240407	} For speedometer cable
		Clip	1	278010	}
		Clip	1	240417	} For speedometer cable
		Clip	1	268883	} 2¼ litre Diesel
		Distance washer	1	4589	} models } at camshaft cover

* Asterisk indicates a new part which has not been used on any previous Rover model

HEAD, SIDE AND TAIL LAMPS. Up to vehicle suffix 'A' inclusive
(Except America Dollar Area)

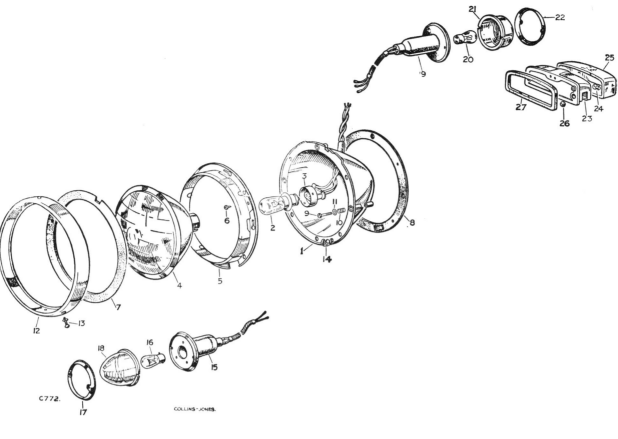

C772.

COLLINS-JONES.

HEAD, SIDE AND TAIL LAMPS. Up to vehicle suffix 'A' inclusive
(Except America Dollar Area)

Plate Ref.	1 2 3 4	DESCRIPTION	Qty	Part No.	REMARKS
		Headlamp, LU 51780 ...	2	272581	RHStg.
		Headlamp, LU 58817 ...	2	605070	LHStg. Except Europe
		Headlamp, LU 58286 ...	2	507198	LHStg. Europe except France
		Headlamp, LU 58287 ...	2	507199	LHStg. France only
1		Body for headlamp, LU 552943 ...	2	274377	RHStg.
2		Bulb for headlamp, LU 414 ...	2	505196	RHStg.
		Bulb for headlamp, LU 415 ...	2	513314	LHStg. Except Europe
		Bulb for headlamp, LU 370 ...	2	244128	LHStg. Europe except France. For early type headlamp 272578, LU 51782 only
		Bulb for headlamp, LU 410 ...	2	505197	LHStg. Europe except France. For late type headlamp 507198, LU 58286
3		Adaptor for bulb, double contact, LU 411 ...	2	262269	LHStg. Europe except France. France only
		Adaptor for bulb, triple contact, LU 554602 ...	2	274378	France only
4		Light unit, LU 553921 ...	2	261794	RHStg.
		Light unit, LU 554447 ...	2	274783	LHStg. Except Europe
		Light unit, LU 553940 ...	2	262342	LHStg. Europe except France. For early type headlamp 272578, LU 51782 only
		Light unit, LU 556452 ...	2	262343	LHStg. Europe except France. For late type headlamp 507198, LU 58286
5		Light unit, LU 553948 ...	2	507804	LHStg. France only
		Rim complete, for light unit, LU 554872 ...	2	262269	
6		Special screw for light unit, LU 186137 ...	6	274379	
7		Rubber gasket for headlamp rim, LU 552906 ...	2	262100	
8		Gasket for body, LU 554279 ...	2	262101	
9		Special screw ⎫Light	6	524872	
10		Spring for screw ⎬unit	6	261940	
11		Cup washer for screw, LU 552818 ⎭adjustment	6	261941	
12		Rim for headlamp, chrome, LU 554439 ...	2	261943	
13		Rim retaining screw, LU 198898 ...	2	272592	
14		Spire nut for rim retaining, LU 187117 ...	2	272594	
		Screw (2 BA x ½" long) ⎫Fixing	6	510469	
		Spring washer ⎬headlamp to	6	71164	
		Nut (2 BA) ⎭front grille	6	3073	
15		Side lamp complete, Sparto 5835 ...	2	2247	
		Rubber body for side lamp, Sparto 12190 ...	2	505144	Sparto type.
16		Bulb for side lamp, LU 207 ...	2	500515	
		Bulbholder, interior, Sparto 12207 ...	2	10211	Alternative to Lucas.
17		Bezel for side lamp, Sparto 12189 ...	2	524497	Check before ordering
		Lens for side lamp, flat type, Sparto 1539 ...	2	500514	
18		Lens for side lamp, domed type ...	2	504130	
		Side lamp complete, LU 52437 ...	2	500513	
		Bulb holder interior, LU 553780 ...	2	510179	
		Bulb for side lamp, LU 989 ...	2	244700	
		Bezel for side lamp, LU 54570407 ...	4	50689	Lucas type.
		Special screw fixing bezel, LU 54127522 ...	4	514148	
		Lens for side lamp, LU 54570404 ...	2	514149	Alternative to Sparto.
		Sleeve for terminal, LU 555910 ...	2	514150	Check before ordering
		Nylon insert ...	4	514145	
				505150	

* Asterisk indicates a new part which has not been used on any previous Rover model

HEAD, SIDE AND TAIL LAMPS. Up to vehicle suffix 'A' inclusive
(Except America Dollar Area)

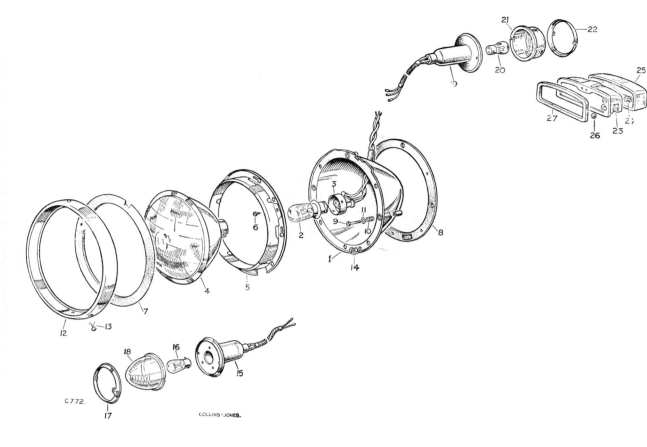

C772.

COLLINS-JONES.

HEAD, SIDE AND TAIL LAMPS. Up to vehicle suffix 'A' inclusive
(Except America Dollar Area)

Plate Ref.	1 2 3 4	DESCRIPTION	Qty	Part No.	REMARKS
		Stop/tail lamp, Sparto 57101	2	276317	Sparto type.
19		Rubber body, Sparto 12173	2	500413	Alternative to Lucas.
20		Bulb for stop/tail lamp, LU 380	2	264590	Check before ordering
21		Lens for stop/tail lamp, Sparto 12198 ...	2	500411	
22		Bezel for glass, Sparto 12172	2	500412	
		Special screw fixing tail lamp to body, Sparto 12210	4	500414	
		Stop/tail lamp, LU 53783	2	510176	
		Lens for stop/tail lamp, LU 54570677 ...	2	514152	
		Special screw fixing lens, LU 576396 ...	4	514143	
		Special nut fixing lens, LU 575219 ...	4	600706	Alternatives
20		Bulb for stop/tail lamp, LU 380	2	264590	
		Bulb holder, interior, LU 573828	2	264782	
		Sleeve for terminal, LU 555910	2	514145	Lucas type.
		Gasket for lamp base, LU 576390	2	514146	Alternative to Sparto.
		Grommet for cable, LU 576401	2	514147	Check before ordering
		Drive screw, fixing tail lamp to body ...	4	78135	
23		Number plate lamp, LU 53876	1	234213	
24		Bulb for number plate lamp, LU 989 ...	1	50689	
25		Lens, LU 54570354	1	601721	
26		Rubber grommet for wire, LU 572039 ...	1	261252	
27		Rubber gasket for lamp, LU 572042 ...	1	261253	

* Asterisk indicates a new part which has not been used on any previous Rover model

K854

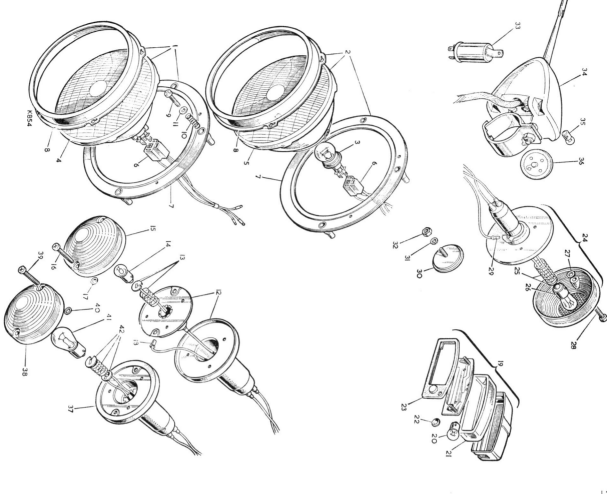

HEAD, SIDE, TAIL AND FLASHER LAMPS. From vehicle suffix 'B' onwards (Except America Dollar Area)

Plate Ref. 1 2 3 4	DESCRIPTION	Qty	Part No.	REMARKS
1	Headlamp complete, sealed beam	2	536006	RHStg
	Headlamp complete, sealed beam	2	536111	LHStg. Except Europe
2	Headlamp complete	2	536112	LHStg. Europe except France and Austria
	Headlamp complete	2	536113	LHStg. France only
	Headlamp complete	2	545107	LHStg. Austria only
3	Bulb for headlamp, LU 410	2	505197	LHStg. Europe except France
	Bulb for headlamp, LU 411	2	505198	LHStg. France only
4	Light unit, sealed beam, LU 54521872	2	542097	RHStg
5	Light unit, sealed beam	2	541537	LHStg. Except Europe
	Light unit, LU 5452r683	2	507804	LHStg. Europe except Austria
	Light unit, LU 54520883	2	545139	LHStg. Austria only
6	Adaptor and leads for headlamp	2	536116	
7	Rim for light unit	2	601999	
8	Rim for headlamp, chrome	2	545149	
9	Screw ⎱ For light unit adjustment and fitting	6	600024	
10	Spring for screw ⎰	6	600025	
11	Fibre washer for screw — headlamp to front grille	6	600026	
	Side lamp complete, Sparto 62402	2	536117	Sparto type.
12	Bulb holder and body complete, Sparto 02313	2	542037	Alternative to Wipac or Lucas.
13	Bulb for side lamp, LU 207	2	10211	Check before ordering
15	Lens for side lamp, Sparto 02301	2	536151	
16	Special screw fixing lens, Sparto 02302/14/15	4	542038	
	Side lamp complete, Wipac WP840	2	531687	Wipac type.
12	Bulb holder and base complete, Wipac 13422	2	542049	Alternative to Sparto or Lucas.
15	Lens for side lamp, Wipac 13431	2	542050	
16	Special screw, fixing lens, Wipac 13429	4	542042	Vehicles for France require Wipac type
	Side lamp complete, LU 52721	2	547344	Lucas type.
12	Bulb holder, interior, LU 553780	2	244700	Alternative to Wipac or Sparto.
14	Bulb for side lamp, LU 989	2	50689	Check before ordering
15	Lens for side lamp, LU 54576127	2	551429	
16	Special screw, fixing lens, LU 54110101	4	600423	
17	Washer for special screw, LU 161385	4	551430	
18	Sleeve for terminal, LU 555910	2	514145	
	Screw (4 BA x ½" long) ⎱ Fixing side	6	77399	
	Spring washer ⎰ lamp to	6	3072	
	Nut (4 BA) — front wing	6	4024	

* Asterisk indicates a new part which has not been used on any previous Rover model!

HEAD, SIDE, TAIL AND FLASHER LAMPS. From vehicle suffix 'B' onwards (Except America Dollar Area)

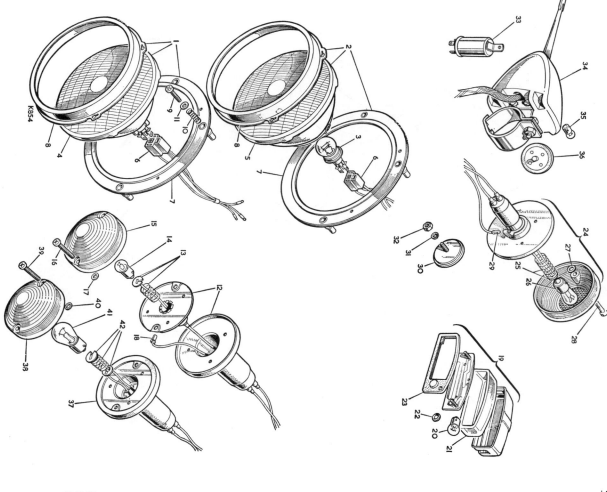

K854

HEAD, SIDE, TAIL AND FLASHER LAMPS. From vehicle suffix 'B' onwards (Except America Dollar Area)

Plate Ref.	1 2 3 4	DESCRIPTION	Qty	Part No.	REMARKS
19		Number plate lamp, LU 53876	1	234213	}
20		Bulb for number plate lamp, LU 989	1	50689	}
21		Lens, LU 54570354	1	607721	} Optional equipment
22		Rubber grommet for wire, LU 572039	1	261252	}
23		Rubber gasket for lamp, LU 572042	1	261253	}
		Stop/tail lamp and rear number plate lamp, Sparto 62401	1	532880	Check before ordering
		Base for lamp, Sparto 02285	2	532881	}
		Bulb holder complete, Sparto 02298	2	532879	} Sparto type.
		Bulb for lamp, LU 380	2	532878	} Alternative to Wipac
		Lens for lamp, Sparto 02280	2	264590	} or Lucas.
		Special screw, fixing lens, Sparto 02282/3/4	4	542036	} Check before ordering
		Drive screw, fixing lamp to rear body	6	77932	}
		Spring washer ⎱ Fixing stop/tail	6	3072	}
		Nut (4 BA) ⎰ lamp to rear body	6	4024	}
		Screw (4 BA x ½" long) Fixing stop/tail lamp to rear body	6	77899	Alternative to drive screw fixing
24		Stop/tail lamp and rear number plate lamp, Wipac WP842	2	531684	Vehicles for France require Wipac type
		Bulb holder and base complete, Wipac 13426	2	542045	}
		Bulb for lamp, LU 380	2	264590	} Wipac type.
		Lens for lamp, Wipac 13428	2	542042	} Check before ordering
		Special screw, fixing lens, Wipac 13430	4	542041	}
		Drive screw, fixing lamp to rear body	6	77932	}
25		Bulb holder, interior, LU 54271A	2	547346	} Lucas type.
26		Bulb for lamp, LU 380	2	264782	} Alternative to Wipac or Sparto.
27		Lens, stop/tail and reflex, LU 54576369	2	551435	} Check before ordering
28		Special screw, fixing lens, LU 54113381	4	551436	}
29		Sleeve for terminal, LU 555910	2	514145	}
		Stop/tail lamp, Sparto 62401/A	2	541520	}
		Base for lamp, Sparto 02285	2	532879	} Sparto type.
		Bulb holder complete, Sparto 02298	2	532878	} Alternative to Wipac.
		Bulb for lamp, LU 380	2	264590	} Check before ordering
		Lens for lamp, Sparto 02280/R	2	541521	}
		Special screw, fixing lens, Sparto 02282/3/4	6	541522	}
		Drive screw, fixing lamp to rear body	6	77932	}
		Stop/tail lamp, Wipac WP891	2	542036	}
		Bulb holder and base complete, Wipac 13426	2	542045	} Wipac type.
		Bulb for lamp, LU 380	2	264590	} Alternative to Sparto.
		Lens for lamp, Wipac 13427	2	542044	} Check before ordering
		Special screw, fixing lens, Wipac 13420	4	542041	}
		Drive screw, fixing lamp to rear body	6	77932	}
30		Red rear reflector, LU 575193	2	551595	}
31		Spring washer ⎱ Fixing reflectors	2	3073	} Optional equipment
32		Nut (2 BA) ⎰	2	2247	}

Asterisk indicates a new part which has not been used on any previous Rover model

HEAD, SIDE, TAIL AND FLASHER LAMPS. From vehicle suffix 'B' onwards (Except America Dollar Area)

HEAD, SIDE, TAIL AND FLASHER LAMPS. From vehicle suffix 'B' onwards (Except America Dollar Area)

Plate Ref.	1 2 3 4	DESCRIPTION	Qty	Part No.	REMARKS
		FLASHING INDICATOR COMPLETE ASSEMBLY	1	541755	Optional equipment up to vehicle suffix 'E' inclusive. Fitted as standard from vehicle suffix 'F' onwards. For Series II models, see Optional Equipment Parts Catalogue
33		Flasher unit, LU FL5 350.0	1	502096	Sparto type.
		Drive screw, fixing flasher unit	1	72628	
34		Self-cancelling switch for flasher, Magnatex	1	519866	Alternative to
35		Bulb for flasher switch, Magnatex GBP V2.2	1	530054	Wipac or Lucas. Check before ordering
36		Wheel for flasher switch	1	522882	
		Feed lead for flasher switch	1	519865	
		Front and rear flasher lamp complete, Sparto 62406	4	536094	Sparto type.
		Lens for flasher lamp, Sparto 02325	4	536152	Alternative to
		Special screw, fixing lens, Sparto 02302/14/15	8	542038	Wipac or Lucas.
		Bulb for flasher lamp, LU 382	8	264591	Check before ordering
		Bulb holder and body complete, Sparto 02364	4	542040	Wipac type.
		Front and rear flasher lamp complete, Wipac 843	4	536148	Wipac type.
		Lens for flasher lamp, Wipac 13423	4	542048	Alternative to
		Special screw, fixing lens, Wipac 13429	8	542042	Sparto or Lucas.
		Bulb for flasher lamp, LU 382	8	264591	Check before ordering
		Bulb holder and base complete, Wipac 13425	4	542046	Check before ordering
37		Front and rear flasher lamp complete, LU 52722A	4	547345	Lucas type.
38		Lens for flasher lamp, LU 54576128	4	551432	Alternative to
39		Special screw, fixing lens LU 54110101	8	600423	Wipac or Sparto.
40		Special washer fixing lens LU 161385	8	551430	Check before ordering
41		Bulb for flasher lamp, LU 382	4	264591	
42		Bulb holder, interior, LU 573832	4	514144	
		Screw (4 BA x ⅞" long) } Fixing front flasher lamp to front wing	6	77899	
		Spring washer	6	3072	
		Nut (4 BA)	6	4024	
		Drive screw fixing flasher lamp to rear body	6	77932	
		Earthing clip on rear chassis crossmember, LU 860019A	2	236366	
		Clip for flasher harness on RH wing valance	2	240406	LH Stg
		Clip, fixing flasher lead to dash	1	233770	
		Clip for lamp and flasher harness to wing valance	1	3621	Diesel

* Asterisk indicates a new part which has not been used on any previous Rover model

HEAD, SIDE, TAIL AND FLASHER LAMPS
(America Dollar Area)

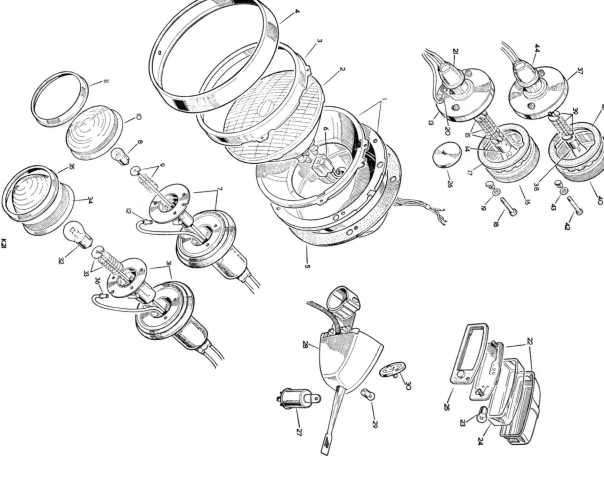

HEAD, SIDE, TAIL AND FLASHER LAMPS
(America Dollar Area)

Plate Ref. 1 2 3 4	DESCRIPTION	Qty	Part No.	REMARKS
1	Headlamp assembly, sealed beam, LU 59191	2	605073	
2	Light unit for headlamp, LU 54522231	2	541537	
3	Rim for light unit, LU 54521913	2	545150	
4	Rim for headlamps, LU 554439	2	272592	
5	Gasket for headlamp, LU 54520505	2	531586	
6	Adaptor and leads for light unit, LU 54931957	2	536116	
7	Screw (2 BA x ½" long) } Fixing headlamp	6	71164	
	Spring washer } to front	6	3073	
	Nut (2 BA) } grille panel	6	2247	
8	Side lamp complete, LU 52602	2	532808	
9	Bulb for side lamp, LU 207	2	10211	
10	Bulb holder, interior, LU 533780	2	244700	
11	Lens, LU 572715 } Fixing	2	271931	
	Rim for lens, LU 572734 } side	2	261640	
12	Sleeve for terminal, LU 555910 } lamp } For side	2	514145	
	Screw (4 BA x ½" long) } Fixing } lamp	6	77899	
	Spring washer } side lamp to	6	3072	
	Nut (4 BA) } front wing	6	4024	
13	Stop/tail lamp complete, LU 54144	2	532809	
14	Bulb for stop/tail lamp, LU 380	2	264590	
15	Bulb holder, interior, LU 573828	2	264782	
16	Lens, LU 54572774	4	551430	
17	Gasket for lens, LU 54572200 } For	4	601719	
18	Special screw, LU 54111111 } Fixing } stop/tail	4	601717	
19	Special washer, LU 161385 } lens } lamp	4	601718	
20	Sleeve for terminal, LU 555910	2	514145	
21	Grommet for cable entry LU 54572202	2	600349	
22	Number plate lamp complete, LU 467	1	502096	
23	Bulb for number plate lamp, LU 989 } For number	1	72628	
24	Lens, LU 54570354 } plate lamp	1	601721	
25	Gasket for lens, LU 54571498 } plate lamp	1	50689	
26	Red rear reflector complete, LU 57047A	2	240542	
27	Flasher unit, LU FL 535020	1	261253	
28	Drive screw, fixing flasher unit	2	519866	
29	Self-cancelling switch for flasher, Magnatex	1	530054	
30	Bulb for flasher switch, Magnatex GBP V2.2	1	522882	
31	Wheel for flasher switch	1	532806	
32	Front flasher lamp, LU 52345	2	264591	
33	Bulb for front flasher lamp, LU 382	2	514144	
34	Bulb holder interior, LU 573832	2	601720	
35	Lens for front flasher lamp, LU 576983	2	601718	
36	Chrome rim for lens, LU 572734	2	261640	
	Sleeve for terminal, LU 555910	2	514145	
	Screw (4 BA x ¼" long) } Fixing front	6	77899	
	Spring washer } flasher lamp	6	3072	
	Nut (4 BA) } to wing	6	4024	
37	Rear flasher lamp complete, LU 691 } flasher lamp	2	532807	
38	Bulb for rear flasher lamp, LU 382	2	264591	
39	Bulb holder interior, LU 573832	2	514144	
40	Lens for rear flasher lamp, LU 54572277	2	601720	
41	Gasket for lens, LU 54572200	2	601717	
42	Special screw, LU 54111111 } Fixing	4	601719	
43	Special washer, LU 161385 } lens	4	551430	
44	Grommet for cable entry LU 54572202	2	600349	

* Asterisk indicates a new part which has not been used on any previous Rover model

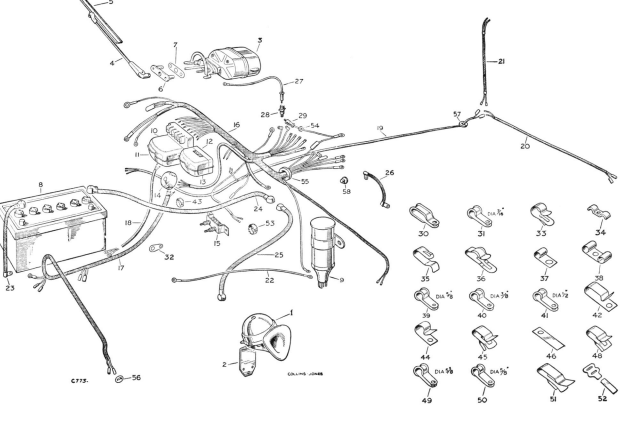

C773.   COLLINS·JONES

GENERAL ELECTRICAL EQUIPMENT, POSITIVE EARTH SYSTEM
Series II and IIA models. Up to vehicle suffix 'C' inclusive

Plate Ref.	1 2 3 4	DESCRIPTION	Qty	Part No.	REMARKS
1		Windtone horn.—Clearhooter 'Alpine' Low Note	1	600890	Clearhooter type. Alternative to Lucas type
		Windtone horn, LU 54068059, Type 9H	1	519871	Lucas type. Alternative to Clearhooter type
		Contact set for windtone horn, LU 702687	1	271500	Lucas type
		Contact set for windtone horn	1	600637	Clearhooter, non-Lucas blade type horn
		Contact set for windtone horn	1	600638	Clearhooter Lucas blade type horn
		Rubber grommet for Lucas type horn	1	234041	
2		Mounting bracket for windtone horn	1	501129	Required when horn is fixed to wing valance
		Bolt (¼" UNF x ¾" long) ⎫ Fixing horn	2	255207	Required when horn is fixed to wing valance
		Spring washer ⎬ to bracket or	2	3074	
		Nut (¼" UNF) ⎭ battery support	2	254810	
		Set bolt (¼" UNF x ⅝" long) ⎫ Fixing horn	3	255006	
		Plain washer ⎬ bracket to	3	3665	
		Spring washer ⎭ toe box	3	3074	
3		Clip, fixing horn lead to bonnet	1	240406	
		Windscreen wiper motor complete, LU 75334	1	277931	Series II models
		Windscreen wiper motor complete	1	519900	Series IIA models
4		Arm for wiper blade, stud fixing	1	26 503	Alternative to box type fixing. Check before ordering
5		Wiper blade, stud fixing, LU 737673	1	272306	Check before ordering
		Arm for wiper blade, box type fixing, LU 5474490	1	530037	Alternative to stud type fixing.
		Wiper blade, LU 741680	1	560941	Check before ordering
6		Escutcheon for windscreen wiper motor	2	330529	
7		Rubber seal for escutcheon	4	302987	
		Screw (2 BA x 1⅝" long) ⎫ Fixing	2	78240	
		Spring washer ⎬ escutcheon	2	3073	
		Nut (2 BA) ⎭ redundant	2	2247	
8		Battery, dry, 12 volt, LU 54027393	1	567882	Petrol models
		Battery, dry, 6 volt, LU 54027345	2	547072	Diesel models
		Ignition coil, LU 45208	1	26 503	
		Rubber boot for ignition coil, LU 54953531	1	272306	
		Acorn nut, LU 408120	1	567977 *	
		Rubber boot for coil acorn nut ⎫ For ignition coil	1	574217 *	For early type coil with terminal nut connection
		Split washer, LU 185015 ⎭	2	214279	Petrol models
		Drive screw ⎫ Fixing coil	2	72628	
		Shakeproof washer ⎬ to dash	2	78114	
10		Voltage regulator box, LU 37182	1	235553	Use with dynamo type C39
		Voltage regulator box, LU 37290	1	514734	Use with dynamo type C40
		Current/voltage regulator box, LU 37475	1	530079	2 litre Diesel models. Use with dynamo type PV5
		Current/voltage regulator box, LU 37471	1	512251	2 litre Diesel models. Use with dynamo type PV6
11		Cover for regulator box, LU 37472	1	530051	2¼ litre Diesel
		Cover for regulator box, LU 391600	1	262859	Petrol models
		Cover for regulator box, LU 335610	1	504444	Diesel models, 2 BA screw type fixing
		Cover for regulator box, LU 54380178	1	513937	Diesel models. Drive screw type fixing

* Asterisk indicates a new part which has not been used on any previous Rover model

GENERAL ELECTRICAL EQUIPMENT, POSITIVE EARTH SYSTEM
Series II and IIA models. Up to vehicle suffix 'C' inclusive

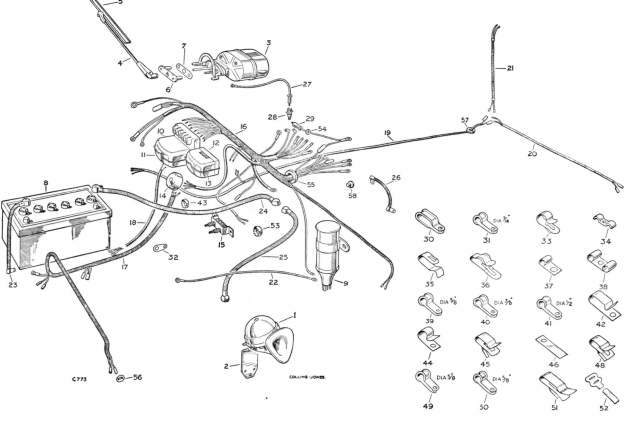

C 773 COLLINS-JONES.

GENERAL ELECTRICAL EQUIPMENT, POSITIVE EARTH SYSTEM
Series II and IIA models. Up to vehicle suffix 'C' inclusive

Plate Ref. 1 2 3 4	DESCRIPTION	Qty	Part No.	REMARKS
	'Lucar' connector earthing blade at voltage regulator box	2	505205	2¼ litre Diesel
12	Screw (2 BA x 1¼″ long)	2	77992	Series II Petrol models
	Spring washer	2	3073	} Series II Petrol models
	Nut (2 BA)	2	2247	
	Drive screw	2	78443	Series II models
	Bolt (¼″ UNF x ¾″ long) } Fixing voltage regulator box	3	255207	Series IIA models
	Shakeproof washer	3	70884	Series IIA Petrol models
	Nut (¼″ UNF)	3	254810	Diesel models
	Riv-nut	3	514929	2 litre } Diesel models / 2¼ litre
13	Fuse box, LU 37260	1	505158	1 off on Series II models
	Fuse box, LU 54038033	1	12738	} Series II models
	Cover for fuse box, LU 291078	1	3279	
	Cover for fuse box, LU 54380114	1	3280	} Series IIA models
	Fuse, 35 amp	2	71164	Series II models
	Special screw, LU 153564 } Fixing fuse box	2	03971	Series IIA models
	Special nut, LU 167530	2		
	Drive screw, fixing fuse box	1		
	Drive screw, locating fuse box	1	529963	Diesel models
14	Junction box, LU 78266A	1	501570	1958 88 Petrol models
	Screw (2 BA x ½″ long) } Fixing junction box	2	501033	Petrol models
	Spring washer	2	3073	
	Nut (2 BA)	2	2247	
15	Resistor for heater plugs, KLG BRQ 3 } resistor	2	545030	Diesel models
	Shakeproof washer } Fixing to dash	2	237119	
	Bolt (2 BA x ¾″ long)	2	311373	
	Nut (2 BA)	2	3823	
16	Dash harness, LU 54940321	1	501034	1958 88 Petrol models
	Dash harness, LU 54940319	1	501032	Petrol models
	Dash harness, LU 54949597	1	519806	1959-1961 Series II 2¼ litre Petrol models
	Dash harness	1	519782	1958-1960 Series II 2¼ litre Petrol models
	Dash harness	1	501033	1958 Diesel models
	Dash harness	1	529965	1959-1961 2 litre Diesel models
17	Headlamp and sidelamp harness, LU 54940428	1	501694	Except LHStg Diesel } Series II models
	Headlamp and sidelamp harness, LU 865028	1	278383	LHStg Diesel
18	Dynamo harness, LU 862980	1	267392	2 litre Petrol models
	Dynamo harness, LU 54940421	1	279074	2 litre Petrol models
	Engine harness	1	519314	1961 Series II 2¼ litre Petrol models
	Dynamo harness	1	526360	Series IIA 2¼ litre Petrol models
	Dynamo harness	1	529965	2¼ litre Diesel models. Early type with separate dash harness. Alternative to main harness. Check before ordering
	Main harness, LU 54934535	1	545218	Late 2¼ litre Diesel models with combined dash harness and dynamo harness. Check before ordering

* Asterisk indicates a new part which has not been used on any previous Rover model

GENERAL ELECTRICAL EQUIPMENT, POSITIVE EARTH SYSTEM
Series II and IIA models. Up to vehicle suffix 'C' inclusive

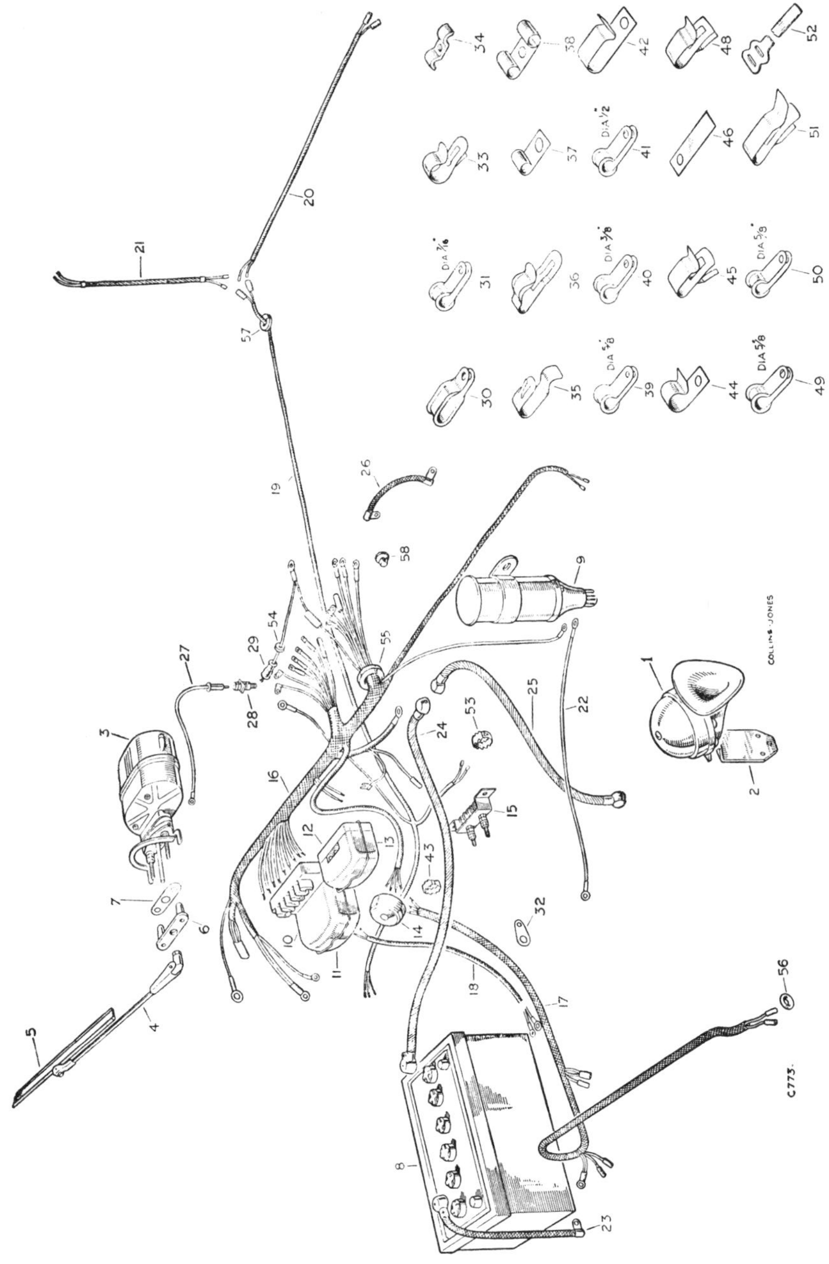

C773
COLLINS-JONES

GENERAL ELECTRICAL EQUIPMENT, POSITIVE EARTH SYSTEM
Series II and IIA models. Up to vehicle suffix 'C' inclusive

Plate Ref.	1 2 3 4 DESCRIPTION	Qty	Part No.	REMARKS
19	Frame harness, LU 54940426	1	501692	88 } Series II models
	Frame harness, LU 835671	1	501693	109 } Series II models
	Frame harness	1	504684	109 Station Wagon } Series II models
	Frame harness, LU 54949642	1	519784	88 } Series IIA models
	Frame harness, LU 54949643	1	526440	109 } Series IIA models
	Frame harness, LU 54949641	1	528900	109 Station Wagon $2\frac{1}{4}$ litre Petrol } Series IIA models
	Frame harness	1	529964	109 Station Wagon $2\frac{1}{4}$ litre Diesel } Series IIA models
20	Rear crossmember harness, LU 820784	1	236137	Series II models
	Rear crossmember harness, LU 54949639	1	519783	Series IIA models
	Extension harness for horn, LU 54946709	1	514895	1960–1961 Series II. When horn is fixed to battery support
	Extension lead for stoplight switch, LU 861080	1	239282	LHStg
	Lead for engine thermostat, LU 54940422	1	279517	$2\frac{1}{4}$ litre Petrol models } 1958–1960 Series II
	Lead for oil pressure switch	1	279518	Petrol models } 1958–1960 Series II
21	Lead, rear number plate lamp, LU 54940414	1	500559	Series II and IIA. Up to vehicle suffix 'A' All America Dollar Area
	Feed lead, dynamo harness to carburetter	1	528931	Series IIA $2\frac{1}{4}$ litre Petrol models. For carburetter **with** starter heater element
	Lead, starter motor to earth	1	531672	$2\frac{1}{4}$ litre Diesel models
	Lead, starter motor to earth, LU 999709	1	219649	Series IIA $2\frac{1}{4}$ litre Petrol models
	Bolt ($\frac{5}{16}$" UNF x $\frac{3}{4}$" long) } Fixing starter earth lead to chassis frame	1	255226	
	Fan disc washer } Fixing starter earth lead to chassis frame	2	510912	
	Nut ($\frac{5}{16}$" UNF) } Fixing starter earth lead to chassis frame	1	254831	
22	Cable, coil to distributor, LU 999152	1	502617	Series II Petrol models
	Cable, coil to distributor, LU 54949599	1	528963	Series IIA $2\frac{1}{4}$ litre Petrol models
23	Cable, battery to earth, LU 812014	1	239558	Petrol models
24	Cable, battery to switch, LU 861081A	1	239559	2 litre Petrol models
	Cable, battery to switch, LU 54940425	1	501690	$2\frac{1}{4}$ litre Petrol models
	Cable, battery to earth, LU 864542	1	275371	2 litre Diesel models
	Cable, battery to earth, LU 54930139	1	530084	$2\frac{1}{4}$ litre Diesel models
	Cable, battery to starter, LU 54941019A	1	503982	} 2 litre Diesel models
	Cable, connecting batteries, LU 54940415	1	501355	} 2 litre Diesel models
	Cable, battery to starter, LU 54930138	1	530085	$2\frac{1}{4}$ litre Diesel models
	Cable, connecting batteries, LU 54931099	1	531695	$2\frac{1}{4}$ litre Diesel models
	Drive screw, fixing battery cables, LU 186111	2	50552	4 off on Diesel models
25	Cable, switch to starter, LU 861082A	1	501691	Petrol models
	Rubber boot for starter cable, LU 861139	2	500980	
	Cable clip, fixing starter cable to flywheel housing	1	12743	$2\frac{1}{4}$ litre Petrol models
26	Cable, gearbox to earth, LU 999709	1	219649	} Series II models
	Bolt ($\frac{1}{4}$" UNF x $\frac{1}{2}$" long) } Fixing gearbox earth cable to frame	1	255204	} Series II models
	Fan disc washer } Fixing gearbox earth cable to frame	2	510170	} Series II models
	Nut ($\frac{1}{4}$" UNF) } Fixing gearbox earth cable to frame	1	254810	} Series II models
	Bolt ($\frac{1}{4}$" UNF x $\frac{7}{8}$" long) } Fixing battery earth cable to frame	1	255208	} Series IIA $2\frac{1}{4}$ litre Diesel models
	Fan disc washer } Fixing battery earth cable to frame	2	510170	} Series IIA $2\frac{1}{4}$ litre Diesel models
	Spring washer } Fixing battery earth cable to frame	1	3074	} Series IIA $2\frac{1}{4}$ litre Diesel models
	Nut ($\frac{1}{4}$" UNF) } Fixing battery earth cable to frame	1	254810	} Series IIA $2\frac{1}{4}$ litre Diesel models
	Clip, battery cable to frame, LU 187051	10	56667	} Series IIA $2\frac{1}{4}$ litre Diesel models
	Drive screw, fixing clip to frame	10	72626	} Series IIA $2\frac{1}{4}$ litre Diesel models
	Grommet, for battery cable at seat base, LU 859856	2	233566	} Series IIA $2\frac{1}{4}$ litre Diesel models

* Asterisk indicates a new part which has not been used on any previous Rover model

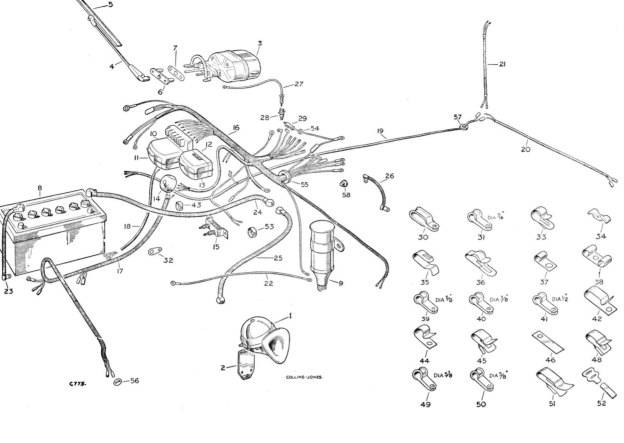

C773.

COLLINS-JONES.

GENERAL ELECTRICAL EQUIPMENT, POSITIVE EARTH SYSTEM
Series II and IIA models. Up to vehicle suffix 'C' inclusive

Plate Ref. 1 2 3 4	DESCRIPTION	Qty	Part No.	REMARKS
	Lead, tank unit to dash harness, LU 863690	1	271225	2 litre Diesel models
	Lead, tank unit to warning lamp, LU 54942032	1	504685	2 litre Diesel, 109 Station Wagon
	Lead, tank unit to dash harness, LU 54992032	1	529966	2¼ litre Diesel models
	Lead for fuel tank unit, LU 54949905	1	528919	Series IIA Petrol models
27	Wiper feed lead, LU 861084	1	239687	Series II models
	Earth lead for wiper motor, LU 861083	1	239688	Series II models
	Earth lead for wiper motor, LU 54949633	1	519899	Series IIA models
28	Plug socket for leads on steering bracket, LU 305514	1	233531	Series IIA models
29	Cap for wiper feed socket, LU 572363	1	240408	Series II models
30	Clip for dynamo harness	1	4082	
31	Clip for battery to switch cable	1	50640	88 Petrol models
32	Mounting plate for cable clips	1	267412	
33	Drive screw, fixing clips to mounting plate	1	72626	
	Cable clip for frame harness, at dash RH	1	237279	
38	Earthing clip on rear crossmember, LU 860019	2	8885	1 off on Diesel models
	Screw (2 BA x ½" long) — Fixing horn lead to dash	2	77869	Not required when horn is fixed to battery support
	Spring washer	2	3073	
	Nut (2 BA)	2	3823	
	Clip for horn lead	1	75454	
	Drive screw, fixing clip to dash	2	233770	
	Fan disc washer	2	3619	
36	Earthing clip for headlamp	2	236366	
	Earthing clip at LH headlamp	1	236366	
35	Cable clip for frame harness, at junction box	2	234603	
	Shakeproof washer	2	71082	
	Clip for rear crossmember harness, LU 860019	2	513282	
	Clip for engine harness	2	2247	
	Cable clip, fixing harness to dash	2	56666	
39	Clip for rear crossmember — clip to rear crossmember	2	219676	1961 Series II 2¼ litre Petrol models
40	Cable clip, fixing harness to dash, RH	1	56667	
41	Cable clip, fixing harness to dash and LH toe box	5	50639	
42	Cable clip on wing valance and toe box	4	56666	3 off on 88
	Cable clip, fixing harness to RH toe box	2	278490	3 off on Series II models
	Cable clip, frame harness to RH toe box	1	56667	Series IIA
	Screw (2BA x ⅜" long) — Fixing clips to dash	10	77846	9 off on 88
	Plain washer	10	2226	9 off on 88
	Spring washer	10	3073	9 off on 88
	Nut (2 BA)	10	3823	9 off on 88
	Screw (¼" UNF x ¾" long) — Fixing clip to LH toe box	2	3621	
	Plain washer	2	255202	
	Spring washer	2	3840	Series IIA
	Nut (¼" UNF)	2	3074	Series II
	Cable cleat, harness to starter solenoid	1	254810	
	Cable clip, fixing harness adjacent to junction box	1	240430	Diesel models
43	Cable clip, fixing harness adjacent to junction box	1	240429	Series II
44	Cable clip, fixing dipper and horn leads to RH accelerator shaft bracket	1	278489	
	Cable clip, for headlamp and horn cable harness	5	233770	
	Cable clip, fixing headlamp harness at front wing	1	50642	Petrol models

* Asterisk indicates a new part which has not been used on any previous Rover model

432

GENERAL ELECTRICAL EQUIPMENT, POSITIVE EARTH SYSTEM
Series II and IIA models. Up to vehicle suffix 'C' inclusive

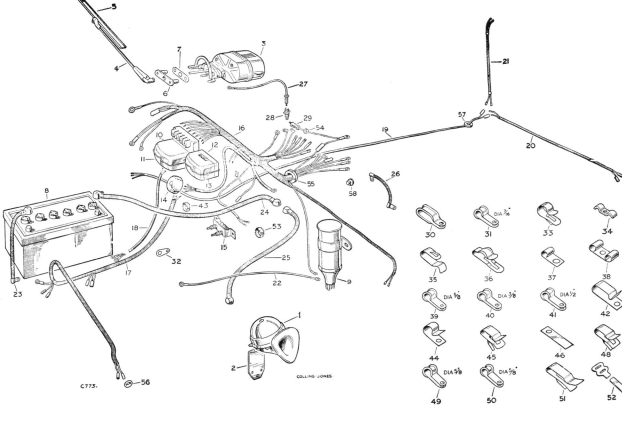

C773.   COLLINS·JONES

Plate Ref. 1 2 3 4	DESCRIPTION	Qty	Part No.	REMARKS
	Cable clip on wing valance	2	233370	
	Cable clip, lamp harness to valance	2	50642	
45	Clip for rear crossmember harness	3	236365	America Dollar Area
	Earthing clip on rear crossmember	2	236366	
	Cable cleat, frame harness to petrol pipe, LU 941834	2	240429	
48	Cable clip, wiper feed lead	1	240406	
	Cable clip, stop lamp lead	1	236365	5 off or LHStg
49	Cable clip on fan cowl top baffle	4	3622	Diesel models
50	Clip, fixing heater switch lead to engine speed control quadrant	3	50637	Diesel models
51	Cable clip at dash, RH	1	233247	2 litre Petrol models
	Clip for tail lamp harness	4	50639	
	Screw (2 BA x 5/8" long) } Fixing clip	4	78177	
	Spring washer } to wheelarch box	4	3073	
	Nut (2 BA)	4	2247	
	Clip, stop/tail lamp leads to rear crossmember	2	56666	
	Screw (2 BA x 1/2" long) } Fixing clip to	2	77869	
	Plain washer } rear crossmember	2	3816	
	Spring washer	2	3073	
	Nut (2 BA)	2	3823	
52	Cable clip, battery to starter cable	1	4200	Series II models
53	Cable cleat for engine harness	2	240430	
	Bowden clip, starter cable to solenoid	2	56667	
	Cable cleat for lead, side lamp harness and dynamo lead to starter cable	3		
	Cable clip harness to dash, upper	1	240429	
	Cleat, coil lead to dash harness	1	240417	2¼ litre Petrol models
	Cleat, dash harness upper, centre and RH	2	240429	2¼ litre Petrol and Diesel models
	Cable clip, starter switch leads	1	240430	
	Cable clip, engine harness leads	1	240431	
	Cleat, harness adjacent to fuse box	1	278490	
	Cable clip, harness to dash, RH lower	1	237121	
	Bolt (2 BA x 5/8" long) } Fixing	1	237127	2¼ litre Petrol models
	Shakeproof washer } clip and earth	2	71082	
	Nut (2 BA) } terminals	2	3823	
	Cable clip, starter switch leads	1	50639	
	Cable clip, engine harness leads	1	278490	
	Screw (2 BA x 1/2" long) } Fixing	2	77869	
	Plain washer } clips to	2	3816	
	Spring washer } dash centre,	2	3073	
	Nut (2 BA) } lower	2	3823	
	Cleat, carburetter heater lead to harness	1	240429	2¼ litre Petrol models with starter heater element is fitted
	Cable clip, starter cable to dash, LH lower	1	3619	
	Bolt (¼" UNF x ½" long) } Fixing clip	1	255204	2¼ litre Petrol models
	Spring washer } to lower heater	1	3074	
	Nut (¼" UNF) } mounting hole	1	254810	
	Clip, fixing dip switch leads to wing stay	1	219676	LHStg
	Cable cleat, fixing starter heater lead to fuel pipe	2	240429	2¼ litre Petrol models when carburetter with starter heater element is fitted

* Asterisk indicates a new part which has not been used on any previous Rover model

GENERAL ELECTRICAL EQUIPMENT, POSITIVE EARTH SYSTEM
Series II and IIA models. Up to vehicle suffix 'C' inclusive

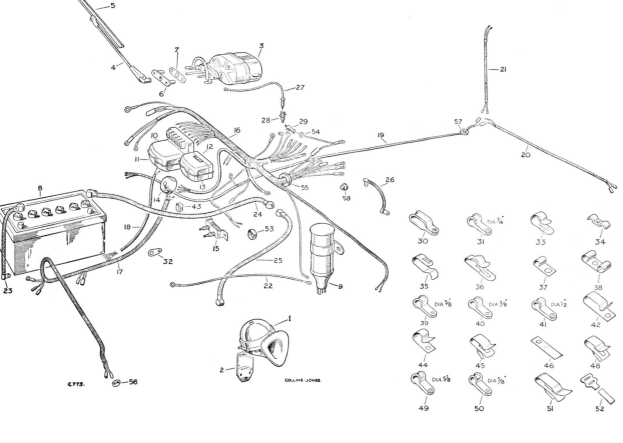

C773.   COLLINS-JONES.

GENERAL ELECTRICAL EQUIPMENT, POSITIVE EARTH SYSTEM
Series II and IIA models. Up to vehicle suffix 'C' inclusive

Plate Ref. 1 2 3 4	DESCRIPTION	Qty	Part No.	REMARKS
	Cable clip, frame harness to RH toe box	2	278490	3 off on Series II
	Cable clip, frame harness to RH toe box	1	513222	Series IIA
	Screw (2 BA x ½" long)	3	77869	Fixing clips to toe box
	Plain washer	3	3816	Fixing clips to toe box
	Spring washer	3	3073	
	Nut (2 BA)	3	3823	
	Cleat, dynamo harness to starter cable and thermo-static switch lead to petrol pipe	6	240429	
	Cleat at dash for engine harness	1	240431	1961 onwards
	Clip, engine leads and battery cable, RH rear side, lower	2	4020	
	Clip for coil and distributor leads at rocker cover rear stud	2	239600	
	Cleat, fixing coil and distributor leads together	2	240429	
	Grommet, cold start cable and coil lead at rear rocker cover	2	06860	
	Grommet for distributor leads	1	06860	
54	Rubber grommet in dash for distributor leads, LU 859341	1	233244	1958–1960
	Rubber grommet in dash for wiper motor lead, LU 859059	2	233243	1961 Series II models
	Rubber grommet, headlamp cable at fan cowl, LU 858063	2	219680	Series II models
55	Rubber grommet in dash for wiper motor lead	2	236389	Series IIA models
	Rubber grommet, main harness, in dash, LU 858063	2	219680	Series IIA models
56	Rubber grommet for side lamp lead, LU 859059	2	219680	
	Rubber grommet for side lamp lead	2	233243	Series II models
	Rubber grommet, tail lamp lead, LU 859059	2	236389	Series IIA models
	Rubber grommet for flasher and horn leads	4	233243	4 off on 109
57	Rubber grommet for frame harness, LU 859060	1	269257	88 Petrol models
	Rubber grommet for frame harness, LU 859060	2	276054	109 Petrol models
	Grommet for frame harness	2	276054	Diesel models
58	Grommet for frame harness	2	276053	
	Rubber plug for redundant hole	1	73198	
59	Rubber cover for battery terminal, LU 862090	2	263158	
	Rubber cover for mixture warning light switch terminals	2	246241	2¼ litre Petrol models
	'Lucar' blade for 2 BA fixing, 14 amp, LU 5419038	As reqd	534337	
	'Lucar' blade for ¼" fixing, 35 amp, LU 5419044	As reqd	534338	
	'Lucar' connector for single lead, 14 amp, LU 5494078	As reqd	534339	
	'Lucar' connector for single lead, 35 amp, LU 5494079	As reqd	534341	Series IIA Land-Rover
	'Lucar' connector for double leads, 14 amp, LU 5494747	As reqd	534340	

* Asterisk indicates a new part which has not been used on any previous Rover model

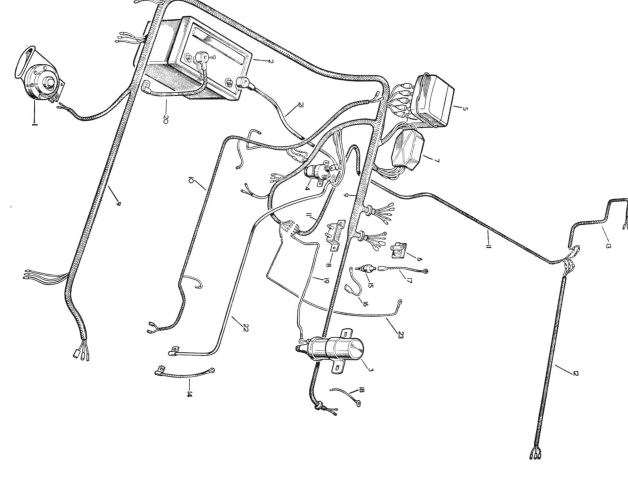

GENERAL ELECTRICAL EQUIPMENT, NEGATIVE EARTH SYSTEM
Series IIA 4-cylinder models. From vehicle suffix 'D' onwards

Plate Ref.	1 2 3 4	DESCRIPTION	Qty	Part No.	REMARKS
1		Windtone horn, Clearhooter 'Alpine' Low Note	1	600890	Clearhooter type ⎫ Alternatives
		Windtone horn, Lucas 54068059, type 9H	1	519871	Lucas type ⎬
		Contact set for windtone horn	1	271500	Lucas type ⎭
		Contact set for windtone horn, LU 702687	1	600638	Clearhooter type
		Bolt (¼" UNF x ¾" long) ⎫ Fixing horn	2	255207	
		Spring washer ⎬ support	2	3074	
		Nut (¼" UNF) ⎭	2	254810	
2		Battery, dry, 12 volt, LU 54027393 ⎫ Fixing coil	1	567882	Petrol models
		Battery, dry, 6 volt, LU 54027345 ⎬ to battery	2	547072	Diesel models
3		Ignition coil, LU 45208	1	567977 *	
		Rubber boot for ignition coil, LU 54953531 ⎫	1	574217 *	
		Acorn nut, LU 408120 ⎬ For ignition coil	1	240102	For early type coil
		Rubber boot for acorn nut ⎭	1	506679	with terminal nut
		Split washer, LU 185015	1	214279	connection
		Drive screw ⎫ Fixing coil	2	72628	
		Shakeproof washer ⎬ to dash	2	78114	
4		Starter solenoid switch, LU 4 ST 76772	1	567969	
		Spring washer, LU 185062 ⎫ For solenoid	2	575014	
		Nut, LU 170578 ⎬ terminal post	2	575015	
		Bolt (10 UNF x ⅝" long) ⎫ Fixing solenoid	2	257019	
		Spring washer ⎬ to dash	2	3073	
		Nut (10 UNF) ⎭	2	257023	
5		Voltage regulator box, LU RB 106/2 37290	1	514734	
		Cover for regulator box, LU 391600	1	262859	
		Drive screw, fixing regulator box to dash	2	78443	
		Current/voltage regulator box, LU RB 340 37387	1	559189	
		Cover for regulator box, LU 54381342	1	600442	
		Screw (10 UNF x 1⅜" long) ⎫ Fixing	3	78477	
		Plain washer ⎬ regulator box	3	3902	
		Riv-nut (10 UNF) ⎭ to dash	3	532848	
6		Voltage regulator for instruments, 10 volt	1	559052	
		Spring washer ⎫ Fixing	2	3101	
		Drive screw ⎬ regulator	2	78001	
7		Fuse box, LU 54038033	1	530047	
		Cover for fuse box, LU 291078	1	261502	
		Fuse, 35 amp	2	12738	
		Drive screw, fixing fuse box	2	78152	
		Drive screw locating fuse box	1	77704	
8		Resistor for heater plugs, KLG	1	545030	
		Bolt (10 UNF x ¾" long) ⎫	2	257015	
		Shakeproof washer ⎬ Fixing resistor to dash	2	311373	
		Nut (10 UNF) ⎭	2	257023	Diesel models
9		Dash harness	1	555776	Petrol models. Up to vehicle suffix 'E' inclusive
10		Dynamo harness	1	555798	Petrol models. Up to vehicle suffix 'E' inclusive
		Dash harness	1	560899 *	Petrol models. From vehicle suffix 'F' onwards
		Engine harness	1	560969 *	Petrol models. From vehicle suffix 'F' onwards
		Dash and dynamo harness	1	559058	Diesel models. Up to vehicle suffix 'E' inclusive
		Dash harness	1	560904 *	Diesel models. From vehicle suffix 'F' onwards

* Asterisk indicates a new part which has not been used on any previous Rover model

GENERAL ELECTRICAL EQUIPMENT, NEGATIVE EARTH SYSTEM
Series IIA 4-cylinder models. From vehicle suffix 'D' onwards

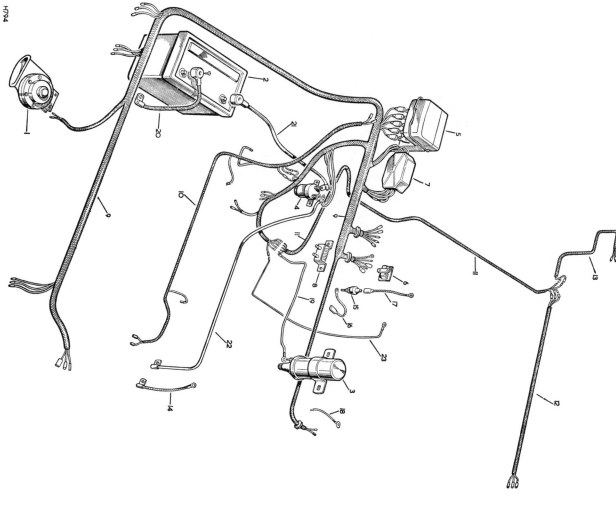

GENERAL ELECTRICAL EQUIPMENT, NEGATIVE EARTH SYSTEM
Series IIA 4-cylinder models. From vehicle suffix 'D' onwards

Plate Ref.	1 2 3 4	DESCRIPTION	Qty	Part No.	REMARKS
11		Frame harness	1	519784	88 models ⎫ Up to vehicle suffix 'E'
		Frame harness	1	526440	109 models ⎬ inclusive
		Frame harness	1	528900	109 Station Wagon, Up to vehicle ⎭ Petrol models
		Frame harness	1	529964	109 Station Wagon, suffix 'E' ⎫ Up to vehicle suffix 'E' ⎬ Diesel models ⎭ inclusive
12		Rear crossmember harness	1	519783	88 models ⎫ Up to vehicle suffix 'E' inclusive
		Frame and rear crossmember harness	1	560903*	88 ⎬
		Frame and rear crossmember harness	1	560898*	109 ⎭
		Frame and rear crossmember harness	1	560901	109 Station Wagon ⎫ Petrol ⎬ models
		Frame and rear crossmember harness	1	560977*	88 ⎫ From vehicle suffix 'F' onwards
		Frame and rear crossmember harness	1	560975*	109 ⎬
		Frame and rear crossmember harness	1	560976*	109 Station Wagon models ⎭
		Frame and rear crossmember harness	1	500559	All America Dollar Area. Optional for ⎫ Diesel ⎬ other areas ⎭ models
13		Lead for rear number plate lamp, LU 54940414	1		
14		Lead, starter motor to earth, LU 999709	1	219649	Petrol models
		Lead, starter motor to earth	1	531672	Up to vehicle suffix 'D' ⎫ inclusive ⎬ Diesel ⎭ models
		Lead, starter motor to earth	1	560979*	From vehicle suffix 'E' ⎫ onwards ⎬
		Fan disc washer for earth lead at starter motor	1	512305	
		Bolt (5/16" UNF x 3/4" long) ⎱ Fixing earth	1	255226	
		Fan disc washer ⎰ lead to	2	510912	
		Nut (5/16" UNF) ⎱ chassis frame	1	254831	
15		Inhibitor socket for wiper motor leads	1	557762	Up to vehicle ⎫
16		Lead, wiper switch to inhibitor socket	1	558800	suffix 'E' ⎬
17		Lead, inhibitor socket to windscreen	1	558801	inclusive ⎭ Except Station Wagon
		Lead, wiper switch to earth	1	558800	Station Wagon. Up to vehicle suffix 'E' inclusive
18		Lead, wiper motor to earth	1	555801	
		Fan disc washer ⎱ Fixing earth	1	519897	Up to vehicle suffix 'E' inclusive
		Drive screw ⎰ lead to dash	1	78001	
19		Cable, switch to wiper motor	1	560965*	From vehicle suffix 'F' onwards
20		Cable, coil to distributor, low tension	1	528963	⎱ Petrol models
21		Cable, battery to earth	1	551318	⎰
		Cable, battery to solenoid switch	1	551319	
		Cable, battery to earth	1	551346	
		Cable, battery to starter	1	551347	Up to vehicle suffix 'D' ⎫ inclusive ⎬ Diesel ⎭ models
		Cable, battery to starter	1	560992*	From vehicle suffix 'E' ⎫ onwards ⎬
		Cable connecting batteries	1	551638	
		Drive screw, fixing battery cables, LU 186111	2	50552	
		Rubber cover for battery terminal, LU 862090	1	263158	4 off on Diesel models

* Asterisk indicates a new part which has not been used on any previous Rover model

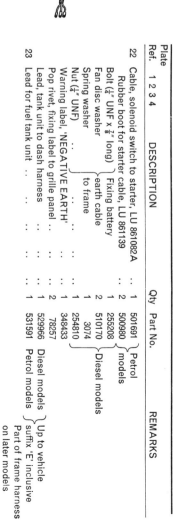

GENERAL ELECTRICAL EQUIPMENT, NEGATIVE EARTH SYSTEM
Series IIA 4-cylinder models. From vehicle suffix 'D' onwards

Plate Ref.	1	2	3	4	DESCRIPTION	Qty	Part No.	REMARKS
22					Cable, solenoid switch to starter, LU 8611082A	1	501691	Petrol
					Rubber boot for starter cable, LU 861139	2	500980	models
					Bolt ($\frac{1}{4}$" UNF x $\frac{7}{8}$" long) ⎫ Fixing battery	1	255208	
					Fan disc washer ⎬ earth cable	2	510170	Diesel models
					Spring washer ⎭ to frame	1	3074	
					Nut ($\frac{1}{4}$" UNF)	1	254810	
					Warning label, 'NEGATIVE EARTH'	1	349433	
					Pop rivet, fixing label to grille panel	2	78257	
23					Lead, tank unit to dash harness	1	529966	Diesel models ⎫ Up to vehicle
					Lead for fuel tank unit	1	531591	Petrol models ⎬ suffix 'E' inclusive
								Part of frame harness
								on later models

GENERAL ELECTRICAL EQUIPMENT, NEGATIVE EARTH SYSTEM
2.6 litre Petrol, 6-cylinder models

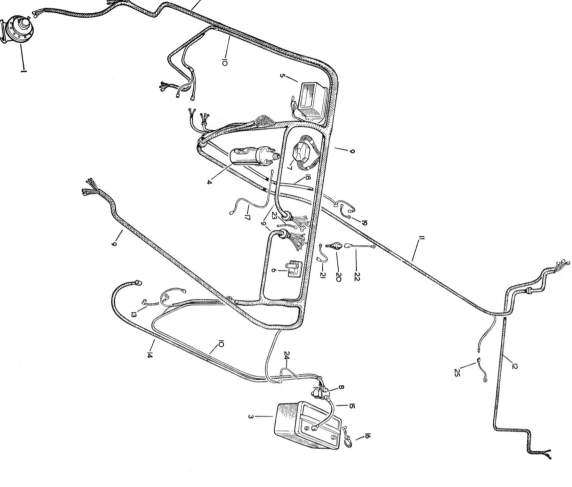

GENERAL ELECTRICAL EQUIPMENT, NEGATIVE EARTH SYSTEM
2.6 litre Petrol, 6-cylinder models

Plate Ref.	1 2 3 4	DESCRIPTION	Qty	Part No.	REMARKS
1		Windtone horn, complete with bracket	1	560749	Lucas type
		Contact set for windtone horn, LU 702687	1	271500	Clearhooter type } Alternatives. Check before ordering
		Contact set for windtone horn	1	600638	
		Rubber grommet for horn	1	234041	
		Bolt (¼" UNF x ⅝" long) } Fixing horn bracket	2	255206	
		Spring washer	2	3074	
		Nut (¼" UNF)	2	254810	
3		Battery, dry, 12 volt, LU 54027393 — to toe box or bridge plate	1	567882	
4		Ignition coil, LU 45208	1	567977	
		Rubber boot, LU 414965 } For ignition coil	1	574217*	
		Rubber boot for ignition coil, LU 54953531	1	510237	
		Terminal nut, LU 54411502	1	214279	For early type coil with terminal nut connection
		Acorn nut	1	506679	
		Bolt (¼" UNF x ¾" long) } Fixing coil to engine	1	240102	
		Plain washer	1	3840	
		Spring washer	1	3074	
		Nut (¼" UNF)	1	254810	
5		Current voltage control box, LU 37517	1	559265	
		Cover for control box, LU 54381342	1	600442	
		Rubber seal for cover	1	507810	
		Screw (10 UNF x 1⅜" long) } Fixing control box	3	78477	
		Plain washer	3	3902	
		Riv-nut (10 UNF)	3	532848	
6		Voltage regulator for instruments	1	559052	
		Drive screw } Fixing regulator to dash	2	78001	
		Spring washer	2	3101	
7		Fuse box, LU 54038033	1	530047	
		Cover for fuse box, LU 54380114	1	505158	
		Fuse, 35 amp	2	12738	
		Drive screw, fixing fuse box	1	78417	
		Drive screw locating fuse box	2	77704	
8		Starter solenoid switch, LU 4ST 76772	1	567969	
		Spring washer, LU 185062 } For solenoid terminal post	2	575014*	
		Nut, LU 170578	2	257019	
		Bolt (10 UNF x ⅝" long) } Fixing solenoid switch	2		
		Spring washer	2	3073	
		Fan disc washer } to seat base	2	513282	
		Nut (10 UNF)	2	257023	
9		Dash harness	1	560754	Up to vehicle suffix 'E' inclusive
		Dash harness	1	560910*	From vehicle suffix 'F' onwards
10		Dynamo harness	1	560532	
11		Frame harness	1	560557	Up to vehicle suffix 'E' } inclusive
12		Rear crossmember harness	1	560555	
		Frame and rear crossmember harness	1	560901*	From vehicle suffix 'F' onwards

GENERAL ELECTRICAL EQUIPMENT, NEGATIVE EARTH SYSTEM
2.6 litre Petrol, 6-cylinder models

Plate Ref.	1 2 3 4	DESCRIPTION		Qty	Part No.	REMARKS
13		Lead, starter motor to earth, LU 999709	...	1	219649	
		Bolt (5/16" UNF x 3/4" long) ⎫ Fixing starter		1	255226	
		Fan disc washer ⎬ earth lead to		2	510912	
		Nut (5/16" UNF) ⎭ chassis frame		1	254831	
14		Cable, solenoid to starter motor	...	1	559140	
15		Cable, battery to starter solenoid	...	1	560567	
16		Cable, battery to earth, negative	...	1	560566	
		Drive screw, fixing battery cables, LU 186111		2	50552	
19		Lead, fuel pump to earth	...	1	559141	From vehicle suffix 'F' onwards
		Cable, fuel pump feed		1	559150*	From vehicle suffix 'F' onwards
18		Cable, fuel pump feed		1	560577	Up to vehicle suffix 'E' inclusive
17		Cable, coil to distributor, low tension		1	528963	From vehicle suffix 'F' onwards
		Cable, coil to distributor, low tension		1	559163	Up to vehicle suffix 'E' inclusive
		Bolt (2 BA x 1/2" long) ⎫ Fixing earth lead		1	234603	
		Fan disc washer ⎬ to fuel pump bracket		1	513282	
		Spring washer ⎭		1	3073	
		Nut (2 BA)		1	3823	
20		Inhibitor socket for wiper motor leads		1	555762	
21		Lead, wiper switch to inhibitor socket		1	555800	Up to vehicle suffix 'E' inclusive
22		Lead, inhibitor socket to windscreen	...	1	555801	
23		Lead, wiper switch to earth	...	1	555800	Station Wagon. Up to vehicle suffix 'E' ⎫ Except inclusive ⎬ Station
24		Earth lead for wiper motor	...	1	555801	⎭ Wagon
		Drive screw ⎫ Fixing earth		1	78001	Up to vehicle suffix 'E' inclusive
		Fan disc washer ⎬ lead to dash		1	519887	
25		Cable, switch to wiper motor	...	1	560965*	From vehicle suffix 'F' onwards
		Earth lead for fuel tank gauge, LU 861083		1	239688	
		Drive screw ⎫ Fixing earth		1	72626	
		Fan disc washer ⎬ lead to body		1	519857	
		Warning label, 'NEGATIVE EARTH'		1	348433	
		Pop rivet, fixing label to grille panel ...		2	78257	

K855

CABLE CLIPS, GROMMETS AND CONNECTORS

CABLE CLIPS, GROMMETS AND CONNECTORS

Plate Ref. 1 2 3 4	DESCRIPTION	Qty	Part No.	REMARKS
1	Cable clip, A—13/64", B—9/32"	As reqd	8885	
	Cable clip, A—13/64", B—7/16"	As reqd	50640	
	Cable clip, A—13/64", B—3/8"	As reqd	50639	
	Cable clip, A—13/64", B—5/8"	As reqd	56666	
	Cable clip, A—13/64", B—5/8"	As reqd	56667	
	Cable clip, A—13/64", B—3/8"	As reqd	513220	
	Cable clip, A—13/64", B—1/16"	As reqd	513221	
	Cable clip, A—13/64", B—3/16"	As reqd	50690	
	Cable clip, A—3/8", B—1 1/32"	As reqd	50637	
	Cable clip, A—7/16", B—1/2"	As reqd	4170	
	Cable clip, A—1/4", B—3/8"	As reqd	514721	
	Cable clip, A—1/16", B—9/16"	As reqd	50637	
	Cable clip, A—1/4", B—5/8"	As reqd	50642	
	Cable clip, A—1/2", B—3/8"	As reqd	3622	
	Cable clip, A—1/8", B—5/8"	As reqd	3621	
	Cable clip, A—1/16", B—5/8"	As reqd	219676	
	Cable clip, A—1/8", B—3/32"	As reqd	513222	
	Cable clip, A—1/8", B—3/32"	As reqd	4020	
	Cable clip, A—1 3/8", B—1 1/16"	As reqd	50641	
	Cable clip, A—1/4", B—1"	As reqd	50638	
2	Cable clip, A—3/8", B—9/32"	As reqd	12743	
3	Cable clip, A—7/16", B—1"	As reqd	4082	
4	Cable clip, A—1 3/16", B—1"	As reqd	278490	
5	Cable clip, A—3/8", B—3/16"	As reqd	239600	
6	Cable clip, A—5/8", B—7/16"	As reqd	278489	
7	Cable clip, A—7/16", B—3/32"	As reqd	560983	
8	Cable clip, A—1/2", B—3/32"	As reqd	233770	
9	Cable clip, A—3/8", B—3/32"	As reqd	240406	
10	Cable clip, A—1/4", B—7/16", C—9/64"	As reqd	240417	
	Cable clip, A—1/4", B—5/16", C—9/64"	As reqd	237279	
11	Cable clip, A—3/8", B—5/16", C—3/32"	As reqd	236365	
	Cable clip, A—3/8", B—1/8", C—6/32"	As reqd	233247	
12	Earthing clip, A—3/8", B—1/8", C—3/32"	As reqd	3619	
13	Bowden clip, A—10", B—5/8"	As reqd	236366	
14	Cable cleat, A—4", B—3/8", C—9/64"	As reqd	4200	
	Cable cleat, A—5", B—3/4"	As reqd	240429	
15	Cable cleat, A—6 1/4", B—3/4"	As reqd	240430	
16	Rubber grommet, A—1 7/16", B—3/32", C—1/16"	As reqd	240431	
17	Rubber grommet, A—3/8", B—1/8", C—7/16"	As reqd	06860	
18	Rubber grommet, A—1 5/16", B—3/4", C—7/16"	As reqd	575042	
	Rubber grommet, black, A—1", B—1/8", C—1/16"	As reqd	219680	
	Rubber grommet, red, A—1", B—1/8", C—1/16"	As reqd	232244	
	Rubber grommet, red, A—1 1/16", B—3/32", C—3/16"	As reqd	236389	
	Rubber grommet, black, A—1", B—3/32", C—4/8"	As reqd	233243	
	Rubber grommet, A—1 1/16", B—1 1/4", C—7/16"	As reqd	233566	
	Rubber grommet, black, A—1 5/8", B—1/4", C—3/16"	As reqd	269257	
	Rubber grommet, black, red, A—1 5/16", B—1/4", C—3/8"	As reqd	276053	
	Rubber grommet, black, A—1/8", B—1/4", C—3/8"	As reqd	276054	
	Rubber grommet, black, A—17/32", B—5/16", C—3/32"	As reqd	233566	
19	Rubber plug, A—5/8", B—5/16", C—13/32"	As reqd	73198	

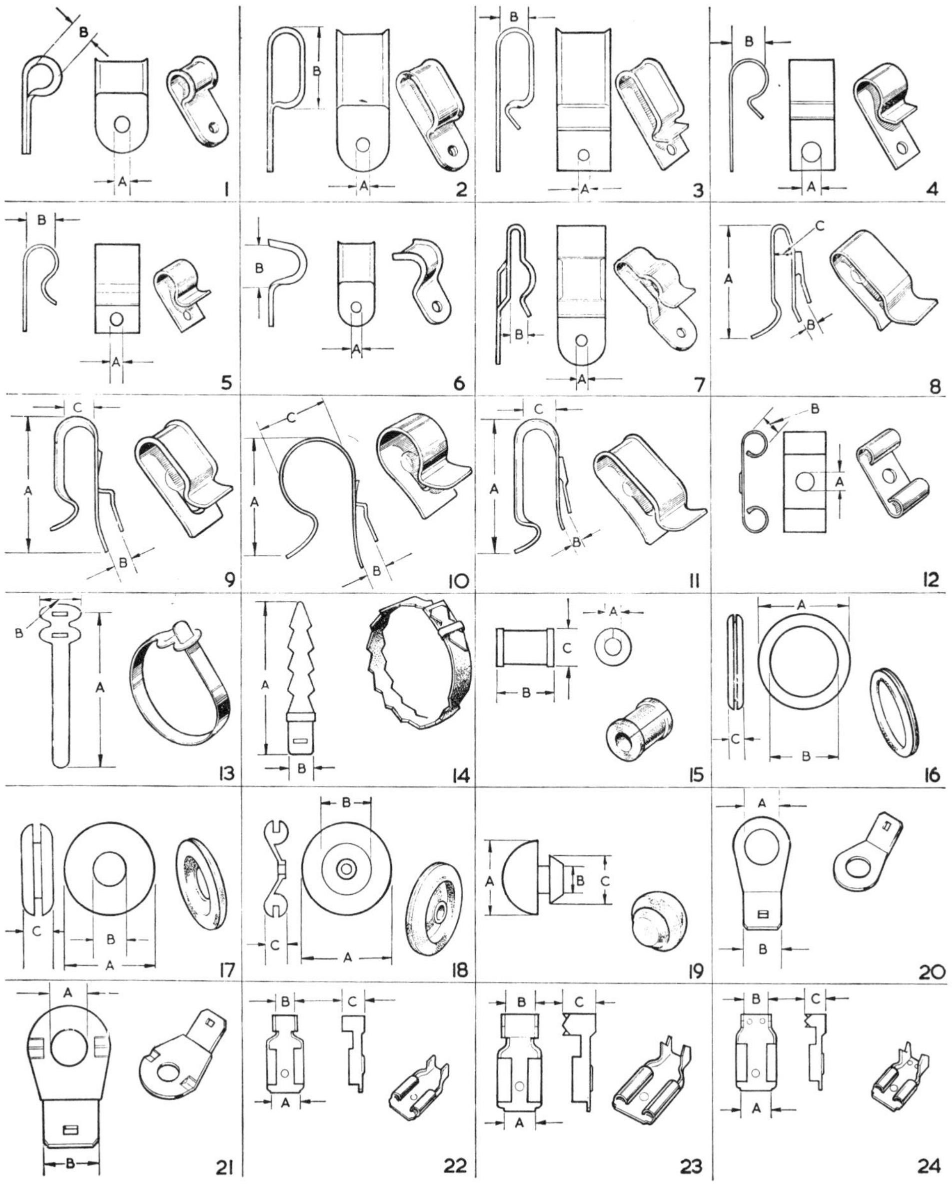

CABLE CLIPS, GROMMETS AND CONNECTORS

21 17 13 9 5 1

22 18 14 10 6 2

23 19 15 11 7 3

24 20 16 12 8 4

K855

CABLE CLIPS, GROMMETS AND CONNECTORS

Plate Ref.	1 2 3 4	DESCRIPTION	Qty	Part No.	REMARKS
20		"Lucar" blade for 2 BA fixing, 14 amp, LU 54190038, $A-\frac{7}{32}$", $B-\frac{19}{64}$" ...	As reqd	534337	
21		'Lucar' blade for $\frac{1}{4}$" fixing, 35 amp, LU 54190044, $A-\frac{17}{64}$", $B-\frac{25}{64}$" ...	As reqd	534338	
22		'Lucar' connector for single lead, 14 amp, LU 5494207B, $A-\frac{15}{64}$", $B-\frac{11}{64}$", $C-\frac{13}{64}$" ...	As reqd	534339	
23		'Lucar' connector for single lead, 35 amp, LU 5494207P, $A-\frac{25}{64}$", $B-\frac{7}{32}$", $C-\frac{23}{64}$" ...	As reqd	534341	
24		'Lucar' connector for double leads, 14 amp, LU 5494274?, $A-\frac{17}{64}$", $B-\frac{17}{64}$", $C-\frac{7}{32}$" ...	As reqd	534340	

* Asterisk indicates a new part which has not been used on any previous Rover model

K856

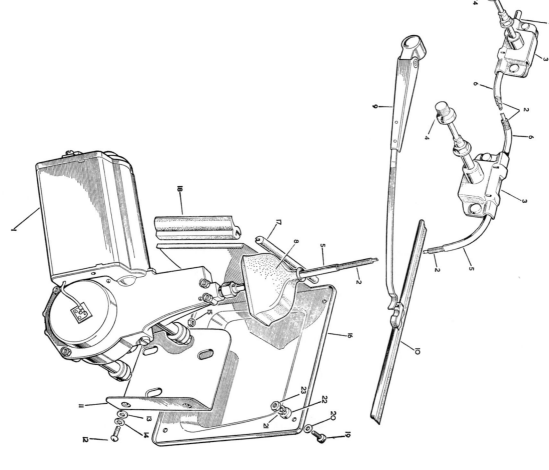

WINDSCREEN WIPER, NEGATIVE EARTH SYSTEM
Series IIA models. From vehicle suffix 'D' to suffix 'E' inclusive

Plate Ref.	1 2 3 4	DESCRIPTION	Qty	Part No.	REMARKS
1		Windscreen wiper motor complete, LU DR 3A	1	551226	
2		Flexible drive cable, LU 743238	1	601701	
3		Wheelbox for wiper, LU 72812	2	551227	
4		Spindle and gear for wheelbox and wiper arm, LU 54700995	2	605904*	
5		Outer casing, motor to wheelbox	1	555754	
6		Outer casing, wheelbox to wheelbox	1	555702	
7		Outer casing, wheelbox end	1	575047	
8		Rubber cover for wiper motor in dash	1	338846	
9		Arm for wiper blade, RH, LU 54711579	1	560944	
		Arm for wiper blade, LH, LU 54711580	1	560945	
10		Wiper blade, LU 741680	2	560941	
11		Mounting bracket for wiper motor	1	338794	
12		Screw (¼" UNF x ½" long) ⎱ Fixing	3	78755	
13		Plain washer ⎰ mounting	3	3912	
14		Spring washer ⎱ bracket to	3	3074	
15		Nut (¼" UNF) ⎰ LH glove box	3	254810	
16		Cover plate for wiper motor	1	345079	
17		Rubber finisher ⎱ For	1	348541	
18		Rubber seal ⎰ cover plate	1	348537	
19		Screw (2 BA x ⅝" long) ⎱ Fixing cover	4	78177	
20		Plain washer ⎱ plate to LH	4	4030	
21		Spring washer ⎰ glove box	4	3073	
22		Distance washer	4	4172	
23		Nut (2 BA)	4	3823	

* Asterisk indicates a new part which has not been used on any previous Rover model

K857

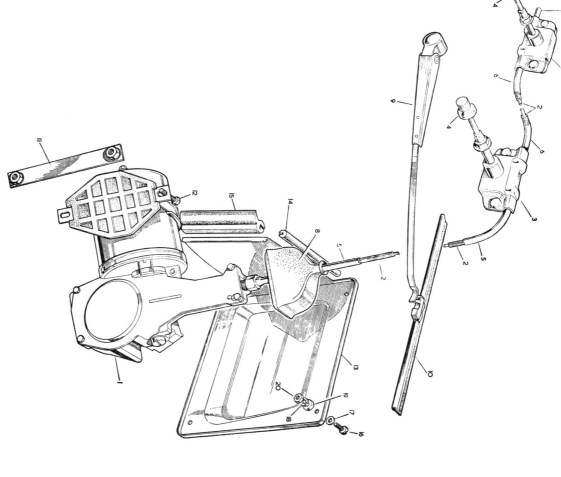

WINDSCREEN WIPER, NEGATIVE EARTH SYSTEM
Series IIA models. From vehicle suffix 'F' onwards

Plate Ref.	1 2 3 4	DESCRIPTION			Qty	Part No.	REMARKS
1		Windscreen wiper motor complete, LU 75664		...	1	606013*	
2		Flexible drive cable, LU 54702585		...	1	606014*	
3		Wheelbox for wiper, LU 72820		...	2	560887*	
4		Spindle and gear for wheel box, LU 72820		...	2	605904*	
5		Outer casing, motor to wheelbox, LU 54700995		...	1	560966*	
6		Outer casing, wheelbox to wheelbox		...	1	560886*	
7		Outer casing, wheelbox end		...	1	575047*	
8		Rubber cover for wiper motor in dash		...	1	338846	
9		Arm for wiper blade, RH, LU 54711579		...	1	560944	
10		Arm for wiper blade, LH, LU 54711580		...	1	560945	
11		Wiper blade, LU 741680		...	2	560941	
12		Nut plate for wiper motor		...	1	560967*	
13		Set screw (¼" UNF x ⅝" long) fixing wiper motor to dash and nut plate		...	2	78709	
14		Cover plate for wiper motor		...	1	345079	
15		Rubber finisher } For cover plate		...	1	348541	
16		Rubber seal } cover plate		...	1	348537	
17		Screw (2 BA x ⅝" long) } Fixing cover plate to LH glove box		...	4	78177	
18		Plain washer } plate to LH		...	4	4030	
19		Spring washer } glove box		...	4	3073	
20		Distance washer		...	4	4172	
		Nut (2 BA)		...	4	3823	

* Asterisk indicates a new part which has not been used on any previous Rover model

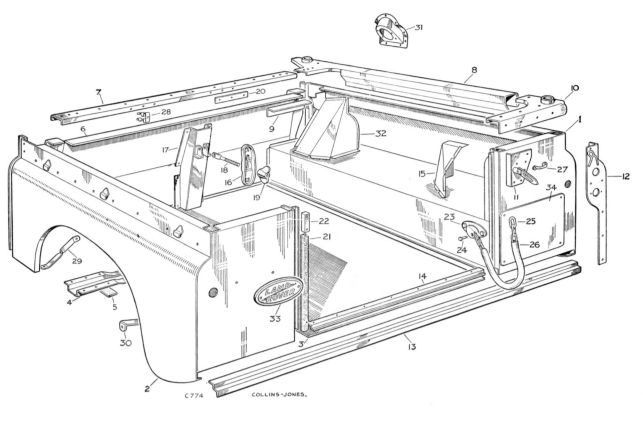

C774   COLLINS-JONES.

BODY, CAPPINGS AND SOCKETS, 88

Plate Ref.	1 2 3 4	DESCRIPTION	Qty	Part No.	REMARKS
		BODY FLOOR AND SIDES ASSEMBLY			
1		Side and wheelarch complete, RH	1	330951	
		Body side panel, RH	1	330214	
		Rear end panel, RH	1	330295	
2		Side and wheelarch complete, LH	1	330248	
		Body side panel, LH	1	330215	
		Rear end panel, LH	1	330296	
3		Rear floor complete	1	330249	
4		Cross-member and pads for rear floor	1	330267	
5		Mounting pad for rear floor cross-member	3	330265	
		Rivet fixing pad to cross-member	6	332582	
		Pop rivet } Fixing cross-member	12	73979	
6		Front panel for rear body	1	300919	
		Rivet } to rear floor	18	301879	
		Rivet fixing front panel to wheelarch	12	330301	
7		Capping for body front panel	4	330789	
8		Capping for body top, side, RH	1	330245	
		Capping for body top, side, LH	1	330832	
9		Corner strengthening angle	1	330833	
		Rivet } Fixing	2	330279	
		Pop rivet } angles	12	302186	
		Rivet } Fixing front	12	300782	
		Pop rivet } capping	8	300783	
		Rivet } Fixing side	8	78248	
		Pop rivet, short } and front	16	300789	
10		Rivet	16	300919	
		Pop rivet, long } corner cappings	2	254831	
		Hood socket complete rear corner, RH	1	330315	
		Hood socket complete rear corner, LH	1	330316	
		Rivet, fixing rear sockets at top	8	300783	
		Pop rivet, short } Fixing rear socket	4	300919	
		Pop rivet, long } at rear	4	300781	
11		Corner bracket and tailboard cotter, RH	1	330233	Except Station Wagon and Hard Top
		Corner bracket and tailboard cotter, LH	1	330234	with rear door
		Corner bracket, plain type, RH	1	330237	} Station Wagon
		Corner bracket, plain type, LH	1	330238	} Also Hard Top when rear door is fitted
		Rivet } Fixing	4	300783	
		Rivet } corner brackets	2	300784	
12		Rear protection angle, RH	1	330224	
		Rear protection angle, LH	1	330225	
		Rivet, long	20	300783	
		Rivet, short } Fixing rear protection angle	1	302186	
13		Rear mounting angle	2	78248	
14		Protecting strip at rear of floor	1	330271	
		Pop rivet	2	338743	
		Drive screw fixing strip	7	77707	

* Asterisk indicates a new part which has not been used on any previous Rover model

BODY, CAPPINGS AND SOCKETS, 88

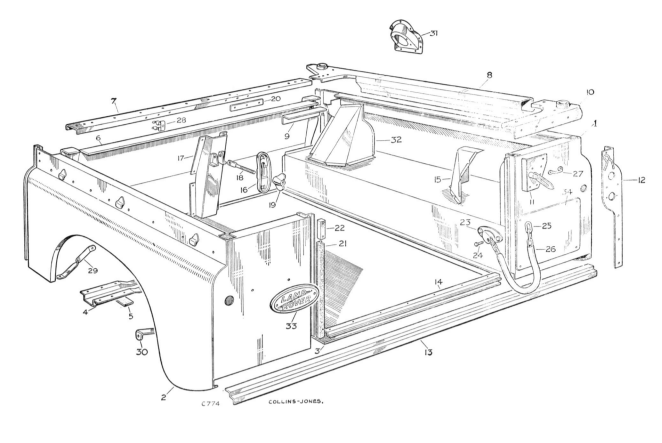

C774   COLLINS-JONES.

BODY, CAPPINGS AND SOCKETS, 88

Plate Ref.	1 2 3 4	DESCRIPTION	Qty	Part No.	REMARKS
15		Cover panel RH for rear lamps	1	330204	
		Cover panel LH for rear lamps	1	330205	
		Screw (2 BA x ½" long) ⎫ Fixing cover panels	6	77869	
		Plain washer ⎬ to rear	6	3816	
		Spring washer ⎪ panel and	6	3073	
		Nut (2 BA) ⎭ wheelarch box	6	2247	
16		Clamp for spare wheel	1	303847	
17		Reinforcement bracket for clamp	1	330208	
		Bolt (¼" UNF x ⅝" long) ⎫ Fixing	8	255206	
		Plain washer ⎬ reinforcement	8	3840	
		Spring washer ⎪ bracket	8	3074	
		Nut (¼" UNF) ⎭ to body	8	254810	
18		Tie bar for spare wheel clamp	1	302933	
		Self-locking nut (¼" UNF) ⎫ Fixing	1	252220	
		Bolt (¼" UNF x 1" long) ⎬ tie bar to	1	255209	
		Plain washer, large ⎪ pivot bracket	1	2284	
		Plain washer, small ⎭	1	3831	
19		Spring washer ⎫ Fixing	1	3842	
		Plain washer ⎬ spare wheel	1	3982	
		Wing nut ⎭ clamp	1	3076	
20		Rubbing strip for spare wheel	2	330212	
		Pop rivet fixing strip to top capping ...	6	300781	Alternatives.
21		Sealing rubber for tailboard, 17" long ...	2	333972	Alternatives.
		Sealing rubber for tailboard, 18⅛" long ...	2	337290	Check before ordering.
22		Rivet fixing sealing rubbers to rear body	10	302818	
		Rubber buffer for tailboard, 3¾" long ...	2	330419	Alternatives
		Rubber buffer for tailboard, 1¼" long ...	2	332146	Check before ordering
		Screw (4 BA x ½" long) ⎫ Fixing buffer	4	77899	
		Plain washer ⎬ to rear body	4	3867	
		Nut (4 BA) ⎭	4	4023	
23		Bracket for tailboard chain	2	302825	
		Bolt (¼" UNF x ⅝" long) ⎫ Fixing bracket	4	255206	
		Spring washer ⎬ to body	4	3074	
		Plain washer ⎪	4	3840	
		Nut (¼" UNF) ⎭	4	254810	
24		Pin ⎫ Fixing tailboard	2	302828	
		Split pin ⎬ chain to bracket	2	3958	
25		Chain for tailboard	2	330399	
26		Sleeve for chain	4	330422	
27		Staple for hood strap	2	300924	
		Screw (2 BA x ¾" long) ⎫ Fixing	4	77758	
		Spring washer ⎬ staples	4	3073	
		Nut (2 BA) ⎭	4	2247	
28		Clip for starting handle and jack handle ...	6	508035	
		Pop rivet fixing clips	6	300919	Alternatives.
		Drive screw fixing clips	6	77932	Check before ordering
		Clip for tyre pump	2	72085	Series 11A 2¼ litre Diesel models
		Pop rivet fixing clips	4	78248	Also Series IIA Petrol models with
		Rubber ring securing tyre pump ...	1	268887	Lucas 2 AC 12 volt AC/DC generator
		Rear wing stay securing tyre pump ...	1	330303	

* Asterisk indicates a new part which has not been used on any previous Rover model

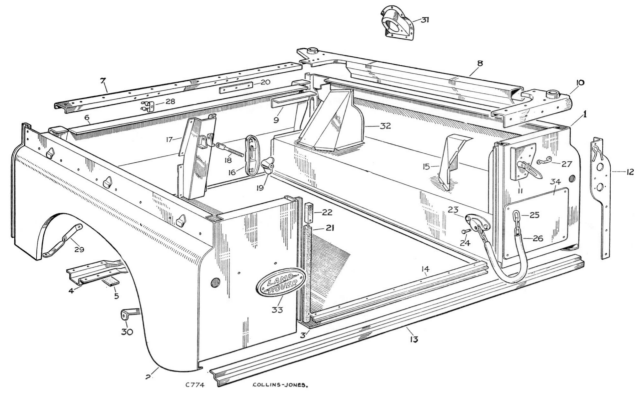

C774   COLLINS-JONES.

BODY, CAPPINGS AND SOCKETS, 88

Plate Ref.	1 2 3 4	DESCRIPTION	Qty	Part No.	REMARKS
29		Rear wing stay LH front	1	330304	
30		Rear wing stay, rear	2	332521	
		Bolt (¼" UNF x ⅝" long)	6	255206	
		Spring washer	6	3074	
		Nut (¼" UNF)	6	254810	
		Bolt (¼" UNF x ⅝" long)	4	255206	Fixing stays to wings, chassis
		Screw (¼" UNF x ⅝" long)	2	78208	and floor
31		Cowl for fuel filler	1	330366	
		Rivet fixing cowl to body side	9	302222	
32		Fuel instruction plate on fuel filler cowl	1	502951	Diesel models
		Cover plate for fuel filler	1	330367	
		Bolt (¼" UNF x ½" long)	4	255204	Alternatives to 4 lines below.
		Plain washer	4	3840	Check before ordering
		Spring washer	4	3074	
		Nut (¼" UNF)	4	254810	
		Screw (2 BA x ½" long)	4	77869	Fixing cover plate to body side
		Plain washer	4	4030	Alternatives to 4 lines above.
		Spring washer	4	3073	Check before ordering
		Nut (2 BA)	4	3823	
		Drive screw	2	78140	
33		'Land-Rover' nameplate, 7 3/16" overall	1	332670	
34		Registration plate	1	239772	
		Pop rivet fixing plates	6	300919	
		Bolt (5/16" UNF x ⅞" long)	12	255227	Fixing plates
		Bolt (5/16" UNF x ¾" long)	2	255226	body to chassis
		Fan disc washer	18	510912	
		Bolt plate	1	336782	
		Nut (5/16" UNF)	22	254831	
		Plastic plug for protection angle	2	330025	For blanking apertures when flasher lamps are not fitted

* Asterisk indicates a new part which has not been used on any previous Rover model

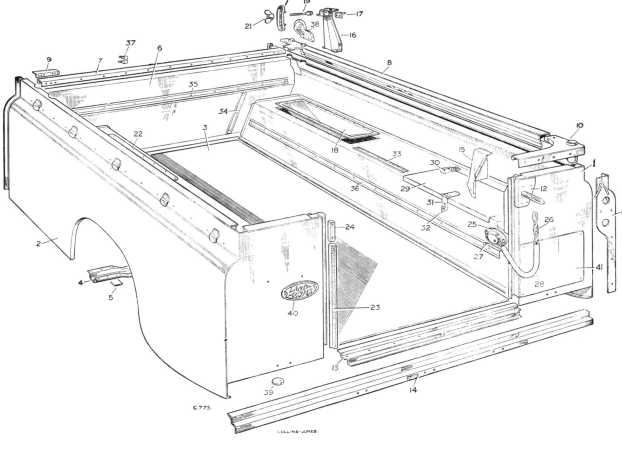

C775.

COLLINS-JONES.

BODY, CAPPINGS AND SOCKETS, 109

Plate Ref.	1	2	3	4	DESCRIPTION	Qty	Part No.	REMARKS
1					BODY COMPLETE ASSEMBLY, REAR ...	1	330958	4 cylinder models
					BODY COMPLETE ASSEMBLY, REAR ...	1	320658	6 cylinder models
1					Side and wheelarch complete, RH ...	1	330663	4 cylinder models
					Side and wheelarch complete, RH	1	348871	6 cylinder models
					Body side panel, RH ...	1	330581	4 cylinder models
					Body side panel, RH ...	1	348872	6 cylinder models
					Rear end panel, RH ...	1	330248	
2					Side and wheelarch complete, LH ...	1	330664	4 cylinder models
					Body side panel, LH ...	1	330582	4 cylinder models
					Body side panel, LH ...	1	348874	6 cylinder models
					Rear end panel, LH ...	1	330249	
3					Rear floor, complete ...	1	330617	
4					Cross-member and pads for rear floor	1	330265	
5					Mounting pad for rear floor cross-member	6	332582	
					Rivet fixing pad to cross-member	12	73979	
6					Pop rivet ⎱ Fixing cross-member	24	300919	
					Rivet ⎰ to rear floor	36	301879	
					Cover plate, rear floor	3	348882	
					Drive screw, fixing plate to rear floor	12	78776	
					Rear mounting angle ...	3	332451	6 cylinder models
					Nut plate for body mounting	12	330243	
6					Front panel for rear body ...	1	330789	
					Rivet fixing front panel to wheelarch	12	300245	
7					Capping for body, front panel ...	1	330840	
8					Capping for body top, side, RH ...	1	330841	
					Capping for body top, side, LH	1	330442	
9					Corner strengthening angle ...	2	300784	
					Rivet ⎱ Fixing	12	300784	
					Pop rivet ⎰ angles	12	78248	
					Pop rivet ⎱ Fixing	8	300783	
					Rivet ⎰ front capping	6	78248	
					Pop rivet ⎱ Fixing side	28	300789	
					Rivet ⎰ and front	28	78248	
					Pop rivet, short ⎱ corner cappings	6	300919	
					Pop rivet, long ⎰	2	78248	
					Bolt (5/16" UNF x 1" long) ⎱ Fixing front	2	255228	
					Plain washer ⎰ corner capping	2	3899	
					Spring washer	2	3075	
					Nut (5/16" UNF)	2	254831	
10					Hood socket complete, rear corner, RH ...	1	330315	
					Hood socket complete, rear corner, LH ...	1	330316	
					Rivet, short ⎱ Fixing	4	300783	
					Rivet, medium ⎰ rear sockets	4	300784	
					Rivet, long	2	302186	
					Pop rivet, short ⎱ at top	4	78248	
					Pop rivet, long ⎰ Fixing rear sockets	4	300919	
					Pop rivet, short at rear	4	78248	
11					Rear protection angle, RH ...	1	330224	
					Rear protection angle, LH ...	1	330225	
					Rivet, short ⎱ Fixing rear	20	300783	
					Rivet, long ⎰ protection	2	302186	
					Pop rivet ⎰ angle	2	78248	

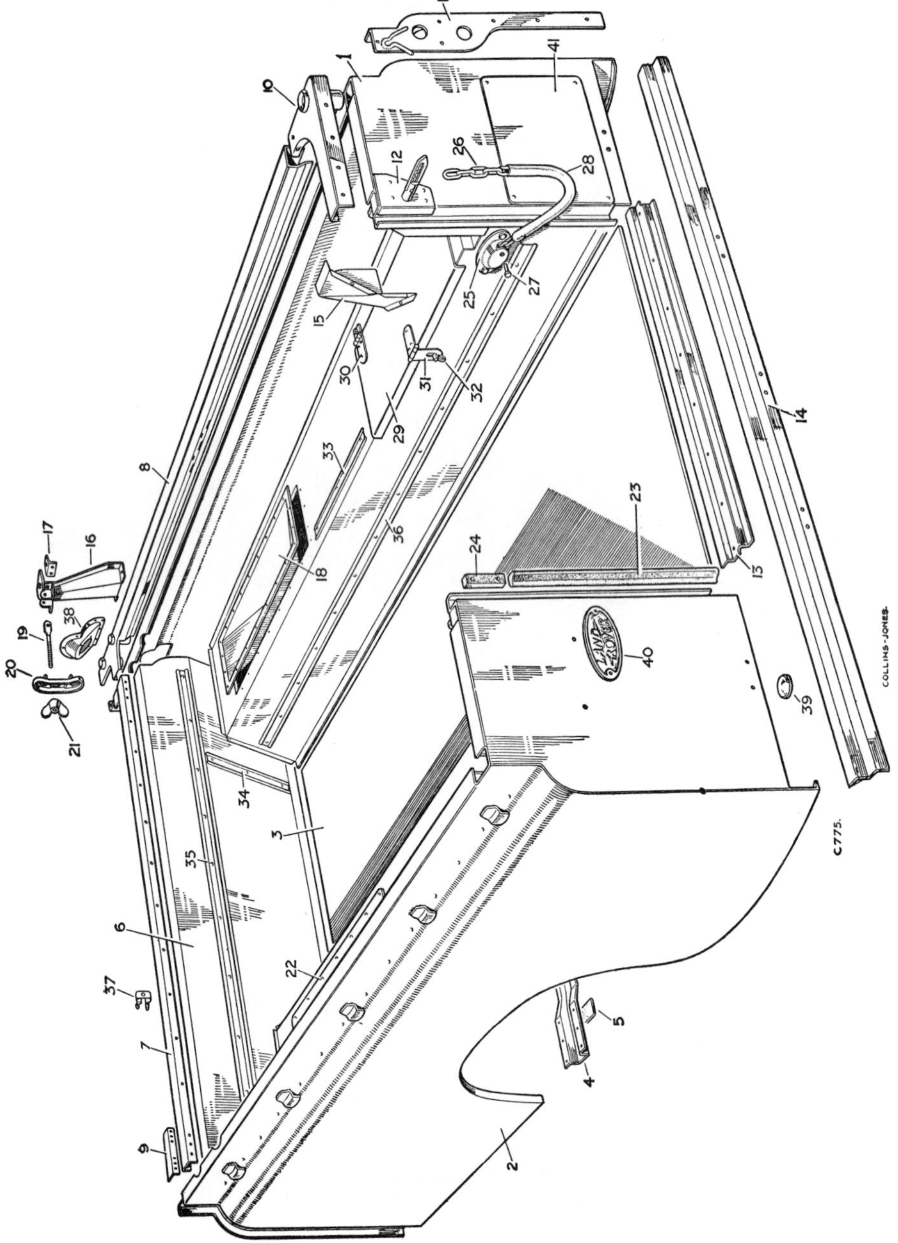
C775.
COLLINS-JONES

Plate Ref.	1 2 3 4	DESCRIPTION	Qty	Part No.	REMARKS
12		Corner bracket and tailboard cotter, RH	1	330233	Except Hard Top with rear door
		Corner bracket and tailboard cotter, LH	1	330234	
		Corner bracket, plain type, RH	1	330237	Hard Top when rear door is fitted
		Corner bracket, plain type, LH	1	330238	
		Rivet } Fixing corner brackets ..	4	300783	
		Rivet	2	300784	
13		Protecting strip at rear of floor	1	330145	
		Drive screw fixing strip	7	77704	
14		Rear mounting angle	1	330468	
15		Cover panel RH for rear lamps	1	330204	
		Cover panel LH for rear lamps	1	330205	
		Screw (2 BA x ½″ long) } Fixing cover panels to rear panel and wheelarch box	6	77869	
		Plain washer	6	3816	
		Spring washer	6	3073	
		Nut (2 BA)	6	2247	
16		Strengthening member for spare wheel mounting	1	330718	
		Bolt (¼″ UNF x ⅝″ long) } Fixing member to side body capping	3	255206	
		Plain washer	3	3900	
		Spring washer	3	3074	
17		Nut plate	1	330824	
18		Spare wheel housing	1	330602	
		Screw (2 BA x ¾″ long) } Fixing spare wheel housing and strengthening member to body side and wheelarch box	22	77887	
		Plain washer	22	3816	
		Spring washer	22	3073	
		Nut (2 BA)	22	2247	
19		Tie bar for spare wheel clamp	1	302933	
		Bolt (¼″ UNF x 1½″ long) } Fixing tie bar to strengthening member	1	256203	
		Plain washer	4	3968	
		Spring washer	2	2284	
		Self-locking nut (¼″ UNF)	1	252220	
20		Clamp for spare wheel	1	303847	
		Plain washer } Fixing spare wheel clamp	1	3982	
21		Wing nut	1	302934	
		Spring washer	1	3076	
		Spare wheel rubbing strip assembly	2	332672	
		Rubbing strip	2	332674	
		Rivet fixing rubbing strip to bracket	4	78248	
		Rivet fixing bracket to capping	4	78248	
22		Cover plate for redundant spare wheel hole ..	1	330616	
		Screw (2 BA x ¾″ long) } Fixing cover plate to wheelarch box	20	77887	
		Plain washer	20	3816	
		Spring washer	20	3073	
		Nut (2 BA)	20	2247	
23		Sealing rubber for tailboard, 17″ long	2	333972	Alternatives. Check before ordering
		Sealing rubber for tailboard, 18⅛″ long	2	337290	
		Rivet fixing sealing rubbers to rear body ..	10	302818	
24		Rubber buffer for tailboard, 3¾″ long	2	330419	Alternatives. Check before ordering
		Rubber buffer for tailboard, 1¼″ long	4	332146	
		Screw (4 BA x ½″ long) } Fixing buffer to rear body	4	77899	
		Plain washer	4	3867	
		Nut (4 BA)	4	4023	

* Asterisk indicates a new part which has not been used on any previous Rover model

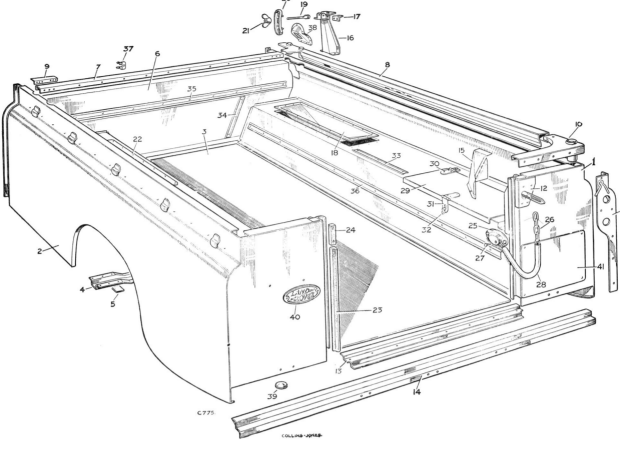

C.775.
COLLINS-JONES.

Plate Ref.	1 2 3 4	DESCRIPTION	Qty	Part No.	REMARKS
25		Tailboard chain bracket	2	302825	
26		Bolt (¼" UNF x ⅝" long) ⎫ Fixing	4	255206	
		Spring washer ⎬ chain bracket	4	3074	
		Plain washer ⎪ to	4	3840	
		Nut (¼" UNF) ⎭ body	4	254810	
26		Chain for tailboard	2	330399	
27		Clevis pin ⎫ Fixing chain	2	302828	
		Split pin ⎭ to bracket	2	4042	
28		Sleeve for chain	2	330422	
29		Locker lid for wheelarch box	2	304401	
30		Hinge for locker lid	4	303875	
		Pop rivet fixing hinge to lid	8	300781	
31		Bolt (¼" UNF x ⅝" long) ⎫ Fixing	4	255206	
		Plain washer ⎬ hinge to	4	3968	
		Spring washer ⎪ wheelarch	4	3074	
		Nut (¼" UNF) ⎭ box	4	254810	
31		Hasp for locker lid	2	303871	For turnbuckle with circular plate ⎫ Alternatives. Check before ordering
		Hasp for locker lid	2	334523	For turnbuckle with semi-circular plate ⎭ Check before ordering
32		Turnbuckle for locker lid	4	300789	With circular plate ⎫ Alternatives.
		Turnbuckle for locker lid	2	300851	⎬ Check before
		Turnbuckle for locker lid	2	334525	With semi-circular plate ⎭ ordering
		Rivet fixing hasp to locker lid	6	77784	
		Nut (2 BA) ⎫ Fixing	6	3073	
		Spring washer ⎬ turnbuckle to	6	2247	
		Screw (2 BA x ½" long) ⎭ locker lid	6	—	
36		Tread plate for rear floor and wheelarch box sides	6	330615	
35		Tread plate, horizontal, front panel	1	308206	
34		Tread plate, vertical, front panel	4	330717	
33		Tread plate, wheelarch box, top	4	304871	Up to early Series IIA models
		Drive screw ⎫ Fixing tread plates	62	78006	
		Rivet, long ⎬ to body	31	300783	
		Rivet, short ⎭ and floor	20	300789	
		Tread plate for locker floor	2	334179	
		Pop rivet fixing tread plate to locker floor	14	78248	
37		Clip for starting handle and jack handle	5	508035	Check before ordering.
		Pop rivet fixing clips	5	300919	
		Drive screw fixing clips	5	77932	Alternatives.
		Clip for tyre pump	5	72085	2¼ litre Series IIA Diesel
		Pop rivet fixing clips	2	78248	
		Rubber ring, securing tyre pump	4	268887	Also Series IIA 2.6 litre Petrol models
		Bracket for starting handle	1	332376	109 Station Wagon
		Bracket for starting handle and jack handle	4	332377	
		Drive screw ⎫ Fixing brackets to base	4	78153	
		Packing washer ⎭ of centre seat	1	4171	
		Rear wing stay	10	332521	
		Stay RH ⎫	8	332453	Alternative to line below
		Stay LH ⎬ Wheelarch panel to body	1	332454	Alternative to two lines above ⎫ Early models with deep type sill channels
		Stay ⎭	2	334895	⎭

* Asterisk indicates a new part which has not been used on any previous Rover model

BODY, CAPPINGS AND SOCKETS, 109

C 775.

COLLINS-JONES.

BODY, CAPPINGS AND SOCKETS, 109

Plate Ref.	1 2 3 4	DESCRIPTION	Qty	Part No.	REMARKS
38		Cover plate for fuel filler ...	1	330366	
		Fuel instruction plate on fuel filler cowl	1	502951	Diesel models
		Rivet fixing cover plate to body side ...	9	302222	
39		Plastic plug for access hole in wheelarch locker ...	2	338027	
		Bracket for squab buffer ...	4	331262	
		Rivet } Fixing bracket to	12	302186	
		Pop rivet } front panel capping	8	78410	
40		'Land-Rover' nameplate	1	332670	
41		Registration plate ...	1	239772	
		Pop rivet fixing plates ...	6	300919	
		Bolt plate ...	1	336782	
		Bolt ($\frac{5}{16}$" UNF x $\frac{7}{8}$" long) } Fixing	16	255227	
		Drive screw } body to	4	69310	
		Fan disc washer } chassis	28	510912	
		Nut ($\frac{5}{16}$" UNF)	18	254831	
		Plastic plug for protection angle	2	338025	For blanking apertures when flasher lamps are not fitted

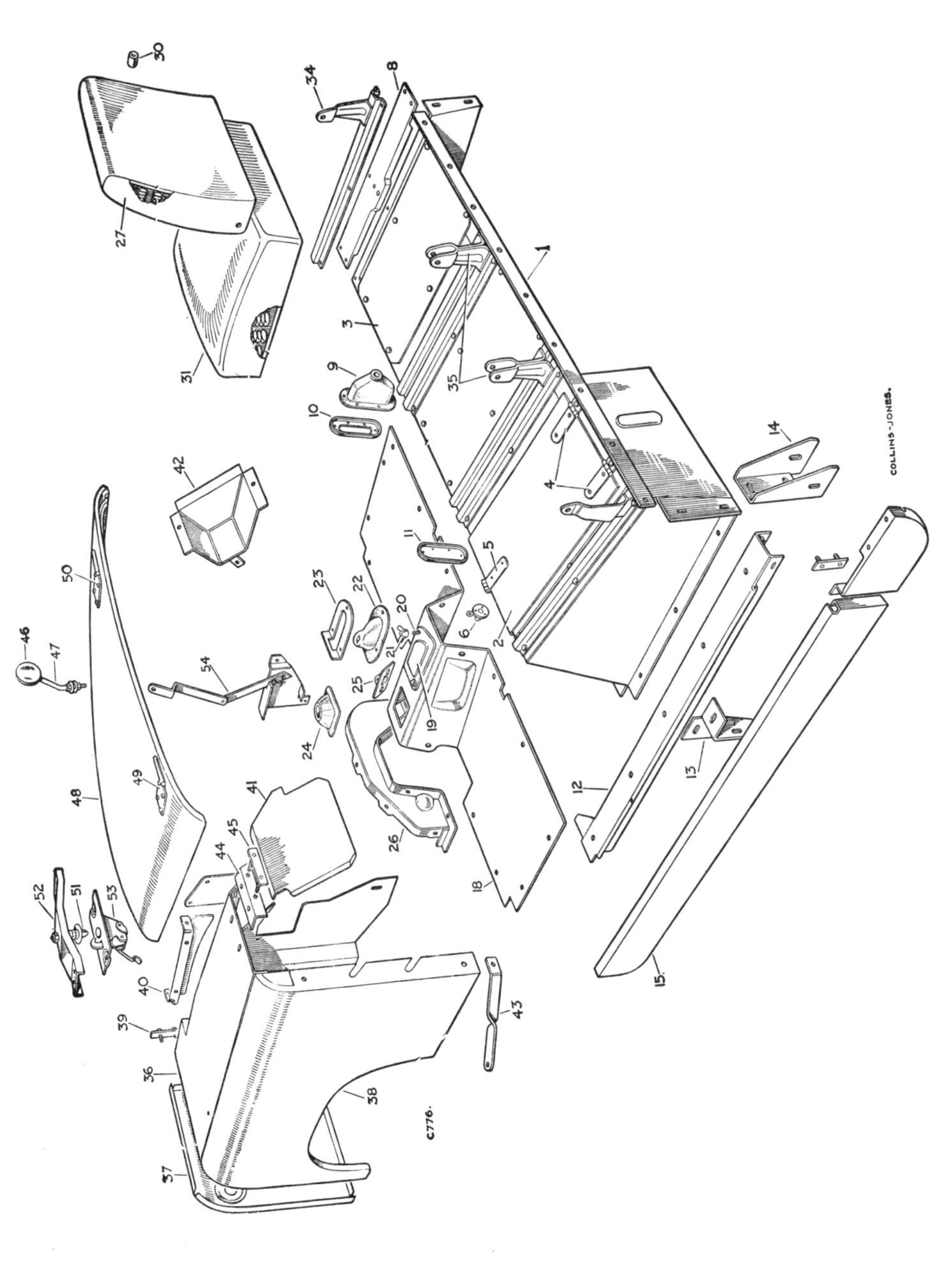

SEAT BASE, SEATS, FRONT FLOOR, FRONT WINGS AND BONNET

Plate Ref. 1 2 3 4	DESCRIPTION	Qty	Part No.	REMARKS
1	Seat base assembly	1	348800	Two locker lid type
	Bolt plate for seat cushion support	1	330908	
	Rivet fixing bolt plate	1	302581	
2	Lid for tool locker	1	330342	For early type one locker lid seat base
	Lid for seat base, centre	1	332522	When circular base turnbuckle is fitted
	Lid for seat base, centre	1	348855	Seriesl IIA 2¼ litre Petrol } When semi-circular base turnbuckle is fitted
	Lid for seat base, centre	1	348854	Series IIA Diesel and 2.6 litre Petrol } When semi-circular base turnbuckle is fitted
	Spire nut } For centre lid	2	302532	Required when centre PTO or hydraulic front winch is fitted to vehicle with two locker lid type seat base
	Special screw } For centre lid	2	320045	Required when centre PTO or hydraulic front winch is fitted to vehicle with two locker lid type seat base
3	Cover panel for petrol tank	1	348859	
4	Hinge for lids	2	303875	Centre lid } When circular base turnbuckle is fitted
	Pop rivet fixing hinges to lid	4	78248	Centre lid } When circular base turnbuckle is fitted
	Set bolt (¼″ UNF x ⅝″ long) } Fixing hinge to seat base	2	255206	Centre lid } When circular base turnbuckle is fitted
	Spring washer } Fixing hinge to seat base	2	3074	Centre lid } When circular base turnbuckle is fitted
	Nut (¼″ UNF) fixing hinge to seat base	2	254810	Centre lid } When circular base turnbuckle is fitted
5	Hasp for locker lid	1	303871	Centre lid } When circular base turnbuckle is fitted
	Rivet fixings hasp to lid	2	78248	Centre lid } When circular base turnbuckle is fitted
6	Turnbuckle for locker lid	1	300851	Centre lid } When circular base turnbuckle is fitted
	Hasp for locker lid, centre	1	334523	When semi-circular base turnbuckle is fitted
	Rivet fixing hasp to lid	2	78248	When semi-circular base turnbuckle is fitted
	Turnbuckle for locker lid, centre	1	334525	When semi-circular base turnbuckle is fitted
	Screw (2 BA x ½″ long) } Fixing turnbuckle to heel board	3	77784	
	Spring washer } Fixing turnbuckle to heel board	3	3073	
	Nut (2 BA) } Fixing turnbuckle to heel board	3	2247	
	Lid for seat base, outer	1	348852	For two locker lid type seat base
	Hasp for locker lid, outer	1	303871	When circular base turnbuckle is fitted } For two locker lid type seat base
	Pop rivet fixing hasp to lid	2	78248	When circular base turnbuckle is fitted } For two locker lid type seat base
	Turnbuckle for locker lid, outer	1	300851	When circular base turnbuckle is fitted } For two locker lid type seat base
	Pop rivet fixing turnbuckle to heelboard	2	78248	When circular base turnbuckle is fitted } For two locker lid type seat base
	Hasp for locker lid, outer	1	334523	When semi-circular base turnbuckle is fitted } For two locker lid type seat base
	Rivet fixing hasp to lid	2	78248	When semi-circular base turnbuckle is fitted } For two locker lid type seat base
	Turnbuckle for locker lid, outer	1	334525	When semi-circular base turnbuckle is fitted } For two locker lid type seat base
	Rivet, fixing turnbuckle to heelboard	3	78248	When semi-circular base turnbuckle is fitted } For two locker lid type seat base
	Grommet for redundant holes in lid end panel	2	334044	For two locker lid type seat base
7	Centre cover panel complete	1	330351	For one locker lid type seat base. Check before ordering
	Centre cover panel complete	1	330881	When centre PTO is fitted } For one locker lid type seat base. Check before ordering
	Set bolt (¼″ UNF x ⅝″ long) } Fixing cover panels to seat base	6	255206	Alternative to spire nut type fixing } 12 off on one locker lid type
	Plain washer } Fixing cover panels to seat base	6	3840	Alternative to spire nut type fixing } 12 off on one locker lid type
	Spring washer } Fixing cover panels to seat base	6	3074	Alternative to spire nut type fixing } 12 off on one locker lid type
	Spire nut } Fixing cover panels to seat base	6	78237	Alternative to set bolt type fixing } 12 off on one locker lid type
	Plain washer } Fixing cover panels to seat base	6	4034	Alternative to set bolt type fixing } 12 off on one locker lid type
	Drive screw } Fixing cover panels to seat base	6	78436	Alternative to set bolt type fixing } 12 off on one locker lid type
	Centre tool tray	1	336512	2¼ litre Diesel
	Centre tool tray	1	348573	2.6 litre Petrol
	Set screw (2 BA x ½″ long) } Fixing tool tray to seat base	4	77869	
	Spring washer } Fixing tool tray to seat base	4	3073	
	Plain washer } Fixing tool tray to seat base	4	3816	

* Asterisk indicates a new part which has not been used on any previous Rover model

SEAT BASE, SEATS, FRONT FLOOR, FRONT WINGS AND BONNET

470

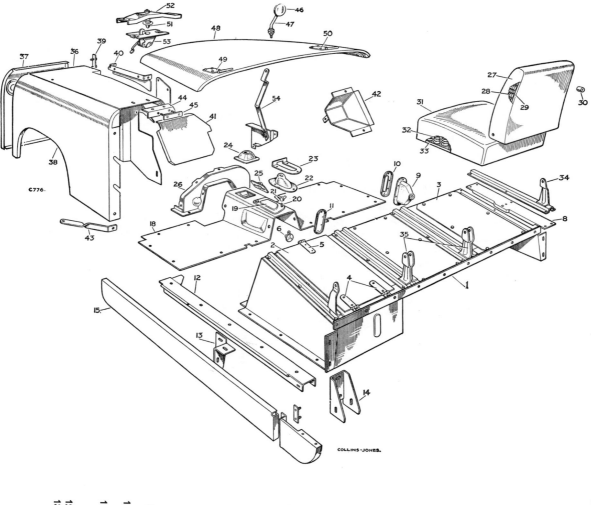

C776.

COLLINS-JONES.

Plate Ref. 1 2 3 4	DESCRIPTION	Qty	Part No.	REMARKS
1	Support strip for centre locker	1	333607	2¼ litre Diesel and 2.6 litre Petrol
	Pop rivet, fixing strip to heelboard	4	78248	
	Reinforcement bracket, front, for battery tray	1	332518	
	Spring washer	3	255206	
	Nut (¼" UNF)	3	3074	
	Bolt (¼" UNF x ⅝" long)	3	254810	2¼ litre Diesel
	Reinforcement bracket, outer, for battery tray	1	332487	
	Bolt (¼" UNF x ⅝" long)	2	255206	
	Spring washer	2	3074	
	Nut (¼" UNF)	2	254810	
	Pivot bracket for battery tie rod	2	336872	2¼ litre Diesel and 2.6 litre Petrol
	Rubber plug for seat base	1	301347	
	Bolt (¼" UNF x ⅝" long)	6	255206	} Fixing pivot
	Spring washer	6	255206	bracket to
	Nut (¼" UNF)	6	3074	locker floor } 2.6 litre Petrol
	Bolt (¼" UNF x ⅝" long)	6	254810	} Fixing stiffening angles,
	Spring washer	8	255206	pivot brackets and
	Nut (¼" UNF)	8	3074	reinforcement brackets
	Nut (¼" UNF)	8	254810	to locker floor
	Bolt (¼" UNF x ⅝" long)	2	255206	} Fixing outer
	Spring washer	2	3074	reinforcement bracket } 2¼ litre Diesel
	Nut (¼" UNF)	2	254810	to rear of
8	Insulating pad, battery stowage, side locker lid	1	334617	
	Heat insulation pad for seat base locker	1	348846	} 2¼ litre Diesel and 2.6 litre Petrol
	Extension panel, RH, at seat base end	1	334966	2 off on one locker lid type
	Extension panel, LH, at seat base end	1	333693	For two locker lid type seat base
	Drive screw	6	78436	} For redundant
	Plain washer	6	4034	holes in
	Spire nut	6	78237	rear stiffener
	Bolt (¼" UNF x ¾" long)	9	255207	} Fixing
	Plain washer	9	3840	seat base
	Spring washer	9	3074	to
	Nut (¼" UNF)	9	254810	rear body
	Screw (¼" UNF x ⅜" long)	2	78216	} Fixing extension panel
	Spring washer	2	3074	
	Nut (¼" UNF)	2	254810	
9	Rubber cover for hand brake	2	330347	Up to vehicle suffix 'C' inclusive
	Rubber cover for handbrake	1	338780	From vehicle suffix 'D' onwards
10	Retainer for rubber cover	1	330348	} Fixing cover
11	Cover plate for hand brake slot	1	78417	to heelboard
	Drive screw	6	330349	
	Drive screw fixing retainer	6	78137	
12	Sill channel RH front	1	330380	
13	Sill channel LH front	1	330381	
	Securing bracket for sill panel	4	330389	For deep type sill panel } Check before ordering
	Securing bracket LH front	4	337774*	6 off on 109 Station Wagon
	Securing bracket for sill panel			For narrow type sill panel

* Asterisk indicates a new part which has not been used on any previous Rover model

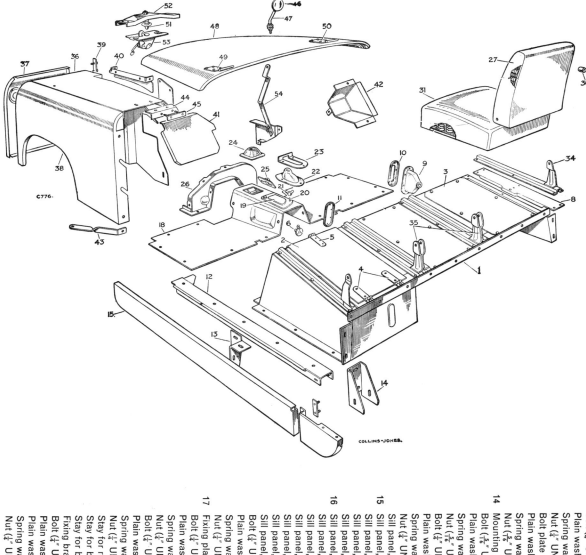

C776.

COLLINS-JONES.

SEAT BASE, SEATS, FRONT FLOOR, FRONT WINGS AND BONNET

Plate Ref. 1 2 3 4	DESCRIPTION		Qty	Part No.	REMARKS
	Bolt ($\frac{1}{4}$" UNF x $\frac{3}{4}$" long)	Fixing bracket to sill channel	8	255207	
	Plain washer		16	3821	
	Spring washer		8	3074	
	Nut ($\frac{1}{4}$" UNF)		8	254810	
	Bolt plate		2	332603	
	Plain washer	Fixing sill channel	4	3841	
	Spring washer		4	3075	
	Nut ($\frac{5}{16}$" UNF)		4	254831	
	Bolt ($\frac{5}{16}$" UNF x $\frac{3}{4}$" long)	Fixing sill channel to dash	4	255207	
	Plain washer		8	330387	
	Spring washer		4	254831	
	Nut ($\frac{5}{16}$" UNF)		4	3840	
14	Mounting bracket for sill channel to rear body		2	254810	
	Bolt ($\frac{5}{16}$" UNF x $\frac{3}{4}$" long)	Fixing mounting bracket to sill channel	8	255226	
	Plain washer		8	3841	
	Spring washer		4	3075	
	Nut ($\frac{5}{16}$" UNF)		4	254831	
	Bolt ($\frac{1}{4}$" UNF x $\frac{3}{4}$" long)	Fixing mounting bracket to body	8	255207	
	Plain washer		8	3840	
	Spring washer		4	3074	
	Nut ($\frac{1}{4}$" UNF)		4	254810	
15	Sill panel, RH front		1	330326	Deep type panel, Depth 4 $\frac{1}{8}$"
	Sill panel, RH front		1	330327	Depth 4 $\frac{1}{8}$"
	Sill panel, LH front		1	337942*	Narrow type panel, Depth 2 $\frac{23}{32}$"
	Sill panel, LH front		1	337943*	Depth 2 $\frac{23}{32}$"
16	Sill panel, RH rear		1	330336	Deep type panel
	Sill panel, RH rear		1	330337	Depth 4 $\frac{15}{16}$"
	Sill panel, RH rear		1	337938*	Narrow type panel, 88
	Sill panel, RH rear		1	337939*	Depth 2 $\frac{23}{32}$", 88
	Sill panel, LH rear		1	330585	Deep type panel, 109
	Sill panel, LH rear		1	330586	Depth 4 $\frac{15}{16}$", 109
	Sill panel, LH rear		1	337932*	Narrow type panel, 109
	Sill panel, LH rear		1	337933*	Depth 2 $\frac{23}{32}$", 109
	Plain washer	Fixing sill panel on sill channel	10	3840	14 off on 109
	Spring washer		20	3074	28 off on 109
	Nut ($\frac{1}{4}$" UNF)		10	254810	14 off on 109
	Bolt ($\frac{1}{4}$" UNF x $\frac{5}{8}$" long)	Fixing sill panels together	10	3821	14 off on 109
	Plain washer		4	3074	
	Spring washer		4	254810	
	Nut ($\frac{1}{4}$" UNF)		4	3821	
17	Fixing plate		2	255206	Alternative fixings
	Fixing plate		2	330333	
	Bolt ($\frac{1}{4}$" UNF x $\frac{5}{8}$" long)	Fixing sill panel to bracket	8	254810	
	Plain washer		8	3074	
	Spring washer		8	3821	
	Nut ($\frac{1}{4}$" UNF)		8	254810	
	Stay for rear sill panels		2	334894	For deep type sill panels
	Stay for body panel and sill panel, rear RH		1	337771*	For narrow type sill panels
	Stay for body panel and sill panel, rear LH		1	337772*	
	Fixing bracket for rear sill stays		2	334893	
	Bolt ($\frac{1}{4}$" UNF x $\frac{5}{8}$" long)	Fixing brackets and stays to chassis and sill panels	8	255206	
	Plain washer, small		10	3840	
	Plain washer, large		4	3817	
	Spring washer		4	3074	
	Nut ($\frac{1}{4}$" UNF)		8	254810	

Check before ordering

* Asterisk indicates a new part which has not been used on any previous Rover model

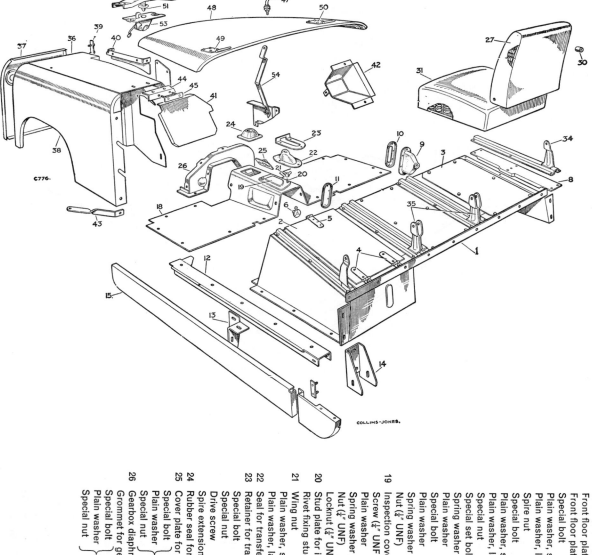

C776.

COLLINS-JONES.

SEAT BASE, SEATS, FRONT FLOOR, FRONT WINGS AND BONNET

Plate Ref.	1 2 3 4	DESCRIPTION	Qty	Part No.	REMARKS
18		Front floor complete	1	332968	4 cylinder models
		Front floor plate, RH	1	348893	6 cylinder
		Front floor plate, LH	1	348894	models
		Special bolt } Fixing front floor to	1	320045	
		Plain washer, small } gearbox diaphragm cover	7	3900	
		Plain washer, large }	4	3817	4 cylinder models
		Spire nut	11	302532	
		Special bolt } Fixing front floor plates to	19	320045	
		Plain washer, small } gearbox diaphragm	15	3900	
		Plain washer, large } cover and heelboard	4	3817	6 cylinder models
		Special nut	19	302532	
		Special set bolt } Fixing	5	78208	
		Spring washer } front floor	5	3074	
		Plain washer } to dash	5	3900	
		Special bolt } Fixing	4	78210	
		Plain washer } front floor	8	3900	
		Spring washer } to sill	4	3074	
		Nut ($\frac{1}{4}$" UNF) } channels	4	254810	
19		Inspection cover for front floor	1	303829	
		Screw ($\frac{1}{4}$" UNF x $\frac{3}{4}$" long) } Fixing	1	78212	
		Spring washer } inspection	3	3912	1958-60
		Plain washer } cover	3	2284	
		Nut ($\frac{1}{4}$" UNF) } to centre	1	254810	
		Locknut ($\frac{1}{4}$" UNF) } cover panel	2	301074	1958-60
20		Stud plate for inspection cover wing nut	1	303824	
		Rivet fixing stud plate	2	3668	
21		Wing nut	1	254810	
		Plain washer, small } Fixing inspection cover	2	3976	
		Plain washer, large	1	3817	
22		Seal for transfer gear lever	1	338871	
23		Retainer for transfer lever seal	1	303817	
		Special bolt } Fixing	3	320045	4 cylinder
		Special bolt } seal to	3	302532	models
		Special nut } floor	3	302532	
		Drive screw	3	78436	6 cylinder
		Spire extension nut	3	78237	models
24		Rubber seal for gear lever	1	301437	
25		Cover plate for operating rod	1	303828	
		Special bolt	2	302533	
		Plain washer } Fixing cover plate	2	3821	
		Special nut	2	302532	
26		Gearbox diaphragm cover complete	1	330069	
		Grommet for gearbox cover	2	334189	
		Special bolt	4	303750	4 cylinder models
		Plain washer } Fixing gearbox cover	4	3900	
		Special nut	4	302532	

* Asterisk indicates a new part which has not been used on any previous Rover model

SEAT BASE, SEATS, FRONT FLOOR, FRONT WINGS AND BONNET

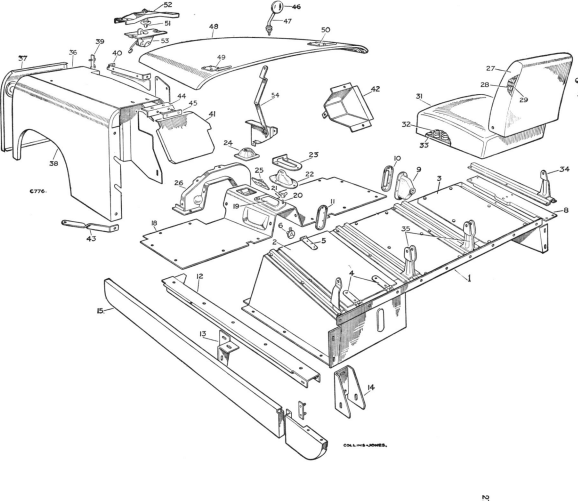

C776.

COLLINS-JONES.

SEAT BASE, SEATS, FRONT FLOOR, FRONT WINGS AND BONNET

Plate Ref.	1 2 3 4	DESCRIPTION	Qty	Part No.	REMARKS
		Gearbox diaphragm cover at front floor	1	348869	
		Gearbox diaphragm cover at dash	1	348868	
		Grommet for gearbox cover at dash	1	345140	
		Grommet for gearbox cover at floor	1	334189	
		Grommet for gearbox cover at dash	1	307388	6 cylinder models
		Special bolt	6	3900	
		Plain washer } Fixing diaphragm covers together	6	302532	
		Spire nut } and to dash	6	78436	
		Drive screw	4	78237	
		Spire extension nut	4		
27		Seat squab, trimmed, outer, GREY	2	349644	
		Seat squab, trimmed, outer, BLACK	2	326908*	
		Seat squab, trimmed, centre, GREY	1	349645	88
		Seat squab, trimmed, centre, BLACK	1	320701*	88
		Seat squab, trimmed, driver's GREY	1	349562	
		Seat squab, trimmed, driver's BLACK	1	320702*	109 Except Station Wagon
		Seat squab, trimmed, passenger's, outer, GREY	1	349644	
		Seat squab, trimmed, passenger's, outer, BLACK	1	326908*	
		Seat squab, trimmed, centre, GREY	1	349645	
		Seat squab, trimmed, centre, BLACK	1	320701*	
		Seat squab, trimmed, driver's GREY	1	349565	
		Seat squab, trimmed, driver's BLACK	1	320708*	109 Station Wagon
		Seat squab, trimmed, passenger's, outer, GREY	1	349563	
		Seat squab, trimmed, passenger's, outer, BLACK	1	320709*	
		Seat squab, trimmed, centre, GREY	1	349564	
		Seat squab, trimmed, centre, BLACK	1	320710*	
		Foam interior for outer squab	2	337873	
		Foam interior for centre squab	1	337880	
		Plain washer } Fixing squab cover	9	4032	
		Drive screw	9	78126	
		Special stud } Fixing squab to supports on non-adjustable seats	6	331170	} 4 off on 109
		Split pin	6	3901	
		Plain washer, thin, 1/4" hole	2	3960	As required
		Special bolt } Fixing squab to supports on adjustable seats	2	349931	
		Plain washer, 5/16" hole	2	331709	
		Plain washer, thin, 1/4" hole	2	3960	
		Plain washer, 5/16" hole	2	349931	
		Plain washer, thick	2	3446	
		Strap retaining front squab	3	331273	88
		Strap retaining front squab, non-adjustable seat	2	331408	109
		Plain washer } Fixing front squab strap	3	4032	} 2 off on 109
		Drive screw	3	77704	For early type squab with spring interior
		Strap retaining front squab, non-adjustable seat	3	331273	2 off on 109 models
		Plain washer, 5/16" hole	3	78441	
		Drive screw	3	4034	
		Plain washer } Fixing strap to squab	3	78237	For squab with foam interior
		Spire nut	3	78237	

* Asterisk indicates a new part which has not been used on any previous Rover model

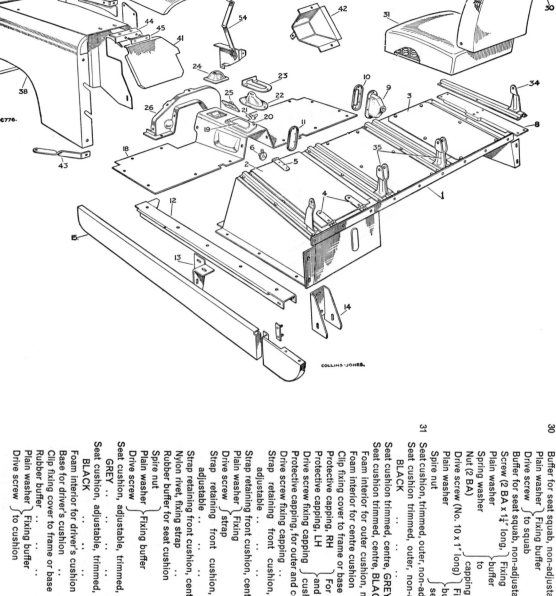

C776.

COLLINS-JONES.

SEAT BASE, SEATS, FRONT FLOOR, FRONT WINGS AND BONNET

Plate Ref. 1 2 3 4	DESCRIPTION	Qty	Part No.	REMARKS
30	Buffer for seat squab, non-adjustable seat	4	312027	6 off on 88 models
	Plain washer } Fixing buffer	4	3886	6 off on 88 models
	Drive screw } to squab	4	77704	For early type squab with
	Buffer for seat squab, non-adjustable seat	4	304125	spring interior
	Screw (2 BA x 1¼" long) } Fixing	4	77993	
	Plain washer } buffer	4	4030	6 off } Alternative
	Spring washer } to	4	3073	on 88 } type } For squab
	Nut (2 BA) } capping	4	3823	models } fixing } with foam
	Drive screw (No. 10 x 1" long) } Fixing	4	78676	interior
	Plain washer } buffer to	4	4030	
	Spire nut } seat squab	4	78237	
31	Seat cushion, trimmed, outer, non-adjustable, GREY	2	349500	Formwood base type. 1 off on 109
	Seat cushion trimmed, outer, non-adjustable, BLACK	2	320700*	Formwood base type. 1 off on 109
	Seat cushion, trimmed, centre, GREY	1	349503	Formwood base type
	Seat cushion trimmed, centre, BLACK	1	320726*	Formwood base type
	Foam interior for outer cushion, non-adjustable	2	337852	1 off on 109
	Foam interior for centre cushion	1	337859	
	Clip fixing cover to frame or base	58	331071	38 off on 109
	Protective capping, RH } For outer	6	331716	4 off
	Protective capping, LH } and centre	6	331717	on 109
	Drive screw fixing capping } cushions	24	78140	16 off on 109
	Protective capping, for outer and centre cushions	12	331909	8 off on 109 } For Formwood base
	Drive screw fixing capping	24	78455	16 off on 109 } type cushion
	Strap retaining front cushion, outer, non-adjustable	2	331272	1 off on 109 } Drive screw type fixing.
	Strap retaining front cushion, centre	1	331273	2 off
	Plain washer } Fixing	3	4203	on 109 } Alternative to nylon rivet
	Drive screw } strap	3	77704	fixing
	Strap retaining front cushion, outer, non-adjustable	2	331973	1 off on 109 } Nylon rivet type fixing.
	Strap retaining front cushion, centre	1	331974	Alternative to drive
	Nylon rivet, fixing strap	3	331971	2 off on 109 } screw fixing
	Rubber buffer for seat cushion	6	304125	For metal base type
	Spire nut	6	78237	4 off on 109 } cushion.
	Plain washer } Fixing buffer	6	4030	
	Drive screw	6	78434	For metal base type cushion. Check before ordering
	Seat cushion, adjustable, trimmed, outer, driver's GREY	1	349506	Formwood base type
	Seat cushion, adjustable, trimmed, outer, driver's BLACK	1	320699*	Formwood base type
	Foam interior for driver's cushion	1	337899	
	Base for driver's cushion	1	331900	
	Clip fixing cover to frame or base	20	331071	
	Rubber buffer	4	3886	
	Plain washer } Fixing buffer	4	312027	
	Drive screw } to cushion	6	69310	For cushion with canvas base

109

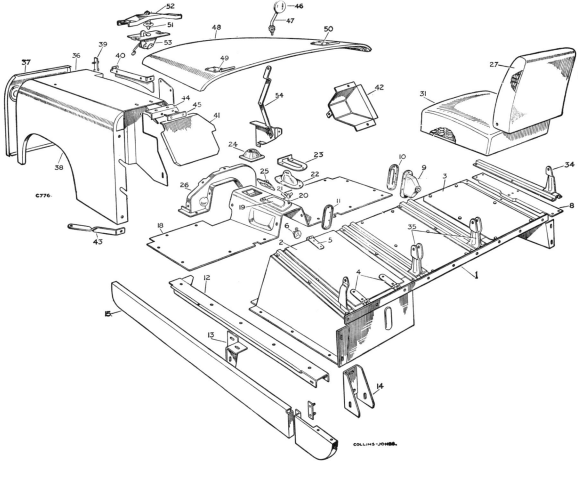

C776

COLLINS-JONES.

SEAT BASE, SEATS, FRONT FLOOR, FRONT WINGS AND BONNET

Plate Ref. 1 2 3 4	DESCRIPTION	Qty	Part No.	REMARKS
	Protective capping, RH front	1	331716	
	Protective capping, LH front	1	331717	} For driver's cushion
	Protective capping, RH rear	1	331718	} For metal base type cushion
	Protective capping, LH rear	1	331719	
	Drive screw fixing capping	8	78140	
	Protective capping, front	2	331909	
	Protective capping, RH	1	331910	} For Formwood base type cushion
	Protective capping, LH	1	331911	For driver's cushion
	Drive screw fixing capping	8	78455	
	Front seat slide with control	1	331102	RHStg
	Front seat slide with control	1	331103	LHStg
	Front seat slide without control	1	331104	
	Screw (¼" UNF x ¾" long)	6	78358	} Fixing slides
	Plain washer	6	3833	
	Spring washer	8	3074	
	Nut (¼" UNF)	8	254810	to seat supports
	Spring washer	8	331466	
	Nut (¼" UNF)	4	254831	
	Frame for adjustable seat cushion	1	3075	
	Distance washer	4	4094	} Fixing frame
	Spring washer	4	3075	} to seat slides
	Nut (⁵⁄₁₆" UNF)	4	2247	
	Screw (2 BA x 1" long)	6	3073	} Fixing studs to seat base and rear body
	Spring washer	6	3073	
	Nut (2 BA)	6	2247	
	Grommet for redundant hole in seat base	1	307220	109
	Rubber stop for squab in frame	2	331083	
	Stud for cushion and squab retaining straps	6	338430	4 off on 109
34	Support for front seat cushion, outer, RH	1	331006	1 off on 109
	Support for front seat cushion, outer, LH	1	331007	
35	Support for front seat, centre	1	331008	
	Retaining bracket for front seat cushion	6	331818	For metal base type cushion. 4 off on 109
	Bolt (¼" UNF x ¾" long)	6	255207	} Fixing seat supports and extension panels to seat base
	Plain washer	28	252207	
	Spring washer	28	3840	17 off on 109
	Nut (¼" UNF)	28	3074	
36	Front wing, top, RH	1	254810	
	Front wing, top, LH	1	330424	
	Front wing, top, RH	1	330425	88
	Front wing, top, LH	1	330426	
37	Front panel and number plate, RH	1	330427	} For front
	Front panel and number plate, LH	1	330531	} wing
	Cover plate at LH wing valance for redundant exhaust hole	1	330436	
	Pop rivet, fixing cover plate	4	330437	
38	Outer panel for front wing, RH	22	330144	1961 models onwards
	Outer panel for front wing, LH	22	78248	
	Bolt (¼" UNF x ⅝" long)	44	255206	} Fixing wing outer panel to wing top
	Plain washer	22	3840	
	Spring washer	22	3074	
	Nut (¼" UNF)	22	254810	

* Asterisk indicates a new part which has not been used on any previous Rover model

SEAT BASE, SEATS, FRONT FLOOR, FRONT WINGS AND BONNET

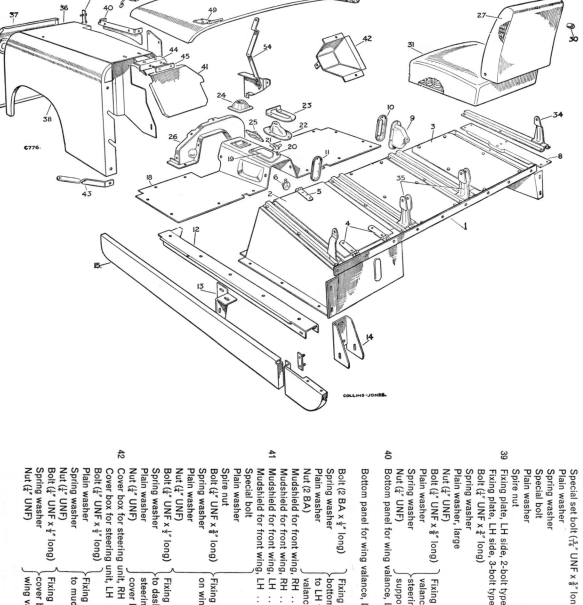

C776.

COLLINS-JONES.

SEAT BASE, SEATS, FRONT FLOOR, FRONT WINGS AND BONNET

Plate Ref.	1 2 3 4	DESCRIPTION	Qty	Part No.	REMARKS
39		Special set bolt (5/16" UNF x 5/8" long) ⎫ Fixing wings to scuttle side pillar	8	78331	Alternatives to three lines below.
		Plain washer	8	2249	Check before ordering
		Spring washer	8	3075	Check before ordering
		Special bolt	8	78392	
		Plain washer	8	3830	Alternatives to two lines above.
		Spire nut ⎭	8	78393	Check before ordering
		Fixing plate, LH side, 2-bolt type	1	330147	Alternatives.
		Fixing plate, LH side, 3-bolt type	1	332589	Check before ordering
		Bolt (1/4" UNF x 3/4" long) ⎫ Fixing wings to grille panel	6	255207	4 off with 3-bolt fixing plate
		Spring washer	6	3074	9 off with 3-bolt fixing plate
		Plain washer, large	14	2215	13 off with 3-bolt fixing plate
		Nut (1/4" UNF)	8	254810	9 off with 3-bolt fixing plate
		Bolt (1/4" UNF x 5/8" long)	2	255206	
		Plain washer	4	2215	
		Spring washer	2	3074	
		Nut (1/4" UNF) ⎭	2	254810	
40		Bottom panel for wing valance, LH	1	330443	All 88 Petrol and Series IIA 88 2¼ litre
		Bottom panel for wing valance, LH	1	330534	All 109 2¼ litre Petrol and Series IIA 109 2¼ litre Diesel models
		Nut (1/4" UNF) ⎫ Fixing wing valance to steering box support bracket	2	254810	
		Spring washer	2	3074	
		Plain washer ⎭	4	2215	
		Bolt (1/4" UNF x 5/8" long) ⎫ Fixing wing valance on wing	2	255206	
		Spring washer	4	3074	
		Plain washer	8	3902	Diesel models
		Nut (2 BA) ⎭	4	3823	
		Bolt (2 BA x 1/4" long) ⎫ Fixing bottom panel to LH wing valance	4	234603	All petrol models and Series IIA 2¼ litre Diesel models
41		Mudshield for front wing, RH	1	330445	RHStg
		Mudshield for front wing, LH	1	330444	LHStg
		Mudshield for front wing, RH	1	330448	RHStg
		Mudshield for front wing, LH	1	330447	LHStg
		Bolt (1/4" UNF x 1/2" long) ⎫ Fixing mudshield to dash and steering unit cover box	8	255204	
		Spring washer	8	3074	
		Plain washer	8	3817	
		Special bolt	6	303750	
		Plain washer	6	3821	
		Spire nut ⎭	6	302532	
		Bolt (1/4" UNF x 5/8" long) ⎫ Fixing mudshield to angle on wing	4	255206	
		Spring washer	4	3074	
		Plain washer	7	3900	
		Nut (1/4" UNF) ⎭	4	254810	
42		Cover box for steering unit, RH	1	330459	RHStg
		Cover box for steering unit, LH	1	330460	LHStg
		Bolt (1/4" UNF x 1/2" long) ⎫ Fixing cover box to mudshield	1	255204	
		Plain washer	1	4071	
		Spring washer	1	3074	
		Nut (1/4" UNF) ⎭	1	254810	
		Bolt (1/4" UNF x 1/2" long) ⎫ Fixing cover box to wing valance	1	3074	
		Spring washer	1	254810	

* Asterisk indicates a new part which has not been used on any previous Rover model

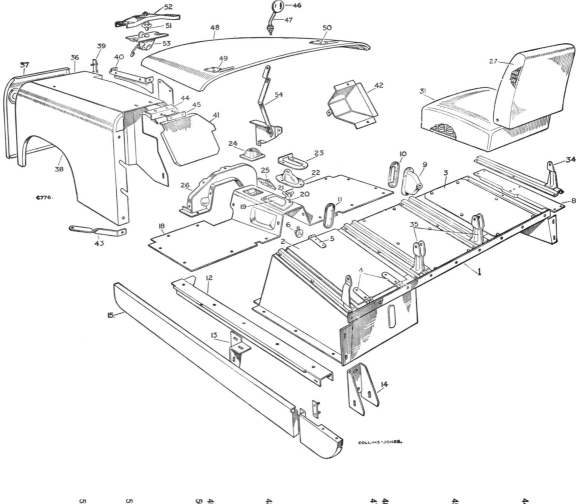

C776.

COLLINS-JONES.

SEAT BASE, SEATS, FRONT FLOOR, FRONT WINGS AND BONNET

Plate Ref.	1 2 3 4	DESCRIPTION	Qty	Part No.	REMARKS
43		Stay for front wing	2	332581	
		Bolt (¼" UNF x ⅝" long) ⎱ Fixing stay to wing and dash	4	255206	
		Spring washer	4	3074	
		Nut (¼" UNF)	4	254810	
44		Bracket, RH ⎱ For rear of wing	1	338652	
		Bracket, LH	1	338653	
		Bolt (¼" UNF x ⅝" long) ⎱ Fixing rear end of front wing to bracket	6	255206	} Alternatives
		Bolt (¼" UNF x ⅞" long)	6	78665	} Alternatives
		Plain washer	12	2215	
		Spring washer	6	254810	
		Nut (¼" UNF)	6	3074	
45		Fixing plate	2	330033	
		Plain washer ⎱ Fixing brackets to dash	4	3840	
		Spring washer	4	3074	
		Nut (¼" UNF)	2	254810	
		Mirror complete, exterior	2	526561	1 off on Export models
46		Mirror only	2	530700	Export models
		Mirror complete, exterior	2	262658	Desmo type ⎱ Alternatives. Check before ordering
47		Mirror only	2	562912*	1 off on Export models. Wingard type
		Arm for mirror	2	605664*	
		Arm for mirror	2	605663*	
		Nut (⅜" BSF) ⎱ Retaining	4	2827	
		Plain washer	4	3076	
		Spring washer ⎱ Fixing mirror to wing	2	2219	
		Plain washer, small	2	2851	
48		Bonnet top panel	1	337963	De-luxe bonnet recessed for spare wheel carrier. Optional on 109
		Bonnet top panel	1	337951	88 Standard model
		Bonnet top panel	1	337957	De-luxe bonnet standard on all 109 models and 88 Station Wagon
49		Hinge for bonnet, LH	1	336476	For early type bonnets with bolted type hinges
50		Hinge for bonnet, RH	1	336474	
		Split pin ⎱ RH hinge	1	4433	
		Plain washer ⎱ Fixing hinges to bonnet panel	1	4432	
		Bolt (¼" UNF x ⅝" long)	6	255206	8 off on early type hinges
		Spring washer	6	3074	
		Nut (¼" UNF)	6	254810	
51		Striker pin for bonnet catch	1	332625	For early type bonnets with bolted type striker bracket
		Striker pin for bonnet catch	1	337969	For late type bonnets with welded type striker bracket
52		Bracket for bonnet striker	1	330548	88 Standard models
		Bracket for bonnet striker	1	330132	For De luxe bonnet
		Bracket for bonnet striker	1	330648	For De luxe recessed bonnet
		Plain washer ⎱ Fixing striker bracket to bonnet	4	3830	For early type bonnets with bolted type bracket
		Spring washer	4	3075	For De luxe bonnets with bolted type bracket
		Set bolt (5/16" UNF x ¾" long)	4	255226	

SEAT BASE, SEATS, FRONT FLOOR, FRONT WINGS AND BONNET

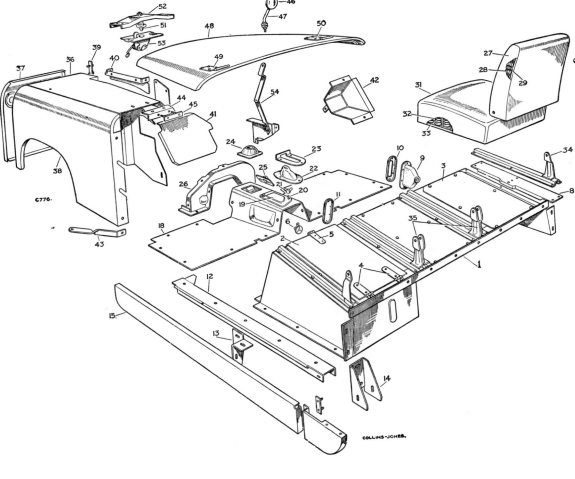

C776.

COLLINS-JONES.

SEAT BASE, SEATS, FRONT FLOOR, FRONT WINGS AND BONNET

Plate Ref.	1 2 3 4	DESCRIPTION	Qty	Part No.	REMARKS
53		Control for bonnet … … …	1	332401	
		Washer plate for bonnet control	1	332400	
		Set bolt (¼" UNF x ⅝" long)	2	255206	} Fixing control to radiator grille
		Plain washer	2	3840	
54		Prop rod for bonnet … …	1	338636	
		Bolt (5⁄16" UNF x 1⅛" long)	1	255229	
		Plain washer	2	3446	
		Plain washer	1	3830	Alternative to pivot pin and clip type fixing
		Spring washer	1	3075	
		Locknut (5⁄16" UNF)	1	254861	} Fixing prop rod
		Retainer clip for pivot pin	2	336536	
		Pivot pin	1	336535	Alternative to bolt and nut type fixing
		Plain washer	1	3446	
		Bolt (¼" UNF x ¾" long)	2	255207	Fixing prop rod to wing valance and grille panel
		Plain washer	4	2215	
		Spring washer	2	3074	
		Nut (¼" UNF)	2	254810	

* Asterisk indicates a new part which has not been used on any previous Rover model

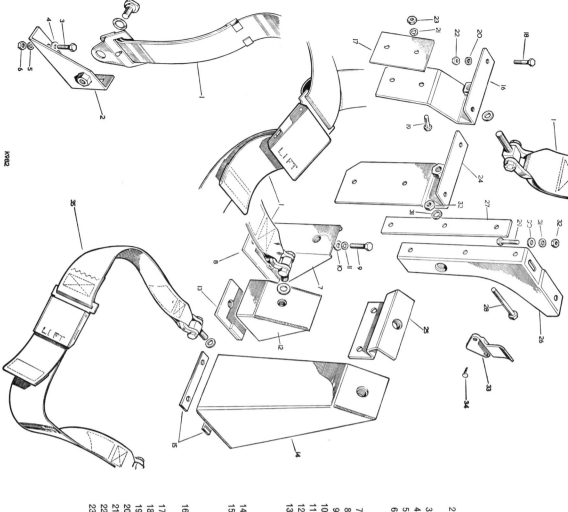

K982

FRONT SEAT SAFETY HARNESS AND ANCHORAGE

Plate Ref.	1 2 3 4	DESCRIPTION	Qty	Part No.	REMARKS
1		Safety harness for front seat, outer ...	2	348776	All models
		ANCHORAGE KIT, RH ...	1	320631	88. Except when
		ANCHORAGE KIT, RH ...	1	320632	cab is fitted
		ANCHORAGE KIT, RH ...	1	320633	88. With
		ANCHORAGE KIT, RH ...	1	320634	truck cab
		ANCHORAGE KIT, RH ...	1	320635	109 with hard top
		ANCHORAGE KIT, RH ...⎫ For safety harness	1	320636	or full length hood
		ANCHORAGE KIT, LH ...⎬	1	320637	109. With
		ANCHORAGE KIT, LH ...⎭	1	320638	truck cab
		ANCHORAGE KIT, LH ...	1	320639	109 Station
		ANCHORAGE KIT, LH ...	1	320640	Wagon
2		Sill gusset, RH ...	1	345100	
		Sill gusset, LH ...	1	345101	
3		Bolt ($\frac{1}{4}$" UNF x $\frac{7}{8}$" long) ⎫ Fixing	2	255208	
4		Plain washer ⎬ sill gussett	2	3821	
5		Spring washer ⎮ to seat	2	3074	
6		Nut ($\frac{1}{4}$" UNF) ⎭ base	2	254810	
		Self-locking nut ($\frac{7}{16}$" UNF) fixing shackle bolt to cab rail ...	1	252163	88. With truck cab
7		Gusset bracket at bulkhead ... ⎫	2	338752	
8		Washer plate complete ⎬ Fixing	2	336567	
9		Set bolt ($\frac{1}{4}$" UNF x $\frac{7}{8}$" long) ⎮ gusset	4	255208	88. All models
10		Plain washer ⎮ bracket	4	3821	
11		Spring washer ⎮ to floor	4	3074	
12		Spring washer ⎭	2	345123	
13		Gusset bracket at bulkhead ... ⎫	2	345125	
		Clamp plate complete ⎬ Fixing	4	255228	109. Except Station
		Set bolt ($\frac{5}{16}$" UNF x 1" long) ⎮ gusset	4	255228	Wagon
14		Spring washer ⎮ bracket	4	3075	
		Plain washer ⎭ to floor	4	3868	
15		Gusset bracket at bulkhead ... ⎫	2	348738	
16		Washer plate ⎬ Fixing gusset bracket	8	336147	109 Station Wagon
		Bolt ($\frac{1}{4}$" UNF x $\frac{7}{8}$" long) ⎮ to bulkhead	8	255208	
		Spring washer ⎮	8	3074	
		Nut ($\frac{1}{4}$" UNF) ⎭	8	254810	
17		Shoulder bracket, RH ...	1	338753	
		Shoulder bracket, LH ...	1	338754	
18		Washer plate ...	2	336566	
19		Bolt ($\frac{1}{4}$" UNF x $\frac{7}{8}$" long) ⎫ Fixing shoulder	4	255227	88. Except when
		Bolt ($\frac{5}{16}$" UNF x $\frac{7}{8}$" long) ⎬ bracket to	4	255208	cab is fitted
20		Spring washer ⎮ bulkhead	4	3074	
21		Spring washer ⎮	4	3075	
22		(Nut $\frac{1}{4}$" UNF) ⎮	4	254810	
23		Nut ($\frac{5}{16}$" UNF) ⎭	4	254831	

Remarks (continued):

Fitted as standard on all Home Market models from vehicle suffix 'D' onwards. Optional on Export models

---

* Asterisk indicates a new part which has not been used on any previous Rover model

FRONT SEAT SAFETY HARNESS AND ANCHORAGE

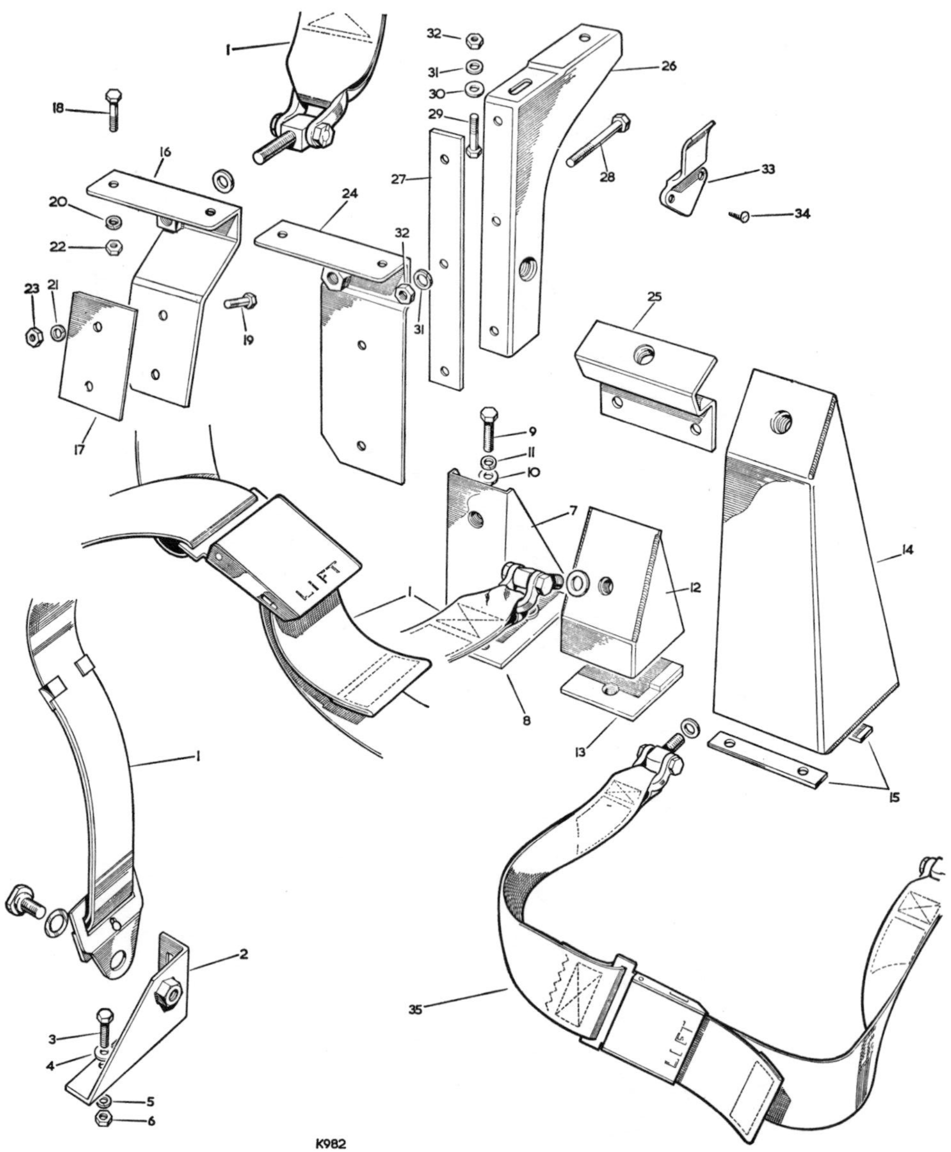

K982

FRONT SEAT SAFETY HARNESS AND ANCHORAGE

Plate Ref.	1 2 3 4 DESCRIPTION		Qty	Part No.	REMARKS	
24	Shoulder bracket, RH		1	338759	109. With hard top or full length hood	Fitted as standard on all Home Market models from vehicle suffix 'D' onwards. Optional on Export models
	Shoulder bracket, LH		1	338760		
	Bolt ($\frac{1}{4}$" UNF x $\frac{7}{8}$" long)	Fixing shoulder bracket to bulkhead	4	255208		
	Bolt ($\frac{5}{16}$" UNF x $\frac{7}{8}$" long)		4	255227		
	Spring washer		4	3074		
	Spring washer		4	3075		
	Nut ($\frac{1}{4}$" UNF)		4	254810		
	Nut ($\frac{5}{16}$" UNF)		4	254831		
25	Shoulder bracket and cab mounting		2	336577	109 with cab	
26	Shoulder bracket, RH		1	348473	109 Station Wagon	
	Shoulder bracket, LH		1	348474		
27	Washer plate	Fixing shoulder bracket to door post and cant rail	2	348447		
28	Bolt ($\frac{1}{4}$" UNF x $2\frac{1}{2}$" long)		6	255219		
29	Bolt ($\frac{1}{4}$" UNF x $\frac{5}{8}$" long)		2	255206		
30	Plain washer		2	3821		
31	Spring washer		8	3074		
32	Nut ($\frac{1}{4}$" UNF)		8	254810		
33	Stowage hook for safety harness		2	345117		
34	Drive screw, fixing hook		4	78126		
35	Front seat safety harness, centre		1	345908*	All models	Optional Equipment
	Gusset bracket at bulkhead		2	338752	88	
	Washer plate complete	Fixing gusset bracket to floor	2	336567		
	Set bolt ($\frac{1}{4}$" UNF x $\frac{7}{8}$" long)		4	255208		
	Plain washer		4	3821		
	Spring washer		4	3074		
	Gusset bracket at bulkhead		2	345123	109 Except Station Wagon †	
	Clamp plate complete		2	345125		
	Set bolt ($\frac{5}{16}$" UNF x 1" long)		4	255228		
	Spring washer		4	3075		
	Plain washer		4	3868		

† **Note:** Anchorage point for centre harness is provided in normal build on late 109 Station Wagons. To convert early models, fit toe panel assembly 345149. See Page 537.

* Asterisk indicates a new part which has not been used on any previous Rover model

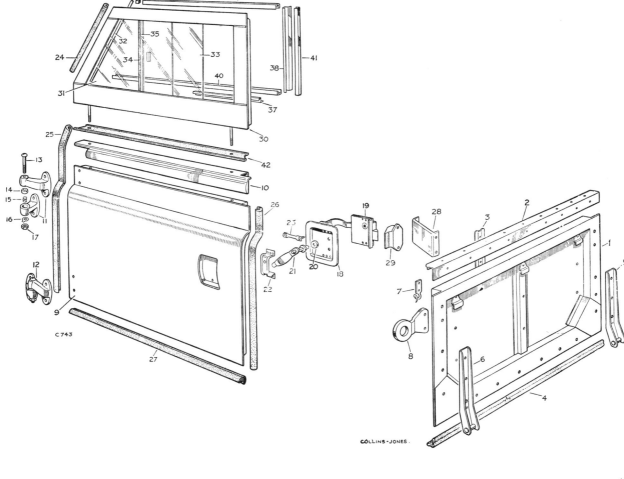

C 743

COLLINS-JONES.

TAILBOARD, DOORS AND SIDE SCREENS

Plate Ref.	1 2 3 4	DESCRIPTION	Qty	Part No.	REMARKS
1		TAILBOARD ASSEMBLY	1	320604	
2		Top capping for tailboard	1	330416	With short flange 1/4" deep
		Top capping for tailboard	1	336411	With short flange 1 7/8" deep — Alternatives. Check before ordering
3		Pop rivet, fixing tailboard top capping	20	78248	
		Tread plate for tailboard	4	304874	
		Rivet, fixing tread plate	16	300789	
4		Sealing rubber, bottom, for tailboard	1	303975	
		Rivet fixing sealing rubber to rear body and tailboard	8	78321	
5		Hinge for tailboard, RH	1	338617	
6		Hinge for tailboard, LH	1	338618	
		Spring washer	1	2289	Not part of tailboard assembly
		Plain washer	4	3895	
		Distance washer	4	4171	Retaining tailboard hinge to chassis
		Set bolt (1/4" UNF x 2 1/8" long)	4	253887	
		Split pin	2	2289	
		Spring washer	6	3074	
		Plain washer	6	3821	
		Nut (1/4" UNF)	6	3896	Fixing hinge to tailboard
		Spring washer	6	3074	
		Plain washer	6	3821	
		Nut (1/4" UNF)	6	3896	Fixing hinge to tailboard
7		Hook for tailboard chain, RH	1	254810	
		Hook for tailboard chain, LH	1	256209	
8		Bolt (1/4" UNF x 2 1/4" long)	4	3074	Fixing locking plate and hook to tailboard
		Nut (1/4" UNF)	4	254810	
		Spring washer	4	3074	
		Locking plate for tailboard	2	336412	
9		FRONT DOOR ASSEMBLY, RH	1	332446	Two-piece door
		FRONT DOOR ASSEMBLY, LH	1	332445	Two-piece door
		FRONT DOOR ASSEMBLY, RH	1	332412	Full length door
		FRONT DOOR ASSEMBLY, LH	1	332411	Full length door. Includes side window glasses and fixings on Page 497
10		Top capping for door, RH	1	337792	Two-piece door only
		Top capping for door, LH	1	337791	
		Top capping for door, RH	1	330122	
		Top capping for door, LH	1	330123	
		Pop rivet fixing cappings	18	78248	
		Pop rivet fixing door outer panel at top	6	300919	
		Waist moulding, RH	1	333717	
		Waist moulding, LH	1	333718	
		Special bolt	8	302533	Fixing moulding to door
		Plain washer	8	302532	
		Special nut	8	2630	
11		Hinge complete, RH upper	1	330015	
		Hinge complete, LH upper	1	330016	
12		Hinge complete, RH lower	1	330017	
		Hinge complete, LH lower	1	330018	
13		Special bolt	1	330953	
14		Cone bush	1	330954	
15		Spring	1	330955	For door hinge
16		Lockwasher	1	330956	
17		Nut	1	330957	

* Asterisk indicates a new part which has not been used on any previous Rover model

TAILBOARD, DOORS AND SIDE SCREENS

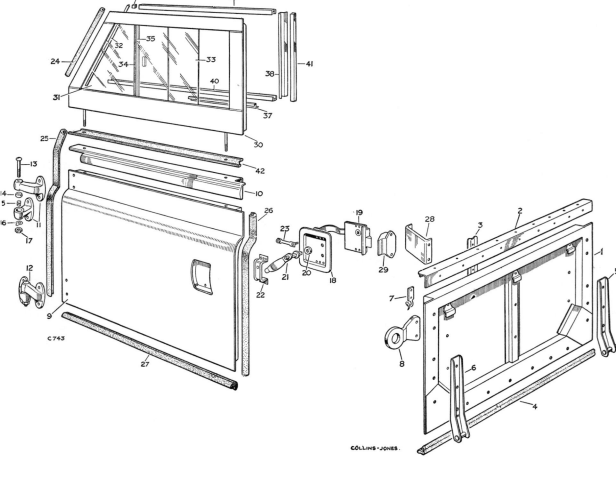

C 743

COLLINS-JONES.

TAILBOARD, DOORS AND SIDE SCREENS

Plate Ref.	1 2 3 4	Description	Qty	Part No.	Remarks
		Bolt (5/16" UNF x 2" long) } Fixing hinges to doors	8	256226	
		Plain washer	8	3966	
		Spring washer	8	3075	
		Nut (5/16" UNF)	8	254831	4 off on 109 Station Wagon
		Nut plate	2	333710	109 Station Wagon
		DOOR LOCK AND MOUNTING ASSEMBLY, RH	1	345433	With non-detachable inner handle
		DOOR LOCK AND MOUNTING ASSEMBLY, LH	1	345434	For early type lock with detachable inner handle. Check before ordering
18		Mounting plate for door lock	1	333980	
19		Door lock, RH	1	330113	
19		Door lock, LH	1	330114	
20		Washer, handle to cover	4	302859	
21		Handle, RH	1	307977	
21		Handle, LH	1	307978	
22		Bracket for door handle	2	302854	
		Screw (2 BA x 1/2" long) } Fixing lock together	12	313858	
		Plain washer	12	3073	
		Spring washer	8	3685	
		Nut (2 BA)	8	2247	
		Rivet	12	302222	
		Screw (1/4" UNF x 1/2" long) } Fixing door lock mounting to door	8	77903	
		Screw retainer	8	254810	
		Nut retainer	2	330536	
		Spring washer	4	337806	4 off when screw retainer is fitted
		Nut (1/4" UNF)	2	337808	2 off when nut retainer is fitted
24		Drive screw, fixing lock assembly to door	8	3074	
25		Seal for door, front upper	8	254810	
		Seal for door, front lower	6	330891	
		Drive screw } Fixing seal to screen	2	330539	
		Rivet	2	330540	
26		Seal for door, front lower, dash	2	330542	
26		Seal for door, rear lower, body RH	2	302818	
26		Seal for door, rear lower, body LH	48	330146	88
27		Seal for door, bottom, sill	1	332484	109
		Rivet fixing lower and bottom seals	1	332485	109
28		Support bracket at door striker	1	255206	
		Support bracket at door striker, RH	8	3074	
		Support bracket at door striker, LH	8	3840	
		Bolt (1/4" UNF x 5/8" long) } Fixing support bracket to body	8	3900	109
		Plain washer, small	8	254810	
		Plain washer, large	8	333140	
		Nut (1/4" UNF)	2	255206	
		Spring washer	4	254810	
29		Striking plate for door lock	4	3840	
		Bolt (1/4" UNF x 5/8" long) } Fixing striking plate to body	4	3074	
		Plain washer	4	254810	
		Spring washer			
		Nut (1/4" UNF)			

* Asterisk indicates a new part which has not been used on any previous Rover model

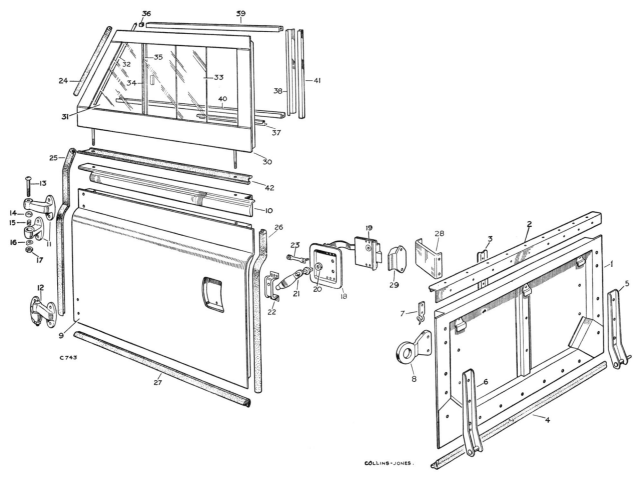

C 743

COLLINS-JONES.

TAILBOARD, DOORS AND SIDE SCREENS

Plate Ref.	1 2 3 4	DESCRIPTION	Qty	Part No.	REMARKS
30		SIDE SCREEN ASSEMBLY, RH	1	330157	Two-piece
30		SIDE SCREEN ASSEMBLY, LH	1	330158	door only
31		Fixed window, front	2	330159	
32		Retainer for window	2	330198	
		Drive screw fixing retainer	10	78140	
33		Sliding window, rear	2	330160	
34		Sealing rubber for front edge of sliding window	2	330660	
35		Channel for sealing rubber	2	330661	
36		Buffer for sliding window at top	2	330162	
		Plain washer ⎫ Fixing buffers and	2	3886	
37		Drive screw ⎭ channels to frame	2	78160	
38		Filler, rear ⎫	4	330202	
38		Filler, top and bottom ⎬ For	2	330203	
39		Channel top ⎪ windows	2	336451	
40		Channel, bottom ⎪	2	336452	
41		Channel, rear ⎭	2	336454	
		Drive screw fixing channel	26	77707	
42		Sealing strip for sidescreen	2	330163	
		Pop rivet ⎫ Fixing	10	78248	
		Plain washer ⎭ sealing strip	10	3912	
		Plain washer ⎫ Fixing	4	3843	Two-piece door only
		Spring washer ⎬ side screen	4	3077	
		Nut (7/16″ UNF) ⎭ to door	4	254813	
		Set bolt (5/16″ UNF x 7/8″ long) ⎫ Fixing door	8	255227	
		Spring washer ⎭ hinges to dash	8	3075	

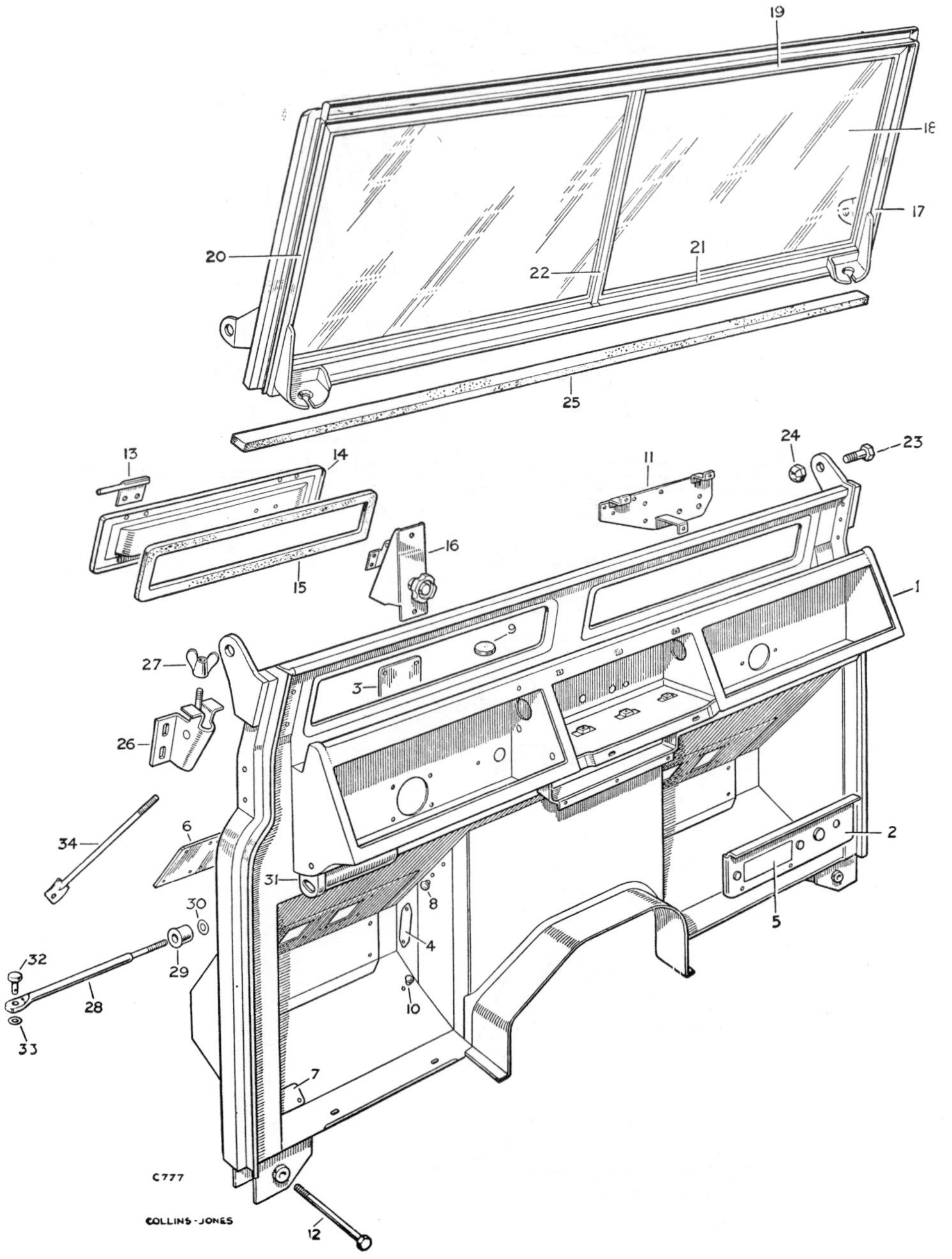
C777
COLLINS-JONES

Plate Ref. 1 2 3 4	DESCRIPTION	Qty	Part No.	REMARKS
1	Dash complete	1	345879*	4 cylinder models
	Dash complete	1	345611	6 cylinder models
	Mounting bracket for control panel	1	345882*	4 cylinder models; Up to vehicle suffix 'C' inclusive
	Pop rivet, fixing bracket to dash	7	78257	4 cylinder models; Up to vehicle suffix 'C' inclusive
2	Panel for controls	1	332725	Petrol models; Up to vehicle suffix 'C' inclusive
	Panel for controls	1	332726	Diesel models; Up to vehicle suffix 'C' inclusive
	Screw (2 BA x ½" long) } Fixing panel to dash	6	77869	Up to vehicle suffix 'C' inclusive
	Spring washer } Fixing panel to dash	6	3073	Up to vehicle suffix 'C' inclusive
	Plain washer } Fixing panel to dash	6	3685	Up to vehicle suffix 'C' inclusive
	Nut (2 BA) } Fixing panel to dash	3	2247	Up to vehicle suffix 'C' inclusive
	Rubber edge finisher, outer } For dash top rail	2	348370*	Up to vehicle suffix 'C' inclusive; 4 cylinder models
	Rubber edge finisher, centre } For dash top rail	1	348371*	Up to vehicle suffix 'C' inclusive; 4 cylinder models
	Finisher for dash top casing	1	345077	From vehicle suffix 'D' to suffix 'E' inclusive
	Finisher for dash top casing	1	345591*	From vehicle suffix 'F' onwards
	Drive screw } Fixing finisher to dash	2	78436	From vehicle suffix 'D' onwards
	Plain washer } Fixing finisher to dash	2	4034	From vehicle suffix 'D' onwards
	Spire extension nut } Fixing finisher to dash	2	78237	From vehicle suffix 'D' onwards
3	Cover panel for steering cutout	1	304111	
	Rivet fixing cover plate	4	302222	
4	Cover plate for accelerator pedal hole	1	303990	
	Bolt (¼" UNF x ⅜" long) } Fixing cover plate to dash	2	255202	
	Spring washer } Fixing cover plate to dash	2	3074	
	Nut (¼" UNF) } Fixing cover plate to dash	2	254810	
5	Cover panel for governor cutout in dash	1	306421	Petrol models; Up to vehicle suffix 'C' inclusive
	Bolt (¼" UNF x ⅝" long) } Fixing cover panel	2	255006	Petrol models; Up to vehicle suffix 'C' inclusive
	Spring washer } Fixing cover panel	2	3074	Petrol models; Up to vehicle suffix 'C' inclusive
	Nut (¼" UNF) } Fixing cover panel	2	254810	Petrol models; Up to vehicle suffix 'C' inclusive
	Cover panel on dash for hand speed control	1	275742	Diesel models; Up to vehicle suffix 'C' inclusive
	Grommet in dash for hand speed control rod ..	1	236389	Diesel models; Up to vehicle suffix 'C' inclusive
	Blanking plate for hand throttle aperture	1	345883*	4 cylinder models From vehicle suffix 'D' onwards
	Pop rivet } Fixing plate to dash	7	78257	4 cylinder models From vehicle suffix 'D' onwards
	Bolt (¼" UNF x ½" long) } Fixing plate to dash	3	255204	4 cylinder models From vehicle suffix 'D' onwards
	Spring washer } Fixing plate to dash	3	3074	4 cylinder models From vehicle suffix 'D' onwards
	Nut (¼" UNF) } Fixing plate to dash	3	254810	4 cylinder models From vehicle suffix 'D' onwards
	Mounting box on dash for hand speed control ..	1	338428	Diesel models. From vehicle suffix 'D' to suffix 'E' inclusive
	Grommet in dash for hand speed control rod ..	1	312937	Diesel models. From vehicle suffix 'D' to suffix 'E' inclusive
	Bolt (¼" UNF x ½" long) } Fixing control and mounting box to dash	4	255204	Diesel models. From vehicle suffix 'D' to suffix 'E' inclusive
	Spring washer } Fixing control and mounting box to dash	4	3074	Diesel models. From vehicle suffix 'D' to suffix 'E' inclusive
	Nut (¼" UNF) } Fixing control and mounting box to dash	4	254810	Diesel models. From vehicle suffix 'D' to suffix 'E' inclusive
6	Cover plate for pedal holes	1	330143	4 cylinder models
	Rivet fixing plate	8	78248	4 cylinder models
	Plate for brake and clutch pedals	1	338086	RHStg
	Plate for brake and clutch pedals	1	338085	LHStg
	Rivet ($\frac{3}{16}$" Imex) fixing pedal plate to dash	12	78697	
	Cover plate for toe panel, LH	1	338091	RHStg
	Cover plate for toe panel, RH	1	338088	LHStg
	Rivet ($\frac{3}{16}$" Imex) fixing cover plate to dash	14	78697	

* Asterisk indicates a new part which has not been used on any previous Rover model

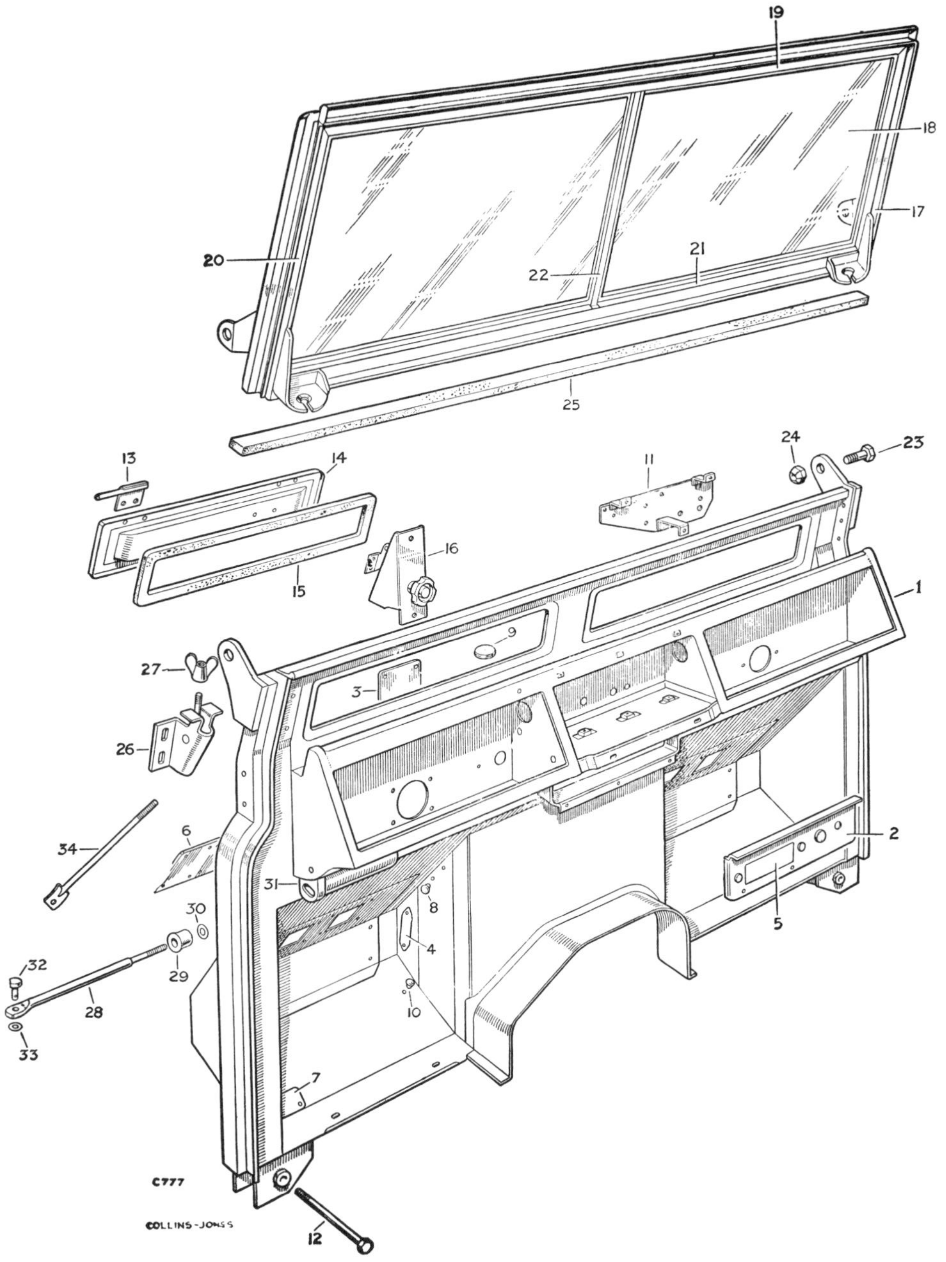

DASH, WINDSCREEN AND VENTILATOR

Plate Ref. 1 2 3 4	DESCRIPTION	Qty	Part No.	REMARKS
7	Cover plate for dipswitch hole	1	307909	1958-60 models
	Screw (2 BA x $\frac{3}{8}$" long) Fixing cover plate to dash	2	77846	1958-60 models
	Spring washer	2	3073	1958-60 models
	Nut (2 BA)	2	3823	1958-60 models
	Drive screw, earth point on dash	1	72628	Series IIA models
8	Rubber plug for redundant accelerator holes	5	307220	
9	Plastic plug, 1$\frac{5}{8}$" dia, for demister holes	2	338027	Alternatives
	Plastic plug, 1$\frac{7}{8}$" dia, for demister holes	4	338009	Alternatives
	Plastic plug, 1$\frac{5}{8}$" dia, for bridge plate	1	338027	
	Plastic plug, $\frac{11}{16}$" dia. For heater pipe holes	2	338019	Alternatives
	Plastic plug, $\frac{3}{4}$" dia.	2	338020	Alternatives
	Plastic plug, $\frac{3}{4}$" dia. for hand throttle hole	1	338020	Petrol models
	Plastic plug $\frac{9}{16}$" dia, for choke control hole	1	338017	6 cylinder Petrol models
	Rubber plug, 3$\frac{5}{32}$" dia, for banjo bolt access hole	1	364684	6 cylinder Petrol models
	Rubber grommet, 1$\frac{1}{2}$" dia, for bridge plate	1	368427	6 cylinder Petrol models
	Plastic plug, $\frac{1}{2}$" dia For redundant windscreen washer holes in dash panel	1	338015	
	Plastic plug, $\frac{3}{4}$" dia	1	338020	
10	Rubber plug, redundant accelerator stop hole	1	73198	
	Rubber grommet for filling redundant clip holes	6	73198	2$\frac{1}{4}$ litre Petrol models. 1 off on Diesel models
	Buffer for bonnet panel on dash front	2	332647	
	Bolt ($\frac{5}{16}$" UNF x $\frac{7}{8}$" long) Fixing dash to steering box support brackets	4	255227	
	Plain washer	10	2249	
	Self-locking nut ($\frac{5}{16}$" UNF)	4	252211	
11	Mounting plate for voltage regulator	1	271863	Series II models
	Drive screw fixing plate at top	2	72628	Series II models
	Bolt ($\frac{1}{4}$" UNF x $\frac{1}{2}$" long) Fixing plate at bottom	1	255204	Series II models
	Plain washer	1	3831	Series II models
	Nut ($\frac{1}{4}$" UNF)	1	254810	Series II models
12	Tie bolt ($\frac{1}{2}$" UNF) Fixing dash to chassis	2	336738	
	Plain washer	4	2253	
	Nut ($\frac{1}{2}$" UNF)	2	255834	
	Ventilator hinge, RH	2	330060	For ventilator lids with hinges fixed by screws and nuts. Check before ordering
13	Ventilator hinge, LH	2	330061	For ventilator lids with hinges fixed by screws and nuts. Check before ordering
	Screw (2 BA x $\frac{1}{2}$" long) Fixing hinges to vent lid	8	78399	For ventilator lids with hinges fixed by screws and nuts. Check before ordering
	Spring washer	8	3073	For ventilator lids with hinges fixed by screws and nuts. Check before ordering
	Plain washer	16	4280	For ventilator lids with hinges fixed by screws and nuts. Check before ordering
	Nut (2 BA)	8	2247	For ventilator lids with hinges fixed by screws and nuts. Check before ordering
	Ventilator lid and hinge for dash, RH	1	336630	
14	Ventilator lid and hinge for dash, LH	1	336631	
	Hinge pin for ventilator lid	4	334121	For ventilator lid with welded hinges. Check before ordering
15	Sealing rubber for ventilator lids	2	330621	
16	Ventilator control mechanism complete	2	337970	Quadrant type
	Knob and fixing screws for ventilator control	2	332327	For quadrant type ventilator control. Check before ordering
	Circlip for ventilator control knob	2	320500	For screw type ventilator control. Check before ordering

* Asterisk indicates a new part which has not been used on any previous Rover model

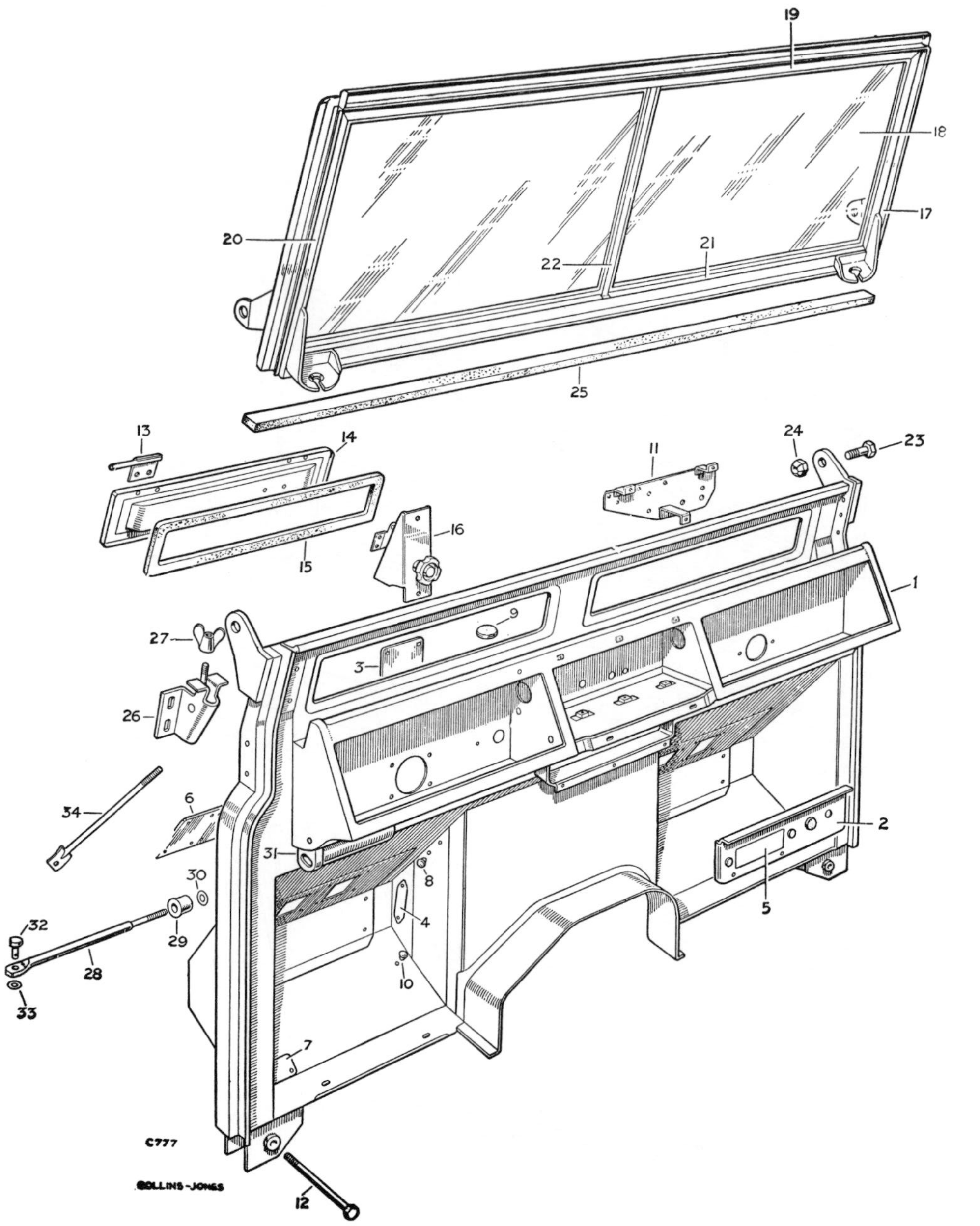
C777
COLLINS-JONES

DASH, WINDSCREEN AND VENTILATOR

Plate Ref.	1 2 3 4 DESCRIPTION	Qty	Part No.	REMARKS
	Drive screw (Fixing ventilator control to dash)	4	78153	For screw type ventilator control. Check before ordering
	Spire nut (Fixing ventilator control to dash)	4	78237	For screw type ventilator control. Check before ordering
	Bolt (2 BA x ½" long) (Fixing ventilator control to dash)	8	234603	For screw type ventilator control. Check before ordering
	Plain washer, small (Fixing ventilator control to dash)	4	3816	For screw type ventilator control. Check before ordering
	Plain washer, medium (Fixing ventilator control to dash)	4	4034	For screw type ventilator control. Check before ordering
	Plain washer, large (Fixing ventilator control to dash)	4	3852	For screw type ventilator control. Check before ordering
	Spring washer (Fixing ventilator control to dash)	8	3073	For screw type ventilator control. Check before ordering
	Nut (2 BA) (Fixing ventilator control to dash)	8	3823	For screw type ventilator control. Check before ordering
	Drive screw (Fixing ventilator control to dash)	4	78436	For quadrant type ventilator. Check before ordering
	Plain washer (Fixing ventilator control to dash)	4	4034	For quadrant type ventilator. Check before ordering
	Spire nut (Fixing ventilator control to dash)	4	78237	For quadrant type ventilator. Check before ordering
	Set bolt (2 BA x ½" long) (Fixing control to ventilator lid)	4	234603	
	Plain washer (Fixing control to ventilator lid)	4	3816	
	Spring washer (Fixing control to ventilator lid)	4	3073	
17	Windscreen complete assembly, toughened glass	1	337643	
	Windscreen complete assembly, laminated glass	1	338845	
18	Glass for windscreen, toughened	2	337644	
	Glass for windscreen, laminated	2	345421	
	Glazing strip for glass	14 ft	78159	
19	Retainer for windscreen glass, RH top	1	330668	
	Retainer for windscreen glass, LH top	1	330669	
20	Retainer for windscreen glass, side	2	330667	
21	Retainer for windscreen glass, RH bottom ..	1	330486	
	Retainer for windscreen glass, LH bottom ..	1	330487	
22	Cover for centre strip	1	330670	Up to vehicle suffix 'C' inclusive
	Cover for centre strip	1	336422	From vehicle suffix 'D' onwards
	Drive screw fixing retainers and cover	37	78126	
23	Special bolt (Fixing windscreen to dash)	2	256280	Bolt and locknut type fixing. Alternatives. Check before ordering
	Plain washer (Fixing windscreen to dash)	4	3829	Bolt and locknut type fixing. Alternatives. Check before ordering
24	Special self-locking nut (Fixing windscreen to dash)	2	250004	Bolt and locknut type fixing. Alternatives. Check before ordering
	Screw (½" UNF x 1⅛" long) (Fixing windscreen to dash)	2	255288	Screw and turret nut type fixing. Alternatives. Check before ordering
	Plain washer (Fixing windscreen to dash)	4	3829	Screw and turret nut type fixing. Alternatives. Check before ordering
	Turret nut (½" UNF), plastic capped (Fixing windscreen to dash)	2	548043	Screw and turret nut type fixing. Alternatives. Check before ordering
25	Rubber sealing strip for windscreen	1	330877	
	End filler for windscreen sealing strip	2	330850	
	Buffer for top of windscreen	1	332550	Only required when hood is fitted
	Pop rivet fixing buffer	4	78410	Only required when hood is fitted
	Fastener for windscreen, RH	1	338194	Turret nut type
26	Fastener for windscreen, LH	1	338195	Turret nut type
27	Wing nut for fastener ($\frac{5}{16}$" BSF)	2	303729	Alternatives. Check before ordering
	Wing nut for fastener ($\frac{5}{16}$" UNF)	2	332098	Alternatives. Check before ordering
	Turret nut ($\frac{5}{16}$" UNF), plastic capped	2	338240	Alternatives. Check before ordering
	Set bolt ($\frac{5}{16}$" UNF x ⅝" long) (Fixing fasteners to dash)	4	255225	
	Plain washer (Fixing fasteners to dash)	4	3841	
	Spring washer (Fixing fasteners to dash)	4	3075	

* Asterisk indicates a new part which has not been used on any previous Rover model

C777

COLLINS-JONES

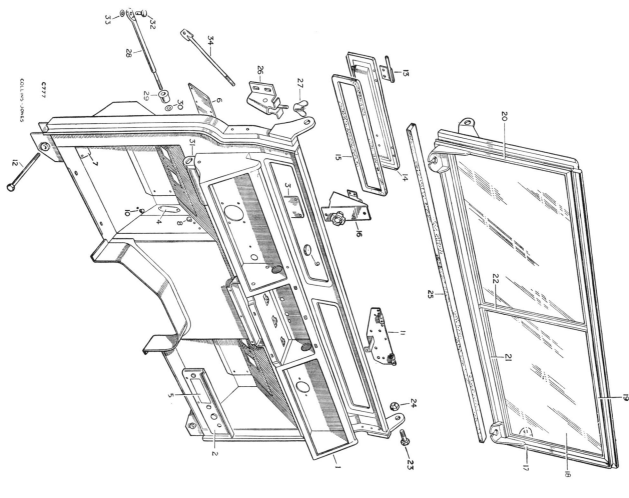

DASH, WINDSCREEN AND VENTILATOR

Plate Ref.	1 2 3 4	DESCRIPTION	Qty	Part No.	REMARKS
28		CHECK STRAP ASSEMBLY	2	306474	
28		Rod for check strap	2	332715	
29		Buffer for check strap	2	306471	
30		Plain washer	4	4035	
		Spring washer	2	3074	Fixing buffer to rod
		Nut (¼" BSF)	4	3819	Alternatives.
		Nut (¼" UNF)	2	254810	Check before ordering
31		Check strap mounting bracket, RH	1	330393	Bolted type
		Check strap mounting bracket, LH	1	330394	
		Bolt (¼" UNF x ⅝" long)	8	255006	Fixing mounting bracket to dash
		Plain washer	8	3946	
		Spring washer	8	3074	
		Plain washer	8	2213	
		Nut (¼" UNF)	8	254810	
32		Clevis pin	2	306564	Fixing check strap
33		Plain washer	4	3815	
		Split pin	2	3901	
34		Tie bar for dash support rod to front door	1	332588	Up to vehicle suffix 'A'
		Tie bar for dash support	1	345549	From vehicle suffix 'B' onwards
		Nut (⅜" BSF) fixing tie bar to dash	2	2827	Alternative fixings
		Nut (⅜" UNF) fixing tie bar to dash	2	254812	
		Bolt (5/16" UNF x ⅞" long)	1	255227	For 5/16" UNF bolt and nut fixing
		Spring washer	1	3075	Up to vehicle suffix 'A'
		Nut (5/16" UNF)	1	254831	
		Bolt (⅜" UNF x ⅞" long)	1	255046	For ⅜" UNF bolt and nut fixing
		Spring washer	1	3076	From vehicle suffix 'B' onwards
		Nut (⅜" UNF)	1	254812	
		Licence holder	1	544573	Plastic type

* Asterisk indicates a new part which has not been used on any previous Rover model

LAND-ROVER

REGISTERED TRADE MARK

88

INTERIM STATION WAGON

GENERAL EXPLANATION

The Interim Station Wagon is mounted on a standard 88 chassis with a modified body. The majority of parts are, therefore, the same as those required for the basic vehicle.

The optional and additional equipment listed below is also required:—

Hard top with side windows
   (less the rear lid and fittings) .. .. .. See optional equipment catalogue
Plain bonnet .. .. See Page 485
Support for spare wheel on bonnet .. .. .. See optional equipment catalogue
Flasher equipment .. .. See page 421
Half-length doors and sidescreens .. .. .. See pages 493 to 497
Locking handles and security catches .. .. .. See optional equipment catalogue
Rear seats .. .. See optional equipment catalogue
Extra driving mirror .. .. See page 485
Floor mats and fixing .. .. See optional equipment catalogue

REGISTERED TRADE MARK

88
STATION WAGON

GENERAL EXPLANATION

The Station Wagon is mounted on a standard 88 chassis with a modified body. The majority of parts are, therefore, the same as those required for the basic vehicle. The following pages 511 to 531 inclusive, list the parts which are peculiar to the Station Wagon.

The optional and additional equipment listed below is also required:—

De-luxe bonnet	See page 485
Flasher equipment	See page 421
Half–length doors and sidescreens	See pages 493 to 497
Locking handles and security catches	See optional equipment catalogue
De-luxe trim for floors and doors	See optional equipment catalogue

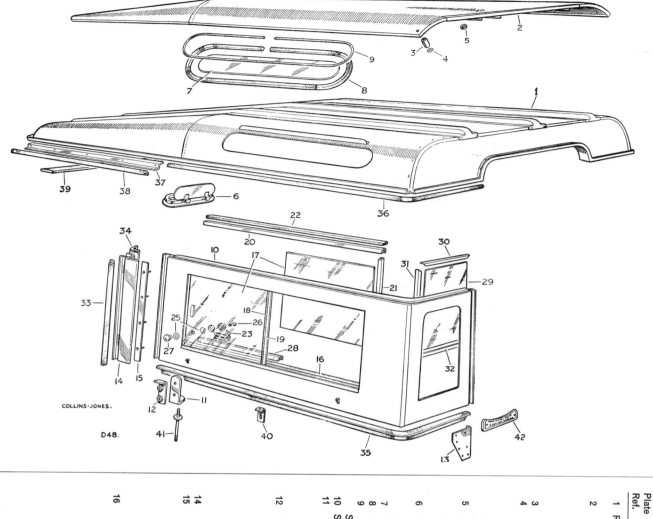

COLLINS-JONES.

D48.

88 STATION WAGON, BODY PANELS AND WINDOWS

Plate Ref.	1 2 3 4	DESCRIPTION		Qty	Part No.	REMARKS
1		ROOF AND TROPICAL ROOF PANEL ASSEMBLY	...	1	332198	
2		Tropical roof panel	...	1	332194	
		Pop rivet	Roof panel to	45	78248	
		Drive screw	stiffeners	45	77704	fixing Alternative
3		Screw (¼" UNF x 2" long)	Fixing	10	78274	
		Plain washer	tropical	20	3900	
4		Distance piece	roof panel	10	336503	
		Rubber washer	to roof	10	302373	
		Spring washer	panel and sides	10	3074	
		Nut (¼" UNF)		10	254810	
		Screw (2 BA x ½" long)	Fixing tropical	8	77869	
		Plain washer	roof panel to	16	3852	
5		Rubber washer	roof at end	8	302371	
		Shakeproof washer	of stiffeners	8	71082	
		Nut (2 BA)		8	3823	
6		Roof ventilator, rear	...	2	333836	
		Roof ventilator, front	...	2	333835	
		Drive screw, fixing ventilator to roof		36	77704	
		Rubber seal for ventilator	...	4	348678	
7		Filler strip for weather strip	...	2	302178	
8		Weather strip for side light	...	2	302177	
9		Side light for roof	...	2	308070	
12		Support bracket at tailboard	...	2	332081	
		Nut plate	to body	2	3075	
		Spring washer	bracket	4	3830	
		Plain washer	mounting	4	255225	
11		Bolt (5/16" UNF x 5/8" long)	Fixing	4	77704	
		Mounting bracket front		2	338554	
10		SIDE PANEL ASSEMBLY, LH		1	332307	
		SIDE PANEL ASSEMBLY, RH		1	332306	
		Nut (¼" UNF)	tailboard corner	4	254810	
		Spring washer	bracket at	8	3074	
		Plain washer	support	8	3840	
		Bolt (¼" UNF x ¾" long)	Fixing	4	255207	
		Screw (¼" UNF x ½" long)	...	2	338552	
		Capping for front door rear seal RH	...	1	332112	
		Capping for front door rear seal LH	...	1	332113	
15		Stud plate	...	2	332065	
14		Screw (2 BA x ½" long)	Fixings cappings	6	78669	
		Plain washer, large	to side	6	3852	
		Plain washer, small	panels	8	3885	
		Spring washer		14	3073	
		Nut (2 BA)		14	3823	
16		Drain channel complete for side windows	...	2	336764	
		Bolt (¼" UNF x 5/8" long)	Fixing	4	255206	
		Plain washer	drain	4	3074	
		Spring washer	channel	4	4071	
		Nut (¼" UNF)	to body	4	254810	

* Asterisk indicates a new part which has not been used on any previous Rover model

88 STATION WAGON, BODY PANELS AND WINDOWS

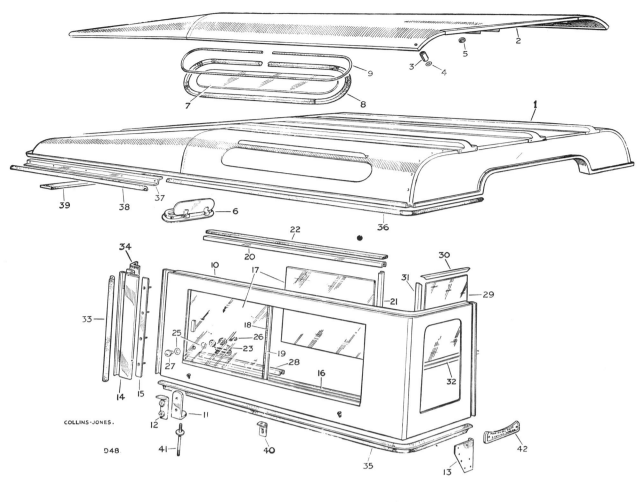

COLLINS-JONES.

948

88 STATION WAGON, BODY PANELS AND WINDOWS

Plate Ref.	1 2 3 4	DESCRIPTION	Qty	Part No.	REMARKS
17		Glass for side window, sliding	4	332217	
18		Sealing rubber for sliding light	2	332230	
19		Channel for sliding light rubber	2	330661	
20		Channel for sliding light—top	4	348396	
		Channel for sliding light—bottom inner	2	348394	
		Channel for sliding light—bottom outer	2	348393	
21		Channel for sliding light—sides	4	336454	
22		Packing strip for top channel	4	332216	
23		Drive screws fixing channels	76	77707	
		Catch for sliding glass, front, overall length $1\frac{1}{2}"$	2	332324	
25		Catch for sliding glass, rear, overall length $1"$	2	332325	
		Washer for catch	16	340391	
26		Screw (2 BA x $\frac{3}{4}"$ long) } Fixing catches	2	78223	2 BA fixing. Alternative to $\frac{1}{4}"$ UNF
		Screw (2 BA x $\frac{7}{8}"$ long) }	2	78260	Check before ordering
		Washer for catch	16	340391	
		Screw ($\frac{1}{4}"$ UNF x $\frac{3}{4}"$ long)	2	78401	$\frac{1}{4}"$ UNF fixing. Alternative to 2 BA
		Screw ($\frac{1}{4}"$ UNF x $\frac{7}{8}"$ long)	2	78402	Check before ordering
27		Tapped plate for catch	4	332329	
28		Runner for sliding catch	4	330848	
		Rivet fixing runner to drain channel	28	78126	
29		Glass for rear end window	2	332200	
30		Retainer for rear end glass upper RH	1	332281	
		Retainer for rear end glass upper LH	1	332280	
31		Retainer for rear end glass inner and outer	4	332282	
32		Retainer for rear end glass lower LH rear	1	332284	
		Retainer for rear end glass lower RH	1	332283	
		Drive screw fixing retainer	36	78126	
33		Seal for front door, upper side	2	302215	
34		Seal at top and bottom for front door, upper side	6	302818	
35		Sealing rubber, lower edge to body	10	332215	
		Rivet fixing rubber to retainer	2	338788	
36		Sealing rubber, RH } Roof to	1	333487	
		Sealing rubber, LH } side	1	334612	
37		Retainer for front door seal top RH	1	330788	
		Retainer for front door seal top LH	1	330789	
38		Seal for front door top	2	330541	
		Rivets, fixing door seal to retainer	10	77869	
		Screw (2 BA x $\frac{1}{2}"$ long) } Fixing	10	3073	
		Spring washer } seal	10	3074	
		Plain washer } retainer	10	3823	
		Nut (2 BA) } to roof	10	3902	
39		Spring washer	10	3074	
		Plain washer	10	3900	
		Rubber seal for canopy	1	330878	
40		Support bracket, centre, body side	2	332201	
		Bolt ($\frac{1}{4}"$ UNF x $\frac{3}{4}"$ long) } Fixing bracket	4	255207	
		Plain washer } to body side	6	3900	
		Rubber buffer, thin } and capping	2	338550	
		Spring washer }	4	3074	
		Nut ($\frac{1}{4}"$ UNF) }	4	254810	

* Asterisk indicates a new part which has not been used on any previous Rover model

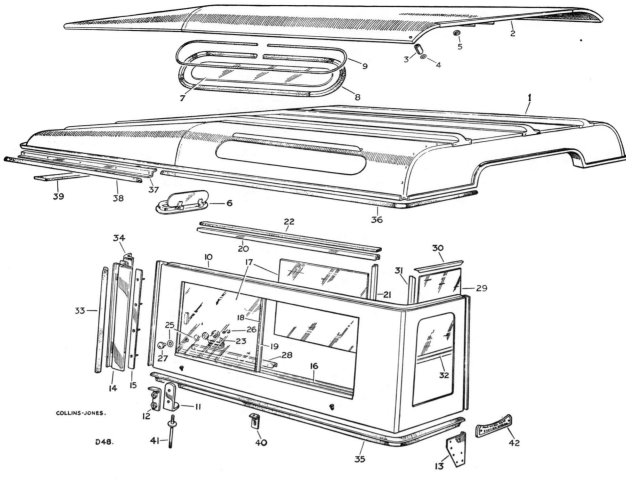

COLLINS-JONES.

D48.

Plate Ref.	1 2 3 4	DESCRIPTION	Qty	Part No.	REMARKS
41		Mounting stud	4	338741	Fixing body
		Plain washer (large)	8	3898	at front and
		Plain washer (small)	2	3830	rear corners
		Special washer	2	332293	
		Rubber buffer, thin	2	338550	
		Rubber buffer, thick	8	338553	
		Spring washer	8	3075	
		Nut (5/16″ UNF)	8	254831	
		Bolt (5/16″ UNF x 7/8″ long)	2	255227	} Alternatives,
		Bolt (5/16″ UNF x 1 1/8″ long)	2	255229	} Check before ordering
		Plain washer	4	3830	Fixing upper to
		Spring washer	2	3075	lower body at rear
		Nut (5/16″ UNF)	2	254831	inner bracket
		Rubber buffer, thick	4	338553	Fixing body at
		Plain washer	4	3840	rear inner
		Spring washer	4	3074	
		Bolt (1/4″ UNF x 5/8″ long)	4	255206	Fixing gusset
		Nut (1/4″ UNF)	4	254810	
		Spring washer	4	3074	
		Bolt (1/4″ UNF x 3/4″ long)	10	255207	Fixing windscreen
		Plain washer	20	3900	to cant
		Spring washer	10	3074	
		Nut (1/4″ UNF)	10	254810	
		Bolt (1/4″ UNF x 3/4″ long)	22	255207	In roof for fixing
		Spring washer	22	3074	rail
		Plain washer	22	3840	at rear corners,
		Spring washer	22	3074	sides and rear
		Nut (1/4″ UNF)	22	254810	
42		'Station Wagon' name plate, front and rear	2	306407	
		Backing plate for name plate, front	1	306408	
		Screw (2 BA x 1/2″ long)	2	77700	Fixing
		Spring washer	2	2260	name plate
		Plain washer	2	3073	to radiator
		Nut (2 BA)	2	2247	grille
		Pop rivet fixing name plate to body	2	30919	
		Rear corner protection angle, RH	1	333253	
		Rear corner protection angle, LH	1	333254	

K9B3

88 STATION WAGON, REAR DOOR

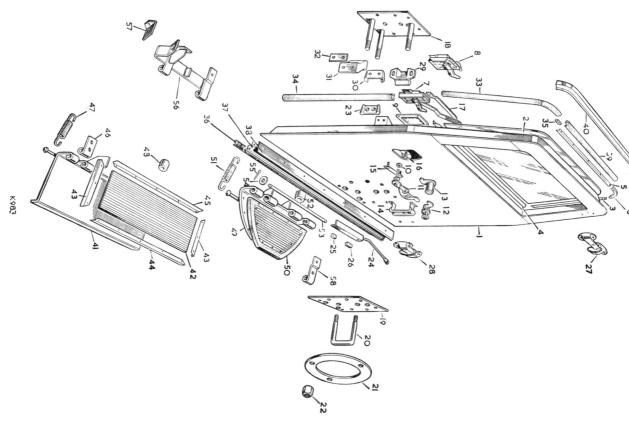

Plate Ref. 1 2 3 4	DESCRIPTION	Qty	Part No.	REMARKS
	REAR DOOR COMPLETE ASSEMBLY ...	1	337780	Incorporating door lock with extended inner handle
	REAR DOOR COMPLETE ASSEMBLY ...	1	337775	Incorporating door lock with short inner handle
1	Rear door ...	1	337381	
2	Glass for rear door ...	1	333031	
3	Retainer for glass, vertical ...	2	333033	
4	Retainer for glass, bottom ...	1	333032	
5	Retainer for glass, top ...	1	333034	
6	Retainer for glass, corners ...	2	333035	
	Drive screw fixing retainers ...	32	78140	
7	DOOR LOCK, MOUNTING AND HANDLE ASSEMBLY ...	1	337789	Cam lever type lock with extended inner handle } Check before ordering
8	DOOR LOCK, MOUNTING AND HANDLE ASSEMBLY ...	1	337801*	Cam lever type lock with short inner handle
9	Mounting plate for door lock ...	1	333080	
10	Washer, handle to cover ...	1	3685	For early type lock with detachable inner handle. Check before ordering
11	Door handle with lock ...	1	332432	For early type lock with detachable inner handle. Check before ordering
		1	306036	Early type lock with hook catch. Check before ordering
12	Barrel lock ...	1	307289	Locking pillar type with plain catch. Cam lever type } Alternatives Check before ordering
13	Barrel lock ...	1	320609	Late type with plain catch. Cam lever type
14	Key for barrel lock ...	1	313858 / 302854	State key number when ordering
	Bracket for door handle	1	302222	
15	Special screw	4	77903	
	Stud plate, door lock, bottom	1	333037	} Fixing door lock
	Spring washer	4	3074	
	Nut (¼" UNF)	4	254810	
16	Locking pillar for catch	1	332434	
	Plain washer	1	2226	} Fixing pillar to door
	Spring washer	1	3073	
	Screw (2 BA x ½" long)	6	2247	
17	Waist rail lock handle protection strip	1	333037	
	Drive screw, fixing lock assembly to door	2	254810	
	Rivet	6	3685	} Fixing lock together
	Nut (2 BA)	4	3073	
	Bolt (¼" UNF x ¾" long)	1	302859	
	Spring washer	4	3074	} Fixing strip to door
	Nut (¼" UNF)	1	78792	
18	Wheel stud plate	1	333435	
19	Clamp plate for spare wheel stud plate ...	2	254810	For lock with extended inner handle
	Screw (¼" UNF x ¾" long)	1	333446	} Fixing stud and clamp plates to rear door
	Spring washer	8	78210	
	Nut (¼" UNF)	8	3074	
20	"U" bolt for spare wheel support ...	1	333447	
	Nut (¼" UNF) fixing 'U' bolt to door	4	254814	
21	Retaining plate for spare wheel	1	333439	
22	Hub nut fixing wheel and retaining plate to 'U' bolt	6	217361	

* Asterisk indicates a new part which has not been used on any previous Rover model

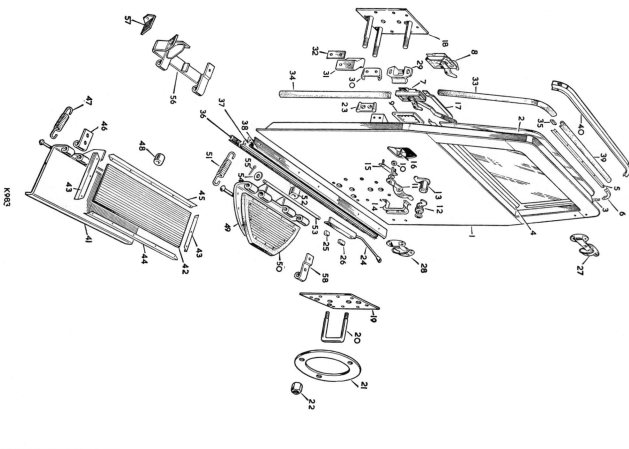

88 STATION WAGON, REAR DOOR

Plate Ref.	1 2 3 4	DESCRIPTION	Qty	Part No.	REMARKS
23		Male dovetail ⎱ For rear end door	1	333438	1958-59
		Male dovetail ⎰	1	332942	1960 onwards
		Screw (¼" UNF x 1" long) ⎱ Fixing male dovetail to door	2	78389	
		Spring washer	2	3074	
		Nut (¼" UNF) ⎰	2	254810	
24		Rod for check strap	1	333041	
25		Buffer for check strap, short	1	306295	
26		Buffer for check strap, long	1	333445	
27		Hinge for rear door, upper	1	333036	
28		Hinge for rear door, lower	1	330117	
		Bolt ($\frac{5}{16}$" UNF x 2" long) ⎱ Fixing hinge to door frame	4	256226	
		Spring washer	4	3075	
		Nut ($\frac{5}{16}$" UNF) ⎰	4	254831	
		Bolt ($\frac{5}{16}$" UNF x $\frac{7}{8}$" long) ⎱ Fixing door hinges to body	4	255227	
		Plain washer	4	3830	
		Spring washer	4	3075	
		Nut ($\frac{5}{16}$" UNF) ⎰	4	254831	
29		Bolt (¼" UNF x 1½" long) ⎱ Fixing check strap rod to floor	4	255213	
		Plain washer, thick	2	3074	
		Plain washer, thin	2	4075	
		Nut (¼" UNF) ⎰	2	3947	
		Striking plate	2	254810	
		Bolt (¼" UNF x 1" long) ⎱ Fixing striking plate to body	2	333140	
		Spring washer ⎰	2	255009	
30		Female dovetail — For rear door	As required		
		Female dovetail ⎱ on body	2	3817	
31		Spacer ⎬ For female dovetail	2	3074	
		Screw (¼" UNF x $\frac{5}{8}$" long) ⎱ fixing dovetail to spacer	2	78233	1958-59
		Screw (¼" UNF x ¾" long) ⎰	2	255207	1960 onwards
32		Shim ⎱ Fixing spacer to LH rear pillar	1	332943	1960 onwards
		Plain washer ⎰	2	305232	
		Spring washer	2	3074	
33		Seal for door sides	3	338788	
34		Seal for door side, LH bottom	1	332563	
35		End filler piece for seals	4	330850	
		Rivet fixing seals to body	22	302818	
36		Rubber seal, door, bottom	1	332564	
37		Seal retainer, door bottom seal	1	332566	
38		Protection strip for retainer	1	332565	
		Drive screw fixing retainer and strip	7	77707	
39		Seal for rear door, top	1	332149	
		Rivet fixing seal	11	302818	
40		Retainer for rear door seal, top ⎱ Fixing retainers to roof panel	1	332151	
		Screw (2 BA x $\frac{3}{8}$" long) ⎬	9	77869	
		Plain washer	9	3902	
		Spring washer ⎭	9	3073	
		Nut (2 BA)	9	3823	

* Asterisk indicates a new part which has not been used on any previous Rover model

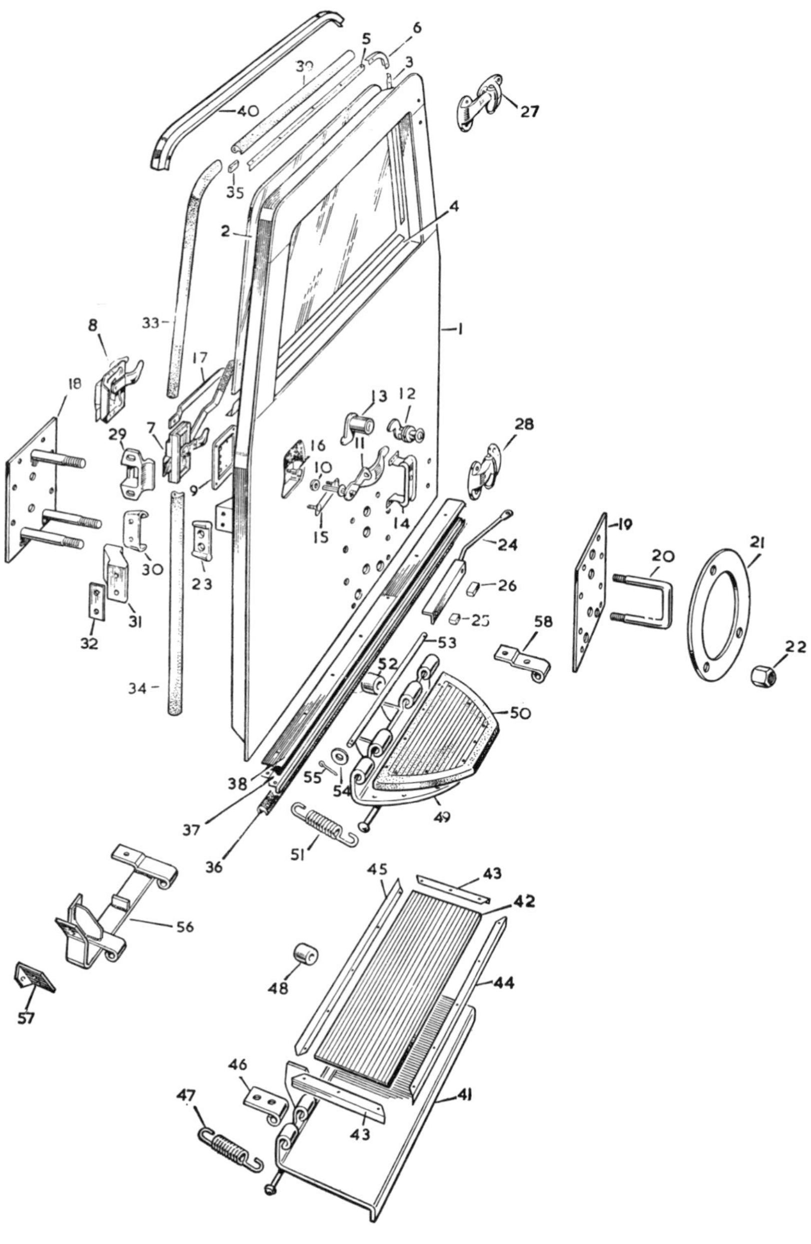

K983

Plate Ref.	1 2 3 4 DESCRIPTION	Qty	Part No.	REMARKS
41	REAR STEP	1	501053	
42	Rubber mat for step	1	245137	
43	Retainer for rear step, side	2	245136	
44	Retainer for rear step mat, front	1	245135	
45	Retainer for rear step mat, rear	1	245134	
	Rivets fixing mat end cappings	16	4022	
46	Hinge centre for rear step	2	245131	
	Bolt ($\frac{3}{8}$" UNF x 4" long) } Fixing hinge centre to rear step	2	256253	
	Locknut ($\frac{3}{8}$" UNF) } Fixing hinge centre to rear step	4	254852	
	Bolt ($\frac{3}{8}$" UNF x $1\frac{1}{8}$" long) } Fixing hinge centre to rear cross member	2	255048	Overall width $19\frac{1}{2}$" Alternative to $11\frac{1}{2}$" step. Check before ordering
	Bolt ($\frac{3}{8}$" UNF x $\frac{7}{8}$" long) } Fixing hinge centre to rear cross member	2	255046	
	Locker } Fixing hinge centre to rear cross member	2	235731	
	Plain washer } Fixing hinge centre to rear cross member	2	2251	
	Packing washer } Fixing hinge centre to rear cross member	2	500588	
	Shakeproof washer } Fixing hinge centre to rear cross member	2	73352	
	Self-locking nut ($\frac{3}{8}$" UNF) } Fixing hinge centre to rear cross member	2	252212	
47	Spring for rear step	2	264362	
48	Buffer for rear step	1	304125	
	Plain washer } Fixing buffer to rear cross member	1	3867	
	Drive screw } Fixing buffer to rear cross member	1	78142	
	REAR STEP	1	508947	Oblong step. Overall width $11\frac{1}{2}$" Alternative to oblong $19\frac{1}{2}$" step or semi-circular step. Check before ordering
	Rubber mat for rear step	1	508957	
	Retainer for rear step mat, side	2	245136	
	Retainer for rear step mat, front	1	508959	
	Retainer for rear step mat, rear	1	508958	
	Rivet fixing mat retainers	14	4022	
	Hinge, centre for rear step RH	1	526471	
	Hinge, centre for rear step LH	1	526472	
	Hinge pin for rear step	1	509120	
	Spring washer } Fixing hinge pin	1	2286	
	Plain washer } Fixing hinge pin	2	3833	
	Split pin } Fixing hinge pin	2	2395	
	Bolt ($\frac{3}{8}$" UNF x $1\frac{5}{8}$" long) } Fixing centre hinges to rear cross-member	2	256042	
	Distance piece } Fixing centre hinges to rear cross-member	2	501070	
	Packing washer } Fixing centre hinges to rear cross-member	2	500588	
	Plain washer } Fixing centre hinges to rear cross-member	2	3933	
	Locknut ($\frac{3}{8}$" UNF) } Fixing centre hinges to rear cross-member	2	252162	
	Support bracket for rear step	1	526473	
	Set bolt ($\frac{3}{8}$" UNF x 1" long) fixing support bracket and centre hinges to rear cross-member	2	255047	
	Spring for rear step	1	264362	
	Buffer for rear step	1	304125	
	Plain washer } Fixing buffer to rear cross-member	1	3867	
	Drive screw } Fixing buffer to rear cross-member	1	78142	
	Buffer for underside of rear step	1	304125	
	Plain washer } Fixing buffer to step	1	3867	
	Screw (2 BA x $\frac{5}{8}$" long) } Fixing buffer to step	1	77758	

* Asterisk indicates a new part which has not been used on any previous Rover model

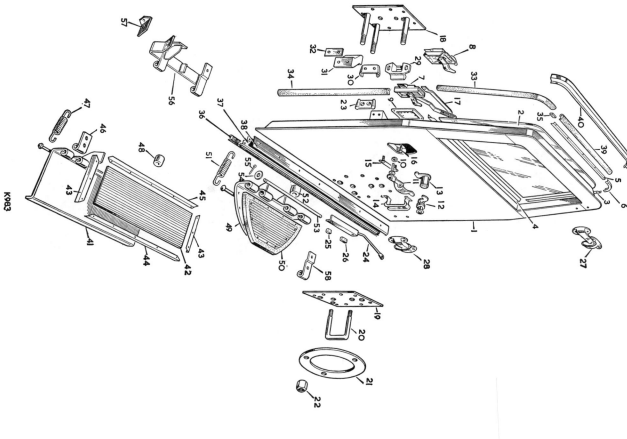

88 STATION WAGON, REAR DOOR

Plate Ref. 1 2 3 4	DESCRIPTION	Qty	Part No.	REMARKS
49	REAR STEP			
	Rubber mat for rear step	1	523589	
	Rivet	7	517977	
	Screw (2 BA x ½" long) } Fixing mat to step	4	307840	
	Spring washer	4	3073	
	Nut (2 BA)	4	3823	
	Hinge, centre for rear step RH	1	526471	
	Hinge, centre for rear step LH	1	526472	
	Bolt (⅜" UNF x 1⅛" long)	2	256042	
	Distance piece	2	501070	
	Plain washer } Fixing centre hinges to rear cross-member	2	500588	Semi-circular step, opposite PTO aperture
	Packing washer	2	3933	
	Locknut (⅜" UNF)	2	252162	
	Support bracket for rear step	1	526473	
	Set bolt (⅜" UNF x 1" long), fixing support bracket and centre hinges to rear cross-member	2	255047	
51	Spring for rear step	1	264362	
52	Buffer for rear step	1	304125	
	Plain washer } Fixing buffer to	1	3867	
53	Drive screw } rear cross-member	1	78142	
	Hinge pin for rear step	1	509120	
	Spring washer	1	2286	
54	Plain washer } Fixing hinge pin	2	3833	
55	Split pin	2	2395	
	REAR STEP			
	Rubber mat for rear step	1	532589	
	Rivet	7	517977	
	Screw (2 BA x ½" long) } Fixing mat	4	307840	Semi-circular step offset to the left of PTO aperture
	Spring washer } to step	4	77869	
	Nut (2 BA)	4	3073	
	Plain washer	4	3823	
	Self-locking nut (5/16" UNF)	1	552132	
	Hinge, centre RH } Fixing RH centre hinge to crossmember	1	552133	
	Bolt (⅜" UNF x 1¼" long)	1	255249	
	Distance piece	1	501070	
	Self-locking nut (⅜" UNF)	1	552129	
	Set bolt (⅜" UNF x 1" long)	1	255247	
	Spring for rear step	1	3076	
	Buffer for rear step	1	304362	
	Drive screw } Fixing buffer to	1	264362	
	Plain washer } rear crossmember	1	3867	
	Hinge pin for rear step	1	509120	
	Spring washer	1	2286	
	Plain washer } Fixing hinge pin	2	3833	
	Split pin	2	2395	
56	Support bracket and hinge centre, LH	1	255029	Alternatives. Check before ordering
57	Anchor bracket for spring } Fixing support bracket and anchor bracket to rear crossmember	2	252212	
	Bolt (5/16" UNF x 1⅛" long)	2	2223	
58	Self-locking nut (5/16" UNF)	2	552132	

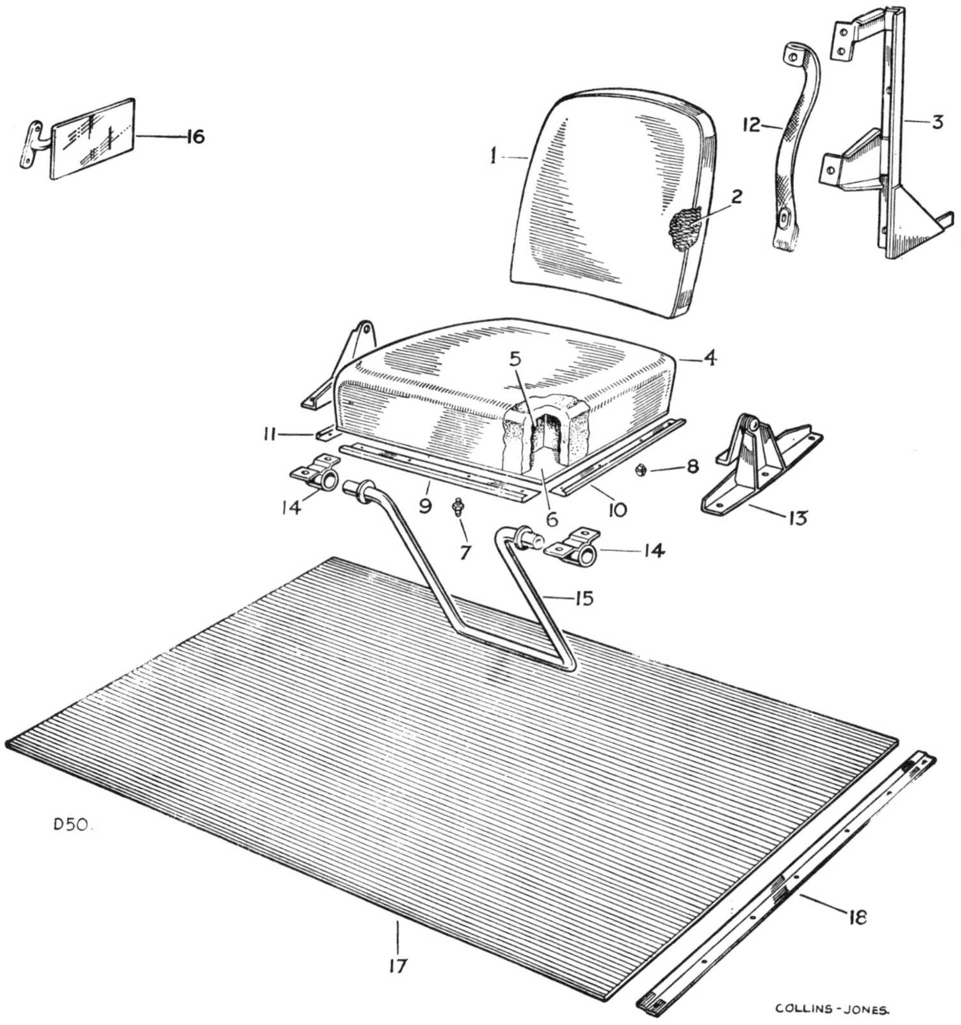

Plate Ref.	1 2 3 4 DESCRIPTION	Qty	Part No.	REMARKS
1	Rear seat backrest complete, GREY	4	331149	
	Rear seat backrest complete, BLACK	4	320703*	
2	Pad for rear seat backrest	4	304054	
	Bolt ($\frac{1}{4}$" UNF x $\frac{1}{2}$" long) } Fixing backrests to supports	20	255204	
	Set bolt ($\frac{1}{4}$" UNF x $\frac{7}{8}$" long) } Fixing backrests to supports	4	255208	
	Plain washer } Fixing backrests to supports	24	3912	
	Spring washer } Fixing backrests to supports	24	3074	
3	Support for backrest RH	4	331478	
	Support for backrest LH	4	331479	
	Bolt ($\frac{1}{4}$" UNF x $\frac{5}{8}$" long) } Fixing backrest support to body capping	6	255206	
	Plain washer, small } Fixing backrest support to body capping	6	3912	
	Plain washer, large } Fixing backrest support to body capping	6	3842	
	Spring washer } Fixing backrest support to body capping	6	3074	
4	Rear seat cushion complete, GREY	4	331148	
	Rear seat cushion complete, BLACK	4	320704*	
5	Rubber interior for seat cushion	4	303809	
6	Seat base panel	4	306316	
7	Fastener stud	4	243618	
	Plain washer } Fixing stud	4	3557	
	Spring washer } Fixing stud	4	3073	
	Nut (2 BA) } Fixing stud	4	2247	
8	Rubber buffer for rear seat cushion	4	312028	
	Woodscrew } Fixing base panel to seat cushion frame	8	20034	
	Cup washer } Fixing base panel to seat cushion frame	8	73962	
9	Front finishers for rear seats	4	307418	
10	Side finishers for rear seats RH	4	307419	
11	Side finishers for rear seats, LH	4	307420	
	Woodscrew } Fixing finishers front and sides	8	20138	
	Woodscrew } Fixing finishers front and sides	16	20147	
12	Retaining straps complete for seat cushion	4	306326	
	Special nut } Fixing backrest and strap to upper body	4	313484	
	Special bolt } Fixing backrest and strap to upper body	4	78296	
13	Cushion pivot bracket complete	8	331122	
14	Support tube bearing } For rear seat cushion	8	306199	
15	Support tube } For rear seat cushion	4	306200	
	Bolt ($\frac{1}{4}$" UNF x $\frac{7}{8}$" long) } Fixing cushion pivot bracket and support tube bearings to seat cushion frame	24	255208	
	Bolt ($\frac{1}{4}$" UNF x $1\frac{1}{4}$" long) } Fixing cushion pivot bracket and support tube bearings to seat cushion frame	8	256201	
	Plain washer } Fixing cushion pivot bracket and support tube bearings to seat cushion frame	32	3912	
	Spring washer } Fixing cushion pivot bracket and support tube bearings to seat cushion frame	32	3074	
	Bolt ($\frac{1}{4}$" UNF x $\frac{5}{8}$" long) } Fixing cushion pivot bracket to wheelarch box	28	255206	
	Plain washer } Fixing cushion pivot bracket to wheelarch box	28	3900	
	Spring washer } Fixing cushion pivot bracket to wheelarch box	28	3074	
	Nut ($\frac{1}{4}$" UNF) } Fixing cushion pivot bracket to wheelarch box	28	254810	
	Packing block for roof lamp	1	332985	
	Mounting pad for roof lamp	1	334111	

* Asterisk indicates a new part which has not been used on any previous Rover model

D50.

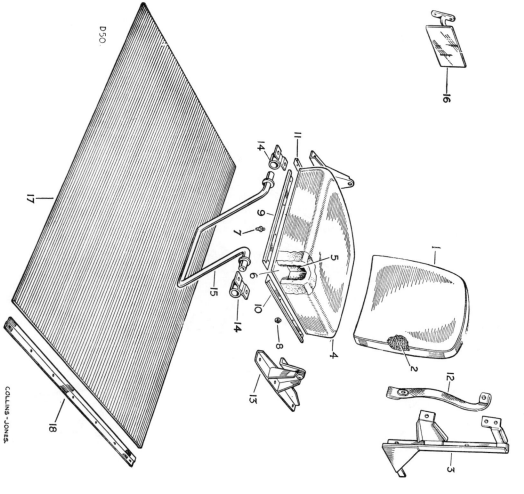

COLLINS - JONES.

88 STATION WAGON, REAR SEATS AND MISCELLANEOUS FITTINGS

Plate Ref.	DESCRIPTION	Qty	Part No.	REMARKS
1 2 3 4	Roof lamp complete	1	265295	
	Lens for roof lamp	1	320608	
	Bulb for roof lamp, LU 382	1	264591	
	Drive screw, fixing lamp and block ...	2	77892	Alternatives.
	Drive screw, fixing lamps and mounting pad	2	77704	Check before ordering
	Lead for roof lamp, LU 861618 ...	1	245386	Alternative to two lines below. Check before ordering
	Lead for roof lamp, roof portion ...	1	265482	
	Lead for roof lamp, chassis portion ...	1	265481	Series II models
	Lead for roof lamp, chassis portion ...	1	519876	Series IIA models. Up to vehicle suffix 'C' inclusive
	Lead for roof lamp, chassis portion ...	1	555879	Series IIA models. From vehicle suffix 'D' onwards
	Grommet for roof lamp lead	1	312856	
	Grommet for lead, LU 859341 ...	2	233244	
	Cable clip	2	41379	
	Drive screw } Fixing roof	2	78155	
	Fuse box, LU/SF5/237131 } lamp lead	1	219078	
	Special screw, LU 153564 } For	2	3279	Series II models
	Special nut, LU 167530 } fuse box	2	3280	
16	Interior mirror	1	345188	
	Support bracket for interior mirror ...	1	338275	From vehicle suffix 'D' onwards
17	Drive screw fixing mirror	2	78495	
18	Rubber mat for rear body floor ...	1	331401	
	Protection strip for floor mat	1	332565	
	Drive screw fixing protection strip ...	7	77707	

* Asterisk indicates a new part which has not been used on any previous Rover model

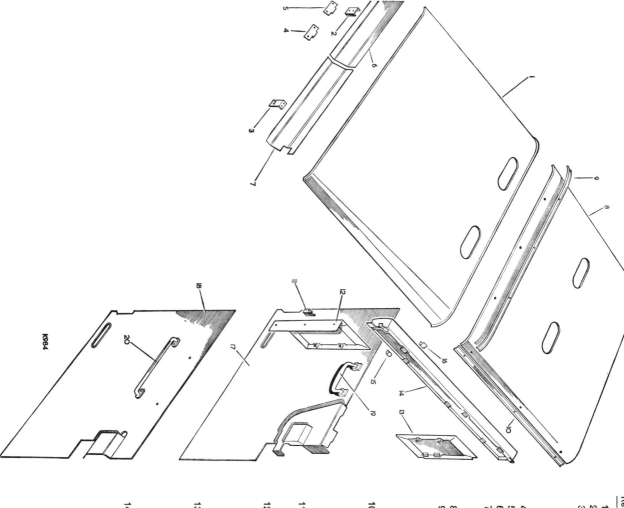

K984

Plate Ref.	1 2 3 4	DESCRIPTION	Qty	Part No.	REMARKS
1		Roof trim, front portion	1	331491	
2		Bracket, cant rail to roof frame	4	331168	
3		Bracket for roof trim, front portion	4	349218	} Alternatives
3		Bracket for roof trim, front portion	4	77704	
		Drive screw (½" No. 6) fixing bracket to cant rail	8	77704	
		Drive screw (½" No. 6) } Fixing	9	3886	
		Plain washer, rear } roof frame	5	3852	
		Plain washer, front } to bracket	4	3852	
7		Canopy trim panel LH	1	331493	
6		Canopy trim panel RH ...	1	331492	
5		Outer bracket, canopy panel ...	1	331390	
4		Centre bracket, canopy panel	1	331391	
		Plain washer } Fixing trim panel	4	3852	
		Drive screw (½" No. 6) } to brackets	4	77704	
8		Head cloth, rear portion	1	331508	
9		Fixing strip, head cloth, front and rear ...	2	331203	
		Plain washer } Fixing	6	77704	
		Drive screw (½" No. 6) } strip to roof	6	3886	
		Retainer for roof trim	4	332240	
		Plain washer } Fixing retainer to vent	4	314394	
		Spire nut	12	77932	
		Drive screw	24	331196	
10		Side rail, head cloth, rear portion	2	3925	
		Plain washer } Fixing	2	77704	
		Drive screw (½" No. 6) } side panel	2	257015	
		Bolt (10 UNF x ⅜" long) } Fixing	2	3902	
		Plain washer } side panel to	2	257023	
		Nut (10 UNF) } roof at rear	2		
11		Fixing bracket, RH, sidelight casing, front	3	331192	
		Fixing bracket, LH, sidelight casing, front ...	3	331193	
		Plain washer	3	331186	
		Drive screw } Fixing casing	6	3886	
		Plain washer } to bracket	6	77704	
12		Sidelight casing, trimmed, LH front } BLACK	1	320683*	
		Sidelight casing, trimmed, RH front } GREY	1	320682*	
		Sidelight casing, trimmed, LH, front	1	331187	
		Sidelight casing, trimmed, RH, front	1	331186	
13		Sidelight casing, trimmed, LH rear } GREY	1	331180	
		Sidelight casing, trimmed, RH rear	1	331179	
		Drive screw } Fixing casing	6	3886	
		Plain washer } to panel	2	77704	
		Drive screw ⅜" No. 6	2	3886	
14		Roll trim for cant rail, RH	1	331497	
		Roll trim for cant rail, LH } GREY	1	331498	
		Roll trim for cant rail, RH	1	320686*	
		Roll trim for cant rail, LH } BLACK	1	320687*	
		Drive screw (½" No. 6) fixing roll trim to body	4	77903	

* Asterisk indicates a new part which has not been used on any previous Rover model

88 STATION WAGON TRIM

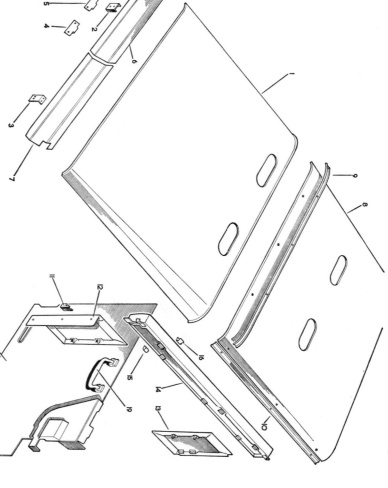

K084

88 STATION WAGON TRIM

Plate Ref.	1 2 3 4	DESCRIPTION	Qty	Part No.	REMARKS
15		Retaining Clip } Fixing	8	331071	
16		Edge clip } roll trim	8	349899	
17		Rear door trim casing, GREY ...	1	331638	
18		Rear door trim casing, GREY ...	1	345426*	When door lock with extended inner handle is fitted } Check before ordering
		Rear door trim casing, BLACK ...	1	320679*	When door lock with short inner handle is fitted }
		Drive screw (¼" No. 6) fixing lock casing to door	4	77704	
19		Door pull handle	1	306460	Alternative to grab handle
		Drive screw (⅝" No. 10) fixing handle to door	2	78167	
20		Door grab handle	1	345450*	Alternative to pull handle
		Set bolt (¼" UNF x ¾" long) } Fixing grab handle to door	2	255207	Check before ordering
		Spring washer	2	3074	

* Asterisk indicates a new part which has not been used on any previous Rover model

LAND-ROVER

REGISTERED TRADE MARK

109
STATION WAGON

GENERAL EXPLANATION

The 109 Station Wagon is mounted on a modified Land-Rover 109 chassis; the majority of parts are, however, the same as those required for the basic vehicle.

The following pages, 535 to 574 inclusive, list the parts which are peculiar to the 109 Station Wagon.

The optional equipment listed below is also fitted:—

Private locks .. .. See optional equipment catalogue

Full-length doors on early models. Later models are fitted with the standard two-piece doors .. See pages 493 to 497

De-luxe bonnet .. .. See page 485

Flasher equipment .. .. See page 421

Front door and floor trim .. .. See optional equipment catalogue

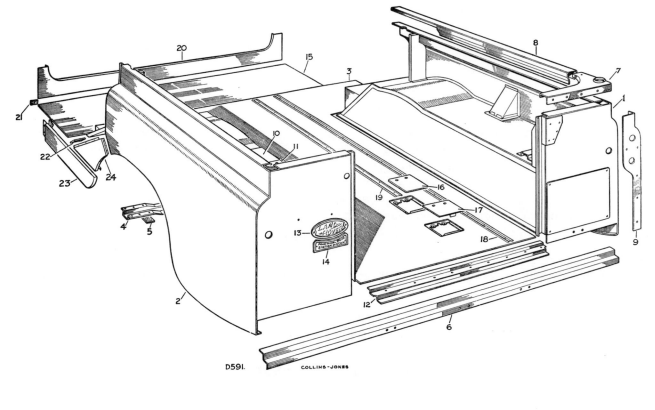

D591.   COLLINS-JONES

Plate Ref.	1 2 3 4	DESCRIPTION	Qty	Part No.	REMARKS
		REAR BODY COMPLETE ASSEMBLY, LOWER	1	332983	
1		Body side and wheelarch complete, RH	1	333562	
		Rear end panel, RH	1	333144	
		Body side panel, RH	1	330248	
2		Body side and wheelarch complete, LH	1	333452	
		Rear end panel, LH	1	333145	
		Body side panel, LH	1	330249	
3		Rear floor	1	333176	
4		Cross-member and pads for rear floor	1	330265	
5		Mounting pad for rear floor cross-member	4	332582	
		Rivet fixing pad to cross-member	16	73979	
6		Rear mounting angle	1	333159	
7		Rear corner capping, RH	1	333266	
		Rear corner capping, LH	1	333267	
		Pop rivet, long	4	78410	
		Pop rivet, short	4	300919	
8		Capping for body top side, RH	1	300783	
		Capping for body top side, LH	1	333234	
		Rivet } Fixing cappings to body	8	333235	
		Pop rivet }	4	300919	
9		Rivet } Fixing cappings to rear body sides	26	300789	
		Pop rivet }	18	333253	
		Rear corner protection angle, RH	1	333254	
		Rear corner protection angle, LH	1	300783	
		Rivet } Fixing protection angle to body	4	300784	
		Rivet, medium }	2	300783	
		Rivet, short }	20	300783	
		Rivet, long }	2	78248	
		Corner bracket, plain type, RH	1	302186	
		Corner bracket, plain type, LH	1	330237	
		Rivet	1	330238	
10		Lid for rear tool locker	2	300783	
11		Hinge complete	2	300784	
		Pop rivet, fixing hinge to lid	4	300401	
		Hasp complete	1	303875	
		Hasp complete	1	303871	For turnbuckle with semi-circular plate } Alternatives. Check before ordering
			1	334523	For turnbuckle with circular plate
		Rivet fixing hasp to lid	2	300789	
		Packing washer	2	3968	
		Bolt (¼" UNF × ⅝" long) } Fixing hinge to wheelarch box	2	255206	
		Spring washer }	2	3074	
		Nut (¼" UNF)	2	254810	
		Turnbuckle for locker lid	1	300851	With circular plate } Alternatives Check before ordering
		Turnbuckle for locker lid	1	334525 } plate	With semi-circular
		Plain washer for turnbuckle	1	3864	
		Rivet fixing turnbuckle to wheelarch box	3	78248	

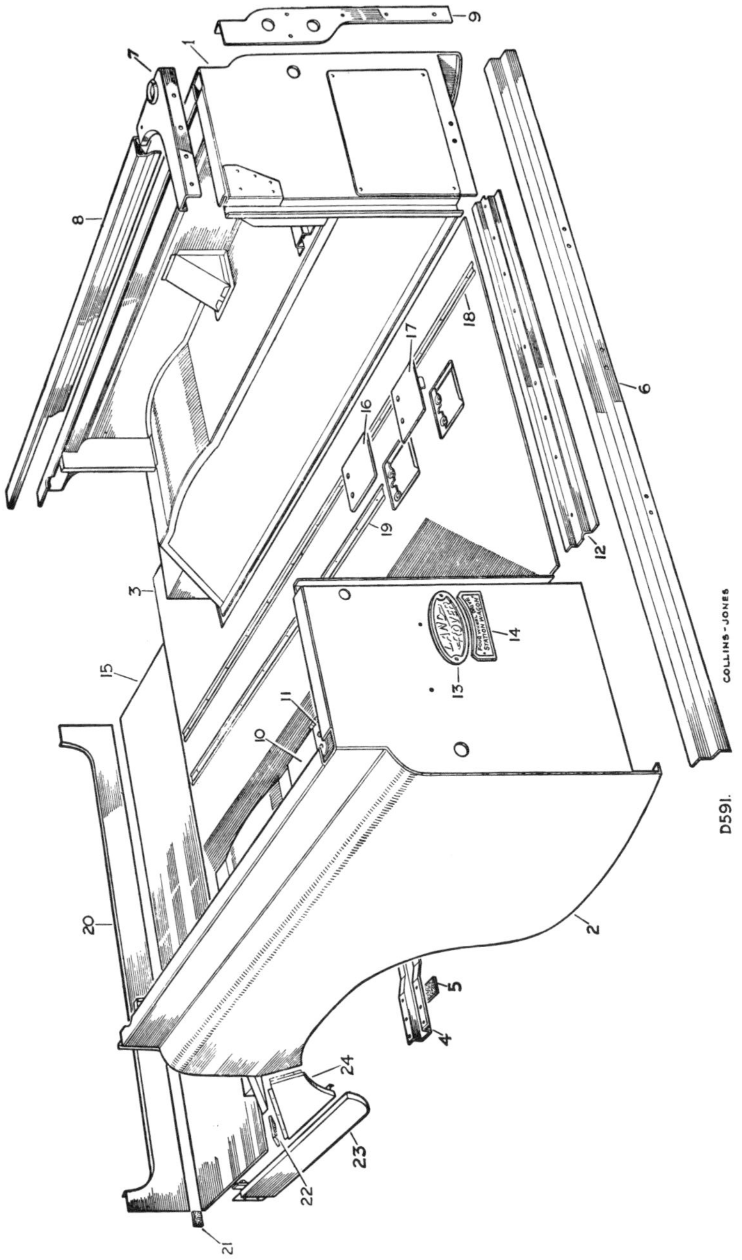

Plate Ref.	1 2 3 4 DESCRIPTION	Qty	Part No.	REMARKS
12	Retainer for floor mat, rear end	1	332756	
	Drive screw fixing retainer	13	77704	
13	'Land-Rover' nameplate	1	332670	
14	'Station Wagon' nameplate	1	306407	Except Twelve-seater Land-Rover
	Pop rivet, fixing nameplates	4	300919	
15	Intermediate floor for rear body	1	333189	4 cylinder models
	Intermediate floor for rear body	1	345120	6 cylinder models
	Special bolt } Fixing floor to rear heelboard and toe panel	10	320045	
	Plain washer } Fixing floor to rear heelboard and toe panel	10	3817	
	Special nut } Fixing floor to rear heelboard and toe panel	10	302532	
	Bolt ($\frac{1}{4}$" UNF x $\frac{3}{4}$" long) } Fixing floor to sill	6	78210	
	Plain washer } Fixing floor to sill	12	3900	
	Spring washer } Fixing floor to sill	6	3074	
	Nut ($\frac{1}{4}$" UNF) } Fixing floor to sill	6	254810	
16	Cover plate for fuel tank, front	1	333201	
17	Cover plate for fuel tank, rear	1	333202	
	Special bolt } Fixing cover plate to floor	4	307339	
	Plain washer } Fixing cover plate to floor	4	3900	
	Special nut } Fixing cover plate to floor	4	302532	
	Bolt ($\frac{5}{16}$" UNF x $1\frac{1}{4}$" long) } Fixing rear body to chassis at rear heelboard	4	255211	
	Fan disc washer } Fixing rear body to chassis at rear heelboard	8	510912	
	Shim } Fixing rear body to chassis at rear heelboard	10	305232	
	Nut ($\frac{5}{16}$" UNF) } Fixing rear body to chassis at rear heelboard	4	254831	
18	Tread plate	4	333183	
	Pop rivet fixing tread plates to floor	36	78248	
	Tread plate for locker floor	1	334179	
	Pop rivet fixing tread plate to locker floor	7	78248	
20	Toe panel complete	1	345149	
	Gusset bracket at toe panel	2	345113*	} Provision for safety harness anchorage, outer and centre
	Backing plate for gusset bracket	2	345115*	} Provision for safety harness anchorage, outer and centre
	Rivet, fixing bracket and plate	12	2986	} Provision for safety harness anchorage, outer and centre
	Bolt ($\frac{1}{4}$" UNF x $\frac{3}{4}$" long) } Fixing toe panel to seat base	2	255207	
	Bolt ($\frac{1}{4}$" UNF x $\frac{5}{8}$" long) } Fixing toe panel to seat base	7	255206	
	Plain washer } Fixing toe panel to seat base	9	3840	
	Spring washer } Fixing toe panel to seat base	9	3074	
	Nut ($\frac{1}{4}$" UNF) } Fixing toe panel to seat base	9	254810	
	Bolt ($\frac{1}{4}$" UNF x $\frac{5}{8}$" long) } Fixing panel gusset to 'BC' post	4	255206	
	Plain washer } Fixing panel gusset to 'BC' post	4	3840	
	Spring washer } Fixing panel gusset to 'BC' post	4	3074	
	Nut ($\frac{1}{4}$" UNF) } Fixing panel gusset to 'BC' post	4	254810	
	Drive screw } Fixing panel gusset to 'BC' post	4	72628	Alternative to nut plate and fixing below
	Nut plate } Fixing panel gusset to 'BC' post	2	333714	} Alternatives to drive screw. Part Number 72628
	Bolt ($\frac{1}{4}$" UNF x $\frac{5}{8}$" long) } Fixing panel gusset to 'BC' post	4	255206	} Alternatives to drive screw. Part Number 72628
	Plain washer } Fixing panel gusset to 'BC' post	4	3840	} Alternatives to drive screw. Part Number 72628
	Spring washer } Fixing panel gusset to 'BC' post	4	3074	} Alternatives to drive screw. Part Number 72628
	Bolt ($\frac{1}{4}$" UNF x $\frac{5}{8}$" long) } Fixing seat box end panel to 'BC' post	4	255206	
	Plain washer } Fixing seat box end panel to 'BC' post	4	3840	
	Spring washer } Fixing seat box end panel to 'BC' post	4	3074	

* Asterisk indicates a new part which has not been used on any previous Rover model

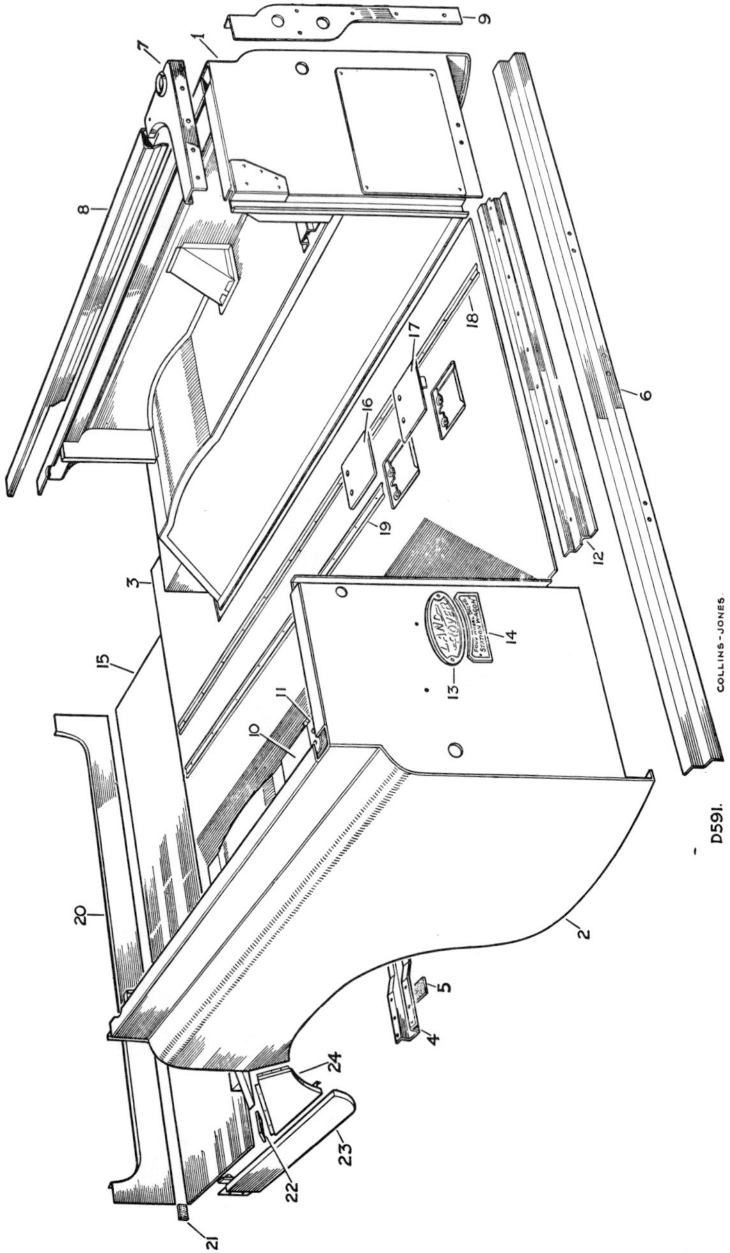

Plate Ref.	1 2 3 4	DESCRIPTION	Qty	Part No.	REMARKS
21		Sealing rubber, 'BC' post to toe panel	2	333745	
22		Sealing rubber, 'D' post to wheelarch front flange	2	333263	
		Sill panel rear RH	1	333160	Deep type panel Depth $4\frac{15}{16}$" — Check before ordering
23		Sill panel, rear LH	1	333161	Deep type panel Depth $4\frac{15}{16}$" — Check before ordering
		Sill panel, rear RH	1	337710*	Narrow type panel Depth $2\frac{27}{32}$" — Check before ordering
		Sill panel, rear LH	1	337711*	Narrow type panel Depth $2\frac{27}{32}$" — Check before ordering
		Bolt ($\frac{1}{4}$" UNF x $\frac{5}{8}$" long) — Fixing sill panel to bracket on sill channel	4	255206	
		Plain washer — Fixing sill panel to bracket on sill channel	8	3821	
		Spring washer — Fixing sill panel to bracket on sill channel	4	3074	
		Nut ($\frac{1}{4}$" UNF) — Fixing sill panel to bracket on sill channel	4	254810	
		Fixing plate — Fixing sill panels together	2	330333	Alternative fixings
		Bolt ($\frac{1}{4}$" UNF x $\frac{5}{8}$" long) — Fixing sill panels together	2	255206	Alternative fixings
		Plain washer — Fixing sill panels together	4	3821	
		Spring washer — Fixing sill panels together	4	3074	
		Nut ($\frac{1}{4}$" UNF) — Fixing sill panels together	4	254810	
		Bolt ($\frac{1}{4}$" UNF x $\frac{5}{8}$" long) — Fixing rear sill panel to body, side	2	255206	
		Plain washer — Fixing rear sill panel to body, side	4	3840	
		Spring washer — Fixing rear sill panel to body, side	2	3074	
		Nut ($\frac{1}{4}$" UNF) — Fixing rear sill panel to body, side	2	254810	
		Stay, sill channel to sill panel	2	334896	Early models with deep type sill panel
		Bolt ($\frac{1}{4}$" UNF x $\frac{5}{8}$" long) — Fixing stay to sill channel	2	255206	Early models with deep type sill panel
		Plain washer — Fixing stay to sill channel	2	3840	Early models with deep type sill panel
		Spring washer — Fixing stay to sill channel	2	3074	Early models with deep type sill panel
		Nut ($\frac{1}{4}$" UNF) — Fixing stay to sill channel	2	254810	Early models with deep type sill panel
		Body side lower front extension, RH	1	333243	
24		Body side lower front extension, LH	1	333244	
		Rivet, panel to rear pillar	18	78248	
		Rivet, front to rear panel	4	78248	
		Bolt ($\frac{1}{4}$" UNF x $\frac{3}{4}$" long) — Fixing rear pillar LH to floor and rear cross-member	4	255207	
		Plain washer — Fixing rear pillar LH to floor and rear cross-member	8	3840	
		Spring washer — Fixing rear pillar LH to floor and rear cross-member	4	3074	
		Nut ($\frac{1}{4}$" UNF) — Fixing rear pillar LH to floor and rear cross-member	4	254810	

* Asterisk indicates a new part which has not been used on any previous Rover model

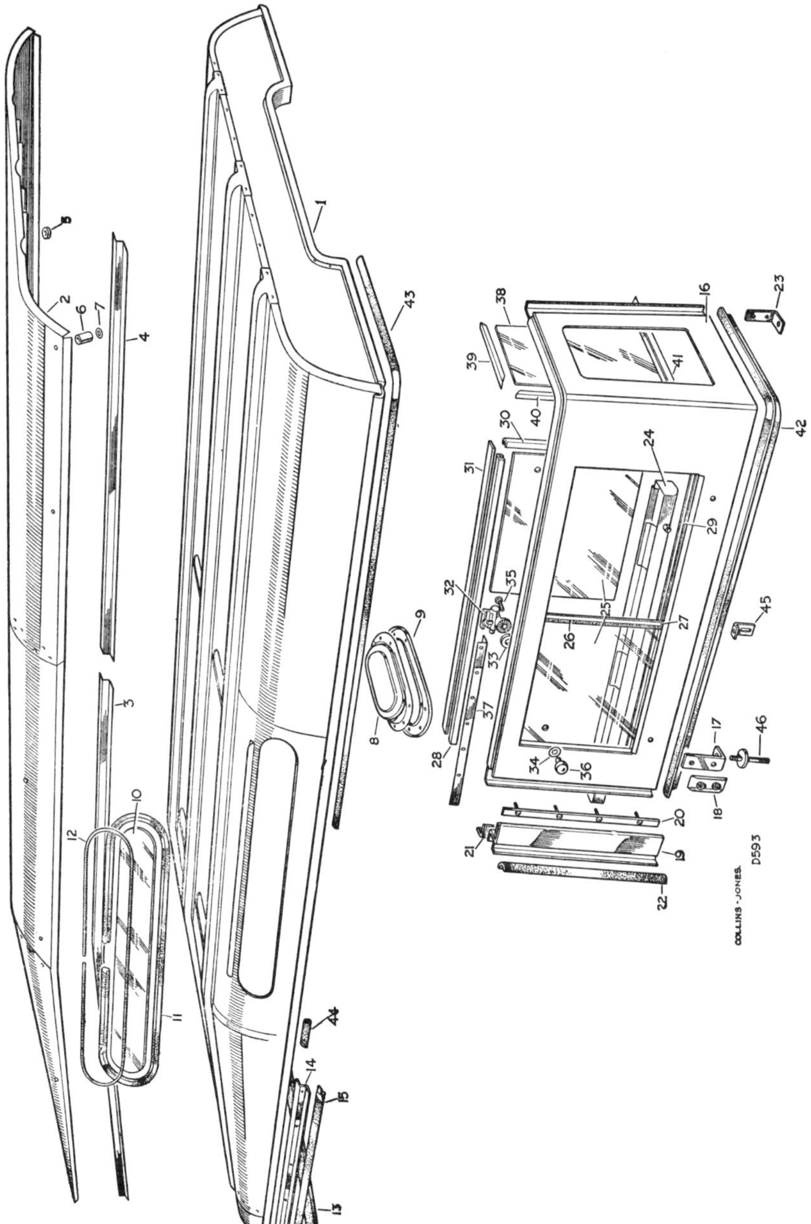
COLLINS - JONES.
D593

109 STATION WAGON, ROOF AND BODY SIDE PANELS

Plate Ref.	1 2 3 4	DESCRIPTION		Qty	Part No.	REMARKS
1		ROOF COMPLETE		1	333211	
2		TROPICAL ROOF PANEL ASSEMBLY		1	332987	
3		Stiffener, front	For roof panel	3	332028	
4		Stiffener, rear	For roof panel	3	333252	
		Drive screw	Fixing stiffener to roof	33	77704	Alternatives. Check before ordering
		Pop rivet	Fixing stiffener to roof	33	78248	Alternatives. Check before ordering
		Screw (2 BA x ½" long)	Fixing panel to roof at end of stiffeners	8	77869	
		Plain washer	Fixing panel to roof at end of stiffeners	16	3852	
5		Rubber washer	Fixing panel to roof at end of stiffeners	8	302371	
		Shakeproof washer	Fixing panel to roof at end of stiffeners	8	71082	
		Nut (2 BA)	Fixing panel to roof at end of stiffeners	8	3823	
		Screw (¼" UNF x 2" long)	Fixing roof to panel at sides	12	78274	
		Screw (¼" UNF x 2¼" long)	Fixing roof to panel at sides	2	78382	
		Plain washer	Fixing roof to panel at sides	28	3900	
6		Distance piece	Fixing roof to panel at sides	14	336503	
7		Rubber washer	Fixing roof to panel at sides	14	302373	
		Spring washer	Fixing roof to panel at sides	14	3074	
		Nut (¼" UNF)	Fixing roof to panel at sides	14	254810	
		Drive screw	Fixing panel to roof stiffener	66	77704	Alternatives. Check before ordering
		Pop rivet	Fixing panel to roof stiffener	66	78248	Alternatives. Check before ordering
8		Roof ventilator		4	333836	
		Rubber seal for ventilator		4	348678	
		Drive screw		32	77704	
9		Retainer for roof trim at ventilator		4	332240	
		Drive screw fixing retainer to vent		24	77932	
10		Roof side light glass		2	308070	
11		Weather strip for glass		2	302177	
12		Filler strip for weather strip		2	302178	
13		Sealing rubber for roof to windscreen		1	330878	
		Bolt (¼" UNF x ¾" long)	Fixing roof to windscreen	10	255207	
		Plain washer	Fixing roof to windscreen	10	3900	
		Spring washer	Fixing roof to windscreen	10	3074	
		Nut (¼" UNF)	Fixing roof to windscreen	10	254810	
		Seal retainer for door top, RH front		1	330788	
14		Seal retainer for door top, LH front		1	330789	
15		Seal for retainer		2	330541	
		Rivet fixing seal to retainer		10	302818	
		Screw (2 BA x ½" long)	Fixing seal retainer to roof	10	77869	
		Plain washer	Fixing seal retainer to roof	10	3852	
		Spring washer	Fixing seal retainer to roof	10	3073	
		Nut (2 BA)	Fixing seal retainer to roof	10	3823	
		Bolt (¼" UNF x ¾" long)	Fixing roof to sides	30	255207	
		Plain washer	Fixing roof to sides	34	3840	
		Spring washer	Fixing roof to sides	30	3074	
		Nut (¼" UNF)	Fixing roof to sides	30	254810	

* Asterisk Indicates a new part which has not been used on any previous Rover model

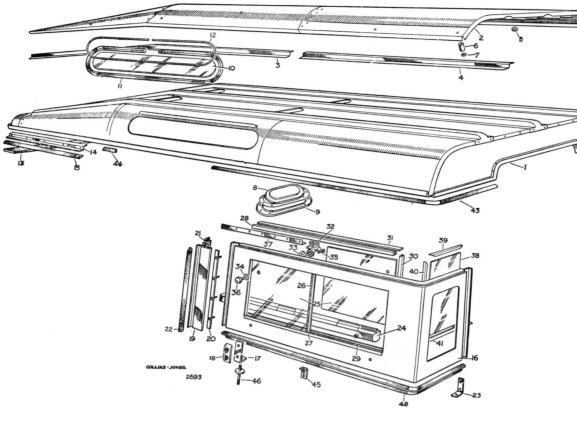

COLLINS-JONES.

D593

109 STATION WAGON, ROOF AND BODY SIDE PANELS

Plate Ref.	1 2 3 4	DESCRIPTION	Qty	Part No.	REMARKS
16		SIDE PANEL ASSEMBLY, RH	1	332306	
		SIDE PANEL ASSEMBLY, LH	1	332307	
17		Mounting bracket front	2	338554	
		Bolt (⅕₆" UNF x ⅝" long) ⎱ Fixing	4	255225	
18		Plain washer ⎰ mounting	4	3830	
		Spring washer ⎱ bracket	4	3075	
		Nut plate ⎰ to body	2	332081	
19		Capping for 'D' post, RH	1	332112	
		Capping for 'D' post, LH	1	332113	
20		Stud plate	2	332065	
		Screw (2 BA x ⅝" long) ⎱ Fixing	6	77869	
		Plain washer, large ⎰ cappings to	6	3852	
		Plain washer, small ⎱ side panels	8	3885	
		Spring washer ⎰	14	3073	
		Nut (2 BA)	14	3823	
21		Sealing rubber at top and bottom, upper side, 'D' post pillar	6	332215	
22		Door sealing rubber at 'D' post, upper side ...	2	338788	
23		Rivet fixing seal	10	78321	
		Support bracket at tailboard	2	338552	
		Bolt (¼" UNF x ¾" long) ⎱ Fixing support	4	255207	
		Plain washer ⎰ bracket at	8	3840	
		Spring washer ⎱ tailboard	4	3074	
		Nut (¼" UNF) ⎰ corner	4	254810	
24		Drain channel complete for side window ...	2	336764	
		Bolt (¼" UNF x ⅝" long) ⎱	4	255206	
		Plain washer ⎰ Fixing drain	4	4071	
		Spring washer ⎱ channel to body	4	3074	
		Nut (¼" UNF) ⎰	4	254810	
25		Glass for side window, sliding ...	4	332217	
26		Sealing rubber for sliding light ...	2	332230	
27		Retainer for sliding light rubber ...	2	330661	
28		Channel for sliding light, top ...	4	348396	
29		Channel for sliding light, bottom inner ...	2	348394	
30		Channel for sliding light, bottom outer ...	2	348393	
31		Channel for sliding light, side ...	2	336454	
		Packing strip for top channel ...	4	332216	
		Drive screw fixing channel	76	77707	
32		Catch for sliding glass, front, overall length 1½"	2	332324	
		Catch for sliding glass, rear, overall length 1" ...	2	332325	
34		Washer for catch	16	340391	
		Screw (2 BA x ⅝" long) ⎱	2	78223	2 BA-type fixing. Alternative to ¼" UNF
35		Screw (2 BA x ⅝" long) ⎰	2	78260	Check before ordering
		Washer for catch ⎱ Fixing catches	16	340391	
		Screw (¼" UNF x ⅝" long) ⎰	2	78401	¼" UNF-type fixing. Alternative to 2 BA
		Screw (¼" UNF x ⅞" long) ⎱	2	78402	Check before ordering
36		Tapped plate for catch ⎰	4	332329	

* Asterisk indicates a new part which has not been used on any previous Rover model

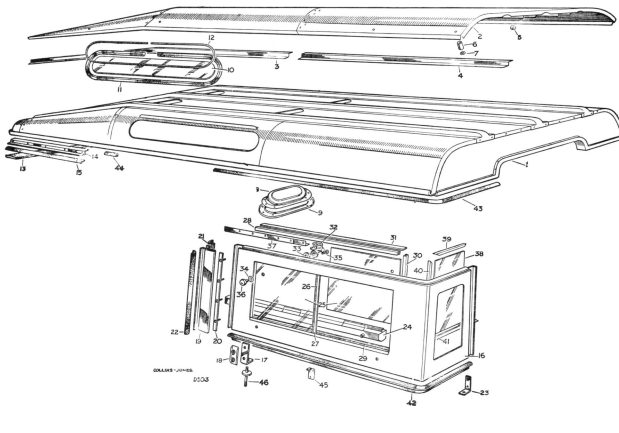

COLLINS-JONES.

D503

109 STATION WAGON, ROOF AND BODY SIDE PANELS

Plate Ref.	1	2	3	4	DESCRIPTION	Qty	Part No.	REMARKS
37					Runner for sliding catch	4	330848	
					Rivet fixing runner to drain channel	28	78248	
38					Glass for rear end window	2	332200	
					Glazing strip for glass	2	332308	
					Retainer for rear end glass, upper RH	1	332280	
39					Retainer for rear end glass, upper LH	1	332281	
40					Retainer for rear end glass, inner and outer	1	332282	
41					Retainer for rear end glass, lower LH	1	332284	
					Retainer for rear end glass, lower RH	1	332283	
					Drive screw fixing retainer	36	78126	
42					Sealing rubber, upper to lower body side	2	333487	
43					Sealing rubber, RH } Body side	2	334612	
					Sealing rubber, LH }	2	334613	
44					Rubber seal, 'BC' post to roof	1	334615	
45					Support bracket, centre body side	2	332201	
					Bolt (¼" UNF x ¾" long)	4	255207	
					Plain washer	6	3900	
					Rubber buffer, thin } Fixing bracket	2	338550	
					Rubber buffer, thick } to body side	2	338553	
					Spring washer } and capping	2	332293	
					Nut (¼" UNF)	4	254810	
46					Mounting stud (⁵⁄₁₆" UNF)	4	338740	
					Plain washer	6	3898	
					Rubber buffer, thin } Fixing upper	2	338550	
					Rubber buffer, thick } to lower body	2	338553	
					Special washer } at front and	2	332293	
					Plain washer } rear corners	2	3830	
					Spring washer	8	3075	
					Nut (⁵⁄₁₆" UNF)	8	254831	
					Bolt (⁵⁄₁₆" UNF x ⁷⁄₈" long)	2	255226 }	Alternatives.
					Bolt (⁵⁄₁₆" UNF x 1⅛" long) } Fixing upper	2	255229 }	Check before ordering
					Plain washer } to lower body at	2	3830	
					Rubber buffer, thick } rear inner bracket	2	338553	
					Spring washer	2	3075	
					Nut (⁵⁄₁₆" UNF)	2	254831	
					Bolt (¼" UNF x ⁵⁄₈" long) } Fixing upper to lower	4	255206	
					Plain washer } body at rear inner	4	3840	
					Spring washer } gusset	4	3074	
					Nut (¼" UNF)	4	254810	

* Asterisk indicates a new part which has not been used on any previous Rover model

N594

COLLINS-JONES

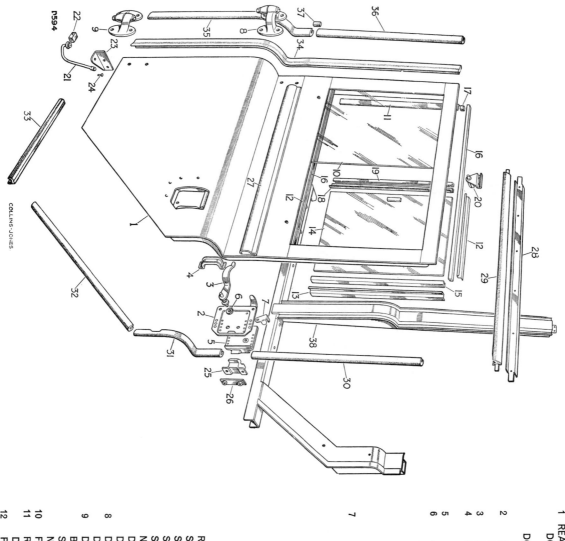

109 STATION WAGON, REAR SIDE DOORS

Plate Ref.	1	2	3	4	DESCRIPTION	Qty	Part No.	REMARKS
1					REAR SIDE DOOR ASSEMBLY, RH	1	345431	With non-detachable inner handle
					REAR SIDE DOOR ASSEMBLY, LH	1	345432	
					DOOR LOCK, MOUNTING AND HANDLE			
					ASSEMBLY, RH	1	345439	
					DOOR LOCK, MOUNTING AND HANDLE			
					ASSEMBLY, LH	1	345440	
2					Mounting plate for door lock	2	333080	
3					Door handle complete, RH	1	307977	
					Door handle complete, LH	1	307978	
4					Bracket for door handle, outer mounting ...	2	302854	
5					Door lock complete, RH	1	330113	For early type lock with detachable
					Door lock complete, LH	1	330114	inner handle. Check before ordering
6					Sealing washer, handle to cover ...	2	302859	
					Screw (2 BA x ¼" long)	8	313858	} Fixing lock,
					Spring washer	12	3073	} brackets and
					Nut (2 BA)	12	2247	} cover together
					Plain washer	8	3685	
7					Screw (2 BA x ¾" long)	4	314203	Models with pivot type catch
					Locking catch, RH	1	333271	} Check before ordering
					Locking catch, LH	1	333272	
					Screw (2 BA x ⅝" long)	2	313860	
					Plain washer, large	2	4034	
					Plain washer, small	2	3902	
					Spring washer	2	3073	
					Nut (2 BA)	4	2247	
					Locknut (2 BA)	2	25996	
					Locking catch, RH	1	345435*	Spring-loaded type catch. } Check
					Locking catch, LH	1	334874	With plunger operating } before
								outside lock case } ordering
					Locking catch, RH	1	345436*	Spring-loaded type catch.
					Locking catch, LH	1	334873	With plunger operating
								inside lock case
					Rivet fixing door lock cover to outer panel	8	302222	
					Screw (¼" UNF x ½" long) ...	8	78792	} Fixing
					Screw retainer, RH	2	337806	} lock assemblies
					Screw retainer, LH	2	337733	} to door
					Spring washer	8	3074	} lock cover
					Nut (¼" UNF)	8	254810	
					Drive screw, fixing lock assembly to door	2	77903	Alternative to 2 lines below
					Door hinge, upper RH	1	330115	Alternative to
					Door hinge, upper LH	1	330116	} line above
8					Door hinge, lower RH	1	330117	
9					Door hinge, lower LH	1	330118	
					Bolt (⅜" UNF x 2" long) ...	8	256226	} Fixing
					Spring washer	8	3075	} door hinge to
					Nut (⅜" UNF)	8	254831	} door frame
10					Fixed window for sidescreen ...	2	333093	
11					Retainer for sidescreen, fixed window	2	333033	
					Drive screw (⅛" No. 6) fixing retainer to frame	10	78140	
12					Filler, top and bottom, for side screen ...	4	333081	
13					Filler, rear, for sidescreen	2	330203	
					Packing strip, for glass channel ...	2	333082	

* Asterisk indicates a new part which has not been used on any previous Rover model

109 STATION WAGON, REAR SIDE DOORS

D594

COLLINS-JOHLS

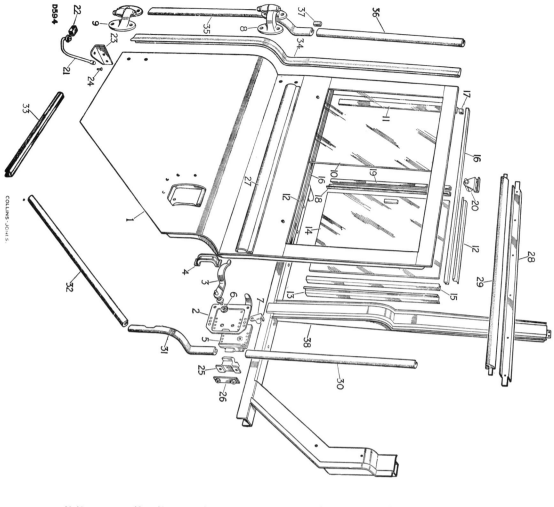

109 STATION WAGON, REAR SIDE DOORS

Plate Ref.	1 2 3 4	DESCRIPTION	Qty	Part No.	REMARKS
14		Sliding window with knob for sidescreen	2	333089	
15		Sliding light channel, rear	2	336454	
16		Sliding light channel, top and bottom	4	336453	
17		Drive screw (¼" No. 6) fixing sliding channel	28	77707	
		Buffer for sidescreen sliding window at top	2	330162	
		Plain washer } Fixing buffer	2	3886	
		Drive screw }	2	78160	
18		Sealing rubber for sliding glass	2	330660	
19		Retainer for sliding glass sealing rubber	2	330661	
20		Sliding window catch	2	332435	
21		Pop rivet fixing catch	4	78248	
		Rod for check strap, RH	1	333041	
		Rod for check strap, LH	1	333204	
22		Buffer for check strap, short	2	306295	
		Buffer for check strap, long	2	333445	
23		Door check bracket, RH, for rear side door	1	333095	
		Door check bracket, LH, for rear side door	1	333096	
		Bolt (5/16" UNF x 1⅛" long)	4	255210	
		Bolt (5/16" UNF x 1" long)	4	255209	
		Shakeproof washer } Fixing brackets to chassis frame	8	72614	
		Shim	8	305232	
		Nut (5/16" UNF)	8	254831	
24		Clevis pin	2	302828	
		Plain washer, small } Fixing door check rod to bracket	4	3947	
		Plain washer, large }	2	3840	
		Split pin	2	4042	
		Bolt, top (5/16" UNF x 2½" long) } Fixing door hinge to 'BC' post at bottom	4	256228	
		Bolt, bottom (5/16" UNF x 3" long) }	4	256230	
		Plain washer	8	3966	
		Spring washer	8	3075	
		Nut (5/16" UNF)	8	254831	
25		Striking plate for rear side door locks	2	333140	
		Plain washer	4	255226	4 off when nut plate is used
		Bolt (¼" UNF x ⅝" long) } Fixing striking plate to 'D' post	4	3840	
		Spring washer }	4	3074	
		Nut plate	2	333137	
26		Nut plate	2	333137	4 off when nut plate is used
27		Waist moulding, rear side door RH	1	333719	
		Waist moulding, rear side door LH	1	333720	
		Special bolt	16	302533	
		Plain washer } Fixing mouldings to doors	16	302532	
		Spire nut	16	2630	
28		Seal retainer for rear side door, RH top	1	333410	
		Seal retainer for rear side door, LH top	1	333411	
		Rubber seal for retainer	2	333228	
29		Rivet fixing seal to retainer	12	78321	
		Screw (2 BA x ½" long)	8	77869	
		Plain washer } Fixing seal and retainer to cant rail	8	3902	
		Spring washer	8	3073	
		Nut (2 BA)	8	3823	

* Asterisk indicates a new part which has not been used on any previous Rover model

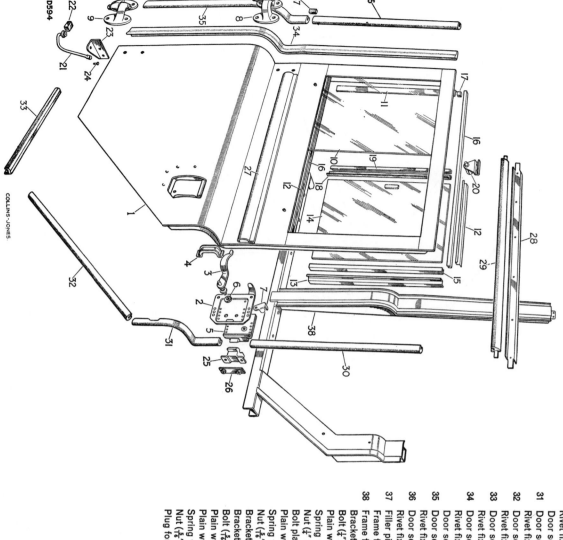

D594

COLLINS-JONES.

109 STATION WAGON, REAR SIDE DOORS

Plate Ref.	1 2 3 4	DESCRIPTION	Qty	Part No.	REMARKS
30		Door sealing rubber for upper vertical 'D' post	2	338788	
		Rivet fixing seal to 'D' post	10	78321	
31		Door sealing rubber for lower vertical 'D' post, RH	1	333261	
		Door sealing rubber for lower vertical 'D' post, LH	1	333262	
		Rivet fixing seal to 'D' post	10	78321	
32		Door sealing rubber for sloping 'D' post	2	338788	
		Rivet fixing seal to 'D' post	8	78321	
33		Door sealing rubber at rear side sills, bottom	2	333233	
		Rivet fixing seal to sill	8	78321	
34		Door sealing rubber at 'C' post	2	333229	
		Rivet fixing seal at 'C' post	22	78321	
35		Door sealing rubber at 'B' post, lower RH	1	330539	
		Door sealing rubber at 'B' post, lower LH	1	330540	
		Rivet fixing seal to 'B' post	12	78321	
36		Door sealing rubber at 'B' post, upper	2	338788	
		Rivet fixing seal to 'B' post	10	78321	
37		Filler piece for 'B' post seal	2	330850	
		Frame for front and rear side doors, RH	1	333100	
		Frame for front and rear side doors, LH	1	333101	
38		Bracket, sill channel to panel	6	330389	
		Bolt (¼" UNF x ¾" long)	12	255207	
		Plain washer	24	3821	
		Spring washer Fixing brackets	12	3074	
		Nut (¼" UNF) to sill channels	12	254810	
		Bolt plate	2	332603	
		Plain washer Fixing sill channels	4	3841	
		Spring washer to dash pillar	4	3075	
		Nut (⁵⁄₁₆" UNF)	4	254831	
		Bracket for sill, rear end, RH	1	333259	
		Bracket for sill, rear end, LH	1	333260	
		Bolt (⁵⁄₁₆" UNF x ⁷⁄₈" long) Fixing bracket	8	255227	
		Plain washer to sill and	16	3841	
		Plain washer chassis frame	2	3898	
		Spring washer	8	3075	
		Nut (⁵⁄₁₆" UNF)	8	254831	
		Plug for redundant hole on 'BC' post, bottom	2	265683	

K0R3

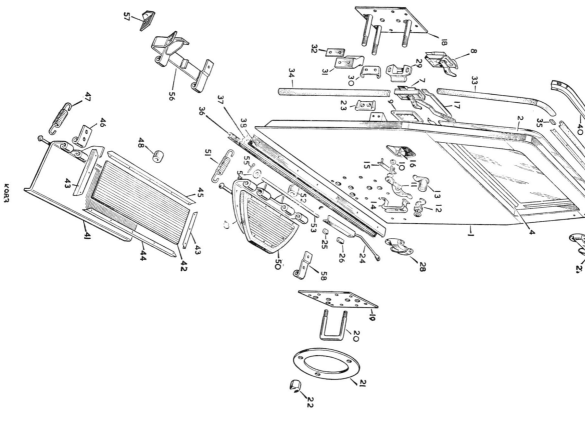

Plate Ref. 1 2 3 4	DESCRIPTION	Qty	Part No.	REMARKS
	REAR DOOR COMPLETE ASSEMBLY ...	1	337780	Incorporating door lock with extended inner handle
			337775*	Incorporating door lock with short inner handle
1	Rear door ...	1	337781	
2	Glass for rear door ...	1	333031	
3	Retainer for glass, vertical ...	2	333033	
4	Retainer for glass, bottom ...	1	333032	
5	Retainer for glass, top ...	1	333034	
6	Retainer for glass, corners ...	2	333035	
7	Drive screw fixing retainers ...	32	78140	
8	DOOR LOCK, MOUNTING AND HANDLE ASSEMBLY ...	1	337789	Cam lever type lock with extended inner handle
			337801*	Cam lever type lock with short inner handle
9	Mounting plate for door lock ...	1	333080	For early type lock with detachable inner handle. Check before ordering
10	Washer, handle to cover ...	1	302859	
11	Door handle with lock ...	1	332432	
12	Barrel lock ...	1	306036	Early type with hook catch. Check before ordering
13	Barrel lock ...	1	320609	With plain catch. Locking pillar type with hook catch. Alternatives. Check before ordering
	Key for barrel lock, Zeni type ...	1	307289	Cam lever type
	Key for barrel lock, Widney type ...	1	320588	State key number when ordering } Alternatives. Check before ordering
14	Bracket for door handle ...	1	302854	
	Screw (2 BA x ½" long) ...	6	313858	For early type lock with detachable inner handle. Check before ordering
	Spring washer ...	6	5073	Fixing lock together
	Plain washer ...	4	3685	
	Nut (2 BA) ...	6	2247	
15	Special screw ...	4	302222	
	Rivet ...	2	78792	2 off when stud plate is fitted
	Stud plate, door lock, bottom ...	1	337806	Fixing door lock to cover
	Spring washer ...	4	3074	
	Nut (¼" UNF) ...	4	254810	
16	Locking pillar for catch ...	1	332434	
	Plain washer ...	1	2226	For early type lock with hook catch. Check before ordering
	Spring washer ...	1	3073	Fixing pillar to door
	Nut (2 BA) ...	1	2247	
17	Waist rail lock handle protection strip ...	1	333037	For lock with extended inner handle
	Bolt (¼" UNF x ¾" long) ...	2	255207	Fixing strip to door
	Spring washer ...	2	3074	
	Nut (¼" UNF) ...	2	254810	
18	Wheel stud plate ...	1	333435	
19	Clamp plate for spare wheel stud plate ...	1	333446	
	Screw (¼" UNF x ¾" long) ...	8	78210	Fixing stud and clamp plates
	Spring washer ...	8	3074	
	Nut (¼" UNF) ...	8	254810	Fixing plates to rear door

* Asterisk indicates a new part which has not been used on any previous Rover model

109 STATION WAGON, REAR DOOR

K983

Plate Ref.	1 2 3 4	DESCRIPTION	Qty	Part No.	REMARKS
20		'U' bolt for spare wheel support ...	1	333447	
21		Nut (1/4" UNF) fixing 'U' bolt to door ...	2	254814	
22		Retaining plate for spare wheel ...	1	333439	
		Hub nut fixing wheel and retaining plate to 'U' bolt	6	217361	
23		Male dovetail } For rear	1	333438	1958-59
		Male dovetail } end door	1	332942	1960 onwards
24		Screw (1/4" UNF x 5/8" long) } Fixing	2	78233	
		Spring washer } male dovetail	2	3074	
		Nut (1/4" UNF) } to door	2	254810	
		Rod for check strap ...	1	333041	
25		Buffer for check strap, short ...	1	306295	
26		Buffer for check strap, long ...	1	333445	
27		Hinge for rear door, upper ...	1	333036	
		Hinge for rear door, lower ...	1	330117	
28		Bolt (5/16" UNF x 2" long) } Fixing	4	256226	
		Spring washer } hinge to	4	3075	
		Nut (5/16" UNF) } door frame	4	254831	
		Bolt (5/16" UNF x 7/8" long) } Fixing door	4	255227	
		Plain washer } hinges to body	4	254831	
		Spring washer	4	3830	
		Nut (5/16" UNF)	4	3075	
		Bolt (1/4" UNF x 1 1/4" long) } Fixing	1	254831	
		Spring washer } check	1	3074	
		Plain washer, thick } strap rod	2	4075	
		Plain washer, thin } to floor	2	3974	
		Nut (1/4" UNF)	2	254810	
29		Striking plate	1	333140	
		Bolt (5/16" UNF x 1" long) } Fixing striking	2	255209	
		Plain washer } plate to rear of body	2	3817	
		Spring washer	2	3074	
		Nut (1/4" UNF)	2	254810	
30		Female dovetail } on body } For rear door	1	333440	1958-59
		Female dovetail }	1	332147	1960 onwards
		Spacer } For female	1	333444	1958-59
		Spacer } dovetail	1	332943	1960 onwards
31		Screw (1/4" UNF x 5/8" long) fixing dovetail to spacer	2	78233	
		Screw (1/4" UNF x 1" long) } Fixing	2	78387	
		Spring washer } male dovetail	2	3074	
		Nut (1/4" UNF) } to door	2	254810	
32		Set bolt (1/4" UNF x 3/4" long) } Fixing spacer to	2	255207	
		Plain washer } LH rear pillar	2	3900	
		Shim	As required	305232	
		Spring washer	2	3074	
33		Seal for door sides ...	3	338788	
34		Seal for door side, LH bottom ...	1	332563	
35		End filler for seals ...	4	330850	
		Rivet fixing seals to body ...	22	302818	
36		Seal for rear door, bottom ...	1	332564	
37		Retainer fixing seal to rear door, bottom ...	1	332756	
38		Protection strip for rear door, bottom ...	1	333203	
		Drive screw (1/4" No. 6) fixing seal and retainer to floor	13	77704	
39		Seal for rear door, top ...	1	332149	
		Rivet fixing seal ...	11	302818	

* Asterisk indicates a new part which has not been used on any previous Rover model

109 STATION WAGON, REAR DOOR

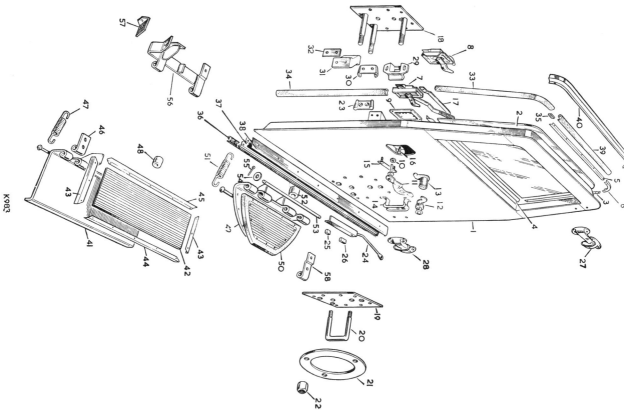

109 STATION WAGON, REAR DOOR

Plate Ref.	1 2 3 4	DESCRIPTION	Qty	Part No.	REMARKS
40		Retainer for rear door seal, top ...	1	332151	
		Screw (2 BA x ½" long) ...	9	77869	} Fixing retainers
		Plain washer ...	9	3902	} to roof panel
		Spring washer ...	9	3073	
		Nut (2 BA) ...	9	3823	
41		REAR STEP			
42		Rubber mat for rear step ...	1	501053	} Oblong step
43		Retainer for rear step, side ...	2	245137	} Overall width 19½"
44		Retainer for rear step mat, front ...	1	245136	
45		Retainer for rear step mat, rear ...	1	245135	
		Rivets fixing mat end cappings ...	16	245134	
46		Hinge, centre for rear step ...	1	4022	
		Locker ...	2	264362	Check before ordering
		Bolt (⅜" UNF x ⅞" long) ...	2	235731	} Fixing hinge centre
		Locknut (⅜" UNF) ...	4	255046	} to rear cross-member
		Bolt (⅜" UNF x 4" long) ...	2	254852	} Fixing hinge centre to rear step
47		Spring for rear step ...	2	256253	
48		Buffer for rear step ...	1	245131	} Alternative to the 11½" step below
		Plain washer ...	1	3867	} Fixing buffer to rear step
		Drive screw ...	1	78142	
		REAR STEP			}Oblong step. Overall width 11½"
		Rubber for rear step ...	1	304125	}Alternative to the oblong 19½" step,
		Retainer for rear step, side ...	2	509120	}or the semi-circular step
		Retainer for rear step mat, front ...	1	77758	Check before ordering
		Retainer for rear step mat, rear ...	1	3867	
		Rivet fixing mat end cappings ...	14	304125	
		Spring for rear step ...	1	508958	
		Buffer for rear step ...	1	508959	
		Plain washer ...	2	508957	} Fixing buffer
		Screw (2 BA x ⅝" long) ...	1	245136	} to step
		Hinge pin for rear step ...	1	245135	
		Plain washer ...	1	3867	} Fixing
		Split pin ...	2	78142	} hinge pin
49		REAR STEP			} Semi-circular step. With hinge brackets
50		Rubber mat for rear step ...	1	307840	} welded to rear cross-member
		Rivet ...	7	517977	
		Screw (2 BA x ½" long) ...	1	523589	} Fixing mat
		Spring washer ...	2	3833	} to step
		Nut (2 BA) ...	2	2395	
51		Spring for rear step ...	1	509120	
52		Buffer for rear step ...	1	78142	
		Plain washer ...	1	3867	} Fixing buffer to
		Drive screw ...	4	3073	} rear cross-member
53		Hinge pin for rear step ...	1	304125	
		Drive screw ...	1	264362	} Fixing buffer to rear cross-member
54		Plain washer ...	1	3867	} Fixing Up to vehicle suffix 'C' inclusive
55		Split pin ...	2	78142	} hinge pin

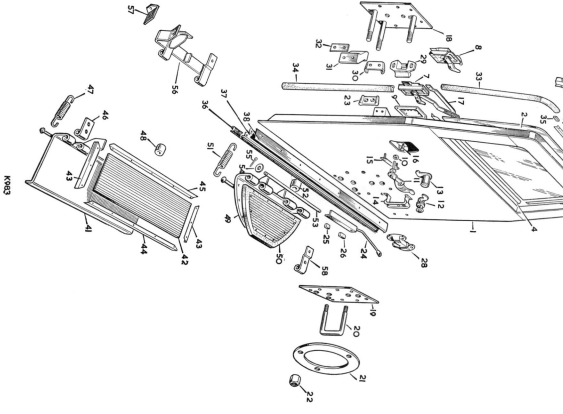

K983

109 STATION WAGON, REAR DOOR

Plate Ref.	1 2 3 4	DESCRIPTION			Qty	Part No.	REMARKS
		REAR STEP					
		Rubber mat for rear step	...	...	1	532589	
		Rivet	...	...	7	517977	
		Screw (2 BA x ½" long)	⎫ Fixing mat		4	307840	
		Spring washer	⎬ to step		4	3073	
		Nut (2 BA)	⎭		4	3823	
56		Support bracket and hinge centre, LH	...	...	1	552132	
57		Anchor bracket for spring	...	...	1	552133	
		Bolt (5/16" UNF x 1⅛" long)	⎫ Fixing support bracket		2	255029	
		Plain washer	⎬ and anchor bracket to		2	2223	
		Self-locking nut (5/16" UNF)	⎭ rear cross-member		2	252211	
58		Hinge, centre RH	...	...	1	552129	Semi-circular step with hinge brackets
		Set bolt (⅜" UNF x 1" long)	⎫ Fixing RH		1	255247	bolted to rear cross-member
		Spring washer	⎬ centre hinge to		1	3076	From vehicle suffix 'D' onwards
			⎭ rear cross-member				
		Spring for rear step	...	...	1	264362	
		Buffer for rear step	...	...	1	312027	
		Drive screw	⎫ Fixing buffer to		1	78417	
		Plain washer	⎬ rear cross-member		1	3867	
		Hinge pin for rear step	...	...	1	509120	
		Spring washer	⎫ Fixing hinge pin		1	2286	
		Plain washer	⎬		2	3833	
		Split pin	⎭		2	2395	

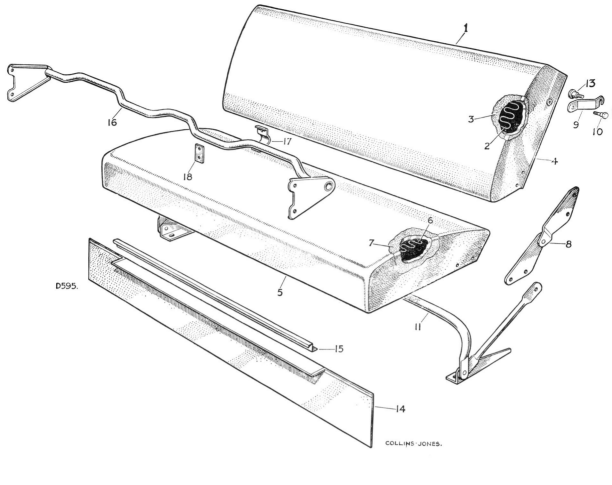

D595.

COLLINS-JONES.

Plate Ref.				DESCRIPTION	Qty	Part No.	REMARKS
	1	2	3	4			
1				CENTRE SEAT COMPLETE ASSEMBLY, GREY ...	1	333526	Early type with cushion 17⅝" wide
				CENTRE SEAT COMPLETE ASSEMBLY, GREY	1	333527	
				CENTRE SEAT COMPLETE ASSEMBLY, BLACK	1	338330	Late type with cushion 15³²⁄₃₂" wide
				CENTRE SEAT COMPLETE ASSEMBLY, BLACK	1	320705*	
1				CENTRE SEAT SQUAB ASSEMBLY, TRIMMED, GREY	1	333330	
				CENTRE SEAT SQUAB ASSEMBLY, TRIMMED, GREY	1	333348	
				CENTRE SEAT SQUAB ASSEMBLY, TRIMMED, BLACK	1	320706*	
2				No-sag seat springs	17	308173	
3				Wire for seat springs	1	308310	
4				Interior for centre squab ...	1	308144	Early type for use with cushion 17⅝" wide
3				Trim board for centre seat squab	1	308156	
				Drive screw ⎫ Fixing trim board	15	77903	
				Plain washer ⎬ to frame	15	3886	
				Trim clip, fixing squab cushion cover	38	315264	
4				Trim board for centre seat squab	1	338353	Late type for use with cushion 15³²⁄₃₂" wide
				Drive screw ⎫ Fixing trim board	15	77903	
				Plain washer ⎬ to frame	15	3886	
				Trim clip, fixing squab cushion cover	38	315264	
5				CENTRE SEAT CUSHION ASSEMBLY, TRIMMED, GREY	1	333528	Early type cushion 17⅝" wide
6				No-sag seat springs	17	308173	
7				Wire for seat springs	1	308310	
				Interior for centre cushion ...	1	308155	
				Trim clip fixing cushion cover ...	38	315264	
5				CENTRE SEAT CUSHION ASSEMBLY, TRIMMED, GREY	1	338349	Late type cushion 15³²⁄₃₂" wide
				CENTRE SEAT CUSHION ASSEMBLY, TRIMMED, BLACK	1	320707*	
6				No-sag seat springs	17	338339	
7				Wire for seat springs	1	308310	
				Interior for centre cushion ...	1	338352	
				Trim clip fixing cushion cover ...	38	315264	
8				Hinge RH ⎫ For seat squad	1	308136	
				Hinge LH ⎬ seat squab	1	308137	
				Set bolt (⁵⁄₁₆" UNF ⎫ Fixing hinge to x ⅝" long) ⎬ cushion and	8	255225	
				Spring washer ⎭ support tube	8	3075	

Check before ordering

109 STATION WAGON, CENTRE SEAT AND FRONT SQUAB

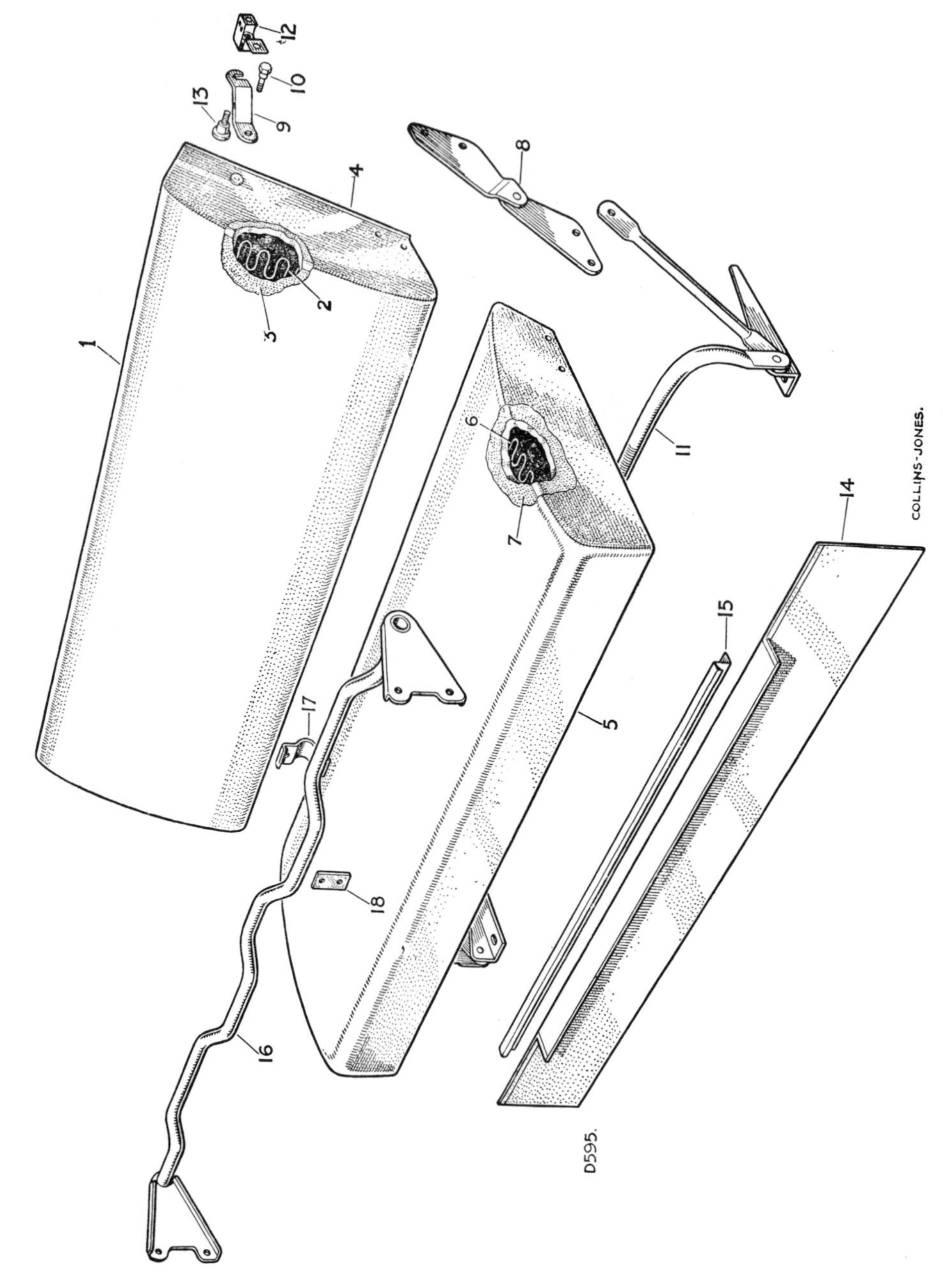

109 STATION WAGON, CENTRE SEAT AND FRONT SQUAB

Plate Ref.	1 2 3 4 DESCRIPTION	Qty	Part No.	REMARKS
	Support link, RH	1	308713	Overall length $2\frac{11}{32}$" each (Check before ordering)
9	Support link, LH	1	308714	
	Support link, RH	1	338332	Overall length $5\frac{9}{32}$" each
	Support link, LH	1	338333	
10	Special set bolt ($\frac{5}{16}$" UNF) } Fixing links to centre squab	2	331708	
	Plain washer	2	3830	
11	Centre seat support tube	1	333507	For use with cushion $17\frac{5}{8}$" wide
	Centre seat support tube	1	338345	For use with cushion $15\frac{17}{32}$" wide
	Set bolt ($\frac{5}{16}$" UNF x $1\frac{1}{2}$" long) } Fixing centre seat to support tube	2	255232	
	Plain washer	2	3899	
	Spring washer	2	3075	
	Bolt ($\frac{5}{16}$" UNF x $\frac{5}{8}$" long) } Fixing centre seat to wheelarch	2	255225	
	Spring washer	2	3075	
	Nut ($\frac{5}{16}$" UNF)	2	254831	
	Bolt ($\frac{5}{16}$" UNF x $\frac{3}{4}$" long) } Fixing centre seat to floor	6	255226	
	Plain washer	6	3830	
	Spring washer	6	3075	
	Nut ($\frac{5}{16}$" UNF)	6	254831	
12	Locking plate and screw for squab support link	2	334174	
13	Locking screw ($\frac{5}{16}$" UNF)	2	334176	
	Bolt ($\frac{1}{4}$" UNF x $1\frac{9}{16}$ long) } Fixing locking plate	4	255205	
	Spring washer	4	3074	
	Nut plate ($\frac{1}{4}$" UNF)	2	334177	
	Strap for rear centre seat	2	331707	
	Stud for strap	2	348430	
	Screw (2 BA x 1" long) } Fixing stud	2	78120	
	Spring washer	2	3073	
	Nut (2 BA)	2	3823	
14	Cover for rear centre toe panel	1	331616	
15	Retainer, cover to seat base	1	345116	
	Drive screw ($\frac{1}{2}$" No. 6) fixing retainer to seat base	5	77704	
16	Grab rail between 'BC' posts in front of centre seat	1	333395	
	Set bolt ($\frac{5}{16}$" UNF x $\frac{3}{4}$" long) } Fixing grab rail to 'BC' post	4	255226	
	Plain washer	4	3830	
	Spring washer	4	3075	
	Front squab, trimmed, driver's, GREY	1	349565	
	Front squab, trimmed, driver's BLACK	1	320708*	
	Front squab, trimmed passenger, GREY	1	349563	
	Front squab, trimmed, passenger, BLACK	1	320709*	
	Front squab, trimmed, centre, GREY	1	349564	
	Front squab, trimmed, centre, BLACK	1	320710*	

* Asterisk indicates a new part which has not been used on any previous Rover model

109 STATION WAGON, CENTRE SEAT AND FRONT SQUAB

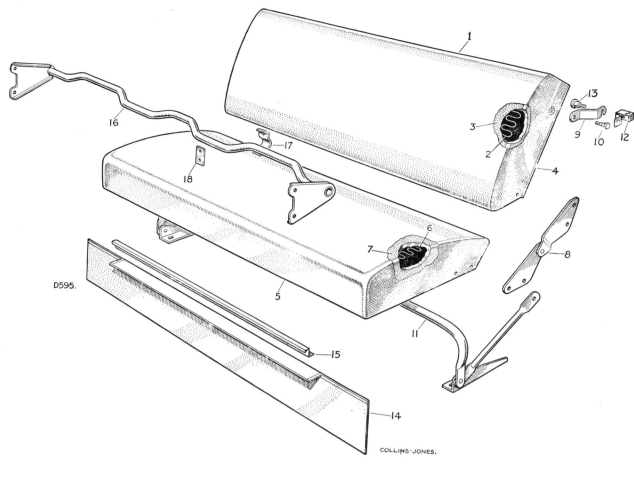

D595.

COLLINS-JONES.

109 STATION WAGON, CENTRE SEAT AND FRONT SQUAB

Plate Ref. 1 2 3 4	DESCRIPTION	Qty	Part No.	REMARKS
17	Retaining clip ⎱ Fixing non-adjustable front seats to grab rail	4	333399	Clip type fixing
18	Packing rubber ⎰	4	339986	Alternative to strap
	Drive screw, to grab rail	8	72628	Check before ordering
	Retaining strap, non-adjustable front seat to grab rail	4	332355	Strap type fixing. Alternative to clip
	Spire nut ⎱ Fixing strap to non-adjustable front seat	8	78237	Check before ordering
	Plain washer ⎰	8	4034	Alternative to three lines below
	Drive screw	8	78438	Check before ordering
	Stud for retaining strap	4	320345	Check before ordering
	Drive screw, fixing stud to grab rail	4	78412	
	Packing rubber	2	334061	
	Plain washer	8	3840	Check before ordering
	Spire nut ⎱ Fixing on adjustable front seat	2	72628	
	Drive screw ⎰	4	78237	
	Plain washer	4	72628	
	Drive screw	2	3885	Alternative to two lines above
	Grab rail behind centre seat, overall length, 45¼"	1	333510	Alternatives
	Grab rail behind centre seat, overall length 21"	1	338334	Alternatives
	Set bolt ($\frac{5}{16}$" UNF x $\frac{5}{8}$" long) ⎱ Fixing grab rail	2	255225	Check before ordering
	Spring washer ⎰	2	3075	
	Clip for starting handle and jack handle	5	508035	
	Bracket for starting handle	4	332376	Alternative to line below
	Bracket for starting handle and jack handle	1	332377	Alternative to line below
	Bracket for starting handle and jack handle	5	332376	Alternative to two lines above
	Rivet fixing clips to brackets	5	78248	
	Drive screw (¼" No. 10) ⎱ Fixing brackets to base of centre seat cushion	10	78153	
	Packing washer ⎰	8	4171	

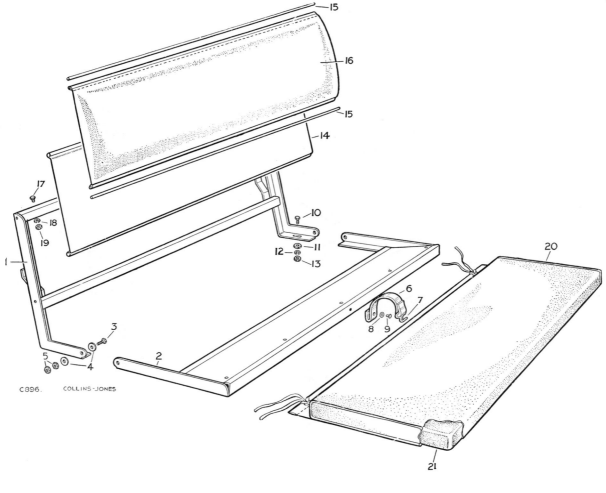

C896. COLLINS-JONES

109 STATION WAGON, REAR SEATS

Plate Ref.	1 2 3 4	DESCRIPTION	Qty	Part No.	REMARKS
		REAR SEAT FRAME ASSEMBLY, RH ...	1	331646	
		REAR SEAT FRAME ASSEMBLY, LH ...	1	331647	
1		Rear seat back frame ...	2	331229	
2		Hinged base for rear seat ...	2	320521	
3		Bolt (¼" UNF x ¾" long) } Fixing seat	4	255207	
4		Plain washer } base to	12	3821	
5		Nut (¼" UNF) } frame back	4	254810	
		Backrest securing clip, RH ...	1	333293	
		Backrest securing clip, LH ...	1	333294	
		Backrest securing clip ...	2	333295	
		Bolt (¼" UNF x ⅝" long) } Fixing	8	255206	
		Spring washer } clips to	8	3074	
		Nut (¼" UNF) } frame	8	254810	
		Buffer ...	2	331699	
		Screw (2 BA x ⅞" long) } Fixing	2	77925	
		Plain washer } buffer to	4	4030	
		Nut (2 BA) } seat frame back	2	3823	
6		Strap for seat ...	2	302648	
7		Hook for strap ...	2	301005	
8		Plain washer } Fixing strap to	4	3816	
9		Pop rivet } hinged base	2	78410	
10		Bolt (¼" UNF x ¾" long) } Fixing seat frame	4	255207	
11		Plain washer } to wheelarch box	4	3821	
12		Spring washer ...	4	3074	
		Packing washer ...	8	2228	
13		Nut (¼" UNF) ...	4	254810	
		REAR SEAT BACKREST TRIMMED ASSEMBLY, GREY ...	2	320084	
		REAR SEAT BACKREST TRIMMED ASSEMBLY, BLACK ...	2	320673*	
14		Back rest panel ...	2	301470	
15		Retaining rod for pad ...	2	301476	
16		Pad for back rest ...	2	302647	
17		Bolt (¼" UNF x ½" long) } Fixing	12	255204	
18		Spring washer } back rest	12	3074	
19		Nut (¼" UNF) } to frame	12	254810	
20		REAR SEAT CUSHION ASSEMBLY, GREY ...	2	331833	
		REAR SEAT CUSHION ASSEMBLY, BLACK ...	2	320674*	
21		Pad for rear cushion ...	2	331835	

* Asterisk indicates a new part which has not been used on any previous Rover model

109 STATION WAGON, BODY TRIM

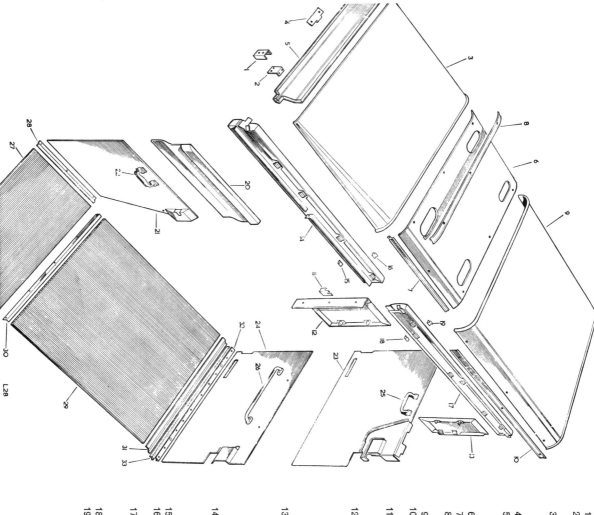

Plate Ref.	1 2 3 4	DESCRIPTION	Qty	Part No.	REMARKS
1		Bracket, cant rail to roof frame	4	331168	
2		Bracket for roof trim, front portion	4	349218	Alternatives
		Drive screw (½" No. 6) fixing bracket to cant rail	8	77704	
3		Roof trim, front portion	1	331491	
		Plain washer	9	3886	
		Drive screw (½" No. 6) fixing roof trim	9	77704	
4		Centre bracket for canopy trim panel	1	331391	
		Drive screw (½" No. 6) fixing bracket to centre bracket	1	331607	
5		Canopy trim panel	1	3852	
		Plain washer	3	3886	
		Drive screw (½" No. 6) fixing trim panel to centre bracket	3	331547	
6		Headcloth, intermediate	1	331605	
		Side strip for rear headcloth	2	331547	
7		Side rail for intermediate headcloth	2	77704	
		Drive screw (1¼" No. 6) fixing strip to roof	5	331216	
8		Fixing strip for intermediate headcloth	2	331203	
9		Headcloth, rear	1	331192	
		Drive screw (½" No. 6) fixing side rail to roof	10	77704	
10		Fixing bracket, RH sidelight casing, front	1	331193	
11		Fixing bracket, LH sidelight casing, front	1	331186	
		Plain washer	6	3886	
		Drive screw (1¼" No. 6)	6	77704	
12		Sidelight casing, front, LH GREY	1	331187	
		Sidelight casing, front, RH	1	320682*	
		Sidelight casing, front, LH BLACK	1	320683*	
		Sidelight casing, front, RH GREY	1	331179	
13		Sidelight casing, rear, LH	1	331180	
		Sidelight casing, rear, RH BLACK	1	320684*	
		Sidelight casing, rear, LH	1	320685*	
		Plain washer to bracket	2	77704	
		Drive screw (1¼" No. 6) to panel	2	3886	
14		Roll trim, cant rail, RH front	1	331071	
		Roll trim, cant rail, LH front GREY	1	320712*	
		Roll trim, cant rail, RH front BLACK	1	331590	
		Roll trim, cant rail, LH front	1	331589	
15		Retaining clip	8	331071	
		Edge clip	8	349899	
16		Drive screw (½" No. 6) fixing roll trim to body	6	77903	
17		Roll trim, cant rail, RH rear	1	331497	
		Roll trim, cant rail, LH rear GREY	1	331498	
		Roll trim, cant rail, RH rear BLACK	1	320686*	
18		Roll trim, cant rail, LH rear	1	320687*	
		Retaining clip	8	331071	
19		Edge clip fixing roll trim	8	349899	
		Drive screw (½" No. 6) fixing roll trim to body	4	77903	

* Asterisk indicates a new part which has not been used on any previous Rover model

109 STATION WAGON, BODY TRIM

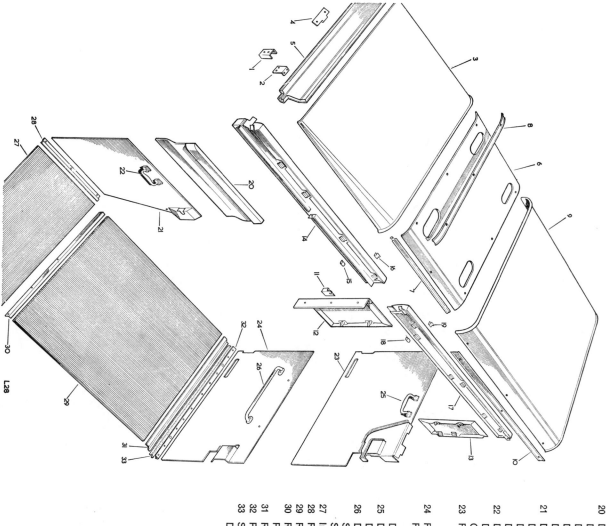

L28

109 STATION WAGON, BODY TRIM

Plate Ref. 1 2 3 4	DESCRIPTION	Qty	Part No.	REMARKS
20	Door trim upper, centre, RH ⎫ GREY	1	339990	
	Door trim upper, centre, LH ⎭	1	339991	
	Door trim upper, centre, RH ⎫ BLACK	1	320713*	
	Door trim upper, centre, LH ⎭	1	320714*	
21	Drive screw (¼″ No. 6) fixing trim to door	10	77704	
	Door trim lower, centre, RH ⎫ GREY	1	349974	
	Door trim lower, centre, LH ⎭	1	349975	
	Door trim lower, centre, RH ⎫ BLACK	1	320715*	
	Door trim lower, centre, LH ⎭	1	320716*	
22	Door pull handle for rear side door	2	306460	
	Drive screw (⅜″ No. 10) fixing handle to door	4	78167	
	Captive nut for drive screw	4	78319	
23	Rear door trim casing, GREY	1	331638	When door lock with extended inner handle is fitted ⎱ Check before ordering
24	Rear door trim casing, GREY ⎫	1	345426*	When door lock with short inner handle is fitted ⎱ Check before ordering
	Rear door trim casing, BLACK ⎭	1	320679*	
25	Drive screw (½″ No. 6) fixing door casing	5	77704	
	Door pull handle for rear door	1	306460	Alternative to grab handle
26	Drive screw (⅝″ No. 10) fixing handle to door	2	78167	
	Door grab handle	1	345450*	Alternative to pull handle
	Set bolt (¼″ UNF x ¾″ long) ⎫ Fixing grab handle to door	2	255207	
	Spring washer ⎭	2	3074	
27	Intermediate floor rubber mat	1	331481	
28	Retainer for floor mat, intermediate	2	331480	
29	Rear floor rubber mat	1	331670	
30	Retainer for rear floor mat, front end	1	331669	
31	Pop rivet fixing retainer	7	78248	
	Retainer for seal and rear floor mat, rear end	1	332756	
32	Protection strip for rear door, bottom	1	333203	
33	Sealing rubber for rear door, bottom	1	332564	
	Drive screw fixing retainer, strip and seal to floor	13	77704	

* Asterisk indicates a new part which has not been used on any previous Rover model

109 STATION WAGON, CHASSIS FRAME, FUEL TANK, EXHAUST SYSTEM, ELECTRICAL EQUIPMENT

Plate Ref. 1 2 3 4	DESCRIPTION	Qty	Part No.	REMARKS
	Chassis frame	1	278665	4 cylinder models. Up to vehicle suffix 'C' inclusive
	Chassis frame	1	562533	4 cylinder models. From vehicle suffix 'D' onwards
	Chassis frame			See footnote ‡
	Cross-member No. 3	1	562575	6 cylinder models
	Cross-member No. 4 ⎱ For chassis frame	1	241130	4 cylinder models
	Cross-member No. 4 ⎰	1	543946	6 cylinder models
	Fuel tank complete	1	552398	Petrol models
	Fuel tank complete	1	543164	Diesel models
	Joint washer for drain plug	1	543165	
	Drain plug	1	515599	
		1	235592	
	Bolt (⅜" UNF x 1⅜" long) ⎱ Fixing fuel tank to frame at front	1	256040	
	Plain washer	2	3109	
	Distance piece	2	508544	
	Mounting rubber	2	508845	
	Spring washer	2	3076	
	Nut (⅜" UNF)	2	254812	
	Bolt (⅜" UNF x 2½" long)	2	256047	
	Distance piece, top	2	501070	
	Plain washer	2	3933	
	Plain washer	1	2265	
	Distance piece ⎱ Distance piece fuel tank to frame at rear	2	500446	
	Mounting rubber	4	500447	
	Distance piece, bottom ⎰	4	500588	
	Self-locking nut (⅜" UNF)	3	252162	
	Spring washer	2	3100	
	Fuel tank gauge unit, SM FT 5300/89	1	501620	Series II Petrol models
	Fuel tank gauge unit, SM FT 5301/18	1	519875	Series IIA Petrol models
	Fuel tank gauge unit, SM FT 5340/06	1	501621	2 litre Diesel models. Up to vehicle suffix 'C' inclusive
	Fuel tank gauge unit, SM FT 5342/00	1	529970	2¼ litre Diesel models
	Fuel tank gauge unit, SM TB 1114/000	1	558846	Series IIA Petrol models. From vehicle suffix 'D' onwards
	Fuel tank gauge unit, SM TB 1214/000	1	558847	Series IIA Diesel models. From vehicle suffix 'D' onwards
	Spring washer ⎱ Fixing gauge unit	6	3101	
	Set screw (3 BA x ½" long) ⎰	6	3100	
	Earth lead for fuel tank gauge unit	1	239688	
	Fan disc washer ⎱ Fixing earth lead	2	519857	
	Drive screw ⎰	1	72626	1961 models onwards
	Fuel tank elbow and pipe, outlet	1	500432	Petrol models
	Fuel tank elbow and pipe, outlet	1	544674	Diesel models
	Fuel tank elbow and pipe, return	1	500818	Diesel models
	Joint washer for feed elbow	1	267837	
	Spring washer ⎱ Fixing elbow to tank	2	3101	
	Set screw (3 BA x ⁷⁄₁₆" long) ⎰	2	3972	
	Fuel pipe, copper, tank to flexible pipe	1	500438	1958-60 Diesel and Petrol models

* Asterisk indicates a new part which has not been used on any previous Rover model

‡ When supplying this chassis for vehicles prior to vehicle suffix 'D', the following must be fitted:
Shaft for handbrake relay, 1 off   Part No. 552746
Self-locking nut (⅜" UNF), 1 off   Part No. 252162

109 STATION WAGON, CHASSIS FRAME, FUEL TANK, EXHAUST SYSTEM, ELECTRICAL EQUIPMENT

Plate Ref. 1 2 3 4	DESCRIPTION	Qty	Part No.	REMARKS
	Fuel pipe (nylon), tank to pump	1	552436	2¼ litre Petrol models, 1961 onwards
	Nut ⎱ Fixing fuel pipe to pump inlet	1	270115	Diesel models, 1961 up to engine suffix 'J' inclusive
	Olive ⎰	1	270105	Diesel models, 1961 up to engine suffix 'J' inclusive
	Fuel pipe, nylon, tank to pump	1	552436	Diesel models without sedimentor
	Olive for fuel pipe, pump end	1	557531	Diesel models with sedimentor
	Fuel pipe, nylon, tank to sedimentor	1	564963	From engine suffix 'K' onwards
	Olive for fuel pipe, sedimentor end	1	557531	From engine suffix 'K' onwards
	Fuel pipe, flex to tank return	1	500821	Diesel model, 1958-60
	Fuel pipe, spill return to tank	1	509910	Diesel models, 1961 onwards
	Bolt (2 BA x ¾" long) ⎱ Fixing spill return pipe	1	250961	
	Fuel pipe, spill return to tank	1	509912	Diesel models, 1961 onwards
	Spring washer	1	3073	Up to engine suffix 'K' inclusive
	Distance piece	1	509412	Petrol models
	Nut (2 BA)	1	552118	Diesel models
	Grommet for spill return pipe ⎱ on injector stud	1	273370	
	Fuel pipe, spill return to tank	1	564962*	Diesel models. From engine suffix 'K' onwards
	Fuel pipe retaining bracket	1	270297	From engine suffix 'K' onwards
	Single pipe clip	1	50183	Petrol model, 1958-60
	Double pipe clip	3	243395	Petrol model, 1958-60
	Pipe clip	3	509415	Diesel model
	Pipe clip	3	500839	Diesel model
	Pipe clip	1	216708	Petrol models
	Pipe clip	1	509412	Petrol models
	Plain washer	1	552118	Diesel models
	Bolt (2 BA x ⅝" long)	1	250960	
	Spring washer	2	2247	
	Nut (2 BA)	1	3073	
	Mounting bracket for fuel pipe clip	1	500445	Petrol model, 1958-60
	Spring washer	1	509413	
	Nut (2 BA)	3	250959	
	Pipe clip at check strap plate	3	3073	
	Double pipe clip	1	2226	
	Pipe clip ⎱ Fixing clip to bracket on No. 3 cross-member	3	2247	
	Pipe clip	3	500833	
	Pipe clip	5	50183	Petrol model, 1958-60
	Drive screw (⅜" No. 6)	1	72626	
	Pipe clip fixing fuel pipe to No. 3 cross-member	3	216708	Petrol model, 1958-60
	Pipe clip fixing fuel pipe to chassis	3	509414	1961 Petrol models onwards
	Drive screw fixing clip to chassis	3	72626	1961 Petrol models onwards
	Intermediate exhaust pipe	1	562785	4 cylinder models
	Intermediate exhaust pipe	1	500289	6 cylinder models
	Exhaust silencer complete	1	500290	4 cylinder models
	Exhaust silencer complete	1	523655	4 cylinder models. Required when hydraulic front winch is fitted
	Exhaust silencer ⎱ 6 cylinder models	1	562737	
	Tail pipe complete ⎰	1	562787	

* Asterisk indicates a new part which has not been used on any previous Rover model

Plate Ref. 1 2 3 4	DESCRIPTION	Qty	Part No.	REMARKS
	Hose, tank to filler tube	1	543782	
	Hose clip, bottom	1	50328	
	Hose clip, top	1	50343	
	Air balance hose for fuel tank	1	543766	
	Breather hose for fuel tank	1	543767	
	Clip fixing breather and tank hose	4	50302	
	Gauge and ammeter panel complete	1	519873	
	Gauge and ammeter panel complete	1	545417	Germany only
	Fuel gauge SM FG 6249/01	1	519874	Germany only
	Ammeter SM 6207/00	1	519840	
	Warning light for headlamp beam, red	1	528950	
	Warning light for headlamp beam, blue	1	545408	Germany only
	Bulb for warning light, LU 987	1	232590	
	Gauge panel complete	1	560744	Series IIA models
	Water temperature gauge SM BT 6106/08	1	560746	From vehicle suffix 'D' onwards
	Fuel gauge SM FG 6106/05	1	555835	Up to vehicle suffix 'C' inclusive
	Warning light, charging, red SM WL 4005/02	1	555837	
	Bulb for warning light, LU 987	1	232590	
	Gauge panel complete	1	504684	Series II models
		1	232590	Series IIA 2¼ litre models
	Frame harness	1	528900	Series II models
	Frame harness	1	529964	Series IIA 2¼ litre models
	Frame harness	1	560557	2.6 litre Petrol models Up to vehicle suffix 'E' inclusive
	Frame harness LU 54949641	1	560901	2¼ litre Petrol and 2.6 litre Diesel models From vehicle suffix 'F' onwards
	Frame and rear cross-member harness	1	504685	2¼ litre Diesel models
		1	560976	2¼ litre Petrol models From vehicle suffix 'E' inclusive
	Frame and rear cross-member harness	2	265482	2 litre Diesel models
	Grommet in chassis side member	1	265481	Up to vehicle suffix 'C' inclusive
	Lead, tank unit to warning lamp LU 54942032	1	519876	Series II models
	Lead for roof lamp, roof portion LU 862546	1	555879	Series IIA models
	Lead for roof lamp, chassis portion LU 862545	1	55877	From vehicle suffix 'D' onwards
	Lead for roof lamp, chassis portion LU 54949635	1	41379	
	Lead for roof lamp, chassis portion LU 54950059	1	78116	
	Cable clip ⎱ Fixing roof lamp lead to dash	1	233244	
	Drive screw ⎰	1	345188	
	Grommet in fuse box for lead	2	78495	
	Interior mirror	1	338275	From vehicle suffix 'D' onwards
	Drive screw fixing mirror	1	240908	Up to vehicle suffix 'C' inclusive
	Support bracket for interior mirror	1	555877	From vehicle suffix 'D' onwards
	Switch for roof lamp	1	560405	From vehicle
	Switch for panel and interior lights LU 34734	1	265295	suffix 'C' inclusive
	Nameplate for switch, 'PANEL/INTERIOR'	1	320608	
	Roof lamp complete LU 56062	1	264591	
	Lens for roof lamp LU 54570209	1	334111	
	Bulb for roof lamp LU 382	1	77892	
	Mounting pad for roof lamp	2		
	Drive screw fixing lamp	1	312856	
	Grommet for roof lamp lead			

LAND-ROVER
REGISTERED TRADE MARK

109 TWELVE-SEATER

SERIES II A

GENERAL EXPLANATION

The 109 Twelve-Seater Land-Rover is based on the normal 109 Station Wagon which has a seating capacity of ten; the majority of parts are therefore identical.

The following pages, 577 to 579 inclusive, list the parts which are peculiar to the 109 Twelve-Seater Land-Rover.

The optional equipment listed below is also fitted:—

Private locks	See
Front door and floor trim	Optional Equipment
Folding steps for side doors	Catalogue
Attachment bracket for towing jaw	
Towing jaw (Export only)	
Interior mirror	See page 574

SPECIAL FEATURES

1. The basic one-piece transverse seat and backrest with all its attachments, is replaced by three separate seats with hinged backrests; these are mounted on tubular steel supports, which pivot on brackets bolted to the floor, and are retained by spring catches, thus enabling the seats to be tipped forward to permit access to the rear portion of the body.

2. The longitudinal rear seats are replaced by longer seats of the same design and they extend over the whole length of the wheelarch boxes, carrying a total of six persons. The rear seats are fixed to body (not tip-up type).

3. The front retainer for the rear floor rubber mat is replaced by three separate retainers.

4. The rear wheelarch locker lid is replaced by a cover plate.

NOTE: 'Station Wagon' nameplate and fixings are not required.

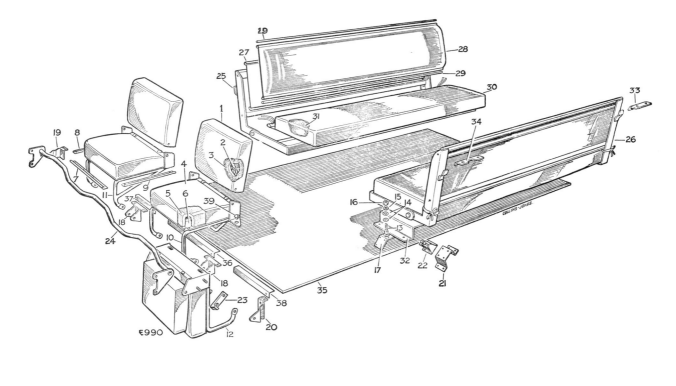

Plate Ref.	1 2 3 4	DESCRIPTION		Qty	Part No.	REMARKS
1		REAR TRANSVERSE SEAT SQUAB ASSEMBLY, TRIMMED GREY		3	334553	
		REAR TRANSVERSE SEAT SQUAB ASSEMBLY, TRIMMED BLACK		3	334562	
2		Spring case for squab		3	334565	
3		Pad for spring case		3	306316	
		Friction washer } Fixing		6	3974	
		Shouldered bolt } squab		6	331709	
4		REAR TRANSVERSE SEAT CUSHION ASSEMBLY, TRIMMED GREY		3	331148	
		REAR TRANSVERSE SEAT CUSHION ASSEMBLY, TRIMMED BLACK		3	320704*	
		Woodscrew } Fixing finishers and seat		12	20147	
		Woodscrew } base at front and sides		6	20138	
5		Seat base panel		3	306316	
6		Front finisher for cushion		3	307418	
7		Side finisher for cushion, RH		3	307419	
8		Side finisher for cushion, LH		3	307420	
9		Woodscrew } Fixing seat base		6	20034	
		Cup washer } at rear		6	73962	
		Spring washer }		6	3074	
		Plain washer } to frame		6	3900	
10		Locating pin for outer transverse seats		2	336296	
		Frame for centre transverse seat		1	334576	
11		Frame for side RH transverse seat		1	334588	
12		Frame for side LH transverse seat		1	334589	
		Set bolt (¼" UNF 1⅞" long) } Fixing		6	256006	
		Spring washer } seat cushion		6	3074	
		Plain washer }		6	3900	
13		Locating pin for outer transverse seats		2	336296	
14		Plain washer } Fixing		4	3900	
15		Spring washer } pin to outer		2	3074	
16		Nut (¼" UNF) } seat frames		2	254810	
		Shouldered bolt } Fixing seat frame		6	331709	
		Plain washer } to brackets		12	2719	
		Nut (¼" UNF) } at pivot point		6	254810	
17		Reinforcing plate for seat support		2	78248	
		Pop rivet, fixing plate to support angle		12	336295	
18		Bracket, between seats } For		2	334573	
19		Bracket RH } rear transverse		1	334574	
20		Bracket LH } Outer seats		1	334575	
		Bolt (¼" UNF x ¾" long) } Fixing brackets		14	255207	
		Plain washer } to body floor		14	3840	
		Spring washer }		14	3074	
		Nut (¼" UNF)		14	254810	
21		Lock support bracket		2	336299	
		Screw (2 BA x ½" long) } Fixing support		8	77869	
		Plain washer } bracket to		8	3851	
		Spring washer } wheel arch		8	3073	
		Nut (2 BA) } front panel		8	3823	

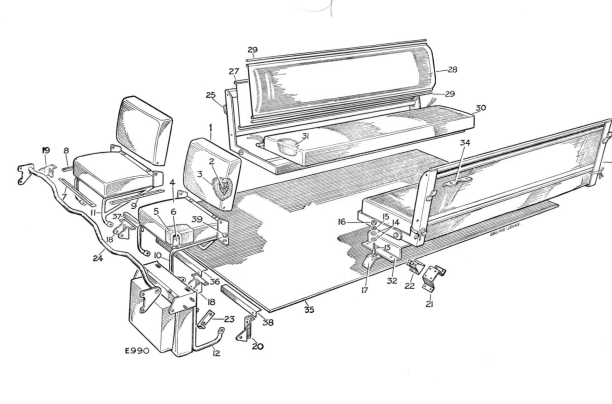

E990

Plate Ref.	1 2 3 4	DESCRIPTION	Qty	Part No.	REMARKS
22		Seat locking catch complete	2	336288	
		Screw (2 BA x ⅜" long) ⎫ Fixing locking	8	77869	
		Plain washer ⎬ catch to	8	4030	
		Spring washer ⎬ support	8	3073	
		Nut (2 BA) ⎭ bracket	8	3823	
23		Striker plate for seat lock	2	336294	
		Plain washer ⎫ Fixing striker	4	3900	
		Spring washer ⎬ plate to	4	3074	
		Nut (¼" UNF) ⎭ seat frames	4	254810	
24		Grab rail between 'BC' posts	1	336285	
25		REAR SEAT FRAME ASSEMBLY, RH	1	334541	
26		REAR SEAT FRAME ASSEMBLY LH	1	334542	
		Bolt (¼" UNF x ½" long) ⎫ Fixing	12	255204	
		Spring washer ⎬ backrest	12	3074	
		Nut (¼" UNF) ⎭ to frame	12	254810	
		Bolt (¼" UNF x ¾" long) ⎫ Fixing frame	8	255207	
		Spring washer ⎬ to wheelarch	8	3074	
		Plain washer ⎬ box and to	8	3821	
		Nut (¼" UNF) ⎭ support angle	8	254810	
		REAR SEAT BACKREST ASSEMBLY, TRIMMED, GREY	2	334530	
		REAR SEAT BACKREST ASSEMBLY, TRIMMED, BLACK	2	334531	
27		Backrest panel	2	320718*	Early type, 48" long ⎫ Alternatives.
27		Backrest panel	2	320719*	Late type, 49" long ⎬ Check before ordering
28		Pad for backrest	2	349523	
29		Retaining rod for pad	2	349522	
30		REAR SEAT CUSHION ASSEMBLY, GREY	2	334601	
30		REAR SEAT CUSHION ASSEMBLY, BLACK	2	333551	
31		Interior for rear cushion	2	78248	
32		Support angle for rear seats	1	349528	
33		Pop rivet fixing angle to wheelarch box, front	12	78248	
34		Retaining bracket for rear seat backrest, rear	2	334601	
		Retaining bracket for rear seat backrest, front	2	331670	
		Pop rivet fixing brackets to side capping	8	73198	
		Plug for redundant holes in wheelarch box	2	331083	
35		Rear floor rubber mat	1	334845	
36		Rear mat retainer, front centre	1	336780	
37		Rear mat retainer, front corner RH	1	336781	
38		Rear mat retainer, front corner LH	1	78248	
		Pop rivet fixing mat retainers	8	334391	
39		Plug for seat and backrest buffer	8	78140	
		Cover plate at rear wheelarch box	1	78248	
		Drive screw fixing cover plate	4		
		Pop rivet for redundant holes in wheelarch	12		

REGISTERED TRADE MARK

FIRE TENDER

1958-61 PETROL MODELS ONLY

GENERAL EXPLANATION

The Fire Tender is mounted on a standard Land-Rover 88 or 109 chassis and body, and the parts are, therefore, the same as those required for the basic models. The following pages, 583 to 597, list the parts which are peculiar to the Fire Tender.

The extras and additional equipment listed below are also required:—

Centre power take-off	.. ..
Spare wheel carrier on bonnet	.. ..
Truck cab	.. ..
Propeller shaft for pump	.. ..
Oil cooler, 2 litre	.. ..
Oil cooler, 2¼ litre	.. ..
Engine governor, 2 litre	.. ..
Engine governor, 2¼ litre	.. ..

See Optional Equipment Catalogue

Full-length doors, 109 models	.. .. Listed on page 493
Rear seat, 1 off, 109 models	.. .. See Optional
Private locks, 109 models	.. .. Equipment Catalogue

But with the following differences:—

Engine governor, 2 litre ..	..	243434 replacing 240628
Governor control quadrant, 2 litre	265346 replacing 240066
Governor control quadrant, 2¼ litre	265346 replacing 508890
Lever for quadrant, 2 litre	..	265348 replacing 235097
Lever for quadrant, 2¼ litre	..	265348 replacing 508692
Washer for lever ..	..	265347 replacing 219659
Support for governor control ..	237549 replacing 240065
LESS	Operating rod	.. .. .. 1 240410
	Ball joint	.. .. .. 2 219065

When universal joint covers are fitted, a modified cover, part number 243547, must be fitted to prevent a foul with the oil cooler return pipe.

All parts normally painted body colour are painted red.

All seats are trimmed in red.

FIRE TENDER PUMP, PETROL MODELS ONLY, 1958-61

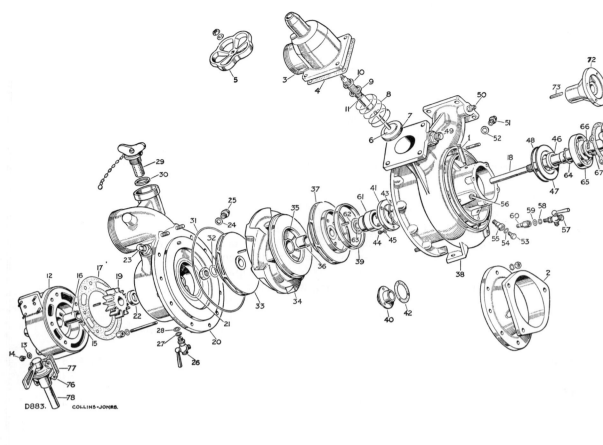

D883. COLLINS-JONES.

FIRE TENDER PUMP, PETROL MODELS ONLY, 1958-61

Plate Ref. 1 2 3 4	DESCRIPTION	Qty	Part No.	REMARKS
1	Pump complete	1	265200	
2	Mounting ring	2	271751	
3	Delivery valve	2	271752	
4	Joint washer for valve	2	271754	
5	Control wheel for valve	2	271755	
6	Spindle and seal complete	2	271756	
7	Rubber seal	2	271757	
8	Rubber seal ⎱ For valve	2	271758	
9	Spring ⎰	2	271759	
10	Circlip ⎱ For valve	2	271760	
11	Seal ⎰	2	271761	
12	Spindle body, for valve	2	271762	
13	Washer ⎱ For priming	1	267623	When drain cock is fitted
14	Plug ⎰	1	271770	2 off when drain cock is fitted
15	Grub screw, securing bush	1	271771	to priming casing
16	Bush for shaft	1	271772	In priming casing on pumps numbered
17	Gasket for priming casing	1	271773	from A745060 onwards
18	Shaft for impellers with washers and keys	1	213959	
19	Priming impeller	1	267620	
20	Suction casing	1	267621	
21	Wearing ring for suction impeller	1	267615	
22	Seal for shaft in suction casing	1	271768	
23	Suction pipe union first aid	2	240041	
24	Washer ⎱ For suction casing	2	271777	to suction casing
25	Plug ⎰	1	271776	
26	Drain cock in suction casing	1	271775	
27	Fibre washer for drain cock	1	271766	
28	Copper washer	1	267621	
29	Cap and filter complete	1	267618	
30	Washer for cap	1	271778	
31	Rubber 'O' ring for priming casing	1	271779	
32	Locking washer for pump shaft nut, suction impeller end	1	275385	
33	Suction impeller	1	267618	
34	Diffuser ring	1	271778	
35	Wearing ring, small pressure impeller	1	271779	
36	Distance piece for impeller	1	267619	
37	Suction and pressure impeller	1	271780	
38	Main casing	1	514574	Up to pump number A745059
39	Main casing with stationary seal	1	271781	From pump number A745060
40	Stationary sealing ring	1	267622	Up to pump number A745059
41	Wearing ring, large pressure impeller	1	271782	From pump number A745060
42	Joint washer for stationary sealing ring	1	514561	Up to pump number A745059
43	Sealing ring for stationary sealing ring	1	514560	From pump number A745060
44	Screw ⎱ For sealing ring	3	514575	From pump number
45	Washer ⎰	3	271783	From pump number A745060
46	Seal for shaft in main casing	1	271782	From pump number A745060
47	Cover for rear seal in main casing	1	271784	2 off from pumps numbered A745060 onwards
48	Sealing ring for rear seal cover	1	514563 ⎱ From pump number 514561 ⎰ A745060	

* Asterisk indicates a new part which has not been used on any previous Rover model

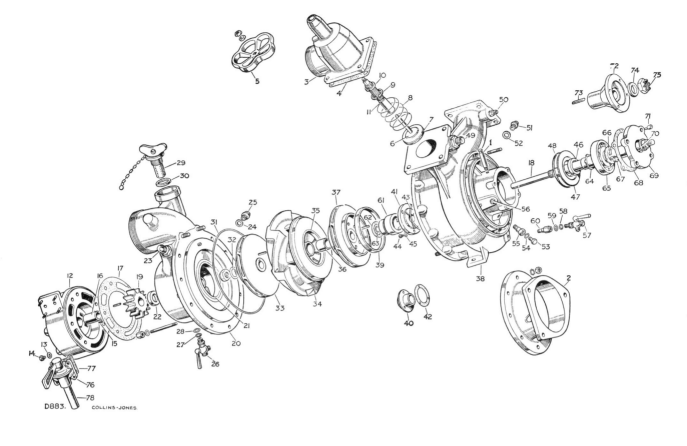

D883. COLLINS-JONES.

FIRE TENDER PUMP, PETROL MODELS ONLY, 1958-61

Plate Ref. 1 2 3 4	DESCRIPTION	Qty	Part No.	REMARKS
49	Union for delivery pipe, first aid	1	240041	
50	Union for pressure gauge pipe	1	3057	
51	Plug for main casing at side	1	271787	
52	Washer for plug	1	271770	
53	Oil drain plug for main casing at bottom	1	271771	
54	Joint washer for plug	1	271771	From pump number A745060
55	Extension tube for plug	1	514564	
56	Grease nipple for main casing	1	271789	
57	Drain cock	1	271772	
58	Fibre washer ⎫ For drain	1	271773	
59	Copper washer ⎬ cock	1	213959	
60	Extension for drain cock	1	514562	From pump number A745060
61	Mechanical seal, for shaft c/w rubber 'O' ring and fibre washer	1	504062	
62	Fibre washer for shaft	1	271790	
63	Rubber 'O' ring for shaft	1	271791	
64	Circlip, front, fixing bearing	1	271793	
65	Ball bearing for shaft	1	267616	
66	Distance piece for bearing	1	271794	
67	Circlip, rear, fixing bearing	1	271793	
68	Joint washer for bearing cover	1	271796	
69	Bearing cover	1	271797	
70	Seal for bearing cover	1	271798	
71	Screw for cover	4	271799	
72	Driving flange	1	271800	
73	Key for flange	1	271801	
74	Washer ⎫ For	1	271802	
75	Nut ⎬ flange	1	271803	
76	Priming cock	1	271804	
77	Joint washer for cock	1	271805	
78	Exhaust pipe	1	271806	
	Bolt (⅜" BSF x 4¾" long)	3	237348	
	Plain washer, small	3	3851	
	Plain washer, large	3	2251	Fixing pump to chassis
	Spring washer	3	3076	
	Nut (⅜" BSF)	3	2827	
	Support plate for pump	1	265051	
	Bolt (⅜" BSF x 2¾" long)	2	237346	Fixing support plate
	Plain washer	As required	2599	
	Self-locking nut (⅜" BSF)	2	50526	to pump
	Bolt (¾" BSF x 2⅛" long)	4	239906	Fixing support plate to
	Plain washer	8	3299	towing plate
	Self-locking nut (¾" BSF)	4	50531	

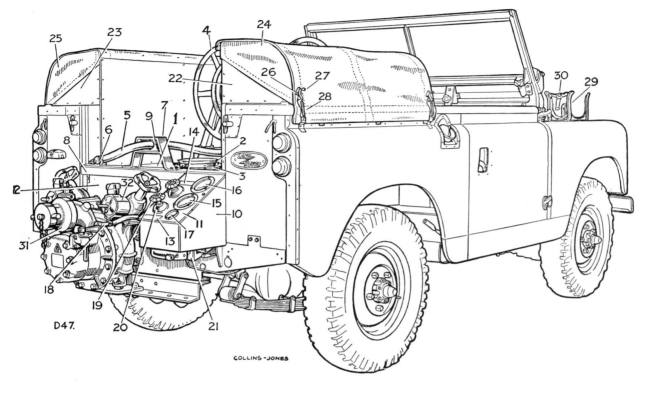

D47.

COLLINS-JONES

FIRE TENDER CONTROLS, PETROL MODELS ONLY, 1958-61

Plate Ref.	1 2 3 4	DESCRIPTION	Qty	Part No.	REMARKS
1		Water tank complete	1	504625	
2		Filler cap for water tank	1	307280	
3		Body for filter	1	307279	
		Filter for water tank	1	303610	
		Bolt (¼" UNF x 1⅛" long)	6	256200	} Fixing filter body
		Plain washer	6	3840	
		Spring washer	6	3074	} to water tank
		Nut (¼" UNF)	6	254810	
		Rubber pad for water tank	4	303661	
		Breather pipe for tank	1	243229	
		Union nut pipe for tank	1	4013	
		Clip for breather pipe	1	50647	
		Strap fixing water tank, long	2	243677	
		Strap fixing water tank, short	2	303612	
		Bolt (¼" UNF x ½" long)	4	255204	} Strap fixing water tank
		Bolt (¼" UNF x ⅝" long)	2	255208	} Fixing strap
		Spring washer	4	3074	} to wheelarch
		Nut (¼" UNF)	8	254810	
4		Hose reel complete	1	277843	
		Hose complete with nozzle	1	303618	
		Special bolt	2	255238	} Fixing
		Nut (5/16" UNF)	2	255249	} strap ends
		Spring washer	12	3076	
		Plain washer	12	4087	
		Bolt (¼" UNF x 2¾" long)	6	253888	} Fixing
		Nut (⅜" UNF)	6	2213	} water tank
		Plain washer	6	3074	
		Spring washer	6	254810	
		Set bolt (2 BA x ½" long)	3	2630	} Fixing clip and
		Spring washer	3	3073	} support to hose
		Plain washer	3	234603	} bobbin bracket
5		Support for hose reel pipe	1	72085	
		Clip for hose reel support	1	302222	
6		Clip for hose nozzle	2	242217	
7		Rivet fixing hose nozzle clips to bobbin support	2	255206	
		Spring washer	8	3074	
		Nut (¼" UNF)	8	254810	
		Bracket for hose bobbin	1	500359	} Fixing
		Bolt (¼" UNF x ⅝" long)	8	503005	} bobbin bracket
		Spring washer	8	254810	} to support
		Nut (⅜" UNF)	8	3074	
8		Support for bobbin bracket and control box	1	243682	
		Support for bobbin bracket and control box	1	243071	
		Packing plate, long	2	255210	} For bobbin
		Packing plate, short	2		} support
		Bolt (¼" UNF x 1⅛" long)	4	3074	} Fixing bobbin
		Spring washer	4	254810	} support to
		Nut (¼" UNF)	4		} wheelarch box

* Asterisk indicates a new part which has not been used on any previous Rover model

FIRE TENDER CONTROLS, PETROL MODELS ONLY, 1958-61

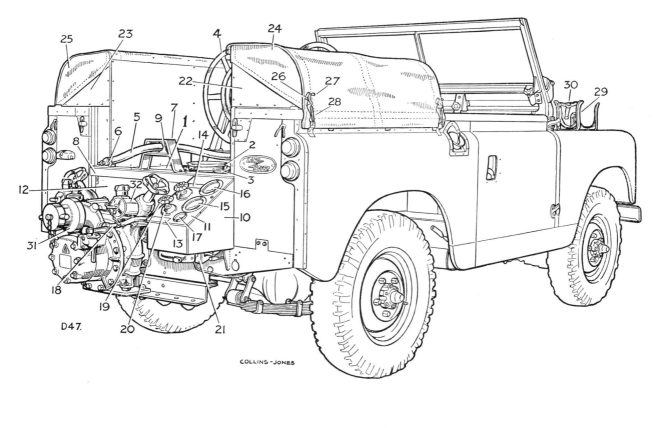

Plate Ref.	1 2 3 4	DESCRIPTION	Qty	Part No.	REMARKS
9		Bobbin for hose, male half	2	268726	4 off on
		Bobbin for hose, female half	2	268727	109
		Special bolt — Fixing bobbin to support	2	503417	
		Castle nut (5/16" UNF) — Fixing bobbin to support	2	254911	4 off on 109
		Split pin — Fixing bobbin to support	2	2555	
		Bracket for hose bobbin	1	502997	
		Bolt (1/4" UNF x 5/8" long) — Fixing bracket to floor and seat base	8	255206	
		Bolt (1/4" UNF x 7/8" long) — Fixing bracket to floor and seat base	8	255208	
		Plain washer — Fixing bracket to floor and seat base	16	2215	
		Spring washer — Fixing bracket to floor and seat base	16	3074	
		Nut (1/4" UNF) — Fixing bracket to floor and seat base	16	254810	
10		Control box at rear of vehicle	1	264798	
		Bolt (1/4" UNF x 5/8" long) — Fixing control box to support	4	255206	
		Spring washer — Fixing control box to support	4	3074	
		Nut (1/4" UNF) — Fixing control box to support	4	254810	
11		Panel for control box	1	264802	
		Screw (1/4" UNF x 3/4" long) — Fixing panel to control box	2	77936	
		Spring washer — Fixing panel to control box	2	3074	
		Plain washer — Fixing panel to control box	2	3840	
12		Panel for tailboard aperture	1	243596	88
		Panel for tailboard aperture	1	503004	109
		Drive screw fixing panel	11	72626	
13		Name plate, suction tank to pump ...	1	240034	
14		Name plate, pressure pump to first aid hose	1	240035	
		Drive screw fixing name plates ...	4	76833	
15		Pressure gauge, 200 lbs sq in. ...	1	271732	
16		Vacuum and pressure gauge, 30" vacuum, 100 lbs sq in. pressure	1	271731	
		Screw (2 BA x 3/4" long) — Fixing gauges to panel	6	78139	
		Spring washer — Fixing gauges to panel	6	3073	
		Nut (2 BA) — Fixing gauges to panel	6	2247	
17		Oil temperature gauge	1	237826	88
		Oil temperature gauge	1	271506	109
		Union for gauge	1	216512	
		Joint washer for union	1	3055	
		Lockwasher for union	1	217321	
		Washer for gauge	1	218751	
		Clip for oil temperature gauge pipe ...	1	41379	
		Drive screw fixing clip	8	72626	
		Tee piece for pressure gauge connection	1	265556	
		Joint washer for pressure gauge tee piece	1	232044	
		Elbow at water valves	2	268666	
		Joint washer—elbow to valves ...	3	231579	
		Distance piece for elbow	2	240109	

FIRE TENDER CONTROLS, PETROL MODELS ONLY, 1958-61

* Asterisk indicates a new part which has not been used on any previous Rover model

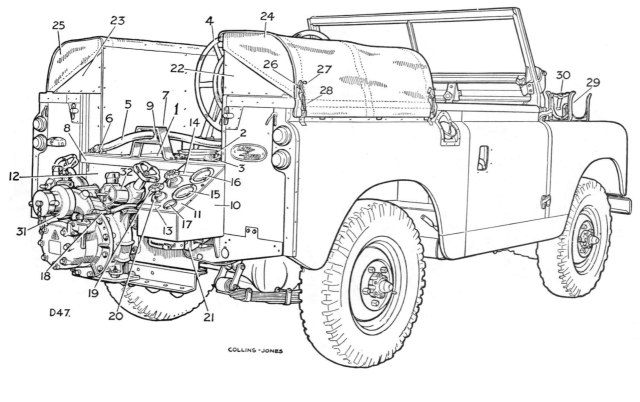

D47.

COLLINS-JONES

FIRE TENDER CONTROLS, PETROL MODELS ONLY, 1958-61

Plate Ref. 1 2 3 4	DESCRIPTION	Qty	Part No.	REMARKS
	Pipe, outlet valve to hose reel	1	277781	‡
	Pipe, outlet valve to hose reel	1	508494	‡‡ 88
	Pipe, hose reel to union	1	508516	
	Double-ended union for pipe	1	240041	109
	Pipe, union to outlet valve	1	503461	
	Pipe, tank to outlet valve	1	264969	88
	Pipe, tank to inlet valve	1	503008	109
	Pipe, outlet valve to inlet valve	1	264963	
18	Pipe, outlet valve to pump	1	264965	
	Pipe, pump to inlet valve	1	264975	
	Pipe complete, tee piece to vacuum gauge	1	264973	
	Double-ended union, for pressure gauge	1	3057	
	Double-ended union, for vacuum gauge pipe	1	213959	
	Joint washer for union	1	240037	
19	Water valve complete	2	240041	
20	Rubber grommet for water valves	2	303663	
	Double ended union ⎱ For water valves,	4	240041	
	Washer for union ⎰ pump and tank	5	243962	
21	Support bracket for governor quadrant	1	237549	
	Bolt (¼″ UNF x ⅝″ long) ⎱ Fixing quadrant	2	255006	
	Spring washer ⎰ to support bracket	2	3074	
	Nut (¼″ UNF)	2	254810	
	Bolt (2 BA x ¾″ long) ⎱ Fixing quadrant	2	234603	
	Spring washer ⎰ to control box	2	3073	
	Plain washer	2	2630	
	Nut (2 BA)	2	2247	
	Governor unloading spring	1	218051	
	Link plate ⎱ For	1	242928	2 litre models
	Wire link ⎰ spring	1	242927	
	Ball joint complete for governor control cable	1	219065	
	Locknut (¼″ BSF) for ball joint	2	2823	
	Spring washer ⎱ Fixing ball joint	1	3074	
	Nut (¼″ BSF) ⎰ to quadrant	1	2823	
	Special screwed adaptor for governor control cable	1	504458	
	Screwed extension rod for governor control cable	1	504459	88
	Screwed extension rod for governor control cable	1	240410	109
	Swivel coupling for control	1	504667	
	Drive screw fixing swivel to frame	2	73892	
	Conduit for control cable, connector to swivel	1	243668	88
	Conduit for control cable, connector to swivel	1	270614	109
	Conduit for control cable, connector to connector	1	233102	
	Conduit for control cable, connector to connector	1	233103	2 litre models
	Conduit for control cable, governor to connector	1	506296	2¼ litre models
	Clamp connector for conduit	2	500569	
	Nipple	1	233093	
	Split pin	1	2388	
	Swivel complete	1	233094	
	Bracket for overslider ⎱ Fixing	1	233100	2 litre models
	Bracket for overslider ⎰ conduit to	1	506284	2¼ litre models
	Shakeproof washer ⎰ overslider cable	1	233017	
	Nut (⅜″ BSF)	1	4373	

* Asterisk indicates a new part which has not been used on any previous Rover model

‡ Up to vehicles numbered:
141003103
142001855 ⎱ 88
144004783

‡‡ From vehicles numbered:
141003104
142001856 ⎱ 88
144004784 ⎰ onwards

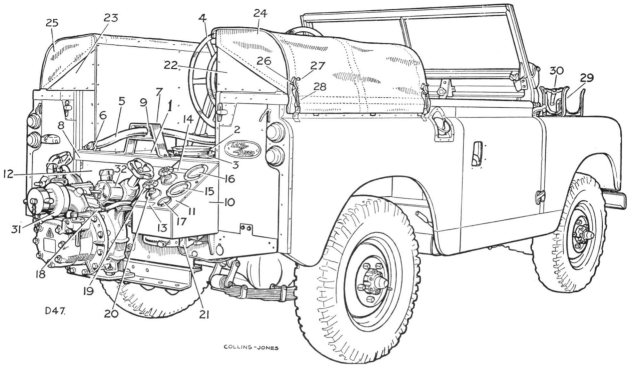

D47.

COLLINS-JONES

FIRE TENDER CONTROLS, PETROL MODELS ONLY, 1958-61

Plate Ref. 1 2 3 4	DESCRIPTION	Qty	Part No.	REMARKS
	Overslider cable	1	244419	
	Bracket for conduit clip at engine	1	237819	2 litre models
	Bolt (5/16" Whit x 19/32" long) } Fixing conduit cap	1	215761	
	Spring washer } bracket to inlet manifold	1	3075	
	Clip for conduit	5	237813	} 4 off on 2¼ litre models
	Drive screw fixing clips	5	72626	
	Nipple, conduit to control cable and swivel	2	233093	
22	Hose box RH	1	508227	} 88
	Hose box LH	1	508228	
23	Hose box RH	1	508283	} 109
	Hose box LH	1	508284	
	Distance piece for body side	2	503410	
	Bolt (1/4" UNF x 5/8" long) } Fixing	18	255206	
	Spring washer } hose boxes	18	3074	
	Nut (1/4" UNF) } to body	18	254810	
	Bolt (1/4" UNF x 7/8" long) } For	6	300781	
	Support bracket RH } hose box	3	506290	} 109
	Support bracket LH }	3	506289	
	Pop rivet fixing support bracket to body capping	18	300781	
	Bolt (1/4" UNF x 7/8" long) } Fixing	6	255208	} 109
	Plain washer } support	6	2760	
	Spring washer } bracket to	6	3074	
	Nut (1/4" UNF) } wheelarch	6	254810	
	Reinforcement plate	2	500566	
	Bolt (1/4" UNF x 1 1/8" long) } Fixing	4	256200	} 88
	Bolt (1/4" UNF x 5/8" long) } hose box to	4	255206	} 109
	Plain washer } wheelarch and	4	3840	
	Spring washer } support brackets	4	3074	} 10 off on
	Nut (1/4" UNF)	4	254810	} 109
	Pop rivet fixing hose box to wheelarch	8	300781	
	Cover plate for spare wheel hole in wheelarch	2	503433	
	Pop rivet fixing cover plate to wheelarch	2	300781	
	Bolt access plate on cover plate	2	503448	} 109
	Drive screw fixing access plate to cover plate	8	72626	
24	Cover for hose box RH	1	508239	
25	Cover for hose box LH	1	508257	
26	Hook assembly for catch	4	503169	
	Top strip for hose box cover	2	500399	
	Bottom strip for hose box cover	2	500400	
	Rivet fixing cover and strip	10	300784	
27	Hook catch } Retaining hose	4	503163	
	Spring } box cover	4	243767	
	Rivet fixing hook catch to body	8	300781	
	Bracket for front cradle	1	303662	
28	Bolt (1/4" UNF x 5/16" long) } Fixing bracket to front	4	255206	
	Spring washer } bumper and radiator	4	3074	
	Nut (1/4" UNF) } top baffle	4	254810	

* Asterisk indicates a new part which has not been used on any previous Rover model

FIRE TENDER CONTROLS, PETROL MODELS ONLY, 1958-61

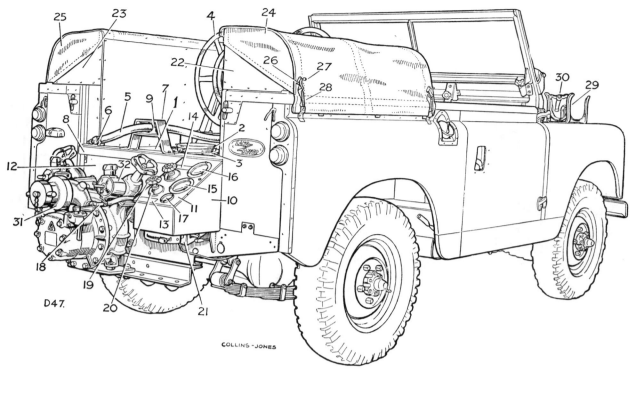

D47.

COLLINS-JONES

Plate Ref. 1 2 3 4	DESCRIPTION	Qty	Part No.	REMARKS
	Front cradle	1	243851	
	Bolt (¼" UNF x ¾" long) } Fixing cradle to bracket	2	255207	
	Spring washer	2	3074	
	Nut (¼" UNF)	2	254810	
29	Support strap for front cradle	2	503015	
	Bolt (¼" UNF x ¾" long) } Fixing straps to cradle bracket	2	255207	
	Spring washer	2	3074	
	Nut (¼" UNF)	2	254810	
	Side cradle	2	303654	
	Bolt (¼" UNF x ⅝" long) } Fixing side cradle to front wings	14	255206	
	Spring washer	14	3074	
	Nut (¼" UNF)	14	254810	
	Stiffener for front wing top	2	303658	
	Pop rivet fixing stiffener	6	300919	
	Support strap, for front cradle	1	243853	
	Rivet } Fixing straps to cradle bracket	2	305665	
	Plain washer	2	3816	
30	Strap for hose cradles, side	2	303650	
	Plain washer } cradle bracket	2	303568	
	Rivet	2	303954	
	Eyelet	2	300785	
	End cap	2	302186	
	Instantaneous buckle	2	3816	
	Rivet } Fixing straps to cradle	24	302186	
31	Suction hose adaptor, 3" diameter output	1	265190	3" type. Alternative to 4" } Check before ordering
	Joint washer for adaptor, 3" diameter	1	265174	Check before ordering
	Suction hose adaptor, 4" diameter output	1	509860	4" type.
	Joint washer for adaptor, 4" diameter output	1	509861	Alternative to 3"
	Cap for adaptor, 4" diameter	1	509866	Check before ordering
	Joint washer for cap, 4" diameter	1	265185	
32	Deliver hose couplings	2	266815	Optional equipment to suit British standard couplings
	Joint washer for delivery hose couplings	4	4017	
	Suction hose spanner	1	273523	
	Support for foam generator RH	1	273529	
	Support for foam generator LH	1	273530	
	Bolt (2 BA x ⅜" long) } Fixing supports	1	273531	
	Plain washer	2	273532	
	Special nut	1	78349	
	Rear step	1	262245	
	Rubber mat for step	1	244687	
	Retainer for mat, rear	1	244688	
	Retainer for mat, front	1	237119	
	Retainer for mat, side	2		
	Rivet fixing mat and retainer	16	2226	
	Bracket for rear seat	2	502998	
	Support bracket for rear seat	2	503002	
	Bolt (¼" UNF x ⅞" long) } Fixing bracket to frame	6	255208	
	Spring washer	6	3074	
	Plain washer	6	2760	
	Nut (¼" UNF)	6	254810	

* Asterisk indicates a new part which has not been used on any previous Rover model

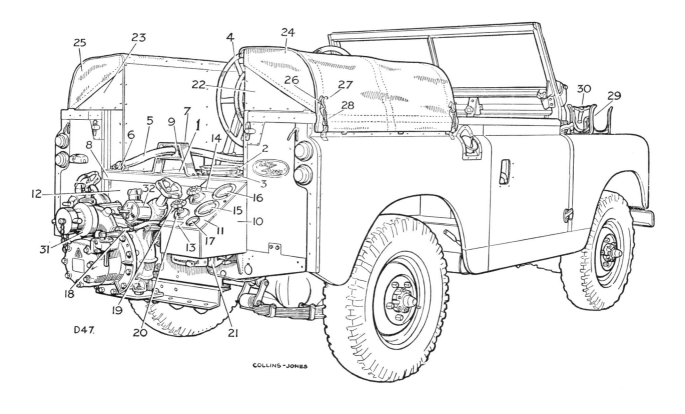

D47.

COLLINS-JONES

FIRE TENDER CONTROLS, PETROL MODELS ONLY, 1958-61

Plate Ref. 1 2 3 4	DESCRIPTION	Qty	Part No.	REMARKS
	Rear seat base complete	1	273979	
	Plain washer for seat base	2	3898	
	Bolt (¼" UNF x ⅞" long) ..	4	255008	
	Plain washer	4	2760	
	Spring washer	4	3074	
	Nut (¼" UNF)	4	254810	
	Rear spring, driver's side	1	272967	
	Rear spring, passenger's side ..	1	272968	
	Main leaf for rear spring	2	537964	88
	Second leaf for rear spring	2	537966	
	Dowel for rear spring	4	279762	
	'U' bolt for rear spring	4	242926	
	Front spring, driver's and passenger's ..	2	264563	2 litre models
	Front spring, driver's side	1	264563	2 litre models
	Front spring, passenger's side	1	265627	88
	'U' bolt, long ⎫ For front	2	265627	2¼ litre models
	'U' bolt, short ⎭ springs	3	265088	88
	Shock absorber, front	2	06694	88
	Shock absorber, rear	2	512102	88
	Tyre (7.00 x 16")	5	512086	88
	Tube (7.00 x 16")	5	242861	88
	Tyre (7.50 x 16")	5	233042	88
	Tube (7.50 x 16")	5	502527	109
		5	270324	

Fixing seat base to support bracket

TRIMMING RAW MATERIALS, ETC.

Plate Ref. 1 2 3 4 — DESCRIPTION	Qty	Part No.	REMARKS
Cotton duck, 72" wide, khaki-green	As required	91957	
Cotton duck, 72" wide, blue	As required	91899	
Sheet wadding, 36" wide open	As required	91341	
Cotton webbing, 3½" x 1" khaki-green	As required	91878	
Cotton webbing, ⅛" x 1", khaki-green	As required	91880	
Cotton webbing, 5/64" x ½" khaki-green	As required	91879	
Cotton webbing, 5/64" x ½", blue	As required	91901	
Cotton webbing, ⅛" x 1", blue	As required	91902	
Cotton webbing, 3½" x 1", blue	As required	91900	
Cotton webbing, 5/64" x 1", black	As required	92088	
Seaming cord	As required	91171	
Thin plain cloth, 38" wide	As required	91829	
Black canvas, 54" wide	As required	91291	
Grey felt	As required	91414	
Black felt, 50" wide	As required	91819	
Leather cloth, grey, 50" wide	As required	91979	
Leather cloth, red	As required	91832	
Leathercloth, black, 50" wide	As required	92473	
Leathercloth, dull black, 50" wide	As required	92175	
White leather cloth, 50" wide	As required	91950	
Piping, green	As required	91857	
Adhepive tape 1" wide	As required	91928	
Plastic cloth, blue check, 54" wide	As required	91912	
Natural felt, 54" wide	As required	91890	
Felted leather cloth, white, 52" wide	As required	91970	
Felted plastic, blue, 52" wide	As required	91909	
Felted plastic, 52" wide, green	As required	91962	
Felted plastic, black	As required	91965	
Black cloth, 72" wide	As required	91913	
Cotton webbing, ... blue	As required	91900	
Fine fluted rubber matting, 36" wide	As required	261577	
Sealing strip, 'Prestik', ¾" x ⅛"	As required	13468	
Sealing strip, 'Prestik', ½" x 1/16"	As required	13296	
Special paint, bronze green, 1 pint tin	As required	262069	
Special paint, bronze green, 1 gallon tin	As required	262072	
Special paint, ivory, 1 pint tin	As required	261887	
Special paint, ivory, 1 gallon tin	As required	261902	
Special paint, blue, 1 pint tin	As required	244272	
Special paint, blue, 1 gallon tin	As required	244275	
Special paint, poppy red, 1 pint tin	As required	244276	
Special paint, poppy red, 1 gallon tin	As required	244279	

TRIMMING RAW MATERIALS, ETC.

Plate Ref. 1 2 3 4 — DESCRIPTION	Qty	Part No.	REMARKS
Special paint, light green, 1 pint tin	As required	503927	
Special paint, light green, 1 gallon tin	As required	503928	
Special paint, sand, 1 pint tin	As required	503929	
Special paint, sand, 1 gallon tin	As required	503930	
Special paint, dark grey, 1 pint tin	As required	503931	
Special paint, dark grey, 1 gallon tin	As required	503932	
Special paint, mid-grey, 1 pint tin	As required	503933	
Special paint, mid-grey, 1 gallon tin	As required	503934	
Special paint, marine blue, 1 pint tin	As required	503935	
Special paint, marine blue, 1 gallon tin	As required	503936	
Special paint, off white, 1 pint tin	As required	503937	
Special paint, off white, 1 gallon tin	As required	503938	
Special paint, limestone, 1 pint tin	As required	510032	
Special paint, limestone, 1 gallon tin	As required	510033	
Thinners, 1 pint tin	As required	261906	
Thinners, 1 gallon tin	As required	261909	
Brake fluid, 32 oz. tin	As required	606021	
Cleaning fluid for wheel cylinder and master cylinder parts, half-pint tin	As required	535642	
⅛" Welding rod (1 lb)	As required	98018	
'Hylomar', sealing compound, 4 oz. tube	As required	534244	
'Bar Seal' sealing pellet for radiator	1	601314	
Anti-freeze mixture, Super Bluecol, quart tin	As required	605532	
Anti-freeze mixture Super Bluecol, gallon tin	As required	605533	
Anti-freeze, mixture Super Bluecol, 5 gallon drum	As required	605534	

OVERHAUL KITS

Plate Ref. 1 2 3 4 — DESCRIPTION	Qty	Part No.	REMARKS
Engine overhaul gasket kit	1	270568	2 litre Petrol
Engine overhaul gasket kit	1	525856	Series II and IIA 2¼ litre Petrol models
Engine overhaul gasket kit	1	605106*	2.6 litre Petrol
Engine overhaul gasket kit	1	272564	2 litre Diesel
Engine overhaul gasket kit	1	525520	2¼ litre Diesel
Gasket set, decarbonising	1	269365	2 litre Petrol
Gasket set, decarbonising	1	525887	Series II and IIA 2¼ litre Petrol models
Gasket set, decarbonising	1	605105*	2.6 litre Petrol
Gasket set, decarbonising	1	272563	2 litre Diesel
Gasket set, decarbonising	1	525521	2¼ litre Diesel
Gearbox gasket kit	1	600603	
Brake wheel cylinder overhaul kit, front	1	275744	88
Brake wheel cylinder overhaul kit, front	1	266684	109 4 cylinder models
Brake wheel cylinder overhaul kit, front GI SP 2189	1	600210	109 6 cylinder models
Brake wheel cylinder overhaul kit, rear	1	266687	88
Brake wheel cylinder overhaul kit, rear, SP 2004	1	266683	109 With steel dust cover } Series II models. Alternatives. Check before ordering
Brake wheel cylinder overhaul kit, rear, GI SP 2103	1	523164	109 With rubber dust cover }
Brake master cylinder overhaul kit, GI SP 2051	1	275744	88 } Check before ordering
Brake master cylinder overhaul kit, GI SP 1980/1	1	502333	109 Series IIA models }
Brake master cylinder overhaul kit GI SP 2385	1	605127	109 4 cylinder models. For CB type master cylinder GI 64067722 } Check before ordering
Brake master cylinder overhaul kit GI SP 2472	1	606023*	109 4 cylinder models. For CV type master cylinder GI 64068750 }
Brake master cylinder overhaul kit, GI SP 1967/4	1	601611	88 For CV type master cylinder GI 64067720 } Check before ordering
Brake master cylinder overhaul kit GI SP 1989	1	503754	109 4 cylinder models. For CB type master cylinder GI 64068893 }
Clutch master cylinder overhaul kit GI SP 1967/4	1	605127	109 6 cylinder models
Clutch slave cylinder overhaul kit GI SP 2029	1	601611	88 For CB type master cylinder
		502335	109 6 cylinder models
Brake servo non-return valve overhaul kit	1	600435	
Brake servo poppet valve overhaul kit	1	600434	Export only
Brake servo 'minor' overhaul kit	1	601981	} 2.6 litre Petrol
Brake servo 'major' overhaul kit	1	601980	}
Clutch slave cylinder overhaul kit GI SP 2029	1	502335	109 6 cylinder models
Repair kit for adjuster unit (transmission brake) GI SP 1856	1	515924	From gearboxes numbered: 14000566, 16000431
Repair kit for expander unit (transmission brake) GI SP 1807	1	515923	151005187 onwards
White brake grease, tube	1	514577	
Red brake rubber grease, tube	1	514578	

*   Asterisk indicates a new part which has not been used on any previous Rover model

OVERHAUL KITS

Plate Ref. 1 2 3 4 — DESCRIPTION	Qty	Part No.	REMARKS
Repair kit for fuel pump, AC 1524507	1	272069	2¼ litre Petrol
Overhaul kit for fuel pump, AC 7950463	1	524684	2¼ litre Petrol models
Repair kit for fuel pump, AC 7950706	1	600904	} Diesel models
Overhaul kit for fuel pump, AC 7950707	1	600905	}
Swivel pin overhaul kit	1	266752	Cone and spring type. Check before ordering
Swivel pin overhaul kit	1	532268	Pin and thrust washer type. Also suitable as conversion kit. Check before ordering
Water pump overhaul kit	1	265255	2 litre Petrol
Water pump overhaul kit	1	530590	Series II and IIA 2¼ litre Petrol and all Diesel models
Carburetter overhaul kit, Zenith MRK 3136	1	605716*	2.6 litre Petrol
Carburetter overhaul kit, Zenith MRK 3082	1	606098*	2.6 litre Petrol } For Zenith carburetter, Check before ordering
Carburetter overhaul kit	1	605092	2¼ litre Petrol } For Solex carburetter, Check before ordering
Carburetter overhaul kit	1	507687	2 litre Petrol }
Carburetter gasket kit	1	266693	2 litre Petrol
Carburetter gasket kit	1	507693	2¼ litre Petrol } For Solex carburetter, Check before ordering
Carburetter gasket kit, Zenith 237Z	1	274895	2¼ litre Petrol } For Zenith carburetter
Carburetter gasket kit, Zenith 235Z	1	605093	2¼ litre Petrol } For Zenith carburetter ordering
Carburetter gasket kit, Zenith 235Z	1	605857*	2.6 litre Petrol } For Zenith carburetter
Bearing set for crankshaft, Std	1	533979	} Series IIA Petrol and Diesel models. Sets include main bearings, thrust washers and connecting rod bearings
Bearing set for crankshaft, .010" U.S.	1	533980	}
Bearing set for crankshaft, .020" U.S.	1	533981	}
Bearing set for crankshaft, .030" U.S.	1	533982	}
Bearing set for crankshaft, .040" U.S.	1	533983	}
Plastigauge strip, box of 24	1	605238*	For checking bearing clearances
'Loctite' retaining compound, 10 cc bottle, Grade AAV	1	606146	
Coolant inhibitor, Marston lubricant SQ 36. 18 oz.	As required	605765*	

*   Asterisk indicates a new part which has not been used on any previous Rover model

TOOLS

Plate Ref. 1 2 3 4	DESCRIPTION	Qty	Part No.	REMARKS
	Wheel brace	1	537179	
	Lifting jack screw type	1	551929	
	Handle for jack, spade end	1	514624	For ratchet
	Handle for jack, round end	1	513072	type jack
	Extension for jack, 35½″ long	1		spade type
	Wooden handle for extension screw type jack	1	261513	For early screw type jack
	Extension for jack, 41½″ long	1	261512	For early
	Extension for jack, 14¾″ long round end with ball retainer	1	543301 / 551929	Use with jack Part No.
	Wooden handle for extension, 14¾″ long	1	543300	For late screw type jack } Alternatives. Check before ordering
	Tyre pump, hand operated	1	524959	
	Connection for tyre pump	1	523638	
	Tyre pump, foot operated	1	535023	Optional
	Grease gun	1	503424	
	Starting handle	1	218508	
	Tool roll	1	219704	
	Combination pliers	1	2703	
	Screwdriver	1	556770	
	Spanner ¾″ x 1″ Whitworth	1	2705	
	Spanner 5/16″ x 7/16″ Whitworth	1	230736	
	Spanner, single end ⅜″ Whitworth	1	277320	
	Spanner 1/16″ x ½″ AF	1	276396	
	Spanner 1/16″ x ¾″ AF	1	277217	
	Spanner ⅝″ x 1/16″ AF	1	276397	
	Spanner 9/16″ x ⅜″ AF	1	549840	
	Adjustable spanner	1	2707	
	Distributor screwdriver	1	240836	For distributor pump bleed screw. Diesel models
	Sparking plug spanner	1	276322	
	Extension for plug spanner	1	276323	Also required for radiator drain plug
	Box spanner	1	52485	
	Tommy bar	1	1403	
	Tyre pressure gauge	1	562019	
	Tyre lever	2	233261	Optional equipment

NUMERICAL INDEX

This list, in part number numerical order, gives the page on which the item will be found

Part No.	Page No.
504032	63
504058	591
504059	591
504062	585
504082	117
504104	337
504105	337
504106	337
504130	413
504135	337
504136	337
504169	109
504233	383
504272	283
504275	283
504276	281
504279	283
504433	263
504438	387
504443	101
504444	425
504485	69
504519	91
504577	289
504588	145
504619	349
504620	351
504621	351
504655	383
504657	383
504667	383
504683	69
504684	429
504685	431
504699	359
504736	65
504765	303
504840	25
504995	99
504997	99
505144	413
505150	413
505158	427
505196	413
505197	413
505198	413
505205	427
505244	383
505597	103
505612	121
505613	121
505675	289
505676	289
505701	71
505786	119
505787	119
505790	291
505796	119
505798	405
505805	163
505826	365
505853	365
505855	365
506046	109
506047	109
506069	51
506210	351
506284	591
506289	593
506290	593
506296	591
506679	425
506679	131
506796	157
506799	41
506812	19
506814	57
506816	103
506817	103
506818	19
507001	85
507025	57
507026	57
507129	403
507198	413
507199	413
507447	249
507687	73
507693	73
507804	413
507810	443
507829	57
508033	330
508034	330
508035	457
508060	381
508148	299
508152	269
508153	269
508175	271
508227	593
508228	593
508239	593
508257	593
508283	593
508284	593
508494	591
508516	591
508544	393
508545	385
508565	299
508566	299
508581	307
508582	311
508895	325
508945	301
508947	521
508957	521
508958	521
508959	521
508962	347
508985	387
509009	157
509011	359
509045	257
509046	259
509120	521
509217	377
509407	347
509412	209
509413	573
509414	573
509415	209
509448	275
509449	275
509463	345
509513	161
509558	347
509582	367
509730	389
509751	257
509767	377
509856	343
509860	595
509861	595
509866	595
509909	163
509910	573
509912	573
509970	333
510032	599
510033	599
510078	147
510170	325
510176	415
510179	413
510237	443
510274	39
510469	413
510489	179
510573	291
510575	295
510730	61
510805	61
510912	429
511036	17
511127	355
511189	237
511203	239
511205	237
511652	119
511680	59
511690	119
511833	9
511878	329
511879	329
511956	65
511957	65
511958	67
511963	169
512018	67
512086	330
512102	330
512106	71
512107	73
512205	57
512206	57
512207	57
512235	341
512237	203
512238	231
512249	177
512251	425
512305	85
512322	277
512359	279
512401	31
512402	69
512412	43
512413	145
512415	291
512416	291
512445	131
512459	117
512460	119
512646	79
512650	341
512651	145
512701	241
512797	177
512798	177
512799	177
512800	177
512806	35
512828	187
512838	299
512839	339
512889	173
513072	602
513171	109
513174	29
513220	447
513221	447
513222	435
513282	431
513314	413
513454	263
513465	65
513506	69
513507	71
513591	25
513617	157
513626	203
513627	203
513639	153
513640	59
513641	59
513694	85
513927	203
513928	203
513929	203
513937	425
513940	301
513991	119
514143	415
514144	421
514145	415
514146	415
514147	415
514149	413
514150	413
514152	415
514189	39
514190	39
514192	39
514193	39
514194	39
514195	39
514224	65
514269	393
514270	393
514451	61
514472	89
514526	45
514527	43
514560	177
514561	583
514562	585
514564	583
514574	583
514575	583
514577	600
514578	600
514580	79
514624	602
514650	249
514721	447
514734	425
514790	279
514827	279
514840	279
514895	429
514929	427
515086	225
515090	213
515291	111
515313	73
515314	73
515321	55
515325	55
515331	365
515332	365
515365	317
515366	317
515405	289
515406	289
515437	157
515465	421
515466	415
515467	317
515470	317
515505	365
515506	365
515508	275
515509	275
515552	203
515573	187
515597	365
515598	367
515599	255
515729	255
515845	277
515848	277
515849	277
515923	317
515924	317
515925	317
515926	317
515927	317
515962	317
515969	99
516028	105
516031	317
516057	347
516059	65
516133	43
516370	25
516410	335
516496	365
516599	291
516699	365
516914	377
516945	365
517026	41
517429	57
517469	365
517502	389
517588	321
517589	321
517613	381
517632	365
517682	163
517684	165
517686	165
517689	163
517690	163
517706	163
517707	163
517711	163
517855	167
517877	283
517878	283
517907	335
517976	157
517977	523
518100	97
518145	105
518146	105
518466	145
518468	49
518473	237
518676	275
518682	337
518719	187
518748	45
518749	45
518750	45
518751	45
518752	45
518753	45
518754	45
518755	45
518757	45
518758	45
518759	45
518760	45
518761	45
518808	143
518809	49
518818	67
519004	173
519006	217
519007	175
519008	57
519009	175
519010	175
519011	175
519054	55
519055	55
519064	11
519206	269
519440	49
519638	279
519740	403
519742	405
519743	403
519744	403
519753	279
519755	279
519775	403
519782	427
519783	429
519784	429
519806	427
519807	335
519814	427
519827	403
519838	385
519840	403
519841	403
519857	572
519864	25
519865	421
519866	421
519871	425
519873	403
519874	403
519875	385
519876	527
519897	439
519899	431
519900	425
520849	335
520940	185
520941	185
520942	185
520943	185
520978	185
521326	239
521328	239
521329	239
521330	239
521543	295
521544	295
521583	99
521584	163
521600	187
522038	287
522318	233
522387	233
522438	205
522593	293
522607	49
522745	59
522756	167
522804	277
522882	421
522913	87
522937	163
522938	163
522939	163
522940	163
523044	295
523084	335
523139	101
523164	295
523181	193
523240	47
523301	13
523302	13
523305	13
523306	9
523307	11
523308	11
523309	11
523310	11
523311	11
523313	11
523314	11
523315	11
523316	11
523317	11
523319	11
523320	45
523321	45
523322	45
523323	45
523324	45
523326	239
523327	45
523328	45
523329	45
523330	45
523331	45
523332	45
523333	45
523334	45
523335	45
523337	45
523338	45
523339	49
523340	49
523341	49
523343	97
523344	97
523345	97
523354	65
523589	523
523637	407
523638	163
523655	367
523695	333
523696	333
523806	423
523916	333
524114	69
524115	71
524151	49
524153	49
524154	49
524155	49

Printed and distributed by

Brooklands Books Ltd. P.O. Box 146,
Cobham, Surrey KT11 1LG, England
Phone: 01932 865051 Fax: 01932 868803

E-mail: sales@brooklands-books.com   www.brooklands-books.com

ISBN 1 85520 2387   Part No 605957   1/7Z3   Printed in England   Ref: B-LR21PH